Introduction to Real Analysis

INTRODUCTION TO REAL ANALYSIS
Third Edition

Robert G. Bartle
Donald R. Sherbert

Eastern Michigan University, Ypsilanti
University of Illinois, Urbana-Champaign

John Wiley & Sons, Inc.

ACQUISITION EDITOR	Barbara Holland
ASSOCIATE EDITOR	Sharon Prendergast
PRODUCTION EDITOR	Ken Santor
PHOTO EDITOR	Nicole Horlacher
ILLUSTRATION COORDINATOR	Gene Aiello

This book was set in Times Roman by Eigentype Compositors, and printed and bound by Hamilton Printing Company. The cover was printed by Phoenix Color Corporation.

This book is printed on acid-free paper.

The paper in this book was manufactured by a mill whose forest management programs include sustained yield harvesting of its timberlands. Sustained yield harvesting principles ensure that the numbers of trees cut each year does not exceed the amount of new growth.

Library of Congress Cataloging in Publication Data:
Bartle, Robert Gardner, 1927–
    Introduction to real analysis / Robert G. Bartle, Donald R., Sherbert. – 3rd ed.
        p.   cm.
    Includes bibliographical references and index.
    ISBN 0-471-32148-6 (alk. paper)
      1. Mathematical analysis.   2. Functions of real variables.
   I. Sherbert, Donald R., 1935– . II. Title.
   QA300.B294   2000
   515–dc21
                                                      99-13829
                                                     CIP

A. M. S. Classification 26-01

Printed in the United States of America

10 9 8 7

PREFACE

The study of real analysis is indispensible for a prospective graduate student of pure or applied mathematics. It also has great value for any undergraduate student who wishes to go beyond the routine manipulations of formulas to solve standard problems, because it develops the ability to think deductively, analyze mathematical situations, and extend ideas to a new context. In recent years, mathematics has become valuable in many areas, including economics and management science as well as the physical sciences, engineering, and computer science. Our goal is to provide an accessible, reasonably paced textbook in the fundamental concepts and techniques of real analysis for students in these areas. This book is designed for students who have studied calculus as it is traditionally presented in the United States. While students find this book challenging, our experience is that serious students at this level are fully capable of mastering the material presented here.

The first two editions of this book were very well received, and we have taken pains to maintain the same spirit and user-friendly approach. In preparing this edition, we have examined every section and set of exercises, streamlined some arguments, provided a few new examples, moved certain topics to new locations, and made revisions. Except for the new Chapter 10, which deals with the generalized Riemann integral, we have not added much new material. While there is more material than can be covered in one semester, instructors may wish to use certain topics as honors projects or extra credit assignments.

It is desirable that the student have had some exposure to proofs, but we do not assume that to be the case. To provide some help for students in analyzing proofs of theorems, we include an appendix on "Logic and Proofs" that discusses topics such as implications, quantifiers, negations, contrapositives, and different types of proofs. We have kept the discussion informal to avoid becoming mired in the technical details of formal logic. We feel that it is a more useful experience to learn how to construct proofs by first watching and then doing than by reading about techniques of proof.

We have adopted a medium level of generality consistently throughout the book: we present results that are general enough to cover cases that actually arise, but we do not strive for maximum generality. In the main, we proceed from the particular to the general. Thus we consider continuous functions on open and closed intervals in detail, but we are careful to present proofs that can readily be adapted to a more general situation. (In Chapter 11 we take particular advantage of the approach.) We believe that it is important to provide students with many examples to aid them in their understanding, and we have compiled rather extensive lists of exercises to challenge them. While we do leave routine proofs as exercises, we do not try to attain brevity by relegating difficult proofs to the exercises. However, in some of the later sections, we do break down a moderately difficult exercise into a sequence of steps.

In Chapter 1 we present a brief summary of the notions and notations for sets and functions that we use. A discussion of Mathematical Induction is also given, since inductive proofs arise frequently. We also include a short section on finite, countable and infinite sets. We recommend that this chapter be covered quickly, or used as background material, returning later as necessary.

Chapter 2 presents the properties of the real number system $\mathbb{R}$. The first two sections deal with the Algebraic and Order Properties and provide some practice in writing proofs of elementary results. The crucial Completeness Property is given in Section 2.3 as the Supremum Property, and its ramifications are discussed throughout the remainder of this chapter.

In Chapter 3 we give a thorough treatment of sequences in $\mathbb{R}$ and the associated limit concepts. The material is of the greatest importance; fortunately, students find it rather natural although it takes some time for them to become fully accustomed to the use of ε. In the new Section 3.7, we give a brief introduction to infinite series, so that this important topic will not be omitted due to a shortage of time.

Chapter 4 on limits of functions and Chapter 5 on continuous functions constitute the heart of the book. Our discussion of limits and continuity relies heavily on the use of sequences, and the closely parallel approach of these chapters reinforces the understanding of these essential topics. The fundamental properties of continuous functions (on intervals) are discussed in Section 5.3 and 5.4. The notion of a "gauge" is introduced in Section 5.5 and used to give alternative proofs of these properties. Monotone functions are discussed in Section 5.6.

The basic theory of the derivative is given in the first part of Chapter 6. This important material is standard, except that we have used a result of Carathéodory to give simpler proofs of the Chain Rule and the Inversion Theorem. The remainder of this chapter consists of applications of the Mean Value Theorem and may be explored as time permits.

Chapter 7, dealing with the Riemann integral, has been completely revised in this edition. Rather than introducing upper and lower integrals (as we did in the previous editions), we here define the integral as a limit of Riemann sums. This has the advantage that it is consistent with the students' first exposure to the integral in calculus and in applications; since it is not dependent on order properties, it permits immediate generalization to complex- and vector-valued functions that students may encounter in later courses. Contrary to popular opinion, this limit approach is no more difficult than the order approach. It also is consistent with the generalized Riemann integral that is discussed in detail in Chapter 10. Section 7.4 gives a brief discussion of the familiar numerical methods of calculating the integral of continuous functions.

Sequences of functions and uniform convergence are discussed in the first two sections of Chapter 8, and the basic transcendental functions are put on a firm foundation in Section 8.3 and 8.4 by using uniform convergence. Chapter 9 completes our discussion of infinite series. Chapters 8 and 9 are intrinsically important, and they also show how the material in the earlier chapters can be applied.

Chapter 10 is completely new; it is a presentation of the generalized Riemann integral (sometimes called the "Henstock-Kurzweil" or the "gauge" integral). It will be new to many readers, and we think they will be amazed that such an apparently minor modification of the definition of the Riemann integral can lead to an integral that is more general than the Lebesgue integral. We believe that this relatively new approach to integration theory is both accessible and exciting to anyone who has studied the basic Riemann integral.

The final Chapter 11 deals with topological concepts. Earlier proofs given for intervals are extended to a more abstract setting. For example, the concept of compactness is given proper emphasis and metric spaces are introduced. This chapter will be very useful for students continuing to graduate courses in mathematics.

Throughout the book we have paid more attention to topics from numerical analysis and approximation theory than is usual. We have done so because of the importance of these areas, and to show that real analysis is not merely an exercise in abstract thought.

We have provided rather lengthy lists of exercises, some easy and some challenging. We have provided "hints" for many of these exercises, to help students get started toward a solution or to check their "answer". More complete solutions of almost every exercise are given in a separate Instructor's Manual, which is available to teachers upon request to the publisher.

It is a satisfying experience to see how the mathematical maturity of the students increases and how the students gradually learn to work comfortably with concepts that initially seemed so mysterious. But there is no doubt that a lot of hard work is required on the part of both the students and the teachers.

In order to enrich the historical perspective of the book, we include brief biographical sketches of some famous mathematicians who contributed to this area. We are particularly indebted to Dr. Patrick Muldowney for providing us with his photograph of Professors Henstock and Kurzweil. We also thank John Wiley & Sons for obtaining photographs of the other mathematicians.

We have received many helpful comments from colleagues at a wide variety of institutions who have taught from earlier editions and liked the book enough to express their opinions about how to improve it. We appreciate their remarks and suggestions, even though we did not always follow their advice. We thank them for communicating with us and wish them well in their endeavors to impart the challenge and excitement of learning real analysis and "real" mathematics. It is our hope that they will find this new edition even more helpful than the earlier ones.

February 24, 1999   **Robert G. Bartle**
Ypsilanti and Urbana   **Donald R. Sherbert**

THE GREEK ALPHABET

A	α	Alpha	N	ν	Nu
B	β	Beta	Ξ	ξ	Xi
Γ	γ	Gamma	O	o	Omicron
Δ	δ	Delta	Π	π	Pi
E	ε	Epsilon	P	ρ	Rho
Z	ζ	Zeta	Σ	σ	Sigma
H	η	Eta	T	τ	Tau
Θ	θ	Theta	Υ	υ	Upsilon
I	ι	Iota	Φ	φ	Phi
K	κ	Kappa	X	χ	Chi
Λ	λ	Lambda	Ψ	ψ	Psi
M	μ	Mu	Ω	ω	Omega

To our wives, Carolyn and Janice,
with our appreciation for their
patience, support, and love.

CONTENTS

CHAPTER 1

PRELIMINARIES

In this initial chapter we will present the background needed for the study of real analysis. Section 1.1 consists of a brief survey of set operations and functions, two vital tools for all of mathematics. In it we establish the notation and state the basic definitions and properties that will be used throughout the book. We will regard the word "set" as synonymous with the words "class", "collection", and "family", and we will not define these terms or give a list of axioms for set theory. This approach, often referred to as "naive" set theory, is quite adequate for working with sets in the context of real analysis.

Section 1.2 is concerned with a special method of proof called Mathematical Induction. It is related to the fundamental properties of the natural number system and, though it is restricted to proving particular types of statements, it is important and used frequently. An informal discussion of the different types of proofs that are used in mathematics, such as contrapositives and proofs by contradiction, can be found in Appendix A.

In Section 1.3 we apply some of the tools presented in the first two sections of this chapter to a discussion of what it means for a set to be finite or infinite. Careful definitions are given and some basic consequences of these definitions are derived. The important result that the set of rational numbers is countably infinite is established.

In addition to introducing basic concepts and establishing terminology and notation, this chapter also provides the reader with some initial experience in working with precise definitions and writing proofs. The careful study of real analysis unavoidably entails the reading and writing of proofs, and like any skill, it is necessary to practice. This chapter is a starting point.

Section 1.1 Sets and Functions

To the reader: In this section we give a brief review of the terminology and notation that will be used in this text. We suggest that you look through quickly and come back later when you need to recall the meaning of a term or a symbol.

If an element x is in a set A, we write

$$x \in A$$

and say that x is a **member** of A, or that x **belongs** to A. If x is *not* in A, we write

$$x \notin A.$$

If every element of a set A also belongs to a set B, we say that A is a **subset** of B and write

$$A \subseteq B \quad \text{or} \quad B \supseteq A$$

We say that a set A is a **proper subset** of a set B if $A \subseteq B$, but there is at least one element of B that is not in A. In this case we sometimes write

$$A \subset B.$$

1.1.1 Definition   Two sets A and B are said to be **equal**, and we write $A = B$, if they contain the same elements.

Thus, to prove that the sets A and B are equal, we must show that

$$A \subseteq B \quad \text{and} \quad B \subseteq A.$$

A set is normally defined by either listing its elements explicitly, or by specifying a property that determines the elements of the set. If P denotes a property that is meaningful and unambiguous for elements of a set S, then we write

$$\{x \in S : P(x)\}$$

for the set of all elements x in S for which the property P is true. If the set S is understood from the context, then it is often omitted in this notation.

Several special sets are used throughout this book, and they are denoted by standard symbols. (We will use the symbol := to mean that the symbol on the left is being *defined* by the symbol on the right.)

- The set of **natural numbers** $\mathbb{N} := \{1, 2, 3, \cdots\}$,
- The set of **integers** $\mathbb{Z} := \{0, 1, -1, 2, -2, \cdots\}$,
- The set of **rational numbers** $\mathbb{Q} := \{m/n : m, n \in \mathbb{Z} \text{ and } n \neq 0\}$,
- The set of **real numbers** $\mathbb{R}$.

The set $\mathbb{R}$ of real numbers is of fundamental importance for us and will be discussed at length in Chapter 2.

1.1.2 Examples   (a)   The set

$$\{x \in \mathbb{N} : x^2 - 3x + 2 = 0\}$$

consists of those natural numbers satisfying the stated equation. Since the only solutions of this quadratic equation are $x = 1$ and $x = 2$, we can denote this set more simply by $\{1, 2\}$.

(b)   A natural number n is **even** if it has the form $n = 2k$ for some $k \in \mathbb{N}$. The set of even natural numbers can be written

$$\{2k : k \in \mathbb{N}\},$$

which is less cumbersome than $\{n \in \mathbb{N} : n = 2k, \ k \in \mathbb{N}\}$. Similarly, the set of **odd** natural numbers can be written

$$\{2k - 1 : k \in \mathbb{N}\}. \qquad \square$$

Set Operations ──

We now define the methods of obtaining new sets from given ones. Note that these set operations are based on the meaning of the words "or", "and", and "not". For the union, it is important to be aware of the fact that the word "or" is used in the *inclusive sense*, allowing the possibility that x may belong to both sets. In legal terminology, this inclusive sense is sometimes indicated by "and/or".

1.1.3 Definition   (a)   The **union** of sets A and B is the set

$$A \cup B := \{x : x \in A \text{ or } x \in B\}.$$

(b) The **intersection** of the sets A and B is the set

$$A \cap B := \{x : x \in A \text{ and } x \in B\}.$$

(c) The **complement of B relative to** A is the set

$$A \backslash B := \{x : x \in A \text{ and } x \notin B\}.$$

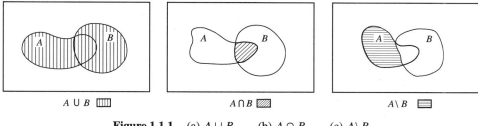

$A \cup B$ ▥ $A \cap B$ ▨ $A \backslash B$ ▤

Figure 1.1.1   (a) $A \cup B$ (b) $A \cap B$ (c) $A \backslash B$

The set that has no elements is called the **empty set** and is denoted by the symbol $\emptyset$. Two sets A and B are said to be **disjoint** if they have no elements in common; this can be expressed by writing $A \cap B = \emptyset$.

To illustrate the method of proving set equalities, we will next establish one of the *DeMorgan laws* for three sets. The proof of the other one is left as an exercise.

1.1.4 Theorem   *If A, B, C are sets, then*

(a) $A \backslash (B \cup C) = (A \backslash B) \cap (A \backslash C)$,

(b) $A \backslash (B \cap C) = (A \backslash B) \cup (A \backslash C)$.

Proof.   To prove (a), we will show that every element in $A \backslash (B \cup C)$ is contained in both $(A \backslash B)$ and $(A \backslash C)$, and conversely.

If x is in $A \backslash (B \cup C)$, then x is in A, but x is not in $B \cup C$. Hence x is in A, but x is neither in B nor in C. Therefore, x is in A but not B, and x is in A but not C. Thus, $x \in A \backslash B$ and $x \in A \backslash C$, which shows that $x \in (A \backslash B) \cap (A \backslash C)$.

Conversely, if $x \in (A \backslash B) \cap (A \backslash C)$, then $x \in (A \backslash B)$ and $x \in (A \backslash C)$. Hence $x \in A$ and both $x \notin B$ and $x \notin C$. Therefore, $x \in A$ and $x \notin (B \cup C)$, so that $x \in A \backslash (B \cup C)$.

Since the sets $(A \backslash B) \cap (A \backslash C)$ and $A \backslash (B \cup C)$ contain the same elements, they are equal by Definition 1.1.1.   Q.E.D.

There are times when it is desirable to form unions and intersections of more than two sets. For a finite collection of sets $\{A_1, A_2, \cdots, A_n\}$, their union is the set A consisting of all elements that belong to *at least one* of the sets A_k, and their intersection consists of all elements that belong to *all* of the sets A_k.

This is extended to an infinite collection of sets $\{A_1, A_2, \cdots, A_n, \cdots\}$ as follows. Their **union** is the set of elements that belong to *at least one* of the sets A_n. In this case we write

$$\bigcup_{n=1}^{\infty} A_n := \{x : x \in A_n \text{ for some } n \in \mathbb{N}\}.$$

Similarly, their **intersection** is the set of elements that belong to *all* of these sets A_n. In this case we write

$$\bigcap_{n=1}^{\infty} A_n := \left\{ x : x \in A_n \text{ for all } n \in \mathbb{N} \right\}.$$

Cartesian Products _____

In order to discuss functions, we define the Cartesian product of two sets.

1.1.5 Definition   If A and B are nonempty sets, then the **Cartesian product** $A \times B$ of A and B is the set of all ordered pairs (a, b) with $a \in A$ and $b \in B$. That is,

$$A \times B := \{(a, b) : a \in A, b \in B\}.$$

Thus if $A = \{1, 2, 3\}$ and $B = \{1, 5\}$, then the set $A \times B$ is the set whose elements are the ordered pairs

$$(1, 1), \quad (1, 5), \quad (2, 1), \quad (2, 5), \quad (3, 1), \quad (3, 5).$$

We may visualize the set $A \times B$ as the set of six points in the plane with the coordinates that we have just listed.

We often draw a diagram (such as Figure 1.1.2) to indicate the Cartesian product of two sets A and B. However, it should be realized that this diagram may be a simplification. For example, if $A := \{x \in \mathbb{R} : 1 \leq x \leq 2\}$ and $B := \{y \in \mathbb{R} : 0 \leq y \leq 1 \text{ or } 2 \leq y \leq 3\}$, then instead of a rectangle, we should have a drawing such as Figure 1.1.3.

We will now discuss the fundamental notion of a *function* or a *mapping*.

To the mathematician of the early nineteenth century, the word "function" meant a definite formula, such as $f(x) := x^2 + 3x - 5$, which associates to each real number x another number $f(x)$. (Here, $f(0) = -5$, $f(1) = -1$, $f(5) = 35$.) This understanding excluded the case of different formulas on different intervals, so that functions could not be defined "in pieces".

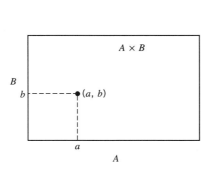

Figure 1.1.2

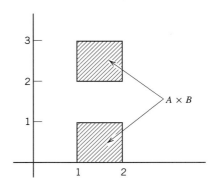

Figure 1.1.3

As mathematics developed, it became clear that a more general definition of "function" would be useful. It also became evident that it is important to make a clear distinction between the function itself and the values of the function. A revised definition might be:

> A function f from a set A into a set B is a rule of correspondence that assigns to each element x in A a uniquely determined element $f(x)$ in B.

But however suggestive this revised definition might be, there is the difficulty of interpreting the phrase "rule of correspondence". In order to clarify this, we will express the definition entirely in terms of sets; in effect, we will define a function to be its **graph**. While this has the disadvantage of being somewhat artificial, it has the advantage of being unambiguous and clearer.

1.1.6 Definition   Let A and B be sets. Then a **function** from A to B is a set f of ordered pairs in $A \times B$ such that for each $a \in A$ there exists a unique $b \in B$ with $(a, b) \in f$. (In other words, if $(a, b) \in f$ and $(a, b') \in f$, then $b = b'$.)

The set A of first elements of a function f is called the **domain** of f and is often denoted by $D(f)$. The set of all second elements in f is called the **range** of f and is often denoted by $R(f)$. Note that, although $D(f) = A$, we only have $R(f) \subseteq B$. (See Figure 1.1.4.)

The essential condition that:

$$(a, b) \in f \qquad \text{and} \qquad (a, b') \in f \qquad \text{implies that} \qquad b = b'$$

is sometimes called the *vertical line test*. In geometrical terms it says every vertical line $x = a$ with $a \in A$ intersects the graph of f exactly once.

The notation

$$f : A \to B$$

is often used to indicate that f is a function from A into B. We will also say that f is a **mapping** of A into B, or that f **maps** A into B. If (a, b) is an element in f, it is customary to write

$$b = f(a) \qquad \text{or sometimes} \qquad a \mapsto b.$$

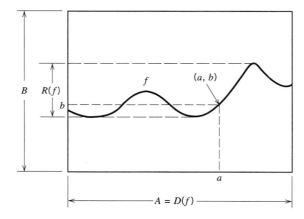

Figure 1.1.4   A function as a graph

If $b = f(a)$, we often refer to b as the **value** of f at a, or as the **image** of a under f.

Transformations and Machines

Aside from using graphs, we can visualize a function as a *transformation* of the set $D(f) = A$ into the set $R(f) \subseteq B$. In this phraseology, when $(a, b) \in f$, we think of f as taking the element a from A and "transforming" or "mapping" it into an element $b = f(a)$ in $R(f) \subseteq B$. We often draw a diagram, such as Figure 1.1.5, even when the sets A and B are not subsets of the plane.

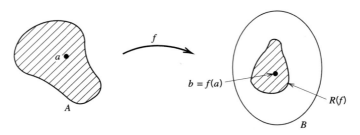

Figure 1.1.5   A function as a transformation

There is another way of visualizing a function: namely, as a *machine* that accepts elements of $D(f) = A$ as *inputs* and produces corresponding elements of $R(f) \subseteq B$ as *outputs*. If we take an element $x \in D(f)$ and put it into f, then out comes the corresponding value $f(x)$. If we put a different element $y \in D(f)$ into f, then out comes $f(y)$ which may or may not differ from $f(x)$. If we try to insert something that does not belong to $D(f)$ into f, we find that it is not accepted, for f can operate only on elements from $D(f)$. (See Figure 1.1.6.)

This last visualization makes clear the distinction between f and $f(x)$: the first is the machine itself, and the second is the output of the machine f when x is the input. Whereas no one is likely to confuse a meat grinder with ground meat, enough people have confused functions with their values that it is worth distinguishing between them notationally.

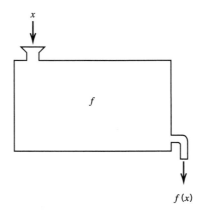

Figure 1.1.6   A function as a machine

Direct and Inverse Images

Let $f : A \to B$ be a function with domain $D(f) = A$ and range $R(f) \subseteq B$.

1.1.7 Definition If E is a subset of A, then the **direct image** of E under f is the subset $f(E)$ of B given by

$$f(E) := \{f(x) : x \in E\}.$$

If H is a subset of B, then the **inverse image** of H under f is the subset $f^{-1}(H)$ of A given by

$$f^{-1}(H) := \{x \in A : f(x) \in H\}.$$

Remark The notation $f^{-1}(H)$ used in this connection has its disadvantages. However, we will use it since it is the standard notation.

Thus, if we are given a set $E \subseteq A$, then a point $y_1 \in B$ is in the direct image $f(E)$ if and only if there exists at least one point $x_1 \in E$ such that $y_1 = f(x_1)$. Similarly, given a set $H \subseteq B$, then a point x_2 is in the inverse image $f^{-1}(H)$ if and only if $y_2 := f(x_2)$ belongs to H. (See Figure 1.1.7.)

1.1.8 Examples **(a)** Let $f : \mathbb{R} \to \mathbb{R}$ be defined by $f(x) := x^2$. Then the direct image of the set $E := \{x : 0 \le x \le 2\}$ is the set $f(E) = \{y : 0 \le y \le 4\}$.

If $G := \{y : 0 \le y \le 4\}$, then the inverse image of G is the set $f^{-1}(G) = \{x : -2 \le x \le 2\}$. Thus, in this case, we see that $f^{-1}(f(E)) \ne E$.

On the other hand, we have $f\left(f^{-1}(G)\right) = G$. But if $H := \{y : -1 \le y \le 1\}$, then we have $f\left(f^{-1}(H)\right) = \{y : 0 \le y \le 1\} \ne H$.

A sketch of the graph of f may help to visualize these sets.

(b) Let $f : A \to B$, and let G, H be subsets of B. We will show that

$$f^{-1}(G \cap H) \subseteq f^{-1}(G) \cap f^{-1}(H).$$

For, if $x \in f^{-1}(G \cap H)$, then $f(x) \in G \cap H$, so that $f(x) \in G$ and $f(x) \in H$. But this implies that $x \in f^{-1}(G)$ and $x \in f^{-1}(H)$, whence $x \in f^{-1}(G) \cap f^{-1}(H)$. Thus the stated implication is proved. [The opposite inclusion is also true, so that we actually have set equality between these sets; see Exercise 13.] $\qquad\square$

Further facts about direct and inverse images are given in the exercises.

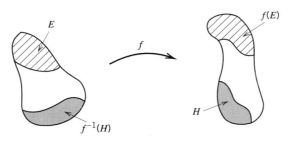

Figure 1.1.7 Direct and inverse images

Special Types of Functions ————————————————————————

The following definitions identify some very important types of functions.

1.1.9 Definition   Let $f : A \to B$ be a function from A to B.

(a) The function f is said to be **injective** (or to be **one-one**) if whenever $x_1 \neq x_2$, then $f(x_1) \neq f(x_2)$. If f is an injective function, we also say that f is an **injection**.

(b) The function f is said to be **surjective** (or to map A **onto** B) if $f(A) = B$; that is, if the range $R(f) = B$. If f is a surjective function, we also say that f is a **surjection**.

(c) If f is both injective and surjective, then f is said to be **bijective**. If f is bijective, we also say that f is a **bijection**.

- In order to prove that a function f is injective, we must establish that:

$$\text{for all } x_1, x_2 \text{ in } A, \text{ if } f(x_1) = f(x_2), \text{ then } x_1 = x_2.$$

   To do this we assume that $f(x_1) = f(x_2)$ and show that $x_1 = x_2$.
   [In other words, the graph of f satisfies the *first horizontal line test:* Every horizontal line $y = b$ with $b \in B$ intersects the graph f in *at most* one point.]

- To prove that a function f is surjective, we must show that for any $b \in B$ there exists at least one $x \in A$ such that $f(x) = b$.
   [In other words, the graph of f satisfies the *second horizontal line test:* Every horizontal line $y = b$ with $b \in B$ intersects the graph f in *at least* one point.]

1.1.10 Example   Let $A := \{x \in \mathbb{R} : x \neq 1\}$ and define $f(x) := 2x/(x-1)$ for all $x \in A$. To show that f is injective, we take x_1 and x_2 in A and assume that $f(x_1) = f(x_2)$. Thus we have

$$\frac{2x_1}{x_1 - 1} = \frac{2x_2}{x_2 - 1},$$

which implies that $x_1(x_2 - 1) = x_2(x_1 - 1)$, and hence $x_1 = x_2$. Therefore f is injective.

To determine the range of f, we solve the equation $y = 2x/(x-1)$ for x in terms of y. We obtain $x = y/(y-2)$, which is meaningful for $y \neq 2$. Thus the range of f is the set $B := \{y \in \mathbb{R} : y \neq 2\}$. Thus, f is a bijection of A onto B.   □

Inverse Functions ————————————————————————

If f is a function from A into B, then f is a special subset of $A \times B$ (namely, one passing the *vertical line test.*) The set of ordered pairs in $B \times A$ obtained by interchanging the members of ordered pairs in f is not generally a function. (That is, the set f may not pass *both* of the *horizontal line tests.*) However, if f is a bijection, then this interchange does lead to a function, called the "inverse function" of f.

1.1.11 Definition   If $f : A \to B$ is a bijection of A onto B, then

$$g := \{(b, a) \in B \times A : (a, b) \in f\}$$

is a function on B into A. This function is called the **inverse function** of f, and is denoted by f^{-1}. The function f^{-1} is also called the **inverse** of f.

We can also express the connection between f and its inverse f^{-1} by noting that $D(f) = R(f^{-1})$ and $R(f) = D(f^{-1})$ and that

$$b = f(a) \qquad \text{if and only if} \qquad a = f^{-1}(b).$$

For example, we saw in Example 1.1.10 that the function

$$f(x) := \frac{2x}{x - 1}$$

is a bijection of $A := \{x \in \mathbb{R} : x \neq 1\}$ onto the set $B := \{y \in \mathbb{R} : y \neq 2\}$. The function inverse to f is given by

$$f^{-1}(y) := \frac{y}{y - 2} \qquad \text{for} \quad y \in B.$$

Remark   We introduced the notation $f^{-1}(H)$ in Definition 1.1.7. It makes sense even if f does not have an inverse function. However, if the inverse function f^{-1} does exist, then $f^{-1}(H)$ is the direct image of the set $H \subseteq B$ under f^{-1}.

Composition of Functions _____

It often happens that we want to "compose" two functions f, g by first finding $f(x)$ and then applying g to get $g(f(x))$; however, this is possible only when $f(x)$ belongs to the domain of g. In order to be able to do this for *all* $f(x)$, we must assume that the range of f is contained in the domain of g. (See Figure 1.1.8.)

1.1.12 Definition   If $f : A \to B$ and $g : B \to C$, and if $R(f) \subseteq D(g) = B$, then the **composite function** $g \circ f$ (note the order!) is the function from A into C defined by

$$(g \circ f)(x) := g(f(x)) \qquad \text{for all} \quad x \in A.$$

1.1.13 Examples   (a)   The order of the composition must be carefully noted. For, let f and g be the functions whose values at $x \in \mathbb{R}$ are given by

$$f(x) := 2x \qquad \text{and} \qquad g(x) := 3x^2 - 1.$$

Since $D(g) = \mathbb{R}$ and $R(f) \subseteq \mathbb{R} = D(g)$, then the domain $D(g \circ f)$ is also equal to $\mathbb{R}$, and the composite function $g \circ f$ is given by

$$(g \circ f)(x) = 3(2x)^2 - 1 = 12x^2 - 1.$$

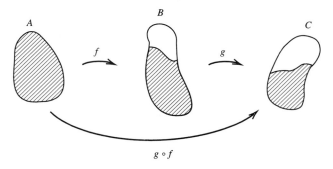

$g \circ f$

Figure 1.1.8   The composition of f and g

On the other hand, the domain of the composite function $f \circ g$ is also $\mathbb{R}$, but

$$(f \circ g)(x) = 2(3x^2 - 1) = 6x^2 - 2.$$

Thus, in this case, we have $g \circ f \neq f \circ g$.

(b) In considering $g \circ f$, some care must be exercised to be sure that the range of f is contained in the domain of g. For example, if

$$f(x) := 1 - x^2 \qquad \text{and} \qquad g(x) := \sqrt{x},$$

then, since $D(g) = \{x : x \geq 0\}$, the composite function $g \circ f$ is given by the formula

$$(g \circ f)(x) = \sqrt{1 - x^2}$$

only for $x \in D(f)$ that satisfy $f(x) \geq 0$; that is, for x satisfying $-1 \leq x \leq 1$.

We note that if we reverse the order, then the composition $f \circ g$ is given by the formula

$$(f \circ g)(x) = 1 - x,$$

but only for those x in the domain $D(g) = \{x : x \geq 0\}$.   □

We now give the relationship between composite functions and inverse images. The proof is left as an instructive exercise.

1.1.14 Theorem   Let $f : A \to B$ and $g : B \to C$ be functions and let H be a subset of C. Then we have

$$(g \circ f)^{-1}(H) = f^{-1}(g^{-1}(H)).$$

Note the *reversal* in the order of the functions.

Restrictions of Functions

If $f : A \to B$ is a function and if $A_1 \subset A$, we can define a function $f_1 : A_1 \to B$ by

$$f_1(x) := f(x) \qquad \text{for} \quad x \in A_1.$$

The function f_1 is called the **restriction of** f **to** A_1. Sometimes it is denoted by $f_1 = f|A_1$.

It may seem strange to the reader that one would ever choose to throw away a part of a function, but there are some good reasons for doing so. For example, if $f : \mathbb{R} \to \mathbb{R}$ is the **squaring function**:

$$f(x) := x^2 \qquad \text{for} \quad x \in \mathbb{R},$$

then f is not injective, so it cannot have an inverse function. However, if we restrict f to the set $A_1 := \{x : x \geq 0\}$, then the restriction $f|A_1$ is a bijection of A_1 onto A_1. Therefore, this restriction has an inverse function, which is the **positive square root function**. (Sketch a graph.)

Similarly, the trigonometric functions $S(x) := \sin x$ and $C(x) := \cos x$ are not injective on all of $\mathbb{R}$. However, by making suitable restrictions of these functions, one can obtain the **inverse sine** and the **inverse cosine** functions that the reader has undoubtedly already encountered.

Exercises for Section 1.1

1. If A and B are sets, show that $A \subseteq B$ if and only if $A \cap B = A$.

2. Prove the second De Morgan Law [Theorem 1.1.4(b)].

3. Prove the Distributive Laws:
    (a) $A \cap (B \cup C) = (A \cap B) \cup (A \cap C)$,
    (b) $A \cup (B \cap C) = (A \cup B) \cap (A \cup C)$.

4. The **symmetric difference** of two sets A and B is the set D of all elements that belong to either A or B but not both. Represent D with a diagram.
    (a)   Show that $D = (A \backslash B) \cup (B \backslash A)$.
    (b)   Show that D is also given by $D = (A \cup B) \backslash (A \cap B)$.

5. For each $n \in \mathbb{N}$, let $A_n = \{(n+1)k : k \in \mathbb{N}\}$.
    (a)   What is $A_1 \cap A_2$?
    (b)   Determine the sets $\bigcup\{A_n : n \in \mathbb{N}\}$ and $\bigcap\{A_n : n \in \mathbb{N}\}$.

6. Draw diagrams in the plane of the Cartesian products $A \times B$ for the given sets A and B.
    (a) $A = \{x \in \mathbb{R} : 1 \le x \le 2 \text{ or } 3 \le x \le 4\}$, $B = \{x \in \mathbb{R} : x = 1 \text{ or } x = 2\}$.
    (b) $A = \{1, 2, 3\}$, $B = \{x \in \mathbb{R} : 1 \le x \le 3\}$.

7. Let $A := B := \{x \in \mathbb{R} : -1 \le x \le 1\}$ and consider the subset $C := \{(x, y) : x^2 + y^2 = 1\}$ of $A \times B$. Is this set a function? Explain.

8. Let $f(x) := 1/x^2$, $x \ne 0$, $x \in \mathbb{R}$.
    (a)   Determine the direct image $f(E)$ where $E := \{x \in \mathbb{R} : 1 \le x \le 2\}$.
    (b)   Determine the inverse image $f^{-1}(G)$ where $G := \{x \in \mathbb{R} : 1 \le x \le 4\}$.

9. Let $g(x) := x^2$ and $f(x) := x + 2$ for $x \in \mathbb{R}$, and let h be the composite function $h := g \circ f$.
    (a)   Find the direct image $h(E)$ of $E := \{x \in \mathbb{R} : 0 \le x \le 1\}$.
    (b)   Find the inverse image $h^{-1}(G)$ of $G := \{x \in \mathbb{R} : 0 \le x \le 4\}$.

10. Let $f(x) := x^2$ for $x \in \mathbb{R}$, and let $E := \{x \in \mathbb{R} : -1 \le x \le 0\}$ and $F := \{x \in \mathbb{R} : 0 \le x \le 1\}$. Show that $E \cap F = \{0\}$ and $f(E \cap F) = \{0\}$, while $f(E) = f(F) = \{y \in \mathbb{R} : 0 \le y \le 1\}$. Hence $f(E \cap F)$ is a proper subset of $f(E) \cap f(F)$. What happens if 0 is deleted from the sets E and F?

11. Let f and E, F be as in Exercise 10. Find the sets $E \backslash F$ and $f(E) \backslash f(F)$ and show that it is *not* true that $f(E \backslash F) \subseteq f(E) \backslash f(F)$.

12. Show that if $f : A \to B$ and E, F are subsets of A, then $f(E \cup F) = f(E) \cup f(F)$ and $f(E \cap F) \subseteq f(E) \cap f(F)$.

13. Show that if $f : A \to B$ and G, H are subsets of B, then $f^{-1}(G \cup H) = f^{-1}(G) \cup f^{-1}(H)$ and $f^{-1}(G \cap H) = f^{-1}(G) \cap f^{-1}(H)$.

14. Show that the function f defined by $f(x) := x/\sqrt{x^2 + 1}$, $x \in \mathbb{R}$, is a bijection of $\mathbb{R}$ onto $\{y : -1 < y < 1\}$.

15. For $a, b \in \mathbb{R}$ with $a < b$, find an explicit bijection of $A := \{x : a < x < b\}$ onto $B := \{y : 0 < y < 1\}$.

16. Give an example of two functions f, g on $\mathbb{R}$ to $\mathbb{R}$ such that $f \ne g$, but such that $f \circ g = g \circ f$.

17. (a)   Show that if $f : A \to B$ is injective and $E \subseteq A$, then $f^{-1}(f(E)) = E$. Give an example to show that equality need not hold if f is not injective.
    (b)   Show that if $f : A \to B$ is surjective and $H \subseteq B$, then $f(f^{-1}(H)) = H$. Give an example to show that equality need not hold if f is not surjective.

18. (a)   Suppose that f is an injection. Show that $f^{-1} \circ f(x) = x$ for all $x \in D(f)$ and that $f \circ f^{-1}(y) = y$ for all $y \in R(f)$.
    (b)   If f is a bijection of A onto B, show that f^{-1} is a bijection of B onto A.

19. Prove that if $f : A \to B$ is bijective and $g : B \to C$ is bijective, then the composite $g \circ f$ is a bijective map of A onto C.

20. Let $f : A \to B$ and $g : B \to C$ be functions.
     (a) Show that if $g \circ f$ is injective, then f is injective.
     (b) Show that if $g \circ f$ is surjective, then g is surjective.

21. Prove Theorem 1.1.14.

22. Let f, g be functions such that $(g \circ f)(x) = x$ for all $x \in D(f)$ and $(f \circ g)(y) = y$ for all $y \in D(g)$. Prove that $g = f^{-1}$.

Section 1.2   Mathematical Induction

Mathematical Induction is a powerful method of proof that is frequently used to establish the validity of statements that are given in terms of the natural numbers. Although its utility is restricted to this rather special context, Mathematical Induction is an indispensable tool in all branches of mathematics. Since many induction proofs follow the same formal lines of argument, we will often state only that a result follows from Mathematical Induction and leave it to the reader to provide the necessary details. In this section, we will state the principle and give several examples to illustrate how inductive proofs proceed.

We shall assume familiarity with the set of natural numbers:

$$\mathbb{N} := \{1, 2, 3, \cdots\},$$

with the usual arithmetic operations of addition and multiplication, and with the meaning of a natural number being less than another one. We will also assume the following fundamental property of $\mathbb{N}$.

1.2.1 Well-Ordering Property of $\mathbb{N}$ *Every nonempty subset of* $\mathbb{N}$ *has a least element.*

A more detailed statement of this property is as follows: If S is a subset of $\mathbb{N}$ and if $S \neq \emptyset$, then there exists $m \in S$ such that $m \leq k$ for all $k \in S$.

On the basis of the Well-Ordering Property, we shall derive a version of the Principle of Mathematical Induction that is expressed in terms of subsets of $\mathbb{N}$.

1.2.2 Principle of Mathematical Induction   *Let S be a subset of $\mathbb{N}$ that possesses the two properties:*

(1) *The number $1 \in S$.*

(2) *For every $k \in \mathbb{N}$, if $k \in S$, then $k + 1 \in S$.*

Then we have $S = \mathbb{N}$.

Proof.   Suppose to the contrary that $S \neq \mathbb{N}$. Then the set $\mathbb{N} \backslash S$ is not empty, so by the Well-Ordering Principle it has a least element m. Since $1 \in S$ by hypothesis (1), we know that $m > 1$. But this implies that $m - 1$ is also a natural number. Since $m - 1 < m$ and since m is the least element in $\mathbb{N}$ such that $m \notin S$, we conclude that $m - 1 \in S$.

We now apply hypothesis (2) to the element $k := m - 1$ in S, to infer that $k + 1 = (m - 1) + 1 = m$ belongs to S. But this statement contradicts the fact that $m \notin S$. Since m was obtained from the assumption that $\mathbb{N} \backslash S$ is not empty, we have obtained a contradiction. Therefore we must have $S = \mathbb{N}$.   Q.E.D.

The Principle of Mathematical Induction is often set forth in the framework of proper-ties or statements about natural numbers. If $P(n)$ is a meaningful statement about $n \in \mathbb{N}$, then $P(n)$ may be true for some values of n and false for others. For example, if $P_1(n)$ is the statement: "$n^2 = n$", then $P_1(1)$ is true while $P_1(n)$ is false for all $n > 1, n \in \mathbb{N}$. On the other hand, if $P_2(n)$ is the statement: "$n^2 > 1$", then $P_2(1)$ is false, while $P_2(n)$ is true for all $n > 1, n \in \mathbb{N}$.

In this context, the Principle of Mathematical Induction can be formulated as follows.

For each $n \in \mathbb{N}$, let $P(n)$ be a statement about n. Suppose that:

(1') $P(1)$ *is true.*
(2') *For every $k \in \mathbb{N}$, if $P(k)$ is true, then $P(k+1)$ is true.*

Then $P(n)$ is true for all $n \in \mathbb{N}$.

The connection with the preceding version of Mathematical Induction, given in 1.2.2, is made by letting $S := \{n \in \mathbb{N} : P(n)$ is true$\}$. Then the conditions (1) and (2) of 1.2.2 correspond exactly to the conditions (1') and (2'), respectively. The conclusion that $S = \mathbb{N}$ in 1.2.2 corresponds to the conclusion that $P(n)$ is true for all $n \in \mathbb{N}$.

In (2') the assumption "if $P(k)$ is true" is called the **induction hypothesis**. In estab-lishing (2'), we are not concerned with the actual truth or falsity of $P(k)$, but only with the validity of the implication "if $P(k)$, then $P(k+1)$". For example, if we consider the statements $P(n)$: "$n = n + 5$", then (2') is logically correct, for we can simply add 1 to both sides of $P(k)$ to obtain $P(k+1)$. However, since the statement $P(1)$: "$1 = 6$" is false, we cannot use Mathematical Induction to conclude that $n = n + 5$ for all $n \in \mathbb{N}$.

It may happen that statements $P(n)$ are false for certain natural numbers but then are true for all $n \geq n_0$ for some particular n_0. The Principle of Mathematical Induction can be modified to deal with this situation. We will formulate the modified principle, but leave its verification as an exercise. (See Exercise 12.)

1.2.3 Principle of Mathematical Induction (second version)   *Let $n_0 \in \mathbb{N}$ and let $P(n)$ be a statement for each natural number $n \geq n_0$. Suppose that:*

(1) *The statement $P(n_0)$ is true.*
(2) *For all $k \geq n_0$, the truth of $P(k)$ implies the truth of $P(k+1)$.*

Then $P(n)$ is true for all $n \geq n_0$.

Sometimes the number n_0 in (1) is called the **base**, since it serves as the starting point, and the implication in (2), which can be written $P(k) \Rightarrow P(k+1)$, is called the **bridge**, since it connects the case k to the case $k+1$.

The following examples illustrate how Mathematical Induction is used to prove asser-tions about natural numbers.

1.2.4 Examples   (a) For each $n \in \mathbb{N}$, the sum of the first n natural numbers is given by

$$1 + 2 + \cdots + n = \tfrac{1}{2}n(n+1).$$

To prove this formula, we let S be the set of all $n \in \mathbb{N}$ for which the formula is true. We must verify that conditions (1) and (2) of 1.2.2 are satisfied. If $n = 1$, then we have $1 = \tfrac{1}{2} \cdot 1 \cdot (1+1)$ so that $1 \in S$, and (1) is satisfied. Next, we *assume* that $k \in S$ and wish to infer from this assumption that $k + 1 \in S$. Indeed, if $k \in S$, then

$$1 + 2 + \cdots + k = \tfrac{1}{2}k(k+1).$$

If we add $k + 1$ to both sides of the assumed equality, we obtain

$$1 + 2 + \cdots + k + (k + 1) = \tfrac{1}{2}k(k + 1) + (k + 1)$$
$$= \tfrac{1}{2}(k + 1)(k + 2).$$

Since this is the stated formula for $n = k + 1$, we conclude that $k + 1 \in S$. Therefore, condition (2) of 1.2.2 is satisfied. Consequently, by the Principle of Mathematical Induction, we infer that $S = \mathbb{N}$, so the formula holds for all $n \in \mathbb{N}$.

(b) For each $n \in \mathbb{N}$, the sum of the squares of the first n natural numbers is given by

$$1^2 + 2^2 + \cdots + n^2 = \tfrac{1}{6}n(n + 1)(2n + 1).$$

To establish this formula, we note that it is true for $n = 1$, since $1^2 = \tfrac{1}{6} \cdot 1 \cdot 2 \cdot 3$. If we assume it is true for k, then adding $(k + 1)^2$ to both sides of the assumed formula gives

$$1^2 + 2^2 + \cdots + k^2 + (k + 1)^2 = \tfrac{1}{6}k(k + 1)(2k + 1) + (k + 1)^2$$
$$= \tfrac{1}{6}(k + 1)(2k^2 + k + 6k + 6)$$
$$= \tfrac{1}{6}(k + 1)(k + 2)(2k + 3).$$

Consequently, the formula is valid for all $n \in \mathbb{N}$.

(c) Given two real numbers a and b, we will prove that $a - b$ is a factor of $a^n - b^n$ for all $n \in \mathbb{N}$.

First we see that the statement is clearly true for $n = 1$. If we now assume that $a - b$ is a factor of $a^k - b^k$, then

$$a^{k+1} - b^{k+1} = a^{k+1} - ab^k + ab^k - b^{k+1}$$
$$= a(a^k - b^k) + b^k(a - b).$$

By the induction hypothesis, $a - b$ is a factor of $a(a^k - b^k)$ and it is plainly a factor of $b^k(a - b)$. Therefore, $a - b$ is a factor of $a^{k+1} - b^{k+1}$, and it follows from Mathematical Induction that $a - b$ is a factor of $a^n - b^n$ for all $n \in \mathbb{N}$.

A variety of divisibility results can be derived from this fact. For example, since $11 - 7 = 4$, we see that $11^n - 7^n$ is divisible by 4 for all $n \in \mathbb{N}$.

(d) The inequality $2^n > 2n + 1$ is false for $n = 1, 2$, but it is true for $n = 3$. If we assume that $2^k > 2k + 1$, then multiplication by 2 gives, when $2k + 2 > 3$, the inequality

$$2^{k+1} > 2(2k + 1) = 4k + 2 = 2k + (2k + 2) > 2k + 3 = 2(k + 1) + 1.$$

Since $2k + 2 > 3$ for all $k \geq 1$, the bridge is valid for all $k \geq 1$ (even though the statement is false for $k = 1, 2$). Hence, with the base $n_0 = 3$, we can apply Mathematical Induction to conclude that the inequality holds for all $n \geq 3$.

(e) The inequality $2^n \leq (n + 1)!$ can be established by Mathematical Induction.

We first observe that it is true for $n = 1$, since $2^1 = 2 = 1 + 1$. If we assume that $2^k \leq (k + 1)!$, it follows from the fact that $2 \leq k + 2$ that

$$2^{k+1} = 2 \cdot 2^k \leq 2(k + 1)! \leq (k + 2)(k + 1)! = (k + 2)!.$$

Thus, if the inequality holds for k, then it also holds for $k + 1$. Therefore, Mathematical Induction implies that the inequality is true for all $n \in \mathbb{N}$.

(f) If $r \in \mathbb{R}$, $r \neq 1$, and $n \in \mathbb{N}$, then

$$1 + r + r^2 + \cdots + r^n = \frac{1 - r^{n+1}}{1 - r}.$$

This is the formula for the sum of the terms in a "geometric progression". It can be established using Mathematical Induction as follows. First, if $n = 1$, then $1 + r = (1 - r^2)/(1 - r)$. If we assume the truth of the formula for $n = k$ and add the term r^{k+1} to both sides, we get (after a little algebra)

$$1 + r + r^k + \cdots + r^{k+1} = \frac{1 - r^{k+1}}{1 - r} + r^{k+1} = \frac{1 - r^{k+2}}{1 - r},$$

which is the formula for $n = k + 1$. Therefore, Mathematical Induction implies the validity of the formula for all $n \in \mathbb{N}$.

[This result can also be proved without using Mathematical Induction. If we let $s_n := 1 + r + r^2 + \cdots + r^n$, then $rs_n = r + r^2 + \cdots + r^{n+1}$, so that

$$(1 - r)s_n = s_n - rs_n = 1 - r^{n+1}.$$

If we divide by $1 - r$, we obtain the stated formula.]

(g) Careless use of the Principle of Mathematical Induction can lead to obviously absurd conclusions. The reader is invited to find the error in the "proof" of the following assertion.

Claim: If $n \in \mathbb{N}$ and if the maximum of the natural numbers p and q is n, then $p = q$.

"Proof."   Let S be the subset of $\mathbb{N}$ for which the claim is true. Evidently, $1 \in S$ since if $p, q \in \mathbb{N}$ and their maximum is 1, then both equal 1 and so $p = q$. Now assume that $k \in S$ and that the maximum of p and q is $k + 1$. Then the maximum of $p - 1$ and $q - 1$ is k. But since $k \in S$, then $p - 1 = q - 1$ and therefore $p = q$. Thus, $k + 1 \in S$, and we conclude that the assertion is true for all $n \in \mathbb{N}$.

(h)   There are statements that are true for *many* natural numbers but that are not true for *all* of them.

For example, the formula $p(n) := n^2 - n + 41$ gives a prime number for $n = 1, 2, \cdots,$ 40. However, $p(41)$ is obviously divisible by 41, so it is not a prime number.   □

Another version of the Principle of Mathematical Induction is sometimes quite useful. It is called the "Principle of Strong Induction", even though it is in fact equivalent to 1.2.2.

1.2.5 Principle of Strong Induction   *Let S be a subset of $\mathbb{N}$ such that*

(1″) $1 \in S$.
(2″) *For every $k \in \mathbb{N}$, if $\{1, 2, \cdots, k\} \subseteq S$, then $k + 1 \in S$.*

Then $S = \mathbb{N}$.

We will leave it to the reader to establish the equivalence of 1.2.2 and 1.2.5.

Exercises for Section 1.2

1. Prove that $1/1 \cdot 2 + 1/2 \cdot 3 + \cdots + 1/n(n + 1) = n/(n + 1)$ for all $n \in \mathbb{N}$.

2. Prove that $1^3 + 2^3 + \cdots + n^3 = \left[\frac{1}{2}n(n + 1)\right]^2$ for all $n \in \mathbb{N}$.

3. Prove that $3 + 11 + \cdots + (8n - 5) = 4n^2 - n$ for all $n \in \mathbb{N}$.

4. Prove that $1^2 + 3^2 + \cdots + (2n - 1)^2 = (4n^3 - n)/3$ for all $n \in \mathbb{N}$.

5. Prove that $1^2 - 2^2 + 3^2 + \cdots + (-1)^{n+1}n^2 = (-1)^{n+1}n(n + 1)/2$ for all $n \in \mathbb{N}$.

6. Prove that $n^3 + 5n$ is divisible by 6 for all $n \in \mathbb{N}$.

7. Prove that $5^{2n} - 1$ is divisible by 8 for all $n \in \mathbb{N}$.

8. Prove that $5^n - 4n - 1$ is divisible by 16 for all $n \in \mathbb{N}$.

9. Prove that $n^3 + (n+1)^3 + (n+2)^3$ is divisible by 9 for all $n \in \mathbb{N}$.

10. Conjecture a formula for the sum $1/1 \cdot 3 + 1/3 \cdot 5 + \cdots + 1/(2n-1)(2n+1)$, and prove your conjecture by using Mathematical Induction.

11. Conjecture a formula for the sum of the first n odd natural numbers $1 + 3 + \cdots + (2n-1)$, and prove your formula by using Mathematical Induction.

12. Prove the Principle of Mathematical Induction 1.2.3 (second version).

13. Prove that $n < 2^n$ for all $n \in \mathbb{N}$.

14. Prove that $2^n < n!$ for all $n \geq 4$, $n \in \mathbb{N}$.

15. Prove that $2n - 3 \leq 2^{n-2}$ for all $n \geq 5$, $n \in \mathbb{N}$.

16. Find all natural numbers n such that $n^2 < 2^n$. Prove your assertion.

17. Find the largest natural number m such that $n^3 - n$ is divisible by m for all $n \in \mathbb{N}$. Prove your assertion.

18. Prove that $1/\sqrt{1} + 1/\sqrt{2} + \cdots + 1/\sqrt{n} > \sqrt{n}$ for all $n \in \mathbb{N}$.

19. Let S be a subset of $\mathbb{N}$ such that (a) $2^k \in S$ for all $k \in \mathbb{N}$, and (b) if $k \in S$ and $k \geq 2$, then $k - 1 \in S$. Prove that $S = \mathbb{N}$.

20. Let the numbers x_n be defined as follows: $x_1 := 1$, $x_2 := 2$, and $x_{n+2} := \frac{1}{2}(x_{n+1} + x_n)$ for all $n \in \mathbb{N}$. Use the Principle of Strong Induction (1.2.5) to show that $1 \leq x_n \leq 2$ for all $n \in \mathbb{N}$.

Section 1.3   Finite and Infinite Sets

When we count the elements in a set, we say "one, two, three,$\cdots$", stopping when we have exhausted the set. From a mathematical perspective, what we are doing is defining a bijective mapping between the set and a portion of the set of natural numbers. If the set is such that the counting does not terminate, such as the set of natural numbers itself, then we describe the set as being infinite.

The notions of "finite" and "infinite" are extremely primitive, and it is very likely that the reader has never examined these notions very carefully. In this section we will define these terms precisely and establish a few basic results and state some other important results that seem obvious but whose proofs are a bit tricky. These proofs can be found in Appendix B and can be read later.

1.3.1 Definition   (a)   The empty set $\emptyset$ is said to have 0 **elements**.

(b)   If $n \in \mathbb{N}$, a set S is said to have n **elements** if there exists a bijection from the set $\mathbb{N}_n := \{1, 2, \cdots, n\}$ onto S.

(c)   A set S is said to be **finite** if it is either empty or it has n elements for some $n \in \mathbb{N}$.

(d)   A set S is said to be **infinite** if it is not finite.

Since the inverse of a bijection is a bijection, it is easy to see that a set S has n elements if and only if there is a bijection from S onto the set $\{1, 2, \cdots, n\}$. Also, since the composition of two bijections is a bijection, we see that a set S_1 has n elements if and only if there is a bijection from S_1 onto another set S_2 that has n elements. Further, a set T_1 is finite if and only if there is a bijection from T_1 onto another set T_2 that is finite.

It is now necessary to establish some basic properties of finite sets to be sure that the definitions do not lead to conclusions that conflict with our experience of counting. From the definitions, it is not entirely clear that a finite set might not have n elements for *more than one* value of n. Also it is conceivably possible that the set $\mathbb{N} := \{1, 2, 3, \cdots\}$ might be a finite set according to this definition. The reader will be relieved that these possibilities do not occur, as the next two theorems state. The proofs of these assertions, which use the fundamental properties of $\mathbb{N}$ described in Section 1.2, are given in Appendix B.

1.3.2 Uniqueness Theorem   *If S is a finite set, then the number of elements in S is a unique number in $\mathbb{N}$.*

1.3.3 Theorem   *The set $\mathbb{N}$ of natural numbers is an infinite set.*

The next result gives some elementary properties of finite and infinite sets.

1.3.4 Theorem   **(a)** *If A is a set with m elements and B is a set with n elements and if $A \cap B = \emptyset$, then $A \cup B$ has $m + n$ elements.*

(b)   *If A is a set with $m \in \mathbb{N}$ elements and $C \subseteq A$ is a set with 1 element, then $A \backslash C$ is a set with $m - 1$ elements.*

(c)   *If C is an infinite set and B is a finite set, then $C \backslash B$ is an infinite set.*

Proof.   (a) Let f be a bijection of $\mathbb{N}_m$ onto A, and let g be a bijection of $\mathbb{N}_n$ onto B. We define h on $\mathbb{N}_{m+n}$ by $h(i) := f(i)$ for $i = 1, \cdots, m$ and $h(i) := g(i - m)$ for $i = m + 1, \cdots, m + n$. We leave it as an exercise to show that h is a bijection from $\mathbb{N}_{m+n}$ onto $A \cup B$.

The proofs of parts (b) and (c) are left to the reader, see Exercise 2.   Q.E.D.

It may seem "obvious" that a subset of a finite set is also finite, but the assertion must be deduced from the definitions. This and the corresponding statement for infinite sets are established next.

1.3.5 Theorem   *Suppose that S and T are sets and that $T \subseteq S$.*

(a)   *If S is a finite set, then T is a finite set.*

(b)   *If T is an infinite set, then S is an infinite set.*

Proof.   (a) If $T = \emptyset$, we already know that T is a finite set. Thus we may suppose that $T \neq \emptyset$. The proof is by induction on the number of elements in S.

If S has 1 element, then the only nonempty subset T of S must coincide with S, so T is a finite set.

Suppose that every nonempty subset of a set with k elements is finite. Now let S be a set having $k + 1$ elements (so there exists a bijection f of $\mathbb{N}_{k+1}$ onto S), and let $T \subseteq S$. If $f(k + 1) \notin T$, we can consider T to be a subset of $S_1 := S \backslash \{f(k + 1)\}$, which has k elements by Theorem 1.3.4(b). Hence, by the induction hypothesis, T is a finite set.

On the other hand, if $f(k + 1) \in T$, then $T_1 := T \backslash \{f(k + 1)\}$ is a subset of S_1. Since S_1 has k elements, the induction hypothesis implies that T_1 is a finite set. But this implies that $T = T_1 \cup \{f(k + 1)\}$ is also a finite set.

(b) This assertion is the contrapositive of the assertion in (a). (See Appendix A for a discussion of the contrapositive.)   Q.E.D.

Countable Sets _____

We now introduce an important type of infinite set.

1.3.6 Definition   **(a)**   A set S is said to be **denumerable** (or **countably infinite**) if there exists a bijection of $\mathbb{N}$ onto S.

(b)   A set S is said to be **countable** if it is either finite or denumerable.

(c)   A set S is said to be **uncountable** if it is not countable.

From the properties of bijections, it is clear that S is denumerable if and only if there exists a bijection of S onto $\mathbb{N}$. Also a set S_1 is denumerable if and only if there exists a bijection from S_1 onto a set S_2 that is denumerable. Further, a set T_1 is countable if and only if there exists a bijection from T_1 onto a set T_2 that is countable. Finally, an infinite countable set is denumerable.

1.3.7 Examples   **(a)**   The set $E := \{2n : n \in \mathbb{N}\}$ of *even* natural numbers is denumerable, since the mapping $f : \mathbb{N} \to E$ defined by $f(n) := 2n$ for $n \in \mathbb{N}$, is a bijection of $\mathbb{N}$ onto E.

Similarly, the set $O := \{2n - 1 : n \in \mathbb{N}\}$ of *odd* natural numbers is denumerable.

(b)   The set $\mathbb{Z}$ of *all* integers is denumerable.

To construct a bijection of $\mathbb{N}$ onto $\mathbb{Z}$, we map 1 onto 0, we map the set of even natural numbers onto the set $\mathbb{N}$ of positive integers, and we map the set of odd natural numbers onto the negative integers. This mapping can be displayed by the enumeration:

$$\mathbb{Z} = \{0, 1, -1, 2, -2, 3, -3, \cdots\}.$$

(c)   The union of two disjoint denumerable sets is denumerable.

Indeed, if $A = \{a_1, a_2, a_3, \cdots\}$ and $B = \{b_1, b_2, b_3, \cdots\}$, we can enumerate the elements of $A \cup B$ as:

$$a_1, b_1, a_2, b_2, a_3, b_3, \cdots. \qquad \square$$

1.3.8 Theorem   *The set $\mathbb{N} \times \mathbb{N}$ is denumerable.*

Informal Proof.   Recall that $\mathbb{N} \times \mathbb{N}$ consists of all ordered pairs (m, n), where $m, n \in \mathbb{N}$. We can enumerate these pairs as:

$$(1, 1), \quad (1, 2), \quad (2, 1), \quad (1, 3), \quad (2, 2), \quad (3, 1), \quad (1, 4), \cdots,$$

according to increasing sum $m + n$, and increasing m. (See Figure 1.3.1.)   Q.E.D.

The enumeration just described is an instance of a "diagonal procedure", since we move along diagonals that each contain finitely many terms as illustrated in Figure 1.3.1. While this argument is satisfying in that it shows exactly what the bijection of $\mathbb{N} \times \mathbb{N} \to \mathbb{N}$ should do, it is not a "formal proof", since it doesn't define this bijection precisely. (See Appendix B for a more formal proof.)

As we have remarked, the construction of an explicit bijection between sets is often complicated. The next two results are useful in establishing the countability of sets, since they do not involve showing that certain mappings are bijections. The first result may seem intuitively clear, but its proof is rather technical; it will be given in Appendix B.

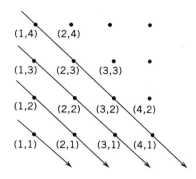

Figure 1.3.1   The set $\mathbb{N} \times \mathbb{N}$

1.3.9 Theorem   *Suppose that S and T are sets and that $T \subseteq S$.*

(a)   *If S is a countable set, then T is a countable set.*

(b)   *If T is an uncountable set, then S is an uncountable set.*

1.3.10 Theorem   *The following statements are equivalent:*

(a)   *S is a countable set.*

(b)   *There exists a surjection of $\mathbb{N}$ onto S.*

(c)   *There exists an injection of S into $\mathbb{N}$.*

Proof.   (a) $\Rightarrow$ (b)   If S is finite, there exists a bijection h of some set $\mathbb{N}_n$ onto S and we define H on $\mathbb{N}$ by

$$H(k) := \begin{cases} h(k) & \text{for} \quad k = 1, \cdots, n, \\ h(n) & \text{for} \quad k > n. \end{cases}$$

Then H is a surjection of $\mathbb{N}$ onto S.

 If S is denumerable, there exists a bijection H of $\mathbb{N}$ onto S, which is also a surjection of $\mathbb{N}$ onto S.

(b) $\Rightarrow$ (c)   If H is a surjection of $\mathbb{N}$ onto S, we define $H_1 : S \to \mathbb{N}$ by letting $H_1(s)$ be the least element in the set $H^{-1}(s) := \{n \in \mathbb{N} : H(n) = s\}$. To see that H_1 is an injection of S into $\mathbb{N}$, note that if $s, t \in S$ and $n_{st} := H_1(s) = H_1(t)$, then $s = H(n_{st}) = t$.

(c) $\Rightarrow$ (a)   If H_1 is an injection of S into $\mathbb{N}$, then it is a bijection of S onto $H_1(S) \subseteq \mathbb{N}$. By Theorem 1.3.9(a), $H_1(S)$ is countable, whence the set S is countable.   Q.E.D.

1.3.11 Theorem   *The set $\mathbb{Q}$ of all rational numbers is denumerable.*

Proof.   The idea of the proof is to observe that the set $\mathbb{Q}^+$ of positive rational numbers is contained in the enumeration:

$$\tfrac{1}{1}, \tfrac{1}{2}, \tfrac{2}{1}, \tfrac{1}{3}, \tfrac{2}{2}, \tfrac{3}{1}, \tfrac{1}{4}, \cdots,$$

which is another "diagonal mapping" (see Figure 1.3.2). However, this mapping is not an injection, since the different fractions $\tfrac{1}{2}$ and $\tfrac{2}{4}$ represent the same rational number.

 To proceed more formally, note that since $\mathbb{N} \times \mathbb{N}$ is countable (by Theorem 1.3.8), it follows from Theorem 1.3.10(b) that there exists a surjection f of $\mathbb{N}$ onto $\mathbb{N} \times \mathbb{N}$. If

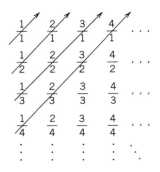

Figure 1.3.2   The set $\mathbb{Q}^+$

$g : \mathbb{N} \times \mathbb{N} \rightarrow \mathbb{Q}^+$ is the mapping that sends the ordered pair (m, n) into the rational number having a representation m/n, then g is a surjection onto $\mathbb{Q}^+$. Therefore, the composition $g \circ f$ is a surjection of $\mathbb{N}$ onto $\mathbb{Q}^+$, and Theorem 1.3.10 implies that $\mathbb{Q}^+$ is a countable set.

Similarly, the set $\mathbb{Q}^-$ of all negative rational numbers is countable. It follows as in Example 1.3.7(b) that the set $\mathbb{Q} = \mathbb{Q}^- \cup \{0\} \cup \mathbb{Q}^+$ is countable. Since $\mathbb{Q}$ contains $\mathbb{N}$, it must be a denumerable set.   Q.E.D.

The next result is concerned with unions of sets. In view of Theorem 1.3.10, we need not be worried about possible overlapping of the sets. Also, we do not have to construct a bijection.

1.3.12 Theorem   *If A_m is a countable set for each $m \in \mathbb{N}$, then the union $A := \bigcup_{m=1}^{\infty} A_m$ is countable.*

Proof.   For each $m \in \mathbb{N}$, let φ_m be a surjection of $\mathbb{N}$ onto A_m. We define $\psi : \mathbb{N} \times \mathbb{N} \rightarrow A$ by

$$\psi(m, n) := \varphi_m(n).$$

We claim that ψ is a surjection. Indeed, if $a \in A$, then there exists a least $m \in \mathbb{N}$ such that $a \in A_m$, whence there exists a least $n \in \mathbb{N}$ such that $a = \varphi_m(n)$. Therefore, $a = \psi(m, n)$.

Since $\mathbb{N} \times \mathbb{N}$ is countable, it follows from Theorem 1.3.10 that there exists a surjection $f : \mathbb{N} \rightarrow \mathbb{N} \times \mathbb{N}$ whence $\psi \circ f$ is a surjection of $\mathbb{N}$ onto A. Now apply Theorem 1.3.10 again to conclude that A is countable.   Q.E.D.

Remark   A less formal (but more intuitive) way to see the truth of Theorem 1.3.12 is to enumerate the elements of $A_m, m \in \mathbb{N}$, as:

$$A_1 = \{a_{11}, a_{12}, a_{13}, \cdots\},$$
$$A_2 = \{a_{21}, a_{22}, a_{23}, \cdots\},$$
$$A_3 = \{a_{31}, a_{32}, a_{33}, \cdots\},$$
$$\cdots \qquad \cdots \qquad \cdots.$$

We then enumerate this array using the "diagonal procedure":

$$a_{11}, a_{12}, a_{21}, a_{13}, a_{22}, a_{31}, a_{14}, \cdots,$$

as was displayed in Figure 1.3.1.

The argument that the set $\mathbb{Q}$ of rational numbers is countable was first given in 1874 by Georg Cantor (1845–1918). He was the first mathematician to examine the concept of infinite set in rigorous detail. In contrast to the countability of $\mathbb{Q}$, he also proved the set $\mathbb{R}$ of real numbers is an uncountable set. (This result will be established in Section 2.5.)

In a series of important papers, Cantor developed an extensive theory of infinite sets and transfinite arithmetic. Some of his results were quite surprising and generated considerable controversy among mathematicians of that era. In a 1877 letter to his colleague Richard Dedekind, he wrote, after proving an unexpected theorem, "I see it, but I do not believe it".

We close this section with one of Cantor's more remarkable theorems.

1.3.13 Cantor's Theorem   *If A is any set, then there is no surjection of A onto the set $\mathcal{P}(A)$ of all subsets of A.*

Proof.   Suppose that $\varphi : A \to \mathcal{P}(A)$ is a surjection. Since $\varphi(a)$ is a subset of A, either a belongs to $\varphi(a)$ or it does not belong to this set. We let

$$D := \{a \in A : a \notin \varphi(a)\}.$$

Since D is a subset of A, if φ is a surjection, then $D = \varphi(a_0)$ for some $a_0 \in A$.

We must have either $a_0 \in D$ or $a_0 \notin D$. If $a_0 \in D$, then since $D = \varphi(a_0)$, we must have $a_0 \in \varphi(a_0)$, contrary to the definition of D. Similarly, if $a_0 \notin D$, then $a_0 \notin \varphi(a_0)$ so that $a_0 \in D$, which is also a contradiction.

Therefore, φ cannot be a surjection.   Q.E.D.

Cantor's Theorem implies that there is an unending progression of larger and larger sets. In particular, it implies that the collection $\mathcal{P}(\mathbb{N})$ of all subsets of the natural numbers $\mathbb{N}$ is uncountable.

Exercises for Section 1.3

1. Prove that a nonempty set T_1 is finite if and only if there is a bijection from T_1 onto a finite set T_2.

2. Prove parts (b) and (c) of Theorem 1.3.4.

3. Let $S := \{1, 2\}$ and $T := \{a, b, c\}$.
   (a) Determine the number of different injections from S into T.
   (b) Determine the number of different surjections from T onto S.

4. Exhibit a bijection between $\mathbb{N}$ and the set of all odd integers greater than 13. $\mathbb{N} \to 13 + 2n$

5. Give an explicit definition of the bijection f from $\mathbb{N}$ onto $\mathbb{Z}$ described in Example 1.3.7(b).

6. Exhibit a bijection between $\mathbb{N}$ and a proper subset of itself. $N \to 2n, \ N \to n+1$

7. Prove that a set T_1 is denumerable if and only if there is a bijection from T_1 onto a denumerable set T_2.

8. Give an example of a countable collection of finite sets whose union is not finite.

9. Prove in detail that if S and T are denumerable, then $S \cup T$ is denumerable.

10. Determine the number of elements in $\mathcal{P}(S)$, the collection of all subsets of S, for each of the following sets:
    (a) $S := \{1, 2\}$,
    (b) $S := \{1, 2, 3\}$,
    (c) $S := \{1, 2, 3, 4\}$.
    Be sure to include the empty set and the set S itself in $\mathcal{P}(S)$.

11. Use Mathematical Induction to prove that if the set S has n elements, then $\mathcal{P}(S)$ has 2^n elements.

12. Prove that the collection $\mathcal{F}(\mathbb{N})$ of all *finite* subsets of $\mathbb{N}$ is countable.

use induction

for each $n \in \mathbb{N}$, there are a countable # of subsets of $\mathbb{N}$ with n elements

assume true for n. Each subset of $\mathbb{N}$ w/ n+1 elements contains n+1 subsets of n elements. Countable ways of adding 1 element to each subset.

CHAPTER 2

THE REAL NUMBERS

In this chapter we will discuss the essential properties of the real number system $\mathbb{R}$. Although it is possible to give a formal construction of this system on the basis of a more primitive set (such as the set $\mathbb{N}$ of natural numbers or the set $\mathbb{Q}$ of rational numbers), we have chosen not to do so. Instead, we exhibit a list of fundamental properties associated with the real numbers and show how further properties can be deduced from them. This kind of activity is much more useful in learning the tools of analysis than examining the logical difficulties of constructing a model for $\mathbb{R}$.

The real number system can be described as a "complete ordered field", and we will discuss that description in considerable detail. In Section 2.1, we first introduce the "algebraic" properties—often called the "field" properties in abstract algebra—that are based on the two operations of addition and multiplication. We continue the section with the introduction of the "order" properties of $\mathbb{R}$ and we derive some consequences of these properties and illustrate their use in working with inequalities. The notion of absolute value, which is based on the order properties, is discussed in Section 2.2.

In Section 2.3, we make the final step by adding the crucial "completeness" property to the algebraic and order properties of $\mathbb{R}$. It is this property, which was not fully understood until the late nineteenth century, that underlies the theory of limits and continuity and essentially all that follows in this book. The rigorous development of real analysis would not be possible without this essential property.

In Section 2.4, we apply the Completeness Property to derive several fundamental results concerning $\mathbb{R}$, including the Archimedean Property, the existence of square roots, and the density of rational numbers in $\mathbb{R}$. We establish, in Section 2.5, the Nested Interval Property and use it to prove the uncountability of $\mathbb{R}$. We also discuss its relation to binary and decimal representations of real numbers.

Part of the purpose of Sections 2.1 and 2.2 is to provide examples of proofs of elementary theorems from explicitly stated assumptions. Students can thus gain experience in writing formal proofs before encountering the more subtle and complicated arguments related to the Completeness Property and its consequences. However, students who have previously studied the axiomatic method and the technique of proofs (perhaps in a course on abstract algebra) can move to Section 2.3 after a cursory look at the earlier sections. A brief discussion of logic and types of proofs can be found in Appendix A at the back of the book.

Section 2.1 The Algebraic and Order Properties of $\mathbb{R}$

We begin with a brief discussion of the "algebraic structure" of the real number system. We will give a short list of basic properties of addition and multiplication from which all other algebraic properties can be derived as theorems. In the terminology of abstract algebra, the system of real numbers is a "field" with respect to addition and multiplication. The basic properties listed in 2.1.1 are known as the *field axioms*. A *binary operation* associates with each pair (a, b) a unique element $B(a, b)$, but we will use the conventional notations of $a + b$ and $a \cdot b$ when discussing the properties of addition and multiplication.

2.1.1 Algebraic Properties of $\mathbb{R}$ On the set $\mathbb{R}$ of real numbers there are two binary operations, denoted by $+$ and $\cdot$ and called **addition** and **multiplication**, respectively. These operations satisfy the following properties:

(A1) $a + b = b + a$ for all a, b in $\mathbb{R}$ (*commutative property of addition*);

(A2) $(a + b) + c = a + (b + c)$ for all a, b, c in $\mathbb{R}$ (*associative property of addition*);

(A3)   there exists an element 0 in $\mathbb{R}$ such that $0 + a = a$ and $a + 0 = a$ for all a in $\mathbb{R}$ (*existence of a zero element*);

(A4)   for each a in $\mathbb{R}$ there exists an element $-a$ in $\mathbb{R}$ such that $a + (-a) = 0$ and $(-a) + a = 0$ (*existence of negative elements*);

(M1) $a \cdot b = b \cdot a$ for all a, b in $\mathbb{R}$ (*commutative property of multiplication*);

(M2) $(a \cdot b) \cdot c = a \cdot (b \cdot c)$ for all a, b, c in $\mathbb{R}$ (*associative property of multiplication*);

(M3)   there exists an element 1 in $\mathbb{R}$ *distinct from* 0 such that $1 \cdot a = a$ and $a \cdot 1 = a$ for all a in $\mathbb{R}$ (*existence of a unit element*);

(M4)   for each $a \neq 0$ in $\mathbb{R}$ there exists an element $1/a$ in $\mathbb{R}$ such that $a \cdot (1/a) = 1$ and $(1/a) \cdot a = 1$ (*existence of reciprocals*);

(D) $a \cdot (b + c) = (a \cdot b) + (a \cdot c)$ and $(b + c) \cdot a = (b \cdot a) + (c \cdot a)$ for all a, b, c in $\mathbb{R}$ (*distributive property of multiplication over addition*).

These properties should be familiar to the reader. The first four are concerned with addition, the next four with multiplication, and the last one connects the two operations. The point of the list is that all the familiar techniques of algebra can be derived from these nine properties, in much the same spirit that the theorems of Euclidean geometry can be deduced from the five basic axioms stated by Euclid in his *Elements*. Since this task more properly belongs to a course in abstract algebra, we will not carry it out here. However, to exhibit the spirit of the endeavor, we will sample a few results and their proofs.

We first establish the basic fact that the elements 0 and 1, whose existence were asserted in (A3) and (M3), are in fact unique. We also show that multiplication by 0 always results in 0.

2.1.2 Theorem   **(a)**   *If z and a are elements in $\mathbb{R}$ with $z + a = a$, then $z = 0$.*

(b)   *If u and $b \neq 0$ are elements in $\mathbb{R}$ with $u \cdot b = b$, then $u = 1$.*

(c)   *If $a \in \mathbb{R}$, then $a \cdot 0 = 0$.*

Proof.   (a)   Using (A3), (A4), (A2), the hypothesis $z + a = a$, and (A4), we get

$$z = z + 0 = z + (a + (-a)) = (z + a) + (-a) = a + (-a) = 0.$$

(b)   Using (M3), (M4), (M2), the assumed equality $u \cdot b = b$, and (M4) again, we get

$$u = u \cdot 1 = u \cdot (b \cdot (1/b)) = (u \cdot b) \cdot (1/b) = b \cdot (1/b) = 1.$$

(c)   We have (why?)

$$a + a \cdot 0 = a \cdot 1 + a \cdot 0 = a \cdot (1 + 0) = a \cdot 1 = a.$$

Therefore, we conclude from (a) that $a \cdot 0 = 0$.   Q.E.D.

We next establish two important properties of multiplication: the uniqueness of reciprocals and the fact that a product of two numbers is zero only when one of the factors is zero.

2.1.3 Theorem   **(a)** *If $a \neq 0$ and b in $\mathbb{R}$ are such that $a \cdot b = 1$, then $b = 1/a$.*
(b)   *If $a \cdot b = 0$, then either $a = 0$ or $b = 0$.*

Proof.   (a) Using (M3), (M4), (M2), the hypothesis $a \cdot b = 1$, and (M3), we have

$$b = 1 \cdot b = ((1/a) \cdot a) \cdot b = (1/a) \cdot (a \cdot b) = (1/a) \cdot 1 = 1/a.$$

(b) It suffices to assume $a \neq 0$ and prove that $b = 0$. (Why?) We multiply $a \cdot b$ by $1/a$ and apply (M2), (M4) and (M3) to get

$$(1/a) \cdot (a \cdot b) = ((1/a) \cdot a) \cdot b = 1 \cdot b = b.$$

Since $a \cdot b = 0$, by 2.1.2(c) this also equals

$$(1/a) \cdot (a \cdot b) = (1/a) \cdot 0 = 0.$$

Thus we have $b = 0$.   Q.E.D.

These theorems represent a small sample of the algebraic properties of the real number system. Some additional consequences of the field properties are given in the exercises.

The operation of **subtraction** is defined by $a - b := a + (-b)$ for a, b in $\mathbb{R}$. Similarly, **division** is defined for a, b in $\mathbb{R}$ with $b \neq 0$ by $a/b := a \cdot (1/b)$. In the following, we will use this customary notation for subtraction and division, and we will use all the familiar properties of these operations. We will ordinarily drop the use of the dot to indicate multiplication and write ab for $a \cdot b$. Similarly, we will use the usual notation for exponents and write a^2 for aa, a^3 for $(a^2)a$; and, in general, we define $a^{n+1} := (a^n)a$ for $n \in \mathbb{N}$. We agree to adopt the convention that $a^1 = a$. Further, if $a \neq 0$, we write $a^0 = 1$ and a^{-1} for $1/a$, and if $n \in \mathbb{N}$, we will write a^{-n} for $(1/a)^n$, when it is convenient to do so. In general, we will freely apply all the usual techniques of algebra without further elaboration.

Rational and Irrational Numbers

We regard the set $\mathbb{N}$ of natural numbers as a subset of $\mathbb{R}$, by identifying the natural number $n \in \mathbb{N}$ with the n-fold sum of the unit element $1 \in \mathbb{R}$. Similarly, we identify $0 \in \mathbb{Z}$ with the zero element of $0 \in \mathbb{R}$, and we identify the n-fold sum of -1 with the integer $-n$. Thus, we consider $\mathbb{N}$ and $\mathbb{Z}$ to be subsets of $\mathbb{R}$.

Elements of $\mathbb{R}$ that can be written in the form b/a where $a, b \in \mathbb{Z}$ and $a \neq 0$ are called **rational numbers**. The set of all rational numbers in $\mathbb{R}$ will be denoted by the standard notation $\mathbb{Q}$. The sum and product of two rational numbers is again a rational number (prove this), and moreover, the field properties listed at the beginning of this section can be shown to hold for $\mathbb{Q}$.

The fact that there are elements in $\mathbb{R}$ that are not in $\mathbb{Q}$ is not immediately apparent. In the sixth century B.C. the ancient Greek society of Pythagoreans discovered that the diagonal of a square with unit sides could not be expressed as a ratio of integers. In view of the Pythagorean Theorem for right triangles, this implies that the square of no rational number can equal 2. This discovery had a profound impact on the development of Greek mathematics. One consequence is that elements of $\mathbb{R}$ that are not in $\mathbb{Q}$ became known as **irrational numbers**, meaning that they are not ratios of integers. Although the word "irrational" in modern English usage has a quite different meaning, we shall adopt the standard mathematical usage of this term.

We will now prove that there does not exist a rational number whose square is 2. In the proof we use the notions of even and odd numbers. Recall that a natural number is **even** if it has the form $2n$ for some $n \in \mathbb{N}$, and it is **odd** if it has the form $2n - 1$ for some $n \in \mathbb{N}$. Every natural number is either even or odd, and no natural number is both even and odd.

2.1.4 Theorem   *There does not exist a rational number r such that $r^2 = 2$.*

Proof.   Suppose, on the contrary, that p and q are integers such that $(p/q)^2 = 2$. We may assume that p and q are positive and have no common integer factors other than 1. (Why?) Since $p^2 = 2q^2$, we see that p^2 is even. This implies that p is also even (because if $p = 2n - 1$ is odd, then its square $p^2 = 2(2n^2 - 2n + 1) - 1$ is also odd). Therefore, since p and q do not have 2 as a common factor, then q must be an odd natural number.

Since p is even, then $p = 2m$ for some $m \in \mathbb{N}$, and hence $4m^2 = 2q^2$, so that $2m^2 = q^2$. Therefore, q^2 is even, and it follows from the argument in the preceding paragraph that q is an even natural number.

Since the hypothesis that $(p/q)^2 = 2$ leads to the contradictory conclusion that q is both even and odd, it must be false.   Q.E.D.

The Order Properties of $\mathbb{R}$

The "order properties" of $\mathbb{R}$ refer to the notions of positivity and inequalities between real numbers. As with the algebraic structure of the system of real numbers, we proceed by isolating three basic properties from which all other order properties and calculations with inequalities can be deduced. The simplest way to do this is to identify a special subset of $\mathbb{R}$ by using the notion of "positivity".

2.1.5 The Order Properties of $\mathbb{R}$ There is a nonempty subset $\mathbb{P}$ of $\mathbb{R}$, called the set of **positive real numbers**, that satisfies the following properties:

(i)   If a, b belong to $\mathbb{P}$, then $a + b$ belongs to $\mathbb{P}$.

(ii)   If a, b belong to $\mathbb{P}$, then ab belongs to $\mathbb{P}$.

(iii)   If a belongs to $\mathbb{R}$, then exactly one of the following holds:

$$a \in \mathbb{P}, \qquad a = 0, \qquad -a \in \mathbb{P}.$$

The first two conditions ensure the compatibility of order with the operations of addition and multiplication, respectively. Condition 2.1.5(iii) is usually called the **Trichotomy Property**, since it divides $\mathbb{R}$ into three distinct types of elements. It states that the set $\{-a : a \in P\}$ of **negative** real numbers has no elements in common with the set $\mathbb{P}$ of positive real numbers, and, moreover, the set $\mathbb{R}$ is the union of three disjoint sets.

If $a \in \mathbb{P}$, we write $a > 0$ and say that a is a **positive** (or a **strictly positive**) real number. If $a \in \mathbb{P} \cup \{0\}$, we write $a \geq 0$ and say that a is a **nonnegative** real number. Similarly, if $-a \in \mathbb{P}$, we write $a < 0$ and say that a is a **negative** (or a **strictly negative**) real number. If $-a \in \mathbb{P} \cup \{0\}$, we write $a \leq 0$ and say that a is a **nonpositive** real number.

The notion of inequality between two real numbers will now be defined in terms of the set $\mathbb{P}$ of positive elements.

2.1.6 Definition   Let a, b be elements of $\mathbb{R}$.

(a)   If $a - b \in \mathbb{P}$, then we write $a > b$ or $b < a$.

(b)   If $a - b \in \mathbb{P} \cup \{0\}$, then we write $a \geq b$ or $b \leq a$.

The Trichotomy Property 2.1.5(iii) implies that for $a, b \in \mathbb{R}$ exactly one of the following will hold:

$$a > b, \qquad a = b, \qquad a < b.$$

Therefore, if both $a \leq b$ and $b \leq a$, then $a = b$.

For notational convenience, we will write

$$a < b < c$$

to mean that both $a < b$ and $b < c$ are satisfied. The other "double" inequalities $a \leq b < c$, $a \leq b \leq c$, and $a < b \leq c$ are defined in a similar manner.

To illustrate how the basic Order Properties are used to derive the "rules of inequalities", we will now establish several results that the reader has used in earlier mathematics courses.

2.1.7 Theorem   *Let a, b, c be any elements of $\mathbb{R}$.*

(a) *If $a > b$ and $b > c$, then $a > c$.*

(b) *If $a > b$, then $a + c > b + c$.*

(c) *If $a > b$ and $c > 0$, then $ca > cb$.*
     If $a > b$ and $c < 0$, then $ca < cb$.

Proof.   (a) If $a - b \in \mathbb{P}$ and $b - c \in \mathbb{P}$, then 2.1.5(i) implies that $(a - b) + (b - c) = a - c$ belongs to $\mathbb{P}$. Hence $a > c$.

(b) If $a - b \in \mathbb{P}$, then $(a + c) - (b + c) = a - b$ is in $\mathbb{P}$. Thus $a + c > b + c$.

(c) If $a - b \in \mathbb{P}$ and $c \in \mathbb{P}$, then $ca - cb = c(a - b)$ is in $\mathbb{P}$ by 2.1.5(ii). Thus $ca > cb$ when $c > 0$.

On the other hand, if $c < 0$, then $-c \in \mathbb{P}$, so that $cb - ca = (-c)(a - b)$ is in $\mathbb{P}$. Thus $cb > ca$ when $c < 0$.   Q.E.D.

It is natural to expect that the natural numbers are positive real numbers. This property is derived from the basic properties of order. The key observation is that the square of any nonzero real number is positive.

2.1.8 Theorem   (a) *If $a \in \mathbb{R}$ and $a \neq 0$, then $a^2 > 0$.*

(b) $1 > 0$.

(c) *If $n \in \mathbb{N}$, then $n > 0$.*

Proof.   (a) By the Trichotomy Property, if $a \neq 0$, then either $a \in \mathbb{P}$ or $-a \in \mathbb{P}$. If $a \in \mathbb{P}$, then by 2.1.5(ii), $a^2 = a \cdot a \in \mathbb{P}$. Also, if $-a \in \mathbb{P}$, then $a^2 = (-a)(-a) \in \mathbb{P}$. We conclude that if $a \neq 0$, then $a^2 > 0$.

(b) Since $1 = 1^2$, it follows from (a) that $1 > 0$.

(c) We use Mathematical Induction. The assertion for $n = 1$ is true by (b). If we suppose the assertion is true for the natural number k, then $k \in \mathbb{P}$, and since $1 \in \mathbb{P}$, we have $k + 1 \in \mathbb{P}$ by 2.1.5(i). Therefore, the assertion is true for all natural numbers.   Q.E.D.

It is worth noting that *no smallest positive real number can exist*. This follows by observing that if $a > 0$, then since $\frac{1}{2} > 0$ (why?), we have that

$$0 < \tfrac{1}{2}a < a.$$

Thus if it is claimed that a is the smallest positive real number, we can exhibit a smaller positive number $\frac{1}{2}a$.

This observation leads to the next result, which will be used frequently as a method of proof. For instance, to prove that a number $a \geq 0$ is actually equal to zero, we see that it suffices to show that a is smaller than an arbitrary positive number.

2.1.9 Theorem *If $a \in \mathbb{R}$ is such that $0 \leq a < \varepsilon$ for every $\varepsilon > 0$, then $a = 0$.*

Proof. Suppose to the contrary that $a > 0$. Then if we take $\varepsilon_0 := \frac{1}{2}a$, we have $0 < \varepsilon_0 < a$. Therefore, it is false that $a < \varepsilon$ for every $\varepsilon > 0$ and we conclude that $a = 0$. Q.E.D.

Remark It is an exercise to show that if $a \in \mathbb{R}$ is such that $0 \leq a \leq \varepsilon$ for every $\varepsilon > 0$, then $a = 0$.

The product of two positive numbers is positive. However, the positivity of a product of two numbers does not imply that each factor is positive. The correct conclusion is given in the next theorem. It is an important tool in working with inequalities.

2.1.10 Theorem *If $ab > 0$, then either*

(i) $a > 0$ and $b > 0$, or

(ii) $a < 0$ and $b < 0$.

Proof. First we note that $ab > 0$ implies that $a \neq 0$ and $b \neq 0$. (Why?) From the Trichotomy Property, either $a > 0$ or $a < 0$. If $a > 0$, then $1/a > 0$ (why?), and therefore $b = (1/a)(ab) > 0$. Similarly, if $a < 0$, then $1/a < 0$, so that $b = (1/a)(ab) < 0$. Q.E.D.

2.1.11 Corollary *If $ab < 0$, then either*

(i) $a < 0$ and $b > 0$, or

(ii) $a > 0$ and $b < 0$.

Inequalities _____

We now show how the Order Properties presented in this section can be used to "solve" certain inequalities. The reader should justify each of the steps.

2.1.12 Examples **(a)** Determine the set A of all real numbers x such that $2x + 3 \leq 6$.
We note that we have[†]

$$ x \in A \quad \Longleftrightarrow \quad 2x + 3 \leq 6 \quad \Longleftrightarrow \quad 2x \leq 3 \quad \Longleftrightarrow \quad x \leq \tfrac{3}{2}. $$

Therefore $A = \left\{ x \in \mathbb{R} : x \leq \frac{3}{2} \right\}$.

(b) Determine the set $B := \left\{ x \in \mathbb{R} : x^2 + x > 2 \right\}$.
We rewrite the inequality so that Theorem 2.1.10 can be applied. Note that

$$ x \in B \quad \Longleftrightarrow \quad x^2 + x - 2 > 0 \quad \Longleftrightarrow \quad (x - 1)(x + 2) > 0. $$

Therefore, we either have (i) $x - 1 > 0$ and $x + 2 > 0$, or we have (ii) $x - 1 < 0$ and $x + 2 < 0$. In case (i) we must have both $x > 1$ and $x > -2$, which is satisfied if and only

---

[†]The symbol $\Longleftrightarrow$ should be read "if and only if".

if $x > 1$. In case (ii) we must have both $x < 1$ and $x < -2$, which is satisfied if and only if $x < -2$.

We conclude that $B = \{x \in \mathbb{R} : x > 1\} \cup \{x \in \mathbb{R} : x < -2\}$.

(c) Determine the set

$$C := \left\{x \in \mathbb{R} : \frac{2x+1}{x+2} < 1\right\}.$$

We note that

$$x \in C \quad\Longleftrightarrow\quad \frac{2x+1}{x+2} - 1 < 0 \quad\Longleftrightarrow\quad \frac{x-1}{x+2} < 0.$$

Therefore we have either (i) $x - 1 < 0$ and $x + 2 > 0$, or (ii) $x - 1 > 0$ and $x + 2 < 0$. (Why?) In case (i) we must have both $x < 1$ and $x > -2$, which is satisfied if and only if $-2 < x < 1$. In case (ii), we must have both $x > 1$ and $x < -2$, which is never satisfied.

We conclude that $C = \{x \in \mathbb{R} : -2 < x < 1\}$.   □

The following examples illustrate the use of the Order Properties of $\mathbb{R}$ in establishing certain inequalities. The reader should verify the steps in the arguments by identifying the properties that are employed.

It should be noted that the existence of square roots of positive numbers has not yet been established; however, we assume the existence of these roots for the purpose of these examples. (The existence of square roots will be discussed in Section 2.4.)

2.1.13 Examples   **(a)** Let $a \geq 0$ and $b \geq 0$. Then

$$(1) \qquad a < b \quad\Longleftrightarrow\quad a^2 < b^2 \quad\Longleftrightarrow\quad \sqrt{a} < \sqrt{b}$$

We consider the case where $a > 0$ and $b > 0$, leaving the case $a = 0$ to the reader. It follows from 2.1.5(i) that $a + b > 0$. Since $b^2 - a^2 = (b-a)(b+a)$, it follows from 2.1.7(c) that $b - a > 0$ implies that $b^2 - a^2 > 0$. Also, it follows from 2.1.10 that $b^2 - a^2 > 0$ implies that $b - a > 0$.

If $a > 0$ and $b > 0$, then $\sqrt{a} > 0$ and $\sqrt{b} > 0$. Since $a = (\sqrt{a})^2$ and $b = (\sqrt{b})^2$, the second implication is a consequence of the first one when a and b are replaced by $\sqrt{a}$ and $\sqrt{b}$, respectively.

We also leave it to the reader to show that if $a \geq 0$ and $b \geq 0$, then

$$(1') \qquad a \leq b \quad\Longleftrightarrow\quad a^2 \leq b^2 \quad\Longleftrightarrow\quad \sqrt{a} \leq \sqrt{b}$$

(b) If a and b are positive real numbers, then their **arithmetic mean** is $\frac{1}{2}(a+b)$ and their **geometric mean** is $\sqrt{ab}$. The **Arithmetic-Geometric Mean Inequality** for a, b is

$$(2) \qquad \sqrt{ab} \leq \tfrac{1}{2}(a+b)$$

with equality occurring if and only if $a = b$.

To prove this, note that if $a > 0$, $b > 0$, and $a \neq b$, then $\sqrt{a} > 0$, $\sqrt{b} > 0$ and $\sqrt{a} \neq \sqrt{b}$. (Why?) Therefore it follows from 2.1.8(a) that $(\sqrt{a} - \sqrt{b})^2 > 0$. Expanding this square, we obtain

$$a - 2\sqrt{ab} + b > 0,$$

whence it follows that

$$\sqrt{ab} < \tfrac{1}{2}(a+b).$$

Therefore (2) holds (with strict inequality) when $a \neq b$. Moreover, if $a = b(> 0)$, then both sides of (2) equal a, so (2) becomes an equality. This proves that (2) holds for $a > 0, b > 0$.

On the other hand, suppose that $a > 0, b > 0$ and that $\sqrt{ab} = \frac{1}{2}(a + b)$. Then, squaring both sides and multiplying by 4, we obtain

$$4ab = (a + b)^2 = a^2 + 2ab + b^2,$$

whence it follows that

$$0 = a^2 - 2ab + b^2 = (a - b)^2.$$

But this equality implies that $a = b$. (Why?) Thus, equality in (2) implies that $a = b$.

Remark   The general Arithmetic-Geometric Mean Inequality for the positive real numbers $a_1, a_2, \cdots, a_n$ is

(3) $$\left(a_1 a_2 \cdots a_n\right)^{1/n} \leq \frac{a_1 + a_2 + \cdots + a_n}{n}$$

with equality occurring if and only if $a_1 = a_2 = \cdots = a_n$. It is possible to prove this more general statement using Mathematical Induction, but the proof is somewhat intricate. A more elegant proof that uses properties of the exponential function is indicated in Exercise 8.3.9 in Chapter 8.

(c)   Bernoulli's Inequality. If $x > -1$, then

(4) $$(1 + x)^n \geq 1 + nx \qquad \text{for all} \quad n \in \mathbb{N}$$

The proof uses Mathematical Induction. The case $n = 1$ yields equality, so the assertion is valid in this case. Next, we assume the validity of the inequality (4) for $k \in \mathbb{N}$ and will deduce it for $k + 1$. Indeed, the assumptions that $(1 + x)^k \geq 1 + kx$ and that $1 + x > 0$ imply (why?) that

$$\begin{aligned}
(1 + x)^{k+1} &= (1 + x)^k \cdot (1 + x) \\
&\geq (1 + kx) \cdot (1 + x) = 1 + (k + 1)x + kx^2 \\
&\geq 1 + (k + 1)x.
\end{aligned}$$

Thus, inequality (4) holds for $n = k + 1$. Therefore, (4) holds for all $n \in \mathbb{N}$.   □

Exercises for Section 2.1

1. If $a, b \in \mathbb{R}$, prove the following.
   (a) If $a + b = 0$, then $b = -a$,
   (b) $-(-a) = a$,
   (c) $(-1)a = -a$,
   (d) $(-1)(-1) = 1$.

2. Prove that if $a, b \in \mathbb{R}$, then
   (a) $-(a + b) = (-a) + (-b)$,
   (b) $(-a) \cdot (-b) = a \cdot b$,
   (c) $1/(-a) = -(1/a)$,
   (d) $-(a/b) = (-a)/b$ if $b \neq 0$.

3. Solve the following equations, justifying each step by referring to an appropriate property or theorem.
   (a) $2x + 5 = 8$,
   (b) $x^2 = 2x$,
   (c) $x^2 - 1 = 3$,
   (d) $(x - 1)(x + 2) = 0$.

4. If $a \in \mathbb{R}$ satisfies $a \cdot a = a$, prove that either $a = 0$ or $a = 1$.

5. If $a \neq 0$ and $b \neq 0$, show that $1/(ab) = (1/a)(1/b)$.

6. Use the argument in the proof of Theorem 2.1.4 to show that there does not exist a rational number s such that $s^2 = 6$.

7. Modify the proof of Theorem 2.1.4 to show that there does not exist a rational number t such that $t^2 = 3$.

8. (a) Show that if x, y are rational numbers, then $x + y$ and xy are rational numbers.
   (b) Prove that if x is a rational number and y is an irrational number, then $x + y$ is an irrational number. If, in addition, $x \neq 0$, then show that xy is an irrational number.

9. Let $K := \{s + t\sqrt{2} : s, t \in \mathbb{Q}\}$. Show that K satisfies the following:
   (a) If $x_1, x_2 \in K$, then $x_1 + x_2 \in K$ and $x_1 x_2 \in K$.
   (b) If $x \neq 0$ and $x \in K$, then $1/x \in K$.
   (Thus the set K is a *subfield* of $\mathbb{R}$. With the order inherited from $\mathbb{R}$, the set K is an ordered field that lies between $\mathbb{Q}$ and $\mathbb{R}$).

10. (a) If $a < b$ and $c \leq d$, prove that $a + c < b + d$.
    (b) If $0 < a < b$ and $0 \leq c \leq d$, prove that $0 \leq ac \leq bd$.

11. (a) Show that if $a > 0$, then $1/a > 0$ and $1/(1/a) = a$.
    (b) Show that if $a < b$, then $a < \frac{1}{2}(a + b) < b$.

12. Let a, b, c, d be numbers satisfying $0 < a < b$ and $c < d < 0$. Give an example where $ac < bd$, and one where $bd < ac$.

13. If $a, b \in \mathbb{R}$, show that $a^2 + b^2 = 0$ if and only if $a = 0$ and $b = 0$.

14. If $0 \leq a < b$, show that $a^2 \leq ab < b^2$. Show by example that it does *not* follow that $a^2 < ab < b^2$.

15. If $0 < a < b$, show that (a) $a < \sqrt{ab} < b$, and (b) $1/b < 1/a$.

16. Find all real numbers x that satisfy the following inequalities.
    (a) $x^2 > 3x + 4$,   (b) $1 < x^2 < 4$,
    (c) $1/x < x$,   (d) $1/x < x^2$.

17. Prove the following form of Theorem 2.1.9: If $a \in \mathbb{R}$ is such that $0 \leq a \leq \varepsilon$ for every $\varepsilon > 0$, then $a = 0$.

18. Let $a, b \in \mathbb{R}$, and suppose that for every $\varepsilon > 0$ we have $a \leq b + \varepsilon$. Show that $a \leq b$.

19. Prove that $\left[\frac{1}{2}(a + b)\right]^2 \leq \frac{1}{2}(a^2 + b^2)$ for all $a, b \in \mathbb{R}$. Show that equality holds if and only if $a = b$.

20. (a) If $0 < c < 1$, show that $0 < c^2 < c < 1$.
    (b) If $1 < c$, show that $1 < c < c^2$.

21. (a) Prove there is no $n \in \mathbb{N}$ such that $0 < n < 1$. (Use the Well-Ordering Property of $\mathbb{N}$.)
    (b) Prove that no natural number can be both even and odd.

22. (a) If $c > 1$, show that $c^n \geq c$ for all $n \in \mathbb{N}$, and that $c^n > c$ for $n > 1$.
    (b) If $0 < c < 1$, show that $c^n \leq c$ for all $n \in \mathbb{N}$, and that $c^n < c$ for $n > 1$.

23. If $a > 0, b > 0$ and $n \in \mathbb{N}$, show that $a < b$ if and only if $a^n < b^n$. [Hint: Use Mathematical Induction].

24. (a) If $c > 1$ and $m, n \in \mathbb{N}$, show that $c^m > c^n$ if and only if $m > n$.
    (b) If $0 < c < 1$ and $m, n \in \mathbb{N}$, show that $c^m < c^n$ if and only if $m > n$.

25. Assuming the existence of roots, show that if $c > 1$, then $c^{1/m} < c^{1/n}$ if and only if $m > n$.

26. Use Mathematical Induction to show that if $a \in \mathbb{R}$ and $m, n, \in \mathbb{N}$, then $a^{m+n} = a^m a^n$ and $(a^m)^n = a^{mn}$.

Section 2.2   Absolute Value and the Real Line

From the Trichotomy Property 2.1.5(iii), we are assured that if $a \in \mathbb{R}$ and $a \neq 0$, then exactly one of the numbers a and $-a$ is positive. The absolute value of $a \neq 0$ is defined to be the positive one of these two numbers. The absolute value of 0 is defined to be 0.

2.2.1 Definition   The **absolute value** of a real number a, denoted by $|a|$, is defined by

$$|a| := \begin{cases} a & \text{if } a > 0, \\ 0 & \text{if } a = 0, \\ -a & \text{if } a < 0. \end{cases}$$

For example, $|5| = 5$ and $|-8| = 8$. We see from the definition that $|a| \geq 0$ for all $a \in \mathbb{R}$, and that $|a| = 0$ if and only if $a = 0$. Also $|-a| = |a|$ for all $a \in \mathbb{R}$. Some additional properties are as follows.

2.2.2 Theorem   **(a)** $|ab| = |a||b|$ for all $a, b \in \mathbb{R}$.
(b) $|a|^2 = a^2$ for all $a \in \mathbb{R}$.
(c)   If $c \geq 0$, then $|a| \leq c$ if and only if $-c \leq a \leq c$.
(d) $-|a| \leq a \leq |a|$ for all $a \in \mathbb{R}$.

Proof.   (a) If either a or b is 0, then both sides are equal to 0. There are four other cases to consider. If $a > 0$, $b > 0$, then $ab > 0$, so that $|ab| = ab = |a||b|$. If $a > 0$, $b < 0$, then $ab < 0$, so that $|ab| = -ab = a(-b) = |a||b|$. The remaining cases are treated similarly.
(b) Since $a^2 \geq 0$, we have $a^2 = |a^2| = |aa| = |a||a| = |a|^2$.
(c) If $|a| \leq c$, then we have both $a \leq c$ and $-a \leq c$ (why?), which is equivalent to $-c \leq a \leq c$. Conversely, if $-c \leq a \leq c$, then we have both $a \leq c$ and $-a \leq c$ (why?), so that $|a| \leq c$.
(d) Take $c = |a|$ in part (c).   Q.E.D.

The following important inequality will be used frequently.

2.2.3 Triangle Inequality   If $a, b \in \mathbb{R}$, then $|a + b| \leq |a| + |b|$.

Proof.   From 2.2.2(d), we have $-|a| \leq a \leq |a|$ and $-|b| \leq b \leq |b|$. On adding these inequalities, we obtain

$$-(|a| + |b|) \leq a + b \leq |a| + |b|.$$

Hence, by 2.2.2(c) we have $|a + b| \leq |a| + |b|$.   Q.E.D.

It can be shown that equality occurs in the Triangle Inequality if and only if $ab > 0$, which is equivalent to saying that a and b have the same sign. (See Exercise 2.)
There are many useful variations of the Triangle Inequality. Here are two.

2.2.4 Corollary   If $a, b \in \mathbb{R}$, then
(a) $\big||a| - |b|\big| \leq |a - b|$,
(b) $|a - b| \leq |a| + |b|$.

Proof.   (a) We write $a = a - b + b$ and then apply the Triangle Inequality to get $|a| = |(a - b) + b| \leq |a - b| + |b|$. Now subtract $|b|$ to get $|a| - |b| \leq |a - b|$. Similarly, from

$|b| = |b - a + a| \le |b - a| + |a|$, we obtain $-|a - b| = -|b - a| \le |a| - |b|$. If we combine these two inequalities, using 2.2.2(c), we get the inequality in (a).

(b) Replace b in the Triangle Inequality by $-b$ to get $|a - b| \le |a| + |-b|$. Since $|-b| = |b|$ we obtain the inequality in (b). Q.E.D.

A straightforward application of Mathematical Induction extends the Triangle Inequality to any finite number of elements of $\mathbb{R}$.

2.2.5 Corollary *If $a_1, a_2, \cdots, a_n$ are any real numbers, then*

$$|a_1 + a_2 + \cdots + a_n| \le |a_1| + |a_2| + \cdots + |a_n|.$$

The following examples illustrate how the properties of absolute value can be used.

2.2.6 Examples (a) Determine the set A of $x \in \mathbb{R}$ such that $|2x + 3| < 7$.

From a modification of 2.2.2(c) for the case of strict inequality, we see that $x \in A$ if and only if $-7 < 2x + 3 < 7$, which is satisfied if and only if $-10 < 2x < 4$. Dividing by 2, we conclude that $A = \{x \in \mathbb{R} : -5 < x < 2\}$.

(b) Determine the set $B := \{x \in \mathbb{R} : |x - 1| < |x|\}$.

One method is to consider cases so that the absolute value symbols can be removed. Here we take the cases

(i) $x \ge 1$, (ii) $0 \le x < 1$, (iii) $x < 0$.

(Why did we choose these three cases?) In case (i) the inequality becomes $x - 1 < x$, which is satisfied without further restriction. Therefore all x such that $x \ge 1$ belong to the set B. In case (ii), the inequality becomes $-(x - 1) < x$, which requires that $x > \frac{1}{2}$. Thus, this case contributes all x such that $\frac{1}{2} < x < 1$ to the set B. In case (iii), the inequality becomes $-(x - 1) < -x$, which is equivalent to $1 < 0$. Since this statement is false, no value of x from case (iii) satisfies the inequality. Forming the union of the three cases, we conclude that $B = \{x \in \mathbb{R} : x > \frac{1}{2}\}$.

There is a second method of determining the set B based on the fact that $a < b$ if and only if $a^2 < b^2$ when both $a \ge 0$ and $b \ge 0$. (See 2.1.13(a).) Thus, the inequality $|x - 1| < |x|$ is equivalent to the inequality $|x - 1|^2 < |x|^2$. Since $|a|^2 = a^2$ for any a by 2.2.2(b), we can expand the square to obtain $x^2 - 2x + 1 < x^2$, which simplifies to $x > \frac{1}{2}$. Thus, we again find that $B = \{x \in \mathbb{R} : x > \frac{1}{2}\}$. This method of squaring can sometimes be used to advantage, but often a case analysis cannot be avoided when dealing with absolute values.

(c) Let the function f be defined by $f(x) := (2x^2 + 3x + 1)/(2x - 1)$ for $2 \le x \le 3$. Find a constant M such that $|f(x)| \le M$ for all x satisfying $2 \le x \le 3$.

We consider separately the numerator and denominator of

$$|f(x)| = \frac{|2x^2 + 3x + 1|}{|2x - 1|}.$$

From the Triangle Inequality, we obtain

$$|2x^2 + 3x + 1| \le 2|x|^2 + 3|x| + 1 \le 2 \cdot 3^2 + 3 \cdot 3 + 1 = 28$$

since $|x| \le 3$ for the x under consideration. Also, $|2x - 1| \ge 2|x| - 1 \ge 2 \cdot 2 - 1 = 3$ since $|x| \ge 2$ for the x under consideration. Thus, $1/|2x - 1| \le 1/3$ for $x \ge 2$. (Why?) Therefore, for $2 \le x \le 3$ we have $|f(x)| \le 28/3$. Hence we can take $M = 28/3$. (Note that we have found one such constant M; evidently any number $H > 28/3$ will also satisfy $|f(x)| \leq H$. It is also possible that $28/3$ is not the smallest possible choice for M.)   □

The Real Line

A convenient and familiar geometric interpretation of the real number system is the real line. In this interpretation, the absolute value $|a|$ of an element a in $\mathbb{R}$ is regarded as the distance from a to the origin 0. More generally, the **distance** between elements a and b in $\mathbb{R}$ is $|a - b|$. (See Figure 2.2.1.)

We will later need precise language to discuss the notion of one real number being "close to" another. If a is a given real number, then saying that a real number x is "close to" a should mean that the distance $|x - a|$ between them is "small". A context in which this idea can be discussed is provided by the terminology of neighborhoods, which we now define.

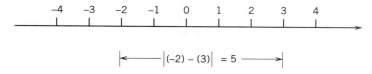

Figure 2.2.1   The distance between $a = -2$ and $b = 3$

2.2.7 Definition   Let $a \in \mathbb{R}$ and $\varepsilon > 0$. Then the ε-**neighborhood** of a is the set $V_\varepsilon(a) := \{x \in \mathbb{R}: |x - a| < \varepsilon\}$.

For $a \in \mathbb{R}$, the statement that x belongs to $V_\varepsilon(a)$ is equivalent to either of the statements (see Figure 2.2.2)

$$-\varepsilon < x - a < \varepsilon \quad \Longleftrightarrow \quad a - \varepsilon < x < a + \varepsilon.$$

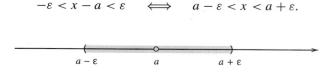

Figure 2.2.2   An ε-neighborhood of a

2.2.8 Theorem   Let $a \in \mathbb{R}$. If x belongs to the neighborhood $V_\varepsilon(a)$ for every $\varepsilon > 0$, then $x = a$.

Proof.   If a particular x satisfies $|x - a| < \varepsilon$ for every $\varepsilon > 0$, then it follows from 2.1.9 that $|x - a| = 0$, and hence $x = a$.   Q.E.D.

2.2.9 Examples   **(a)**   Let $U := \{x : 0 < x < 1\}$. If $a \in U$, then let ε be the smaller of the two numbers a and $1 - a$. Then it is an exercise to show that $V_\varepsilon(a)$ is contained in U. Thus each element of U has some ε-neighborhood of it contained in U.

(b)   If $I := \{x : 0 \leq x \leq 1\}$, then for any $\varepsilon > 0$, the ε-neighborhood $V_\varepsilon(0)$ of 0 contains points not in I, and so $V_\varepsilon(0)$ is not contained in I. For example, the number $x_\varepsilon := -\varepsilon/2$ is in $V_\varepsilon(0)$ but not in I.

(c)   If $|x - a| < \varepsilon$ and $|y - b| < \varepsilon$, then the Triangle Inequality implies that

$$|(x + y) - (a + b)| = |(x - a) + (y - b)|$$
$$\leq |x - a| + |y - b| < 2\varepsilon.$$

Thus if x, y belong to the ε-neighborhoods of a, b, respectively, then $x + y$ belongs to the 2ε-neighborhood of $a + b$ (but not necessarily to the ε-neighborhood of $a + b$).   □

Exercises for Section 2.2

1. If $a, b \in \mathbb{R}$ and $b \neq 0$, show that:
    (a) $|a| = \sqrt{a^2}$,
    (b) $|a/b| = |a|/|b|$.

2. If $a, b \in \mathbb{R}$, show that $|a + b| = |a| + |b|$ if and only if $ab \geq 0$.

3. If $x, y, z \in \mathbb{R}$ and $x \leq z$, show that $x \leq y \leq z$ if and only if $|x - y| + |y - z| = |x - z|$. Interpret this geometrically.

4. Show that $|x - a| < \varepsilon$ if and only if $a - \varepsilon < x < a + \varepsilon$.

5. If $a < x < b$ and $a < y < b$, show that $|x - y| < b - a$. Interpret this geometrically.

6. Find all $x \in \mathbb{R}$ that satisfy the following inequalities:
    (a) $|4x - 5| \leq 13$,
    (b) $|x^2 - 1| \leq 3$.

7. Find all $x \in \mathbb{R}$ that satisfy the equation $|x + 1| + |x - 2| = 7$.

8. Find all $x \in \mathbb{R}$ that satisfy the following inequalities.
    (a) $|x - 1| > |x + 1|$,
    (b) $|x| + |x + 1| < 2$.

9. Sketch the graph of the equation $y = |x| - |x - 1|$.

10. Find all $x \in \mathbb{R}$ that satisfy the inequality $4 < |x + 2| + |x - 1| < 5$.

11. Find all $x \in \mathbb{R}$ that satisfy both $|2x - 3| < 5$ and $|x + 1| > 2$ simultaneously.

12. Determine and sketch the set of pairs (x, y) in $\mathbb{R} \times \mathbb{R}$ that satisfy:
    (a) $|x| = |y|$,
    (b) $|x| + |y| = 1$,
    (c) $|xy| = 2$,
    (d) $|x| - |y| = 2$.

13. Determine and sketch the set of pairs (x, y) in $\mathbb{R} \times \mathbb{R}$ that satisfy:
    (a) $|x| \leq |y|$,
    (b) $|x| + |y| \leq 1$,
    (c) $|xy| \leq 2$,
    (d) $|x| - |y| \geq 2$.

14. Let $\varepsilon > 0$ and $\delta > 0$, and $a \in \mathbb{R}$. Show that $V_\varepsilon(a) \cap V_\delta(a)$ and $V_\varepsilon(a) \cup V_\delta(a)$ are γ-neighborhoods of a for appropriate values of γ.

15. Show that if $a, b \in \mathbb{R}$, and $a \neq b$, then there exist ε-neighborhoods U of a and V of b such that $U \cap V = \emptyset$.

16. Show that if $a, b \in \mathbb{R}$ then
    (a) $\max\{a, b\} = \frac{1}{2}(a + b + |a - b|)$ and $\min\{a, b\} = \frac{1}{2}(a + b - |a - b|)$.
    (b) $\min\{a, b, c\} = \min\{\min\{a, b\}, c\}$.

17. Show that if $a, b, c \in \mathbb{R}$, then the "middle number" is $\mathrm{mid}\{a, b, c\} = \min\{\max\{a, b\}, \max\{b, c\}, \max\{c, a\}\}$.

Section 2.3   The Completeness Property of $\mathbb{R}$

Thus far, we have discussed the algebraic properties and the order properties of the real number system $\mathbb{R}$. In this section we shall present one more property of $\mathbb{R}$ that is often called the "Completeness Property". The system $\mathbb{Q}$ of rational numbers also has the algebraic and order properties described in the preceding sections, but we have seen that $\sqrt{2}$ *cannot* be represented as a rational number; therefore $\sqrt{2}$ does not belong to $\mathbb{Q}$. This observation shows the necessity of an additional property to characterize the real number system. This additional property, the Completeness (or the Supremum) Property, is an essential property of $\mathbb{R}$, and we will say that $\mathbb{R}$ is a *complete ordered field*. It is this special property that permits us to define and develop the various limiting procedures that will be discussed in the chapters that follow.

There are several different ways to describe the Completeness Property. We choose to give what is probably the most efficient approach by assuming that each nonempty bounded subset of $\mathbb{R}$ has a supremum.

Suprema and Infima

We now introduce the notions of upper bound and lower bound for a set of real numbers. These ideas will be of utmost importance in later sections.

2.3.1 Definition   Let S be a nonempty subset of $\mathbb{R}$.

(a)   The set S is said to be **bounded above** if there exists a number $u \in \mathbb{R}$ such that $s \leq u$ for all $s \in S$. Each such number u is called an **upper bound** of S.

(b)   The set S is said to be **bounded below** if there exists a number $w \in \mathbb{R}$ such that $w \leq s$ for all $s \in S$. Each such number w is called a **lower bound** of S.

(c)   A set is said to be **bounded** if it is both bounded above and bounded below. A set is said to be **unbounded** if it is not bounded.

For example, the set $S := \{x \in \mathbb{R} : x < 2\}$ is bounded above; the number 2 and any number larger than 2 is an upper bound of S. This set has no lower bounds, so that the set is not bounded below. Thus it is unbounded (even though it is bounded above).

If a set has one upper bound, then it has infinitely many upper bounds, because if u is an upper bound of S, then the numbers $u + 1, u + 2, \cdots$ are also upper bounds of S. (A similar observation is valid for lower bounds.)

In the set of upper bounds of S and the set of lower bounds of S, we single out their least and greatest elements, respectively, for special attention in the following definition. (See Figure 2.3.1.)

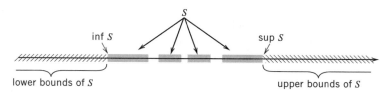

Figure 2.3.1   inf S and   sup S

2.3.2 Definition   Let S be a nonempty subset of $\mathbb{R}$.

(a)   If S is bounded above, then a number u is said to be a **supremum** (or a **least upper bound**) of S if it satisfies the conditions:

    (1) u is an upper bound of S, and

    (2)   if v is any upper bound of S, then $u \leq v$.

(b) If S is bounded below, then a number w is said to be an **infimum** (or a **greatest lower bound**) of S if it satisfies the conditions:

(1′) w is a lower bound of S, and

(2′) if t is any lower bound of S, then $t \leq w$.

It is not difficult to see that *there can be only one supremum of a given subset S of $\mathbb{R}$.* (Then we can refer to *the* supremum of a set instead of *a* supremum.) For, suppose that u_1 and u_2 are both suprema of S. If $u_1 < u_2$, then the hypothesis that u_2 is a supremum implies that u_1 cannot be an upper bound of S. Similarly, we see that $u_2 < u_1$ is not possible. Therefore, we must have $u_1 = u_2$. A similar argument can be given to show that the infimum of a set is uniquely determined.

If the supremum or the infimum of a set S exists, we will denote them by

$$\sup S \quad \text{and} \quad \inf S.$$

We also observe that if u' is an arbitrary upper bound of a nonempty set S, then $\sup S \leq u'$. This is because $\sup S$ is the least of the upper bounds of S.

First of all, it needs to be emphasized that in order for a nonempty set S in $\mathbb{R}$ to have a supremum, it must have an upper bound. Thus, not every subset of $\mathbb{R}$ has a supremum; similarly, not every subset of $\mathbb{R}$ has an infimum. Indeed, there are four possibilities for a nonempty subset S of $\mathbb{R}$: it can

(i)   have both a supremum and an infimum,
(ii)   have a supremum but no infimum,
(iii)   have a infimum but no supremum,
(iv)   have neither a supremum nor an infimum.

We also wish to stress that in order to show that $u = \sup S$ for some nonempty subset S of $\mathbb{R}$, we need to show that *both* (1) and (2) of Definition 2.3.2(a) hold. It will be instructive to reformulate these statements. First the reader should see that the following two statements about a number u and a set S are equivalent:

(1) u is an upper bound of S,
(1′) $s \leq u$ for all $s \in S$.

Also, the following statements about an upper bound u of a set S are equivalent:

(2)   if v is any upper bound of S, then $u \leq v$,
(2′)   if $z < u$, then z is not an upper bound of S,
(2″)   if $z < u$, then there exists $s_z \in S$ such that $z < s_z$,
(2‴)   if $\varepsilon > 0$, then there exists $s_\varepsilon \in S$ such that $u - \varepsilon < s_\varepsilon$.

Therefore, we can state two alternate formulations for the supremum.

2.3.3 Lemma   *A number u is the supremum of a nonempty subset S of $\mathbb{R}$ if and only if u satisfies the conditions:*

(1) $s \leq u$ for all $s \in S$,
(2) *if $v < u$, then there exists $s' \in S$ such that $v < s'$.*

We leave it to the reader to write out the details of the proof.

2.3.4 Lemma   *An upper bound u of a nonempty set S in $\mathbb{R}$ is the supremum of S if and only if for every $\varepsilon > 0$ there exists an $s_\varepsilon \in S$ such that $u - \varepsilon < s_\varepsilon$.*

Proof.   If u is an upper bound of S that satisfies the stated condition and if $v < u$, then we put $\varepsilon := u - v$. Then $\varepsilon > 0$, so there exists $s_\varepsilon \in S$ such that $v = u - \varepsilon < s_\varepsilon$. Therefore, v is not an upper bound of S, and we conclude that $u = \sup S$.

Conversely, suppose that $u = \sup S$ and let $\varepsilon > 0$. Since $u - \varepsilon < u$, then $u - \varepsilon$ is not an upper bound of S. Therefore, some element s_ε of S must be greater than $u - \varepsilon$; that is, $u - \varepsilon < s_\varepsilon$. (See Figure 2.3.2.)

Q.E.D.

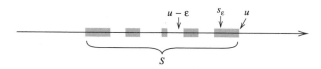

Figure 2.3.2 $u = \sup S$

It is important to realize that the supremum of a set may or may not be an element of the set. Sometimes it is and sometimes it is not, depending on the particular set. We consider a few examples.

2.3.5 Examples   **(a)**   If a nonempty set S_1 has a finite number of elements, then it can be shown that S_1 has a largest element u and a least element w. Then $u = \sup S_1$ and $w = \inf S_1$, and they are both members of S_1. (This is clear if S_1 has only one element, and it can be proved by induction on the number of elements in S_1; see Exercises 11 and 12.)

(b)   The set $S_2 := \{x : 0 \le x \le 1\}$ clearly has 1 for an upper bound. We prove that 1 is its supremum as follows. If $v < 1$, there exists an element $s' \in S_2$ such that $v < s'$. (Name one such element s'.) Therefore v is not an upper bound of S_2 and, since v is an arbitrary number $v < 1$, we conclude that $\sup S_2 = 1$. It is similarly shown that $\inf S_2 = 0$. Note that both the supremum and the infimum of S_2 are contained in S_2.

(c)   The set $S_3 := \{x : 0 < x < 1\}$ clearly has 1 for an upper bound. Using the same argument as given in (b), we see that $\sup S_3 = 1$. In this case, the set S_3 does *not* contain its supremum. Similarly, $\inf S_3 = 0$ is not contained in S_3. $\square$

The Completeness Property of $\mathbb{R}$

It is not possible to prove on the basis of the field and order properties of $\mathbb{R}$ that were discussed in Section 2.1 that every nonempty subset of $\mathbb{R}$ that is bounded above has a supremum in $\mathbb{R}$. However, it is a deep and fundamental property of the real number system that this is indeed the case. We will make frequent and essential use of this property, especially in our discussion of limiting processes. The following statement concerning the existence of suprema is our final assumption about $\mathbb{R}$. Thus, we say that $\mathbb{R}$ is a *complete ordered field*.

2.3.6 The Completeness Property of $\mathbb{R}$ *Every nonempty set of real numbers that has an upper bound also has a supremum in $\mathbb{R}$.*

This property is also called the **Supremum Property** of $\mathbb{R}$. The analogous property for infima can be deduced from the Completeness Property as follows. Suppose that S is a nonempty subset of $\mathbb{R}$ that is bounded below. Then the nonempty set $\overline{S} := \{-s : s \in S\}$ is bounded above, and the Supremum Property implies that $u := \sup \overline{S}$ exists in $\mathbb{R}$. The reader should verify in detail that $-u$ is the infimum of S.

Exercises for Section 2.3

1. Let $S_1 := \{x \in \mathbb{R} : x \geq 0\}$. Show in detail that the set S_1 has lower bounds, but no upper bounds. Show that inf $S_1 = 0$.

2. Let $S_2 = \{x \in \mathbb{R} : x > 0\}$. Does S_2 have lower bounds? Does S_2 have upper bounds? Does inf S_2 exist? Does sup S_2 exist? Prove your statements.

3. Let $S_3 = \{1/n : n \in \mathbb{N}\}$. Show that sup $S_3 = 1$ and inf $S_3 \geq 0$. (It will follow from the Archimedean Property in Section 2.4 that inf $S_3 = 0$.)

4. Let $S_4 := \{1 - (-1)^n/n : n \in \mathbb{N}\}$. Find inf S_4 and sup S_4.

5. Let S be a nonempty subset of $\mathbb{R}$ that is bounded below. Prove that inf $S = -\sup\{-s : s \in S\}$.

6. If a set $S \subseteq \mathbb{R}$ contains one of its upper bounds, show that this upper bound is the supremum of S.

7. Let $S \subseteq \mathbb{R}$ be nonempty. Show that $u \in \mathbb{R}$ is an upper bound of S if and only if the conditions $t \in \mathbb{R}$ and $t > u$ imply that $t \notin S$.

8. Let $S \subseteq \mathbb{R}$ be nonempty. Show that if $u = \sup S$, then for every number $n \in \mathbb{N}$ the number $u - 1/n$ is not an upper bound of S, but the number $u + 1/n$ is an upper bound of S. (The converse is also true; see Exercise 2.4.3.)

9. Show that if A and B are bounded subsets of $\mathbb{R}$, then $A \cup B$ is a bounded set. Show that $\sup(A \cup B) = \sup\{\sup A, \sup B\}$.

10. Let S be a bounded set in $\mathbb{R}$ and let S_0 be a nonempty subset of S. Show that inf $S \leq$ inf $S_0 \leq$ sup $S_0 \leq$ sup S.

11. Let $S \subseteq \mathbb{R}$ and suppose that $s^* := \sup S$ belongs to S. If $u \notin S$, show that $\sup(S \cup \{u\}) = \sup\{s^*, u\}$.

12. Show that a nonempty finite set $S \subseteq \mathbb{R}$ contains its supremum. [Hint: Use Mathematical Induction and the preceding exercise.]

13. Show that the assertions (1) and (1') before Lemma 2.3.3 are equivalent.

14. Show that the assertions (2), (2'), (2''), and (2''') before Lemma 2.3.3 are equivalent.

15. Write out the details of the proof of Lemma 2.3.3.

Section 2.4   Applications of the Supremum Property

We will now discuss how to work with suprema and infima. We will also give some very important applications of these concepts to derive fundamental properties of $\mathbb{R}$. We begin with examples that illustrate useful techniques in applying the ideas of supremum and infimum.

2.4.1 Example   (a)   It is an important fact that taking suprema and infima of sets is compatible with the algebraic properties of $\mathbb{R}$. As an example, we present here the compatibility of taking suprema and addition.

Let S be a nonempty subset of $\mathbb{R}$ that is bounded above, and let a be any number in $\mathbb{R}$. Define the set $a + S := \{a + s : s \in S\}$. We will prove that

$$\sup(a + S) = a + \sup S.$$

If we let $u := \sup S$, then $x \leq u$ for all $x \in S$, so that $a + x \leq a + u$. Therefore, $a + u$ is an upper bound for the set $a + S$; consequently, we have $\sup(a + S) \leq a + u$.

Now if v is *any* upper bound of the set $a + S$, then $a + x \leq v$ for all $x \in S$. Consequently $x \leq v - a$ for all $x \in S$, so that $v - a$ is an upper bound of S. Therefore, $u = \sup S \leq v - a$, which gives us $a + u \leq v$. Since v is any upper bound of $a + S$, we can replace v by $\sup(a + S)$ to get $a + u \leq \sup(a + S)$.

Combining these inequalities, we conclude that

$$\sup(a + S) = a + u = a + \sup S.$$

For similar relationships between the suprema and infima of sets and the operations of addition and multiplication, see the exercises.

(b) If the suprema or infima of two sets are involved, it is often necessary to establish results in two stages, working with one set at a time. Here is an example.

Suppose that A and B are nonempty subsets of $\mathbb{R}$ that satisfy the property:

$$a \leq b \qquad \text{for all } a \in A \text{ and all } b \in B.$$

We will prove that

$$\sup A \leq \inf B.$$

For, given $b \in B$, we have $a \leq b$ for all $a \in A$. This means that b is an upper bound of A, so that $\sup A \leq b$. Next, since the last inequality holds for all $b \in B$, we see that the number $\sup A$ is a lower bound for the set B. Therefore, we conclude that $\sup A \leq \inf B$. □

Functions

The idea of upper bound and lower bound is applied to functions by considering the range of a function. Given a function $f : D \to \mathbb{R}$, we say that f is **bounded above** if the set $f(D) = \{f(x) : x \in D\}$ is bounded above in $\mathbb{R}$; that is, there exists $B \in \mathbb{R}$ such that $f(x) \leq B$ for all $x \in D$. Similarly, the function f is **bounded below** if the set $f(D)$ is bounded below. We say that f is **bounded** if it is bounded above and below; this is equivalent to saying that there exists $B \in \mathbb{R}$ such that $|f(x)| \leq B$ for all $x \in D$.

The following example illustrates how to work with suprema and infima of functions.

2.4.2 Example Suppose that f and g are real-valued functions with common domain $D \subseteq \mathbb{R}$. We assume that f and g are bounded.

(a) If $f(x) \leq g(x)$ for all $x \in D$, then $\sup f(D) \leq \sup g(D)$, which is sometimes written:

$$\sup_{x \in D} f(x) \leq \sup_{x \in D} g(x).$$

We first note that $f(x) \leq g(x) \leq \sup g(D)$, which implies that the number $\sup g(D)$ is an upper bound for $f(D)$. Therefore, $\sup f(D) \leq \sup g(D)$.

(b) We note that the hypothesis $f(x) \leq g(x)$ for all $x \in D$ in part (a) does not imply any relation between $\sup f(D)$ and $\inf g(D)$.

For example, if $f(x) := x^2$ and $g(x) := x$ with $D = \{x : 0 \leq x \leq 1\}$, then $f(x) \leq g(x)$ for all $x \in D$. However, we see that $\sup f(D) = 1$ and $\inf g(D) = 0$. Since $\sup g(D) = 1$, the conclusion of (a) holds.

(c) If $f(x) \le g(y)$ for all $x, y \in D$, then we may conclude that $\sup f(D) \le \inf g(D)$, which we may write as:

$$\sup_{x \in D} f(x) \le \inf_{y \in D} g(y).$$

(Note that the functions in (b) do not satisfy this hypothesis.)

The proof proceeds in two stages as in Example 2.4.1(b). The reader should write out the details of the argument. □

Further relationships between suprema and infima of functions are given in the exercises.

The Archimedean Property

Because of your familiarity with the set $\mathbb{R}$ and the customary picture of the real line, it may seem obvious that the set $\mathbb{N}$ of natural numbers is *not* bounded in $\mathbb{R}$. How can we prove this "obvious" fact? In fact, we cannot do so by using only the Algebraic and Order Properties given in Section 2.1. Indeed, we must use the Completeness Property of $\mathbb{R}$ as well as the Inductive Property of $\mathbb{N}$ (that is, if $n \in \mathbb{N}$, then $n + 1 \in \mathbb{N}$).

The absence of upper bounds for $\mathbb{N}$ means that given any real number x there exists a natural number n (depending on x) such that $x < n$.

2.4.3 Archimedean Property If $x \in \mathbb{R}$, then there exists $n_x \in \mathbb{N}$ such that $x < n_x$.

Proof. If the assertion is false, then $n \le x$ for all $n \in \mathbb{N}$; therefore, x is an upper bound of $\mathbb{N}$. Therefore, by the Completeness Property, the nonempty set $\mathbb{N}$ has a supremum $u \in \mathbb{R}$. Subtracting 1 from u gives a number $u - 1$ which is smaller than the supremum u of $\mathbb{N}$. Therefore $u - 1$ is not an upper bound of $\mathbb{N}$, so there exists $m \in \mathbb{N}$ with $u - 1 < m$. Adding 1 gives $u < m + 1$, and since $m + 1 \in \mathbb{N}$, this inequality contradicts the fact that u is an upper bound of $\mathbb{N}$. Q.E.D.

2.4.4 Corollary If $S := \{1/n : n \in \mathbb{N}\}$, then $\inf S = 0$.

Proof. Since $S \ne \emptyset$ is bounded below by 0, it has an infimum and we let $w := \inf S$. It is clear that $w \ge 0$. For any $\varepsilon > 0$, the Archimedean Property implies that there exists $n \in \mathbb{N}$ such that $1/\varepsilon < n$, which implies $1/n < \varepsilon$. Therefore we have

$$0 \le w \le 1/n < \varepsilon.$$

But since $\varepsilon > 0$ is arbitrary, it follows from Theorem 2.1.9 that $w = 0$. Q.E.D.

2.4.5 Corollary If $t > 0$, there exists $n_t \in \mathbb{N}$ such that $0 < 1/n_t < t$.

Proof. Since $\inf\{1/n : n \in \mathbb{N}\} = 0$ and $t > 0$, then t is not a lower bound for the set $\{1/n : n \in \mathbb{N}\}$. Thus there exists $n_t \in \mathbb{N}$ such that $0 < 1/n_t < t$. Q.E.D.

2.4.6 Corollary If $y > 0$, there exists $n_y \in \mathbb{N}$ such that $n_y - 1 \le y < n_y$.

Proof.   The Archimedean Property ensures that the subset $E_y := \{m \in \mathbb{N} : y < m\}$ of $\mathbb{N}$ is not empty. By the Well-Ordering Property 1.2.1, E_y has a least element, which we denote by n_y. Then $n_y - 1$ does not belong to E_y, and hence we have $n_y - 1 \le y < n_y$.   Q.E.D.

Collectively, the Corollaries 2.4.4–2.4.6 are sometimes referred to as the Archimedean Property of $\mathbb{R}$.

The Existence of $\sqrt{2}$

The importance of the Supremum Property lies in the fact that it guarantees the existence of real numbers under certain hypotheses. We shall make use of it in this way many times. At the moment, we shall illustrate this use by proving the existence of a positive real number x such that $x^2 = 2$; that is, the positive square root of 2. It was shown earlier (see Theorem 2.1.4) that such an x cannot be a rational number; thus, we will be deriving the existence of at least one irrational number.

2.4.7 Theorem   *There exists a positive real number x such that $x^2 = 2$.*

Proof.   Let $S := \{s \in \mathbb{R} : 0 \le s, s^2 < 2\}$. Since $1 \in S$, the set is not empty. Also, S is bounded above by 2, because if $t > 2$, then $t^2 > 4$ so that $t \notin S$. Therefore the Supremum Property implies that the set S has a supremum in $\mathbb{R}$, and we let $x := \sup S$. Note that $x > 1$.

We will prove that $x^2 = 2$ by ruling out the other two possibilities: $x^2 < 2$ and $x^2 > 2$.

First assume that $x^2 < 2$. We will show that this assumption contradicts the fact that $x = \sup S$ by finding an $n \in \mathbb{N}$ such that $x + 1/n \in S$, thus implying that x is not an upper bound for S. To see how to choose n, note that $1/n^2 \le 1/n$ so that

$$\left(x + \frac{1}{n}\right)^2 = x^2 + \frac{2x}{n} + \frac{1}{n^2} \le x^2 + \frac{1}{n}(2x + 1).$$

Hence if we can choose n so that

$$\frac{1}{n}(2x + 1) < 2 - x^2,$$

then we get $(x + 1/n)^2 < x^2 + (2 - x^2) = 2$. By assumption we have $2 - x^2 > 0$, so that $(2 - x^2)/(2x + 1) > 0$. Hence the Archimedean Property (Corollary 2.4.5) can be used to obtain $n \in \mathbb{N}$ such that

$$\frac{1}{n} < \frac{2 - x^2}{2x + 1}.$$

These steps can be reversed to show that for this choice of n we have $x + 1/n \in S$, which contradicts the fact that x is an upper bound of S. Therefore we cannot have $x^2 < 2$.

Now assume that $x^2 > 2$. We will show that it is then possible to find $m \in \mathbb{N}$ such that $x - 1/m$ is also an upper bound of S, contradicting the fact that $x = \sup S$. To do this, note that

$$\left(x - \frac{1}{m}\right)^2 = x^2 - \frac{2x}{m} + \frac{1}{m^2} > x^2 - \frac{2x}{m}.$$

Hence if we can choose m so that

$$\frac{2x}{m} < x^2 - 2,$$

then $(x - 1/m)^2 > x^2 - (x^2 - 2) = 2$. Now by assumption we have $x^2 - 2 > 0$, so that $(x^2 - 2)/2x > 0$. Hence, by the Archimedean Property, there exists $m \in \mathbb{N}$ such that

$$\frac{1}{m} < \frac{x^2 - 2}{2x}.$$

These steps can be reversed to show that for this choice of m we have $(x - 1/m)^2 > 2$. Now if $s \in S$, then $s^2 < 2 < (x - 1/m)^2$, whence it follows from 2.1.13(a) that $s < x - 1/m$. This implies that $x - 1/m$ is an upper bound for S, which contradicts the fact that $x = \sup S$. Therefore we cannot have $x^2 > 2$.

Since the possibilities $x^2 < 2$ and $x^2 > 2$ have been excluded, we must have $x^2 = 2$.
                                                                              Q.E.D.

By slightly modifying the preceding argument, the reader can show that if $a > 0$, then there is a unique $b > 0$ such that $b^2 = a$. We call b the **positive square root** of a and denote it by $b = \sqrt{a}$ or $b = a^{1/2}$. A slightly more complicated argument involving the binomial theorem can be formulated to establish the existence of a unique **positive nth root** of a, denoted by $\sqrt[n]{a}$ or $a^{1/n}$, for each $n \in \mathbb{N}$.

Remark   If in the proof of Theorem 2.4.7 we replace the set S by the set of rational numbers $T := \{r \in \mathbb{Q} : 0 \le r, r^2 < 2\}$, the argument then gives the conclusion that $y := \sup T$ satisfies $y^2 = 2$. Since we have seen in Theorem 2.1.4 that y cannot be a rational number, it follows that the set T that consists of rational numbers does not have a supremum belonging to the set $\mathbb{Q}$. Thus the ordered field $\mathbb{Q}$ of rational numbers does *not* possess the Completeness Property.

Density of Rational Numbers in $\mathbb{R}$

We now know that there exists at least one irrational real number, namely $\sqrt{2}$. Actually there are "more" irrational numbers than rational numbers in the sense that the set of rational numbers is countable (as shown in Section 1.3), while the set of irrational numbers is uncountable (see Section 2.5). However, we next show that in spite of this apparent disparity, the set of rational numbers is "dense" in $\mathbb{R}$ in the sense that given any two real numbers there is a rational number between them (in fact, there are infinitely many such rational numbers).

2.4.8 The Density Theorem   *If x and y are any real numbers with $x < y$, then there exists a rational number $r \in \mathbb{Q}$ such that $x < r < y$.*

Proof.   It is no loss of generality (why?) to assume that $x > 0$. Since $y - x > 0$, it follows from Corollary 2.4.5 that there exists $n \in \mathbb{N}$ such that $1/n < y - x$. Therefore, we have $nx + 1 < ny$. If we apply Corollary 2.4.6 to $nx > 0$, we obtain $m \in \mathbb{N}$ with $m - 1 \le nx < m$. Therefore, $m \le nx + 1 < ny$, whence $nx < m < ny$. Thus, the rational number $r := m/n$ satisfies $x < r < y$.   Q.E.D.

To round out the discussion of the interlacing of rational and irrational numbers, we have the same "betweenness property" for the set of irrational numbers.

2.4.9 Corollary   *If x and y are real numbers with $x < y$, then there exists an irrational number z such that $x < z < y$.*

Proof.   If we apply the Density Theorem 2.4.8 to the real numbers $x/\sqrt{2}$ and $y/\sqrt{2}$, we obtain a rational number $r \neq 0$ (why?) such that

$$\frac{x}{\sqrt{2}} < r < \frac{y}{\sqrt{2}}.$$

Then $z := r\sqrt{2}$ is irrational (why?) and satisfies $x < z < y$.   Q.E.D.

Exercises for Section 2.4

1. Show that $\sup\{1 - 1/n : n \in \mathbb{N}\} = 1$.

2. If $S := \{1/n - 1/m : n, m \in \mathbb{N}\}$, find inf S and sup S.

3. Let $S \subseteq \mathbb{R}$ be nonempty. Prove that if a number u in $\mathbb{R}$ has the properties: (i) for every $n \in \mathbb{N}$ the number $u - 1/n$ is not an upper bound of S, and (ii) for every number $n \in \mathbb{N}$ the number $u + 1/n$ is an upper bound of S, then $u = \sup S$. (This is the converse of Exercise 2.3.8.)

4. Let S be a nonempty bounded set in $\mathbb{R}$.
    (a) Let $a > 0$, and let $aS := \{as : s \in S\}$. Prove that

    $$\inf(aS) = a \inf S, \qquad \sup(aS) = a \sup S.$$

    (b) Let $b < 0$ and let $bS = \{bs : s \in S\}$. Prove that

    $$\inf(bS) = b \sup S, \qquad \sup(bS) = b \inf S.$$

5. Let X be a nonempty set and let $f : X \to \mathbb{R}$ have bounded range in $\mathbb{R}$. If $a \in \mathbb{R}$, show that Example 2.4.1(a) implies that

    $$\sup\{a + f(x) : x \in X\} = a + \sup\{f(x) : x \in X\}$$

    Show that we also have

    $$\inf\{a + f(x) : x \in X\} = a + \inf\{f(x) : x \in X\}$$

6. Let A and B be bounded nonempty subsets of $\mathbb{R}$, and let $A + B := \{a + b : a \in A, b \in B\}$. Prove that $\sup(A + B) = \sup A + \sup B$ and $\inf(A + B) = \inf A + \inf B$.

7. Let X be a nonempty set, and let f and g be defined on X and have bounded ranges in $\mathbb{R}$. Show that

    $$\sup\{f(x) + g(x) : x \in X\} \leq \sup\{f(x) : x \in X\} + \sup\{g(x) : x \in X\}$$

    and that

    $$\inf\{f(x) : x \in X\} + \inf\{g(x) : x \in X\} \leq \inf\{f(x) + g(x) : x \in X\}.$$

    Give examples to show that each of these inequalities can be either equalities or strict inequalities.

8. Let $X = Y := \{x \in \mathbb{R} : 0 < x < 1\}$. Define $h : X \times Y \to \mathbb{R}$ by $h(x, y) := 2x + y$.
    (a) For each $x \in X$, find $f(x) := \sup\{h(x, y) : y \in Y\}$; then find $\inf\{f(x) : x \in X\}$.
    (b) For each $y \in Y$, find $g(y) := \inf\{h(x, y) : x \in X\}$; then find $\sup\{g(y) : y \in Y\}$. Compare with the result found in part (a).

9. Perform the computations in (a) and (b) of the preceding exercise for the function $h : X \times Y \to \mathbb{R}$ defined by

    $$h(x, y) := \begin{cases} 0 & \text{if } x < y, \\ 1 & \text{if } x \geq y. \end{cases}$$

10. Let X and Y be nonempty sets and let $h : X \times Y \to \mathbb{R}$ have bounded range in $\mathbb{R}$. Let $f : X \to \mathbb{R}$ and $g : Y \to \mathbb{R}$ be defined by

    $$f(x) := \sup\{h(x, y) : y \in Y\}, \qquad g(y) := \inf\{h(x, y) : x \in X\}.$$

Prove that

$$\sup\{g(y) : y \in Y\} \le \inf\{f(x) : x \in X\}$$

We sometimes express this by writing

$$\sup_{y} \inf_{x} h(x, y) \le \inf_{x} \sup_{y} h(x, y).$$

Note that Exercises 8 and 9 show that the inequality may be either an equality or a strict inequality.

11. Let X and Y be nonempty sets and let $h : X \times Y \to \mathbb{R}$ have bounded range in $\mathbb{R}$. Let $F: X \to \mathbb{R}$ and $G : Y \to \mathbb{R}$ be defined by

$$F(x) := \sup\{h(x, y) : y \in Y\}, \qquad G(y) := \sup\{h(x, y) : x \in X\}.$$

Establish the **Principle of the Iterated Suprema**:

$$\sup\{h(x, y) : x \in X, y \in Y\} = \sup\{F(x) : x \in X\} = \sup\{G(y) : y \in Y\}$$

We sometimes express this in symbols by

$$\sup_{x,y} h(x, y) = \sup_{x} \sup_{y} h(x, y) = \sup_{y} \sup_{x} h(x, y).$$

12. Given any $x \in \mathbb{R}$, show that there exists a *unique* $n \in \mathbb{Z}$ such that $n - 1 \le x < n$.

13. If $y > 0$, show that there exists $n \in \mathbb{N}$ such that $1/2^n < y$.

14. Modify the argument in Theorem 2.4.7 to show that there exists a positive real number y such that $y^2 = 3$.

15. Modify the argument in Theorem 2.4.7 to show that if $a > 0$, then there exists a positive real number z such that $z^2 = a$.

16. Modify the argument in Theorem 2.4.7 to show that there exists a positive real number u such that $u^3 = 2$.

17. Complete the proof of the Density Theorem 2.4.8 by removing the assumption that $x > 0$.

18. If $u > 0$ is any real number and $x < y$, show that there exists a rational number r such that $x < ru < y$. (Hence the set $\{ru : r \in \mathbb{Q}\}$ is dense in $\mathbb{R}$.)

Section 2.5   Intervals

The Order Relation on $\mathbb{R}$ determines a natural collection of subsets called "intervals". The notations and terminology for these special sets will be familiar from earlier courses. If $a, b \in \mathbb{R}$ satisfy $a < b$, then the **open interval** determined by a and b is the set

$$(a, b) := \{x \in \mathbb{R} : a < x < b\}.$$

The points a and b are called the **endpoints** of the interval; however, the endpoints are not included in an open interval. If both endpoints are adjoined to this open interval, then we obtain the **closed interval** determined by a and b; namely, the set

$$[a, b] := \{x \in \mathbb{R} : a \le x \le b\}.$$

The two **half-open** (or **half-closed**) intervals determined by a and b are $[a, b)$, which includes the endpoint a, and $(a, b]$, which includes the endpoint b.

Each of these four intervals is bounded and has **length** defined by $b - a$. If $a = b$, the corresponding open interval is the empty set $(a, a) = \emptyset$, whereas the corresponding closed interval is the singleton set $[a, a] = \{a\}$.

There are five types of unbounded intervals for which the symbols ∞ (or $+\infty$) and $-\infty$ are used as notational convenience in place of the endpoints. The **infinite open intervals** are the sets of the form

$$(a, \infty) := \{x \in \mathbb{R} : x > a\} \qquad \text{and} \qquad (-\infty, b) := \{x \in \mathbb{R} : x < b\}.$$

The first set has no upper bounds and the second one has no lower bounds. Adjoining endpoints gives us the **infinite closed intervals**:

$$[a, \infty) := \{x \in \mathbb{R} : a \leq x\} \qquad \text{and} \qquad (-\infty, b] := \{x \in \mathbb{R} : x \leq b\}.$$

It is often convenient to think of the entire set $\mathbb{R}$ as an infinite interval; in this case, we write $(-\infty, \infty) := \mathbb{R}$. No point is an endpoint of $(-\infty, \infty)$.

Warning It must be emphasized that ∞ and $-\infty$ are *not* elements of $\mathbb{R}$, but only convenient symbols.

Characterization of Intervals

An obvious property of intervals is that if two points x, y with $x < y$ belong to an interval I, then any point lying between them also belongs to I. That is, if $x < t < y$, then the point t belongs to the same interval as x and y. In other words, if x and y belong to an interval I, then the interval $[x, y]$ is contained in I. We now show that a subset of $\mathbb{R}$ possessing this property must be an interval.

2.5.1 Characterization Theorem *If S is a subset of $\mathbb{R}$ that contains at least two points and has the property*

(1) $$\text{if} \quad x, y \in S \quad \text{and} \quad x < y, \quad \text{then} \quad [x, y] \subseteq S,$$

then S is an interval.

Proof. There are four cases to consider: (i) S is bounded, (ii) S is bounded above but not below, (iii) S is bounded below but not above, and (iv) S is neither bounded above nor below.

Case (i): Let $a := \inf S$ and $b := \sup S$. Then $S \subseteq [a, b]$ and we will show that $(a, b) \subseteq S$.

If $a < z < b$, then z is not a lower bound of S, so there exists $x \in S$ with $x < z$. Also, z is not an upper bound of S, so there exists $y \in S$ with $z < y$. Therefore $z \in [x, y]$, so property (1) implies that $z \in S$. Since z is an arbitrary element of (a, b), we conclude that $(a, b) \subseteq S$.

Now if $a \in S$ and $b \in S$, then $S = [a, b]$. (Why?) If $a \notin S$ and $b \notin S$, then $S = (a, b)$. The other possibilities lead to either $S = (a, b]$ or $S = [a, b)$.

Case (ii): Let $b := \sup S$. Then $S \subseteq (-\infty, b]$ and we will show that $(-\infty, b) \subseteq S$. For, if $z < b$, then there exist $x, y \in S$ such that $z \in [x, y] \subseteq S$. (Why?) Therefore $(-\infty, b) \subseteq S$. If $b \in S$, then $S = (-\infty, b]$, and if $b \notin S$, then $S = (-\infty, b)$.

Cases (iii) and (iv) are left as exercises. Q.E.D.

Nested Intervals

We say that a sequence of intervals I_n, $n \in \mathbb{N}$, is **nested** if the following chain of inclusions holds (see Figure 2.5.1):

$$I_1 \supseteq I_2 \supseteq \cdots \supseteq I_n \supseteq I_{n+1} \supseteq \cdots$$

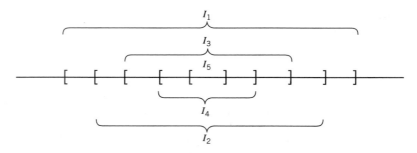

Figure 2.5.1   Nested intervals

For example, if $I_n := [0, 1/n]$ for $n \in \mathbb{N}$, then $I_n \supseteq I_{n+1}$ for each $n \in \mathbb{N}$ so that this sequence of intervals is nested. In this case, the element 0 belongs to all I_n and the Archimedean Property 2.4.5 can be used to show that 0 is the only such common point. (Prove this.) We denote this by writing $\bigcap_{n=1}^{\infty} I_n = \{0\}$.

It is important to realize that, in general, a nested sequence of intervals need *not* have a common point. For example, if $J_n := (0, 1/n)$ for $n \in \mathbb{N}$, then this sequence of intervals is nested, but there is no common point, since for every given $x > 0$, there exists (why?) $m \in \mathbb{N}$ such that $1/m < x$ so that $x \notin J_m$. Similarly, the sequence of intervals $K_n := (n, \infty)$, $n \in \mathbb{N}$, is nested but has no common point. (Why?)

However, it is an important property of $\mathbb{R}$ that every nested sequence of *closed, bounded* intervals does have a common point, as we will now prove. Notice that the completeness of $\mathbb{R}$ plays an essential role in establishing this property.

2.5.2 Nested Intervals Property   *If $I_n = \left[a_n, b_n\right]$, $n \in \mathbb{N}$, is a nested sequence of closed bounded intervals, then there exists a number $\xi \in \mathbb{R}$ such that $\xi \in I_n$ for all $n \in \mathbb{N}$.*

Proof.   Since the intervals are nested, we have $I_n \subseteq I_1$ for all $n \in \mathbb{N}$, so that $a_n \leq b_1$ for all $n \in \mathbb{N}$. Hence, the nonempty set $\{a_n : n \in \mathbb{N}\}$ is bounded above, and we let ξ be its supremum. Clearly $a_n \leq \xi$ for all $n \in \mathbb{N}$.

We claim also that $\xi \leq b_n$ for all n. This is established by showing that for any particular n, the number b_n is an upper bound for the set $\{a_k : k \in \mathbb{N}\}$. We consider two cases. (i) If $n \leq k$, then since $I_n \supseteq I_k$, we have $a_k \leq b_k \leq b_n$. (ii) If $k < n$, then since $I_k \supseteq I_n$, we have $a_k \leq a_n \leq b_n$. (See Figure 2.5.2.) Thus, we conclude that $a_k \leq b_n$ for all k, so that b_n is an upper bound of the set $\{a_k : k \in \mathbb{N}\}$. Hence, $\xi \leq b_n$ for each $n \in \mathbb{N}$. Since $a_n \leq \xi \leq b_n$ for all n, we have $\xi \in I_n$ for all $n \in \mathbb{N}$.   Q.E.D.

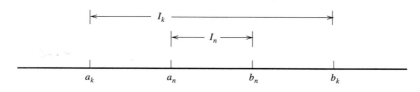

Figure 2.5.2   If $k < n$, then $I_n \subseteq I_k$

2.5.3 Theorem *If $I_n := [a_n, b_n]$, $n \in \mathbb{N}$, is a nested sequence of closed, bounded intervals such that the lengths $b_n - a_n$ of I_n satisfy*

$$\inf\{b_n - a_n : n \in \mathbb{N}\} = 0,$$

then the number ξ contained in I_n for all $n \in \mathbb{N}$ is unique.

Proof. If $\eta := \inf\{b_n : n \in \mathbb{N}\}$, then an argument similar to the proof of 2.5.2 can be used to show that $a_n \leq \eta$ for all n, and hence that $\xi \leq \eta$. In fact, it is an exercise (see Exercise 10) to show that $x \in I_n$ for all $n \in \mathbb{N}$ if and only if $\xi \leq x \leq \eta$. If we have $\inf\{b_n - a_n : n \in \mathbb{N}\} = 0$, then for any $\varepsilon > 0$, there exists an $m \in \mathbb{N}$ such that $0 \leq \eta - \xi \leq b_m - a_m < \varepsilon$. Since this holds for all $\varepsilon > 0$, it follows from Theorem 2.1.9 that $\eta - \xi = 0$. Therefore, we conclude that $\xi = \eta$ is the only point that belongs to I_n for every $n \in \mathbb{N}$.   Q.E.D.

The Uncountability of $\mathbb{R}$

The concept of a countable set was discussed in Section 1.3 and the countability of the set $\mathbb{Q}$ of rational numbers was established there. We will now use the Nested Interval Property to prove that the set $\mathbb{R}$ is an *uncountable* set. The proof was given by Georg Cantor in 1874 in the first of his papers on infinite sets. He later published a proof that used decimal representations of real numbers, and that proof will be given later in this section.

2.5.4 Theorem   *The set $\mathbb{R}$ of real numbers is not countable.*

Proof.   We will prove that the unit interval $I := [0, 1]$ is an uncountable set. This implies that the set $\mathbb{R}$ is an uncountable set, for if $\mathbb{R}$ were countable, then the subset I would also be countable. (See Theorem 1.3.9(a).)

The proof is by contradiction. If we assume that I is countable, then we can enumerate the set as $I = \{x_1, x_2, \cdots, x_n, \cdots\}$. We first select a closed subinterval I_1 of I such that $x_1 \notin I_1$, then select a closed subinterval I_2 of I_1 such that $x_2 \notin I_2$, and so on. In this way, we obtain nonempty closed intervals

$$I_1 \supseteq I_2 \supseteq \cdots \supseteq I_n \supseteq \cdots$$

such that $I_n \subseteq I$ and $x_n \notin I_n$ for all n. The Nested Intervals Property 2.5.2 implies that there exists a point $\xi \in I$ such that $\xi \in I_n$ for all n. Therefore $\xi \neq x_n$ for all $n \in \mathbb{N}$, so the enumeration of I is not a complete listing of the elements of I, as claimed. Hence, I is an uncountable set.   Q.E.D.

The fact that the set $\mathbb{R}$ of real numbers is uncountable can be combined with the fact that the set $\mathbb{Q}$ of rational numbers is countable to conclude that the set $\mathbb{R}\backslash\mathbb{Q}$ of irrational numbers is uncountable. Indeed, since the union of two countable sets is countable (see 1.3.7(c)), if $\mathbb{R}\backslash\mathbb{Q}$ is countable, then since $\mathbb{R} = \mathbb{Q} \cup (\mathbb{R}\backslash\mathbb{Q})$, we conclude that $\mathbb{R}$ is also a countable set, which is a contradiction. Therefore, the set of irrational numbers $\mathbb{R}\backslash\mathbb{Q}$ is an uncountable set.

[†]Binary Representations

We will digress briefly to discuss informally the binary (and decimal) representations of real numbers. It will suffice to consider real numbers between 0 and 1, since the representations for other real numbers can then be obtained by adding a positive or negative number.

[†]The remainder of this section can be omitted on a first reading.

If $x \in [0, 1]$, we will use a repeated bisection procedure to associate a sequence (a_n) of 0s and 1s as follows. If $x \neq \frac{1}{2}$ belongs to the left subinterval $[0, \frac{1}{2}]$ we take $a_1 := 0$, while if x belongs to the right subinterval $[\frac{1}{2}, 1]$ we take $a_1 = 1$. If $x = \frac{1}{2}$, then we may take a_1 to be either 0 or 1. In any case, we have

$$\frac{a_1}{2} \leq x \leq \frac{a_1 + 1}{2}.$$

We now bisect the interval $[\frac{1}{2}a_1, \frac{1}{2}(a_1 + 1)]$. If x is not the bisection point and belongs to the left subinterval we take $a_2 := 0$, and if x belongs to the right subinterval we take $a_2 := 1$. If $x = \frac{1}{4}$ or $x = \frac{3}{4}$, we can take a_2 to be either 0 or 1. In any case, we have

$$\frac{a_1}{2} + \frac{a_2}{2^2} \leq x \leq \frac{a_1}{2} + \frac{a_2 + 1}{2^2}.$$

We continue this bisection procedure, assigning at the nth stage the value $a_n := 0$ if x is not the bisection point and lies in the left subinterval, and assigning the value $a_n := 1$ if x lies in the right subinterval. In this way we obtain a sequence (a_n) of 0s or 1s that correspond to a nested sequence of intervals containing the point x. For each n, we have the inequality

$$(2) \qquad \frac{a_1}{2} + \frac{a_2}{2^2} + \cdots + \frac{a_n}{2^n} \leq x \leq \frac{a_1}{2} + \frac{a_2}{2^2} + \cdots + \frac{a_n + 1}{2^n}.$$

If x is the bisection point at the nth stage, then $x = m/2^n$ with m odd. In this case, we may choose either the left or the right subinterval; however, once this subinterval is chosen, then all subsequent subintervals in the bisection procedure are determined. [For instance, if we choose the left subinterval so that $a_n = 0$, then x is the right endpoint of all subsequent subintervals, and hence $a_k = 1$ for all $k \geq n + 1$. On the other hand, if we choose the right subinterval so that $a_n = 1$, then x is the left endpoint of all subsequent subintervals, and hence $a_k = 0$ for all $k \geq n + 1$. For example, if $x = \frac{3}{4}$, then the two possible sequences for x are $1, 0, 1, 1, 1, \cdots$ and $1, 1, 0, 0, 0, \cdots$.]

To summarize: *If $x \in [0, 1]$, then there exists a sequence (a_n) of 0s and 1s such that inequality (2) holds for all $n \in \mathbb{N}$.* In this case we write

$$(3) \qquad x = (.a_1 a_2 \cdots a_n \cdots)_2,$$

and call (3) a **binary representation** of x. This representation is unique except when $x = m/2^n$ for m odd, in which case x has the two representations

$$x = (.a_1 a_2 \cdots a_{n-1} 1000 \cdots)_2 = (.a_1 a_2 \cdots a_{n-1} 0111 \cdots)_2,$$

one ending in 0s and the other ending in 1s.

Conversely, each sequence of 0s and 1s is the binary representation of a unique real number in $[0, 1]$. The inequality corresponding to (2) determines a closed interval with length $1/2^n$ and the sequence of these intervals is nested. Therefore, Theorem 2.5.3 implies that there exists a unique real number x satisfying (2) for every $n \in \mathbb{N}$. Consequently, x has the binary representation $(.a_1 a_2 \cdots a_n \cdots)_2$.

Remark   The concept of binary representation is extremely important in this era of digital computers. A number is entered in a digital computer on "bits", and each bit can be put in one of two states—either it will pass current or it will not. These two states correspond to the values 1 and 0, respectively. Thus, the binary representation of a number can be stored in a digital computer on a string of bits. Of course, in actual practice, since only finitely many bits can be stored, the binary representations must be truncated. If n binary digits are used for a number $x \in [0, 1]$, then the accuracy is at most $1/2^n$. For example, to assure four-decimal accuracy, it is necessary to use at least 15 binary digits (or 15 bits).

Decimal Representations

Decimal representations of real numbers are similar to binary representations, except that we subdivide intervals into *ten* equal subintervals instead of two.

Thus, given $x \in [0, 1]$, if we subdivide $[0, 1]$ into ten equal subintervals, then x belongs to a subinterval $[b_1/10, (b_1 + 1)/10]$ for some integer b_1 in $\{0, 1, \cdots, 9\}$. Proceeding as in the binary case, we obtain a sequence (b_n) of integers with $0 \le b_n \le 9$ for all $n \in \mathbb{N}$ such that x satisfies

$$(4) \qquad \frac{b_1}{10} + \frac{b_2}{10^2} + \cdots + \frac{b_n}{10^n} \le x \le \frac{b_1}{10} + \frac{b_2}{10^2} + \cdots + \frac{b_n + 1}{10^n}.$$

In this case we say that x has a **decimal representation** given by

$$x = .b_1 b_2 \cdots b_n \cdots .$$

If $x \ge 1$ and if $B \in \mathbb{N}$ is such that $B \le x < B + 1$, then $x = B.b_1 b_2 \cdots b_n \cdots$ where the decimal representation of $x - B \in [0, 1]$ is as above. Negative numbers are treated similarly.

The fact that each decimal determines a unique real number follows from Theorem 2.5.3, since each decimal specifies a nested sequence of intervals with lengths $1/10^n$.

The decimal representation of $x \in [0, 1]$ is unique except when x is a subdivision point at some stage, which can be seen to occur when $x = m/10^n$ for some $m, n \in \mathbb{N}$, $1 \le m \le 10^n$. (We may also assume that m is not divisible by 10.) When x is a subdivision point at the nth stage, one choice for b_n corresponds to selecting the left subinterval, which causes all subsequent digits to be 9, and the other choice corresponds to selecting the right subinterval, which causes all subsequent digits to be 0. [For example, if $x = \frac{1}{2}$ then $x = .4999 \cdots = .5000 \cdots$, and if $y = 38/100$ then $y = .37999 \cdots = .38000 \cdots$.]

Periodic Decimals

A decimal $B.b_1 b_2 \cdots b_n \cdots$ is said to be **periodic** (or to be **repeating**), if there exist $k, n \in \mathbb{N}$ such that $b_n = b_{n+m}$ for all $n \ge k$. In this case, the block of digits $b_k b_{k+1} \cdots b_{k+m-1}$ is repeated once the kth digit is reached. The smallest number m with this property is called the **period** of the decimal. For example, $19/88 = .2159090 \cdots 90 \cdots$ has period $m = 2$ with repeating block 90 starting at $k = 4$. A **terminating decimal** is a periodic decimal where the repeated block is simply the digit 0.

We will give an informal proof of the assertion: A *positive real number is rational if and only if its decimal representation is periodic.*

For, suppose that $x = p/q$ where $p, q \in \mathbb{N}$ have no common integer factors. For convenience we will also suppose that $0 < p < q$. We note that the process of "long division" of q into p gives the decimal representation of p/q. Each step in the division process produces a remainder that is an integer from 0 to $q - 1$. Therefore, after at most q steps, some remainder will occur a second time and, at that point, the digits in the quotient will begin to repeat themselves in cycles. Hence, the decimal representation of such a rational number is periodic.

Conversely, if a decimal is periodic, then it represents a rational number. The idea of the proof is best illustrated by an example. Suppose that $x = 7.31414 \cdots 14 \cdots$. We multiply by a power of 10 to move the decimal point to the first repeating block; here obtaining $10x = 73.1414 \cdots$. We now multiply by a power of 10 to move one block to the left of the decimal point; here getting $1000x = 7314.1414 \cdots$. We now subtract to obtain an integer; here getting $1000x - 10x = 7314 - 73 = 7241$, whence $x = 7241/990$, a rational number.

Cantor's Second Proof

We will now give Cantor's second proof of the uncountability of $\mathbb{R}$. This is the elegant "diagonal" argument based on decimal representations of real numbers.

2.5.5 Theorem   *The unit interval* $[0, 1] := \{x \in \mathbb{R} : 0 \le x \le 1\}$ *is not countable.*

Proof.   The proof is by contradiction. We will use the fact that every real number $x \in [0, 1]$ has a decimal representation $x = 0.b_1 b_2 b_3 \cdots$, where $b_i = 0, 1 \cdots, 9$. Suppose that there is an enumeration $x_1, x_2, x_3 \cdots$ of all numbers in $[0, 1]$, which we display as:

$$x_1 = 0.b_{11} b_{12} b_{13} \cdots b_{1n} \cdots,$$
$$x_2 = 0.b_{21} b_{22} b_{23} \cdots b_{2n} \cdots,$$
$$x_3 = 0.b_{31} b_{32} b_{33} \cdots b_{3n} \cdots,$$
$$\cdots \quad \cdots$$
$$x_n = 0.b_{n1} b_{n2} b_{n3} \cdots b_{nn} \cdots,$$
$$\cdots \quad \cdots$$

We now define a real number $y := 0.y_1 y_2 y_3 \cdots y_n \cdots$ by setting $y_1 := 2$ if $b_{11} \ge 5$ and $y_1 := 7$ if $b_{11} \le 4$; in general, we let

$$y_n := \begin{cases} 2 & \text{if } b_{nn} \ge 5, \\ 7 & \text{if } b_{nn} \le 4. \end{cases}$$

Then $y \in [0, 1]$. Note that the number y is not equal to any of the numbers with two decimal representations, since $y_n \ne 0, 9$ for all $n \in \mathbb{N}$. Further, since y and x_n differ in the nth decimal place, then $y \ne x_n$ for any $n \in \mathbb{N}$. Therefore, y is not included in the enumeration of $[0, 1]$, contradicting the hypothesis.   Q.E.D.

Exercises for Section 2.5

1. If $I := [a, b]$ and $I' := [a', b']$ are closed intervals in $\mathbb{R}$, show that $I \subseteq I'$ if and only if $a' \le a$ and $b \le b'$.

2. If $S \subseteq \mathbb{R}$ is nonempty, show that S is bounded if and only if there exists a closed bounded interval I such that $S \subseteq I$.

3. If $S \subseteq \mathbb{R}$ is a nonempty bounded set, and $I_S := [\inf S, \sup S]$, show that $S \subseteq I_S$. Moreover, if J is any closed bounded interval containing S, show that $I_S \subseteq J$.

4. In the proof of Case (ii) of Theorem 2.5.1, explain why x, y exist in S.

5. Write out the details of the proof of case (iv) in Theorem 2.5.1.

6. If $I_1 \supseteq I_2 \supseteq \cdots \supseteq I_n \supseteq \cdots$ is a nested sequence of intervals and if $I_n = [a_n, b_n]$, show that $a_1 \le a_2 \le \cdots \le a_n \le \cdots$ and $b_1 \ge b_2 \ge \cdots \ge b_n \ge \cdots$.

7. Let $I_n := [0, 1/n]$ for $n \in \mathbb{N}$. Prove that $\bigcap_{n=1}^{\infty} I_n = \{0\}$.

8. Let $J_n := (0, 1/n)$ for $n \in \mathbb{N}$. Prove that $\bigcap_{n=1}^{\infty} J_n = \emptyset$.

9. Let $K_n := (n, \infty)$ for $n \in \mathbb{N}$. Prove that $\bigcap_{n=1}^{\infty} K_n = \emptyset$.

10. With the notation in the proofs of Theorems 2.5.2 and 2.5.3, show that we have $\eta \in \bigcap_{n=1}^{\infty} I_n$. Also show that $[\xi, \eta] = \bigcap_{n=1}^{\infty} I_n$.

11. Show that the intervals obtained from the inequalities in (2) form a nested sequence.

12. Give the two binary representations of $\frac{3}{8}$ and $\frac{7}{16}$.

13. (a) Give the first four digits in the binary representation of $\frac{1}{3}$.
    (b) Give the complete binary representation of $\frac{1}{3}$.

14. Show that if $a_k, b_k \in \{0, 1, \cdots, 9\}$ and if

$$\frac{a_1}{10} + \frac{a_2}{10^2} + \cdots + \frac{a_n}{10^n} = \frac{b_1}{10} + \frac{b_2}{10^2} + \cdots + \frac{b_m}{10^m} \neq 0,$$

then $n = m$ and $a_k = b_k$ for $k = 1, \cdots, n$.

15. Find the decimal representation of $-\frac{2}{7}$.

16. Express $\frac{1}{7}$ and $\frac{2}{19}$ as periodic decimals.

17. What rationals are represented by the periodic decimals $1.25137 \cdots 137 \cdots$ and $35.14653 \cdots 653 \cdots$?

CHAPTER 3

SEQUENCES AND SERIES

Now that the foundations of the real number system $\mathbb{R}$ have been laid, we are prepared to pursue questions of a more analytic nature, and we will begin with a study of the convergence of sequences. Some of the early results may be familiar to the reader from calculus, but the presentation here is intended to be rigorous and will lead to certain more profound theorems than are usually discussed in earlier courses.

We will first introduce the meaning of the convergence of a sequence of real numbers and establish some basic, but useful, results about convergent sequences. We then present some deeper results concerning the convergence of sequences. These include the Monotone Convergence Theorem, the Bolzano-Weierstrass Theorem, and the Cauchy Criterion for convergence of sequences. It is important for the reader to learn both the theorems and how the theorems apply to special sequences.

Because of the linear limitations inherent in a book it is necessary to decide where to locate the subject of infinite series. It would be reasonable to follow this chapter with a full discussion of infinite series, but this would delay the important topics of continuity, differentiation, and integration. Consequently, we have decided to compromise. A brief introduction to infinite series is given in Section 3.7 at the end of this chapter, and a more extensive treatment is given later in Chapter 9. Thus readers who want a fuller discussion of series at this point can move to Chapter 9 after completing this chapter.

Augustin-Louis Cauchy

Augustin-Louis Cauchy (1789–1857) was born in Paris just after the start of the French Revolution. His father was a lawyer in the Paris police department, and the family was forced to flee during the Reign of Terror. As a result, Cauchy's early years were difficult and he developed strong anti-revolutionary and pro-royalist feelings. After returning to Paris, Cauchy's father became secretary to the newly-formed Senate, which included the mathematicians Laplace and Lagrange. They were impressed by young Cauchy's mathematical talent and helped him begin his career.

He entered the École Polytechnique in 1805 and soon established a reputation as an exceptional mathematician. In 1815, the year royalty was restored, he was appointed to the faculty of the École Polytechnique, but his strong political views and his uncompromising standards in mathematics often resulted in bad relations with his colleagues. After the July revolution of 1830, Cauchy refused to sign the new loyalty oath and left France for eight years in self-imposed exile. In 1838, he accepted a minor teaching post in Paris, and in 1848 Napoleon III reinstated him to his former position at the École Polytechnique, where he remained until his death.

Cauchy was amazingly versatile and prolific, making substantial contributions to many areas, including real and complex analysis, number theory, differential equations, mathematical physics and probability. He published eight books and 789 papers, and his collected works fill 26 volumes. He was one of the most important mathematicians in the first half of the nineteenth century.

Section 3.1   Sequences and Their Limits

A sequence in a set S is a function whose domain is the set $\mathbb{N}$ of natural numbers, and whose range is contained in the set S. In this chapter, we will be concerned with sequences in $\mathbb{R}$ and will discuss what we mean by the convergence of these sequences.

3.1.1 Definition   A **sequence of real numbers** (or a **sequence in** $\mathbb{R}$) is a function defined on the set $\mathbb{N} = \{1, 2, \cdots\}$ of natural numbers whose range is contained in the set $\mathbb{R}$ of real numbers.

In other words, a sequence in $\mathbb{R}$ assigns to each natural number $n = 1, 2, \cdots$ a uniquely determined real number. If $X : \mathbb{N} \to \mathbb{R}$ is a sequence, we will usually denote the value of X at n by the symbol x_n rather than using the function notation $X(n)$. The values x_n are also called the **terms** or the **elements** of the sequence. We will denote this sequence by the notations

$$X, \qquad (x_n), \qquad (x_n : n \in \mathbb{N}).$$

Of course, we will often use other letters, such as $Y = (y_k)$, $Z = (z_i)$, and so on, to denote sequences.

We purposely use parentheses to emphasize that the ordering induced by the natural order of $\mathbb{N}$ is a matter of importance. Thus, we distinguish notationally between the sequence $(x_n : n \in \mathbb{N})$, whose infinitely many terms have an ordering, and the set of values $\{x_n : n \in \mathbb{N}\}$ in the range of the sequence which are not ordered. For example, the sequence $X := ((-1)^n : n \in \mathbb{N})$ has infinitely many terms that alternate between -1 and 1, whereas the set of values $\{(-1)^n : n \in \mathbb{N}\}$ is equal to the set $\{-1, 1\}$, which has only two elements.

Sequences are often defined by giving a formula for the nth term x_n. Frequently, it is convenient to list the terms of a sequence in order, stopping when the rule of formation seems evident. For example, we may define the sequence of reciprocals of the even numbers by writing

$$X := \left(\frac{1}{2}, \frac{1}{4}, \frac{1}{6}, \frac{1}{8}, \cdots \right),$$

though a more satisfactory method is to specify the formula for the general term and write

$$X := \left(\frac{1}{2n} : n \in \mathbb{N} \right)$$

or more simply $X = (1/2n)$.

Another way of defining a sequence is to specify the value of x_1 and give a formula for x_{n+1} $(n \geq 1)$ in terms of x_n. More generally, we may specify x_1 and give a formula for obtaining x_{n+1} from $x_1, x_2, \cdots, x_n$. Sequences defined in this manner are said to be **inductively** (or **recursively**) defined.

3.1.2 Examples   **(a)**   If $b \in \mathbb{R}$, the sequence $B := (b, b, b, \cdots)$, all of whose terms equal b, is called the **constant sequence** b. Thus the constant sequence 1 is the sequence $(1, 1, 1, \cdots)$, and the constant sequence 0 is the sequence $(0, 0, 0, \cdots)$.

(b) If $b \in \mathbb{R}$, then $B := (b^n)$ is the sequence $B = (b, b^2, b^3, \cdots, b^n, \cdots)$. In particular, if $b = \frac{1}{2}$, then we obtain the sequence

$$\left(\frac{1}{2^n} : n \in \mathbb{N} \right) = \left(\frac{1}{2}, \frac{1}{4}, \frac{1}{8}, \cdots, \frac{1}{2^n}, \cdots \right).$$

(c) The sequence of $(2n : n \in \mathbb{N})$ of even natural numbers can be defined inductively by

$$x_1 := 2, \quad x_{n+1} := x_n + 2,$$

or by the definition

$$y_1 := 2, \quad y_{n+1} := y_1 + y_n.$$

(d) The celebrated **Fibonacci sequence** $F := (f_n)$ is given by the inductive definition

$$f_1 := 1, \quad f_2 := 1, \quad f_{n+1} := f_{n-1} + f_n \quad (n \geq 2).$$

Thus each term past the second is the sum of its two immediate predecessors. The first ten terms of F are seen to be $(1, 1, 2, 3, 5, 8, 13, 21, 34, 55, \cdots)$.   □

The Limit of a Sequence

There are a number of different limit concepts in real analysis. The notion of limit of a sequence is the most basic, and it will be the focus of this chapter.

3.1.3 Definition   A sequence $X = (x_n)$ in $\mathbb{R}$ is said to **converge** to $x \in \mathbb{R}$, or x is said to be a **limit** of (x_n), if for every $\varepsilon > 0$ there exists a natural number $K(\varepsilon)$ such that for all $n \geq K(\varepsilon)$, the terms x_n satisfy $|x_n - x| < \varepsilon$.

If a sequence has a limit, we say that the sequence is **convergent**; if it has no limit, we say that the sequence is **divergent**.

Note   The notation $K(\varepsilon)$ is used to emphasize that the choice of K depends on the value of ε. However, it is often convenient to write K instead of $K(\varepsilon)$. In most cases, a "small" value of ε will usually require a "large" value of K to guarantee that the distance $|x_n - x|$ between x_n and x is less than ε for all $n \geq K = K(\varepsilon)$.

When a sequence has limit x, we will use the notation

$$\lim X = x \quad \text{or} \quad \lim(x_n) = x.$$

We will sometimes use the symbolism $x_n \to x$, which indicates the intuitive idea that the values x_n "approach" the number x as $n \to \infty$.

3.1.4 Uniqueness of Limits   *A sequence in $\mathbb{R}$ can have at most one limit.*

Proof.   Suppose that x' and x'' are both limits of (x_n). For each $\varepsilon > 0$ there exist K' such that $|x_n - x'| < \varepsilon/2$ for all $n \geq K'$, and there exists K'' such that $|x_n - x''| < \varepsilon/2$ for all $n \geq K''$. We let K be the larger of K' and K''. Then for $n \geq K$ we apply the Triangle Inequality to get

$$|x' - x''| = |x' - x_n + x_n - x''|$$
$$\leq |x' - x_n| + |x_n - x''| < \varepsilon/2 + \varepsilon/2 = \varepsilon.$$

Since $\varepsilon > 0$ is an arbitrary positive number, we conclude that $x' - x'' = 0$.   Q.E.D.

For $x \in \mathbb{R}$ and $\varepsilon > 0$, recall that the ε-neighborhood of x is the set

$$V_\varepsilon(x) := \{u \in \mathbb{R} : |u - x| < \varepsilon\}.$$

(See Section 2.2.) Since $u \in V_\varepsilon(x)$ is equivalent to $|u - x| < \varepsilon$, the definition of convergence of a sequence can be formulated in terms of neighborhoods. We give several different ways of saying that a sequence x_n converges to x in the following theorem.

3.1.5 Theorem   *Let $X = (x_n)$ be a sequence of real numbers, and let $x \in \mathbb{R}$. The following statements are equivalent.*

(a)   *X converges to x.*

(b)   *For every $\varepsilon > 0$, there exists a natural number K such that for all $n \geq K$, the terms x_n satisfy $|x_n - x| < \varepsilon$.*

(c)   *For every $\varepsilon > 0$, there exists a natural number K such that for all $n \geq K$, the terms x_n satisfy $x - \varepsilon < x_n < x + \varepsilon$.*

(d)   *For every ε-neighborhood $V_\varepsilon(x)$ of x, there exists a natural number K such that for all $n \geq K$, the terms x_n belong to $V_\varepsilon(x)$.*

Proof.   The equivalence of (a) and (b) is just the definition. The equivalence of (b), (c), and (d) follows from the following implications:

$$|u - x| < \varepsilon \quad \Longleftrightarrow \quad -\varepsilon < u - x < \varepsilon \quad \Longleftrightarrow \quad x - \varepsilon < u < x + \varepsilon \quad \Longleftrightarrow \quad u \in V_\varepsilon(x).$$

Q.E.D.

With the language of neighborhoods, one can describe the convergence of the sequence $X = (x_n)$ to the number x by saying: *for each ε-neighborhood $V_\varepsilon(x)$ of x, all but a finite number of terms of X belong to $V_\varepsilon(x)$.* The finite number of terms that may not belong to the ε-neighborhood are the terms $x_1, x_2, \cdots, x_{K-1}$.

Remark   The definition of the limit of a sequence of real numbers is used to verify that a proposed value x is indeed the limit. It does *not* provide a means for initially determining what that value of x might be. Later results will contribute to this end, but quite often it is necessary in practice to arrive at a conjectured value of the limit by direct calculation of a number of terms of the sequence. Computers can be helpful in this respect, but since they can calculate only a finite number of terms of a sequence, such computations do not in any way constitute a proof of the value of the limit.

The following examples illustrate how the definition is applied to prove that a sequence has a particular limit. In each case, a positive ε is given and we are required to find a K, depending on ε, as required by the definition.

3.1.6 Examples   **(a)** $\lim(1/n) = 0$.

If $\varepsilon > 0$ is given, then $1/\varepsilon > 0$. By the Archimedean Property 2.4.5, there is a natural number $K = K(\varepsilon)$ such that $1/K < \varepsilon$. Then, if $n \geq K$, we have $1/n \leq 1/K < \varepsilon$. Consequently, if $n \geq K$, then

$$\left| \frac{1}{n} - 0 \right| = \frac{1}{n} < \varepsilon.$$

Therefore, we can assert that the sequence $(1/n)$ converges to 0.

(b) $\lim(1/(n^2 + 1)) = 0$.

Let $\varepsilon > 0$ be given. To find K, we first note that if $n \in \mathbb{N}$, then

$$\frac{1}{n^2 + 1} < \frac{1}{n^2} \leq \frac{1}{n}.$$

Now choose K such that $1/K < \varepsilon$, as in (a) above. Then $n \geq K$ implies that $1/n < \varepsilon$, and therefore

$$\left| \frac{1}{n^2 + 1} - 0 \right| = \frac{1}{n^2 + 1} < \frac{1}{n} < \varepsilon.$$

Hence, we have shown that the limit of the sequence is zero.

(c) $\lim \left(\dfrac{3n + 2}{n + 1} \right) = 3$.

Given $\varepsilon > 0$, we want to obtain the inequality

(1)
$$\left| \frac{3n + 2}{n + 1} - 3 \right| < \varepsilon$$

when n is sufficiently large. We first simplify the expression on the left:

$$\left| \frac{3n + 2}{n + 1} - 3 \right| = \left| \frac{3n + 2 - 3n - 3}{n + 1} \right| = \left| \frac{-1}{n + 1} \right| = \frac{1}{n + 1} < \frac{1}{n}.$$

Now if the inequality $1/n < \varepsilon$ is satisfied, then the inequality (1) holds. Thus if $1/K < \varepsilon$, then for any $n \geq K$, we also have $1/n < \varepsilon$ and hence (1) holds. Therefore the limit of the sequence is 3.

(d) If $0 < b < 1$, then $\lim(b^n) = 0$.

We will use elementary properties of the natural logarithm function. If $\varepsilon > 0$ is given, we see that

$$b^n < \varepsilon \quad \Longleftrightarrow \quad n \ln b < \ln \varepsilon \quad \Longleftrightarrow \quad n > \ln \varepsilon / \ln b.$$

(The last inequality is reversed because $\ln b < 0$.) Thus if we choose K to be a number such that $K > \ln \varepsilon / \ln b$, then we will have $0 < b^n < \varepsilon$ for all $n \geq K$. Thus we have $\lim(b^n) = 0$.

For example, if $b = .8$, and if $\varepsilon = .01$ is given, then we would need $K > \ln .01 / \ln .8 \approx 20.6377$. Thus $K = 21$ would be an appropriate choice for $\varepsilon = .01$. $\square$

Remark   The $K(\varepsilon)$ Game   In the notion of convergence of a sequence, one way to keep in mind the connection between the ε and the K is to think of it as a game called the $K(\varepsilon)$ Game. In this game, Player A asserts that a certain number x is the limit of a sequence (x_n). Player B challenges this assertion by giving Player A a specific value for $\varepsilon > 0$. Player A must respond to the challenge by coming up with a value of K such that $|x_n - x| < \varepsilon$ for all $n > K$. If Player A can always find a value of K that works, then he wins, and the sequence is convergent. However, if Player B can give a specific value of $\varepsilon > 0$ for which Player A cannot respond adequately, then Player B wins, and we conclude that the sequence does not converge to x.

In order to show that a sequence $X = (x_n)$ does *not* converge to the number x, it is enough to produce one number $\varepsilon_0 > 0$ such that no matter what natural number K is chosen, one can find a particular n_K satisfying $n_K \geq K$ such that $|x_{n_K} - x| \geq \varepsilon_0$. (This will be discussed in more detail in Section 3.4.)

3.1.7 Example   The sequence $(0, 2, 0, 2, \cdots, 0, 2, \cdots)$ does *not* converge to the number 0.

If Player A asserts that 0 is the limit of the sequence, he will lose the $K(\varepsilon)$ Game when Player B gives him a value of $\varepsilon < 2$. To be definite, let Player B give Player A the value $\varepsilon_0 = 1$. Then no matter what value Player A chooses for K, his response will not be adequate, for Player B will respond by selecting an even number $n > K$. Then the corresponding value is $x_n = 2$ so that $|x_n - 0| = 2 > 1 = \varepsilon_0$. Thus the number 0 is not the limit of the sequence.   □

Tails of Sequences ──

It is important to realize that the convergence (or divergence) of a sequence $X = (x_n)$ depends only on the "ultimate behavior" of the terms. By this we mean that if, for any natural number m, we drop the first m terms of the sequence, then the resulting sequence X_m converges if and only if the original sequence converges, and in this case, the limits are the same. We will state this formally after we introduce the idea of a "tail" of a sequence.

3.1.8 Definition   If $X = (x_1, x_2, \cdots, x_n, \cdots)$ is a sequence of real numbers and if m is a given natural number, then the m-**tail** of X is the sequence

$$X_m := (x_{m+n} : n \in \mathbb{N}) = (x_{m+1}, x_{m+2}, \cdots)$$

For example, the 3-tail of the sequence $X = (2, 4, 6, 8, 10, \cdots, 2n, \cdots)$, is the sequence $X_3 = (8, 10, 12, \cdots, 2n + 6, \cdots)$.

3.1.9 Theorem   *Let $X = (x_n : n \in \mathbb{N})$ be a sequence of real numbers and let $m \in \mathbb{N}$. Then the m-tail $X_m = (x_{m+n} : n \in \mathbb{N})$ of X converges if and only if X converges. In this case, $\lim X_m = \lim X$.*

Proof.   We note that for any $p \in \mathbb{N}$, the pth term of X_m is the $(p + m)$th term of X. Similarly, if $q > m$, then the qth term of X is the $(q - m)$th term of X_m.

Assume X converges to x. Then given any $\varepsilon > 0$, if the terms of X for $n \geq K(\varepsilon)$ satisfy $|x_n - x| < \varepsilon$, then the terms of X_m for $k \geq K(\varepsilon) - m$ satisfy $|x_k - x| < \varepsilon$. Thus we can take $K_m(\varepsilon) = K(\varepsilon) - m$, so that X_m also converges to x.

Conversely, if the terms of X_m for $k \geq K_m(\varepsilon)$ satisfy $\left|x_k - x\right| < \varepsilon$, then the terms of X for $n \geq K(\varepsilon) + m$ satisfy $\left|x_n - x\right| < \varepsilon$. Thus we can take $K(\varepsilon) = K_m(\varepsilon) + m$.

Therefore, X converges to x if and only if X_m converges to x.   Q.E.D.

We shall sometimes say that a sequence X *ultimately* has a certain property if some tail of X has this property. For example, we say that the sequence $(3, 4, 5, 5, 5, \cdots, 5, \cdots)$ is "ultimately constant". On the other hand, the sequence $(3, 5, 3, 5, \cdots, 3, 5, \cdots)$ is not ultimately constant. The notion of convergence can be stated using this terminology: A sequence X converges to x if and only if the terms of X are ultimately in every ε-neighborhood of x. Other instances of this "ultimate terminology" will be noted below.

Further Examples ───

In establishing that a number x is the limit of a sequence (x_n), we often try to simplify the difference $|x_n - x|$ before considering an $\varepsilon > 0$ and finding a $K(\varepsilon)$ as required by the definition of limit. This was done in some of the earlier examples. The next result is a more formal statement of this idea, and the examples that follow make use of this approach.

3.1.10 Theorem   *Let* (x_n) *be a sequence of real numbers and let* $x \in \mathbb{R}$. *If* (a_n) *is a sequence of positive real numbers with* $\lim(a_n) = 0$ *and if for some constant* $C > 0$ *and some* $m \in \mathbb{N}$ *we have*

$$|x_n - x| \leq C a_n \qquad \text{for all} \quad n \geq m,$$

then it follows that $\lim(x_n) = x$.

Proof.   If $\varepsilon > 0$ is given, then since $\lim(a_n) = 0$, we know there exists $K = K(\varepsilon/C)$ such that $n \geq K$ implies

$$a_n = |a_n - 0| < \varepsilon/C.$$

Therefore it follows that if both $n \geq K$ and $n \geq m$, then

$$|x_n - x| \leq C a_n < C(\varepsilon/C) = \varepsilon.$$

Since $\varepsilon > 0$ is arbitrary, we conclude that $x = \lim(x_n)$.   Q.E.D.

3.1.11 Examples   **(a)**   If $a > 0$, then $\lim \left(\dfrac{1}{1 + na} \right) = 0$.

Since $a > 0$, then $0 < na < 1 + na$, and therefore $0 < 1/(1 + na) < 1/(na)$. Thus we have

$$\left| \frac{1}{1 + na} - 0 \right| \leq \left(\frac{1}{a} \right) \frac{1}{n} \qquad \text{for all} \quad n \in \mathbb{N}.$$

Since $\lim(1/n) = 0$, we may invoke Theorem 3.1.10 with $C = 1/a$ and $m = 1$ to infer that $\lim(1/(1 + na)) = 0$.

(b)   If $0 < b < 1$, then $\lim(b^n) = 0$.

This limit was obtained earlier in Example 3.1.6(d). We will give a second proof that illustrates the use of Bernoulli's Inequality (see Example 2.1.13(c)).

Since $0 < b < 1$, we can write $b = 1/(1 + a)$, where $a := (1/b) - 1$ so that $a > 0$. By Bernoulli's Inequality, we have $(1 + a)^n \geq 1 + na$. Hence

$$0 < b^n = \frac{1}{(1 + a)^n} \leq \frac{1}{1 + na} < \frac{1}{na}.$$

Thus from Theorem 3.1.10 we conclude that $\lim(b^n) = 0$.

In particular, if $b = .8$, so that $a = .25$, and if we are given $\varepsilon = .01$, then the preceding inequality gives us $K(\varepsilon) = 4/(.01) = 400$. Comparing with Example 3.1.6(d), where we obtained $K = 25$, we see this method of estimation does not give us the "best" value of K. However, for the purpose of establishing the limit, the size of K is immaterial.

(c)   If $c > 0$, then $\lim(c^{1/n}) = 1$.

The case $c = 1$ is trivial, since then $(c^{1/n})$ is the constant sequence $(1, 1, \cdots)$, which evidently converges to 1.

If $c > 1$, then $c^{1/n} = 1 + d_n$ for some $d_n > 0$. Hence by Bernoulli's Inequality 2.1.13(c),

$$c = (1 + d_n)^n \geq 1 + n d_n \qquad \text{for} \quad n \in \mathbb{N}.$$

Therefore we have $c - 1 \geq n d_n$, so that $d_n \leq (c - 1)/n$. Consequently we have

$$\left| c^{1/n} - 1 \right| = d_n \leq (c - 1)\frac{1}{n} \qquad \text{for} \quad n \in \mathbb{N}.$$

We now invoke Theorem 3.1.10 to infer that $\lim(c^{1/n}) = 1$ when $c > 1$.

Now suppose that $0 < c < 1$; then $c^{1/n} = 1/(1+h_n)$ for some $h_n > 0$. Hence Bernoulli's Inequality implies that

$$c = \frac{1}{(1+h_n)^n} \le \frac{1}{1+nh_n} < \frac{1}{nh_n},$$

from which it follows that $0 < h_n < 1/nc$ for $n \in \mathbb{N}$. Therefore we have

$$0 < 1 - c^{1/n} = \frac{h_n}{1+h_n} < h_n < \frac{1}{nc}$$

so that

$$\left|c^{1/n} - 1\right| < \left(\frac{1}{c}\right)\frac{1}{n} \qquad \text{for} \quad n \in \mathbb{N}.$$

We now apply Theorem 3.1.10 to infer that $\lim(c^{1/n}) = 1$ when $0 < c < 1$.

(d) $\lim(n^{1/n}) = 1$

Since $n^{1/n} > 1$ for $n > 1$, we can write $n^{1/n} = 1 + k_n$ for some $k_n > 0$ when $n > 1$. Hence $n = (1+k_n)^n$ for $n > 1$. By the Binomial Theorem, if $n > 1$ we have

$$n = 1 + nk_n + \tfrac{1}{2}n(n-1)k_n^2 + \cdots \ge 1 + \tfrac{1}{2}n(n-1)k_n^2,$$

whence it follows that

$$n - 1 \ge \tfrac{1}{2}n(n-1)k_n^2.$$

Hence $k_n^2 \le 2/n$ for $n > 1$. If $\varepsilon > 0$ is given, it follows from the Archimedean Property that there exists a natural number N_ε such that $2/N_\varepsilon < \varepsilon^2$. It follows that if $n \ge \sup\{2, N_\varepsilon\}$ then $2/n < \varepsilon^2$, whence

$$0 < n^{1/n} - 1 = k_n \le (2/n)^{1/2} < \varepsilon.$$

Since $\varepsilon > 0$ is arbitrary, we deduce that $\lim(n^{1/n}) = 1$. $\qquad\qquad\square$

Exercises for Section 3.1

1. The sequence (x_n) is defined by the following formulas for the nth term. Write the first five terms in each case:
   (a) $x_n := 1 + (-1)^n$,
   (b) $x_n := (-1)^n/n$,
   (c) $x_n := \dfrac{1}{n(n+1)}$,
   (d) $x := \dfrac{1}{n^2+2}$.

2. The first few terms of a sequence (x_n) are given below. Assuming that the "natural pattern" indicated by these terms persists, give a formula for the nth term x_n.
   (a) $5, 7, 9, 11, \cdots$,
   (b) $1/2, -1/4, 1/8, -1/16, \cdots$,
   (c) $1/2, 2/3, 3/4, 4/5, \cdots$,
   (d) $1, 4, 9, 16, \cdots$.

3. List the first five terms of the following inductively defined sequences.
   (a) $x_1 := 1$, $\quad x_{n+1} = 3x_n + 1$,
   (b) $y_1 := 2$, $\quad y_{n+1} = \tfrac{1}{2}(y_n + 2/y_n)$,
   (c) $z_1 := 1$, $\quad z_2 := 2$, $\quad z_{n+2} := (z_{n+1} + z_n)/(z_{n+1} - z_n)$,
   (d) $s_1 = 3$, $\quad s_2 := 5$, $\quad s_{n+2} := s_n + s_{n+1}$.

4. For any $b \in \mathbb{R}$, prove that $\lim(b/n) = 0$.

5. Use the definition of the limit of a sequence to establish the following limits.

(a) $\lim \left(\dfrac{n}{n^2 + 1} \right) = 0,$ (b) $\lim \left(\dfrac{2n}{n + 1} \right) = 2,$

(c) $\lim \left(\dfrac{3n + 1}{2n + 5} \right) = \dfrac{3}{2},$ (d) $\lim \left(\dfrac{n^2 - 1}{2n^2 + 3} \right) = \dfrac{1}{2}.$

6. Show that

(a) $\lim \left(\dfrac{1}{\sqrt{n + 7}} \right) = 0,$ (b) $\lim \left(\dfrac{2n}{n + 2} \right) = 2,$

(c) $\lim \left(\dfrac{\sqrt{n}}{n + 1} \right) = 0,$ (d) $\lim \left(\dfrac{(-1)^n n}{n^2 + 1} \right) = 0.$

7. Let $x_n := 1/\ln(n + 1)$ for $n \in \mathbb{N}$.
   (a) Use the definition of limit to show that $\lim(x_n) = 0$.
   (b) Find a specific value of $K(\varepsilon)$ as required in the definition of limit for each of (i) $\varepsilon = 1/2$, and (ii) $\varepsilon = 1/10$.

8. Prove that $\lim(x_n) = 0$ if and only if $\lim(|x_n|) = 0$. Give an example to show that the convergence of $(|x_n|)$ need not imply the convergence of (x_n).

9. Show that if $x_n \geq 0$ for all $n \in \mathbb{N}$ and $\lim(x_n) = 0$, then $\lim \left(\sqrt{x_n} \right) = 0$.

10. Prove that if $\lim(x_n) = x$ and if $x > 0$, then there exists a natural number M such that $x_n > 0$ for all $n \geq M$.

11. Show that $\lim \left(\dfrac{1}{n} - \dfrac{1}{n + 1} \right) = 0.$

12. Show that $\lim(1/3^n) = 0$.

13. Let $b \in \mathbb{R}$ satisfy $0 < b < 1$. Show that $\lim(nb^n) = 0$. [*Hint*: Use the Binomial Theorem as in Example 3.1.11(d).]

14. Show that $\lim \left((2n)^{1/n} \right) = 1$.

15. Show that $\lim(n^2/n!) = 0$.

16. Show that $\lim(2^n/n!) = 0$. [*Hint*: If $n \geq 3$, then $0 < 2^n/n! \leq 2 \left(\frac{2}{3} \right)^{n-2}$.]

17. If $\lim(x_n) = x > 0$, show that there exists a natural number K such that if $n \geq K$, then $\frac{1}{2}x < x_n < 2x$.

Section 3.2   Limit Theorems

In this section we will obtain some results that enable us to evaluate the limits of certain sequences of real numbers. These results will expand our collection of convergent sequences rather extensively. We begin by establishing an important property of convergent sequences that will be needed in this and later sections.

3.2.1 Definition   A sequence $X = (x_n)$ of real numbers is said to be **bounded** if there exists a real number $M > 0$ such that $|x_n| \leq M$ for all $n \in \mathbb{N}$.

Thus, the sequence (x_n) is bounded if and only if the set $\{x_n : n \in \mathbb{N}\}$ of its values is a bounded subset of $\mathbb{R}$.

3.2.2 Theorem   *A convergent sequence of real numbers is bounded.*

Proof. Suppose that $\lim(x_n) = x$ and let $\varepsilon := 1$. Then there exists a natural number $K = K(1)$ such that $|x_n - x| < 1$ for all $n \geq K$. If we apply the Triangle Inequality with $n \geq K$ we obtain

$$|x_n| = |x_n - x + x| \leq |x_n - x| + |x| < 1 + |x|.$$

If we set

$$M := \sup \left\{ |x_1|, |x_2|, \cdots, |x_{K-1}|, 1 + |x| \right\},$$

then it follows that $|x_n| \leq M$ for all $n \in \mathbb{N}$.   Q.E.D.

We will now examine how the limit process interacts with the operations of addition, subtraction, multiplication, and division of sequences. If $X = (x_n)$ and $Y = (y_n)$ are sequences of real numbers, then we define their **sum** to be the sequence $X + Y := (x_n + y_n)$, their **difference** to be the sequence $X - Y := (x_n - y_n)$, and their **product** to be the sequence $X \cdot Y := (x_n y_n)$. If $c \in \mathbb{R}$, we define the **multiple** of X by c to be the sequence $cX := (cx_n)$. Finally, if $Z = (z_n)$ is a sequence of real numbers with $z_n \neq 0$ for all $n \in \mathbb{N}$, then we define the **quotient** of X and Z to be the sequence $X/Z := (x_n/z_n)$.

For example, if X and Y are the sequences

$$X := (2, 4, 6, \cdots 2n, \cdots), \quad Y := \left(\frac{1}{1}, \frac{1}{2}, \frac{1}{3}, \cdots, \frac{1}{n}, \cdots \right),$$

then we have

$$X + Y = \left(\frac{3}{1}, \frac{9}{2}, \frac{19}{3}, \cdots, \frac{2n^2 + 1}{n}, \cdots \right),$$

$$X - Y = \left(\frac{1}{1}, \frac{7}{2}, \frac{17}{3}, \cdots, \frac{2n^2 - 1}{n}, \cdots \right),$$

$$X \cdot Y = (2, 2, 2, \cdots, 2, \cdots),$$

$$3X = (6, 12, 18, \cdots, 6n, \cdots),$$

$$X/Y = (2, 8, 18, \cdots, 2n^2, \cdots).$$

We note that if Z is the sequence

$$Z := (0, 2, 0, \cdots, 1 + (-1)^n, \cdots),$$

then we can define $X + Z$, $X - Z$ and $X \cdot Z$, but X/Z is not defined since some of the terms of Z are zero.

We now show that sequences obtained by applying these operations to convergent sequences give rise to new sequences whose limits can be predicted.

3.2.3 Theorem   **(a)** Let $X = (x_n)$ and $Y = (y_n)$ be sequences of real numbers that converge to x and y, respectively, and let $c \in \mathbb{R}$. Then the sequences $X + Y$, $X - Y$, $X \cdot Y$, and cX converge to $x + y$, $x - y$, xy, and cx, respectively.

(b)   If $X = (x_n)$ converges to x and $Z = (z_n)$ is a sequence of nonzero real numbers that converges to z and if $z \neq 0$, then the quotient sequence X/Z converges to x/z.

Proof.   (a) To show that $\lim(x_n + y_n) = x + y$, we need to estimate the magnitude of $|(x_n + y_n) - (x + y)|$. To do this we use the Triangle Inequality 2.2.3 to obtain

$$|(x_n + y_n) - (x + y)| = |(x_n - x) + (y_n - y)|$$
$$\leq |x_n - x| + |y_n - y|.$$

By hypothesis, if $\varepsilon > 0$ there exists a natural number K_1 such that if $n \geq K_1$, then $|x_n - x| < \varepsilon/2$; also there exists a natural number K_2 such that if $n \geq K_2$, then $|y_n - y| < \varepsilon/2$. Hence if $K(\varepsilon) := \sup\{K_1, K_2\}$, it follows that if $n \geq K(\varepsilon)$ then

$$|(x_n + y_n) - (x + y)| \leq |x_n - x| + |y_n - y|$$
$$< \tfrac{1}{2}\varepsilon + \tfrac{1}{2}\varepsilon = \varepsilon.$$

Since $\varepsilon > 0$ is arbitrary, we infer that $X + Y = (x_n + y_n)$ converges to $x + y$.

Precisely the same argument can be used to show that $X - Y = (x_n - y_n)$ converges to $x - y$.

To show that $X \cdot Y = (x_n y_n)$ converges to xy, we make the estimate

$$|x_n y_n - xy| = |(x_n y_n - x_n y) + (x_n y - xy)|$$
$$\leq |x_n(y_n - y)| + |(x_n - x)y|$$
$$= |x_n||y_n - y| + |x_n - x||y|.$$

According to Theorem 3.2.2 there exists a real number $M_1 > 0$ such that $|x_n| \leq M_1$ for all $n \in \mathbb{N}$ and we set $M := \sup\{M_1, |y|\}$. Hence we have the estimate

$$|x_n y_n - xy| \leq M|y_n - y| + M|x_n - x|.$$

From the convergence of X and Y we conclude that if $\varepsilon > 0$ is given, then there exist natural numbers K_1 and K_2 such that if $n \geq K_1$ then $|x_n - x| < \varepsilon/2M$, and if $n \geq K_2$ then $|y_n - y| < \varepsilon/2M$. Now let $K(\varepsilon) = \sup\{K_1, K_2\}$; then, if $n \geq K(\varepsilon)$ we infer that

$$|x_n y_n - xy| \leq M|y_n - y| + M|x_n - x|$$
$$< M(\varepsilon/2M) + M(\varepsilon/2M) = \varepsilon.$$

Since $\varepsilon > 0$ is arbitrary, this proves that the sequence $X \cdot Y = (x_n y_n)$ converges to xy.

The fact that $cX = (cx_n)$ converges to cx can be proved in the same way; it can also be deduced by taking Y to be the constant sequence $(c, c, c, \cdots)$. We leave the details to the reader.

(b) We next show that if $Z = (z_n)$ is a sequence of nonzero numbers that converges to a nonzero limit z, then the sequence $(1/z_n)$ of reciprocals converges to $1/z$. First let $\alpha := \tfrac{1}{2}|z|$ so that $\alpha > 0$. Since $\lim(z_n) = z$, there exists a natural number K_1 such that if $n \geq K_1$ then $|z_n - z| < \alpha$. It follows from Corollary 2.2.4(a) of the Triangle Inequality that $-\alpha \leq -|z_n - z| \leq |z_n| - |z|$ for $n \geq K_1$, whence it follows that $\tfrac{1}{2}|z| = |z| - \alpha \leq |z_n|$ for $n \geq K_1$. Therefore $1/|z_n| \leq 2/|z|$ for $n \geq K_1$ so we have the estimate

$$\left| \frac{1}{z_n} - \frac{1}{z} \right| = \left| \frac{z - z_n}{z_n z} \right| = \frac{1}{|z_n z|}|z - z_n|$$
$$\leq \frac{2}{|z|^2}|z - z_n| \qquad \text{for all} \quad n \geq K_1.$$

Now, if $\varepsilon > 0$ is given, there exists a natural number K_2 such that if $n \geq K_2$ then $|z_n - z| < \tfrac{1}{2}\varepsilon|z|^2$. Therefore, it follows that if $K(\varepsilon) = \sup\{K_1, K_2\}$, then

$$\left| \frac{1}{z_n} - \frac{1}{z} \right| < \varepsilon \qquad \text{for all} \quad n > K(\varepsilon).$$

Since $\varepsilon > 0$ is arbitrary, it follows that

$$\lim\left(\frac{1}{z_n} \right) = \frac{1}{z}.$$

The proof of (b) is now completed by taking Y to be the sequence $(1/z_n)$ and using the fact that $X \cdot Y = (x_n/z_n)$ converges to $x(1/z) = x/z$.   Q.E.D.

Some of the results of Theorem 3.2.3 can be extended, by Mathematical Induction, to a finite number of convergent sequences. For example, if $A = (a_n)$, $B = (b_n)$, $\cdots$, $Z = (z_n)$ are convergent sequences of real numbers, then their sum $A + B + \cdots + Z = (a_n + b_n + \cdots + z_n)$ is a convergent sequence and

(1) $\lim(a_n + b_n + \cdots + z_n) = \lim(a_n) + \lim(b_n) + \cdots + \lim(z_n).$

Also their product $A \cdot B \cdots Z := (a_n b_n \cdots z_n)$ is a convergent sequence and

(2) $\lim(a_n b_n \cdots z_n) = \big(\lim(a_n)\big)\big(\lim(b_n)\big) \cdots \big(\lim(z_n)\big).$

Hence, if $k \in \mathbb{N}$ and if $A = (a_n)$ is a convergent sequence, then

(3) $\lim(a_n^k) = \big(\lim(a_n)\big)^k.$

We leave the proofs of these assertions to the reader.

3.2.4 Theorem   If $X = (x_n)$ is a convergent sequence of real numbers and if $x_n \geq 0$ for all $n \in \mathbb{N}$, then $x = \lim(x_n) \geq 0$.

Proof.   Suppose the conclusion is not true and that $x < 0$; then $\varepsilon := -x$ is positive. Since X converges to x, there is a natural number K such that

$$x - \varepsilon < x_n < x + \varepsilon \qquad \text{for all} \quad n \geq K.$$

In particular, we have $x_K < x + \varepsilon = x + (-x) = 0$. But this contradicts the hypothesis that $x_n \geq 0$ for all $n \in \mathbb{N}$. Therefore, this contradiction implies that $x \geq 0$.   Q.E.D.

We now give a useful result that is formally stronger than Theorem 3.2.4.

3.2.5 Theorem   If $X = (x_n)$ and $Y = (y_n)$ are convergent sequences of real numbers and if $x_n \leq y_n$ for all $n \in \mathbb{N}$, then $\lim(x_n) \leq \lim(y_n)$.

Proof.   Let $z_n := y_n - x_n$ so that $Z := (z_n) = Y - X$ and $z_n \geq 0$ for all $n \in \mathbb{N}$. It follows from Theorems 3.2.4 and 3.2.3 that

$$0 \leq \lim Z = \lim(y_n) - \lim(x_n),$$

so that $\lim(x_n) \leq \lim(y_n)$.   Q.E.D.

The next result asserts that if all the terms of a convergent sequence satisfy an inequality of the form $a \leq x_n \leq b$, then the limit of the sequence satisfies the same inequality. Thus if the sequence is convergent, one may "pass to the limit" in an inequality of this type.

3.2.6 Theorem   If $X = (x_n)$ is a convergent sequence and if $a \leq x_n \leq b$ for all $n \in \mathbb{N}$, then $a \leq \lim(x_n) \leq b$.

Proof.   Let Y be the constant sequence $(b, b, b, \cdots)$. Theorem 3.2.5 implies that $\lim X \leq \lim Y = b$. Similarly one shows that $a \leq \lim X$.   Q.E.D.

The next result asserts that if a sequence Y is squeezed between two sequences that converge to the *same limit,* then it must also converge to this limit.

3.2.7 Squeeze Theorem   *Suppose that $X = (x_n)$, $Y = (y_n)$, and $Z = (z_n)$ are sequences of real numbers such that*

$$x_n \le y_n \le z_n \qquad \text{for all} \quad n \in \mathbb{N},$$

and that $\lim(x_n) = \lim(z_n)$. *Then* $Y = (y_n)$ *is convergent and*

$$\lim(x_n) = \lim(y_n) = \lim(z_n).$$

Proof.   Let $w := \lim(x_n) = \lim(z_n)$. If $\varepsilon > 0$ is given, then it follows from the convergence of X and Z to w that there exists a natural number K such that if $n \ge K$ then

$$|x_n - w| < \varepsilon \qquad \text{and} \qquad |z_n - w| < \varepsilon.$$

Since the hypothesis implies that

$$x_n - w \le y_n - w \le z_n - w \qquad \text{for all} \quad n \in \mathbb{N},$$

it follows (why?) that

$$-\varepsilon < y_n - w < \varepsilon$$

for all $n \ge K$. Since $\varepsilon > 0$ is arbitrary, this implies that $\lim(y_n) = w$.   Q.E.D.

Remark   Since any tail of a convergent sequence has the same limit, the hypotheses of Theorems 3.2.4, 3.2.5, 3.2.6, and 3.2.7 can be weakened to apply to the tail of a sequence. For example, in Theorem 3.2.4, if $X = (x_n)$ is "ultimately positive" in the sense that there exists $m \in \mathbb{N}$ such that $x_n \ge 0$ for all $n \ge m$, then the same conclusion that $x \ge 0$ will hold. Similar modifications are valid for the other theorems, as the reader should verify.

3.2.8 Examples   **(a)**   The sequence (n) is divergent.
It follows from Theorem 3.2.2 that if the sequence $X := (n)$ is convergent, then there exists a real number $M > 0$ such that $n = |n| < M$ for all $n \in \mathbb{N}$. But this violates the Archimedean Property 2.4.3.

(b)   The sequence $((-1)^n)$ is divergent.
This sequence $X = ((-1)^n)$ is bounded (take $M := 1$), so we cannot invoke Theorem 3.2.2. However, assume that $a := \lim X$ exists. Let $\varepsilon := 1$ so that there exists a natural number K_1 such that

$$|(-1) - a| < 1 \qquad \text{for all} \quad n \ge K_1.$$

If n is an odd natural number with $n \ge K_1$, this gives $|-1 - a| < 1$, so that $-2 < a < 0$. (Why?) On the other hand, if n is an even natural number with $n \ge K_1$, this inequality gives $|1 - a| < 1$ so that $0 < a < 2$. Since a cannot satisfy both of these inequalities, the hypothesis that X is convergent leads to a contradiction. Therefore the sequence X is divergent.

(c) $\lim \left(\dfrac{2n + 1}{n} \right) = 2$.
If we let $X := (2)$ and $Y := (1/n)$, then $((2n + 1)/n) = X + Y$. Hence it follows from Theorem 3.2.3(a) that $\lim(X + Y) = \lim X + \lim Y = 2 + 0 = 2$.

(d) $\lim \left(\dfrac{2n + 1}{n + 5} \right) = 2$.

Since the sequences $(2n + 1)$ and $(n + 5)$ are not convergent (why?), it is not possible to use Theorem 3.2.3(b) directly. However, if we write

$$\frac{2n + 1}{n + 5} = \frac{2 + 1/n}{1 + 5/n},$$

we can obtain the given sequence as one to which Theorem 3.2.3(b) applies when we take $X := (2 + 1/n)$ and $Z := (1 + 5/n)$. (Check that all hypotheses are satisfied.) Since $\lim X = 2$ and $\lim Z = 1 \neq 0$, we deduce that $\lim\big((2n + 1)/(n + 5)\big) = 2/1 = 2$.

(e) $\lim \left(\dfrac{2n}{n^2 + 1} \right) = 0.$

Theorem 3.2.3(b) does not apply directly. (Why?) We note that

$$\frac{2n}{n^2 + 1} = \frac{2}{n + 1/n},$$

but Theorem 3.2.3(b) does not apply here either, because $(n + 1/n)$ is not a convergent sequence. (Why not?) However, if we write

$$\frac{2n}{n^2 + 1} = \frac{2/n}{1 + 1/n^2},$$

then we can apply Theorem 3.2.3(b), since $\lim(2/n) = 0$ and $\lim(1 + 1/n^2) = 1 \neq 0$. Therefore $\lim(2n/(n^2 + 1)) = 0/1 = 0$.

(f) $\lim \left(\dfrac{\sin n}{n} \right) = 0.$

We cannot apply Theorem 3.2.3(b) directly, since the sequence (n) is not convergent [neither is the sequence $(\sin n)$]. It does not appear that a simple algebraic manipulation will enable us to reduce the sequence into one to which Theorem 3.2.3 will apply. However, if we note that $-1 \leq \sin n \leq 1$, then it follows that

$$-\frac{1}{n} \leq \frac{\sin n}{n} \leq \frac{1}{n} \qquad \text{for all} \quad n \in \mathbb{N}.$$

Hence we can apply the Squeeze Theorem 3.2.7 to infer that $\lim(n^{-1} \sin n) = 0$. (We note that Theorem 3.1.10 could also be applied to this sequence.)

(g) Let $X = (x_n)$ be a sequence of real numbers that converges to $x \in \mathbb{R}$. Let p be a polynomial; for example, let

$$p(t) := a_k t^k + a_{k-1} t^{k-1} + \cdots + a_1 t + a_0,$$

where $k \in \mathbb{N}$ and $a_j \in \mathbb{R}$ for $j = 0, 1, \cdots, k$. It follows from Theorem 3.2.3 that the sequence $(p(x_n))$ converges to $p(x)$. We leave the details to the reader as an exercise.

(h) Let $X = (x_n)$ be a sequence of real numbers that converges to $x \in \mathbb{R}$. Let r be a rational function (that is, $r(t) := p(t)/q(t)$, where p and q are polynomials). Suppose that $q(x_n) \neq 0$ for all $n \in \mathbb{N}$ and that $q(x) \neq 0$. Then the sequence $(r(x_n))$ converges to $r(x) = p(x)/q(x)$. We leave the details to the reader as an exercise. $\qquad \square$

We conclude this section with several results that will be useful in the work that follows.

3.2.9 Theorem *Let the sequence $X = (x_n)$ converge to x. Then the sequence $(|x_n|)$ of absolute values converges to $|x|$. That is, if $x = \lim(x_n)$, then $|x| = \lim(|x_n|)$.*

Proof.   It follows from the Triangle Inequality (see Corollary 2.2.4(a)) that

$$\left| |x_n| - |x| \right| \le |x_n - x| \qquad \text{for all} \quad n \in \mathbb{N}.$$

The convergence of $(|x_n|)$ to $|x|$ is then an immediate consequence of the convergence of (x_n) to x.   Q.E.D.

3.2.10 Theorem   *Let $X = (x_n)$ be a sequence of real numbers that converges to x and suppose that $x_n \ge 0$. Then the sequence $(\sqrt{x_n})$ of positive square roots converges and $\lim (\sqrt{x_n}) = \sqrt{x}$.*

Proof.   It follows from Theorem 3.2.4 that $x = \lim(x_n) \ge 0$ so the assertion makes sense. We now consider the two cases: (i) $x = 0$ and (ii) $x > 0$.

   Case (i)   If $x = 0$, let $\varepsilon > 0$ be given. Since $x_n \to 0$ there exists a natural number K such that if $n \ge K$ then

$$0 \le x_n = x_n - 0 < \varepsilon^2.$$

Therefore [see Example 2.1.13(a)], $0 \le \sqrt{x_n} < \varepsilon$ for $n \ge K$. Since $\varepsilon > 0$ is arbitrary, this implies that $\sqrt{x_n} \to 0$.

   Case (ii)   If $x > 0$, then $\sqrt{x} > 0$ and we note that

$$\sqrt{x_n} - \sqrt{x} = \frac{\left(\sqrt{x_n} - \sqrt{x}\right)\left(\sqrt{x_n} + \sqrt{x}\right)}{\sqrt{x_n} + \sqrt{x}} = \frac{x_n - x}{\sqrt{x_n} + \sqrt{x}} \; .$$

Since $\sqrt{x_n} + \sqrt{x} \ge \sqrt{x} > 0$, it follows that

$$\left| \sqrt{x_n} - \sqrt{x} \right| \le \left(\frac{1}{\sqrt{x}} \right) |x_n - x|.$$

The convergence of $\sqrt{x_n} \to \sqrt{x}$ follows from the fact that $x_n \to x$.   Q.E.D.

   For certain types of sequences, the following result provides a quick and easy "ratio test" for convergence. Related results can be found in the exercises.

3.2.11 Theorem   *Let (x_n) be a sequence of positive real numbers such that $L := \lim(x_{n+1}/x_n)$ exists. If $L < 1$, then (x_n) converges and $\lim(x_n) = 0$.*

Proof.   By 3.2.4 it follows that $L \ge 0$. Let r be a number such that $L < r < 1$, and let $\varepsilon := r - L > 0$. There exists a number $K \in \mathbb{N}$ such that if $n \ge K$ then

$$\left| \frac{x_{n+1}}{x_n} - L \right| < \varepsilon.$$

It follows from this (why?) that if $n \ge K$, then

$$\frac{x_{n+1}}{x_n} < L + \varepsilon = L + (r - L) = r.$$

Therefore, if $n \ge K$, we obtain

$$0 < x_{n+1} < x_n r < x_{n-1} r^2 < \cdots < x_K r^{n-K+1}.$$

If we set $C := x_K / r^K$, we see that $0 < x_{n+1} < C r^{n+1}$ for all $n \ge K$. Since $0 < r < 1$, it follows from 3.1.11(b) that $\lim(r^n) = 0$ and therefore from Theorem 3.1.10 that $\lim(x_n) = 0$.   Q.E.D.

As an illustration of the utility of the preceding theorem, consider the sequence (x_n) given by $x_n := n/2^n$. We have

$$\frac{x_{n+1}}{x_n} = \frac{n+1}{2^{n+1}} \cdot \frac{2^n}{n} = \frac{1}{2}\left(1 + \frac{1}{n}\right),$$

so that $\lim(x_{n+1}/x_n) = \frac{1}{2}$. Since $\frac{1}{2} < 1$, it follows from Theorem 3.2.11 that $\lim(n/2^n) = 0$.

Exercises for Section 3.2

1. For x_n given by the following formulas, establish either the convergence or the divergence of the sequence $X = (x_n)$.

    (a) $x_n := \dfrac{n}{n+1}$,

    (b) $x_n := \dfrac{(-1)^n n}{n+1}$,

    (c) $x_n := \dfrac{n^2}{n+1}$,

    (d) $x_n := \dfrac{2n^2 + 3}{n^2 + 1}$.

2. Give an example of two divergent sequences X and Y such that:
    (a) their sum $X + Y$ converges,
    (b) their product XY converges.

3. Show that if X and Y are sequences such that X and $X + Y$ are convergent, then Y is convergent.

4. Show that if X and Y are sequences such that X converges to $x \neq 0$ and XY converges, then Y converges.

5. Show that the following sequences are not convergent.
    (a) (2^n),
    (b) $((-1)^n n^2)$.

6. Find the limits of the following sequences:

    (a) $\lim\left((2 + 1/n)^2\right)$,

    (b) $\lim\left(\dfrac{(-1)^n}{n+2}\right)$,

    (c) $\lim\left(\dfrac{\sqrt{n}-1}{\sqrt{n}+1}\right)$,

    (d) $\lim\left(\dfrac{n+1}{n\sqrt{n}}\right)$.

7. If (b_n) is a bounded sequence and $\lim(a_n) = 0$, show that $\lim(a_n b_n) = 0$. Explain why Theorem 3.2.3 *cannot* be used.

8. Explain why the result in equation (3) before Theorem 3.2.4 *cannot* be used to evaluate the limit of the sequence $\left((1 + 1/n)^n\right)$.

9. Let $y_n := \sqrt{n+1} - \sqrt{n}$ for $n \in \mathbb{N}$. Show that (y_n) and $(\sqrt{n}\,y_n)$ converge. Find their limits.

10. Determine the following limits.
    (a) $\lim\left((3\sqrt{n})^{1/2n}\right)$,
    (b) $\lim\left((n+1)^{1/\ln(n+1)}\right)$.

11. If $0 < a < b$, determine $\lim\left(\dfrac{a^{n+1} + b^{n+1}}{a^n + b^n}\right)$.

12. If $a > 0, b > 0$, show that $\lim\left(\sqrt{(n+a)(n+b)} - n\right) = (a+b)/2$.

13. Use the Squeeze Theorem 3.2.7 to determine the limits of the following.
    (a) $\left(n^{1/n^2}\right)$,
    (b) $\left((n!)^{1/n^2}\right)$.

14. Show that if $z_n := (a^n + b^n)^{1/n}$ where $0 < a < b$, then $\lim(z_n) = b$.

15. Apply Theorem 3.2.11 to the following sequences, where a, b satisfy $0 < a < 1, b > 1$.
    (a) (a^n),
    (b) $(b^n/2^n)$,
    (c) (n/b^n),
    (d) $(2^{3n}/3^{2n})$.

68 CHAPTER 3 SEQUENCES AND SERIES

16. (a) Give an example of a convergent sequence (x_n) of positive numbers with $\lim(x_{n+1}/x_n) = 1$.
    (b) Give an example of a divergent sequence with this property. (Thus, this property cannot be used as a test for convergence.)

17. Let $X = (x_n)$ be a sequence of positive real numbers such that $\lim(x_{n+1}/x_n) = L > 1$. Show that X is not a bounded sequence and hence is not convergent.

18. Discuss the convergence of the following sequences, where a, b satisfy $0 < a < 1, b > 1$.
    (a) $(n^2 a^n)$,   (b) (b^n/n^2),
    (c) $(b^n/n!)$,   (d) $(n!/n^n)$.

19. Let (x_n) be a sequence of positive real numbers such that $\lim(x_n^{1/n}) = L < 1$. Show that there exists a number r with $0 < r < 1$ such that $0 < x_n < r^n$ for all sufficiently large $n \in \mathbb{N}$. Use this to show that $\lim(x_n) = 0$.

20. (a) Give an example of a convergent sequence (x_n) of positive numbers with $\lim(x_n^{1/n}) = 1$.
    (b) Give an example of a divergent sequence (x_n) of positive numbers with $\lim(x_n^{1/n}) = 1$. (Thus, this property cannot be used as a test for convergence.)

21. Suppose that (x_n) is a convergent sequence and (y_n) is such that for any $\varepsilon > 0$ there exists M such that $|x_n - y_n| < \varepsilon$ for all $n \geq M$. Does it follow that (y_n) is convergent?

22. Show that if (x_n) and (y_n) are convergent sequences, then the sequences (u_n) and (v_n) defined by $u_n := \max\{x_n, y_n\}$ and $v_n := \min\{x_n, y_n\}$ are also convergent. (See Exercise 2.2.16.)

23. Show that if (x_n), (y_n), (z_n) are convergent sequences, then the sequence (w_n) defined by $w_n := \mathrm{mid}\{x_n, y_n, z_n\}$ is also convergent. (See Exercise 2.2.17.)

Section 3.3 Monotone Sequences

Until now, we have obtained several methods of showing that a sequence $X = (x_n)$ of real numbers is convergent:

(i) We can use Definition 3.1.3 or Theorem 3.1.5 directly. This is often (but not always) difficult to do.

(ii) We can dominate $|x_n - x|$ by a multiple of the terms in a sequence (a_n) known to converge to 0, and employ Theorem 3.1.10.

(iii) We can identify X as a sequence obtained from other sequences that are known to be convergent by taking tails, algebraic combinations, absolute values, or square roots, and employ Theorems 3.1.9, 3.2.3, 3.2.9, or 3.2.10.

(iv) We can "squeeze" X between two sequences that converge to the same limit and use Theorem 3.2.7.

(v) We can use the "ratio test" of Theorem 3.2.11.

Except for (iii), all of these methods require that we already know (or at least suspect) the value of the limit, and we then verify that our suspicion is correct.

There are many instances, however, in which there is no obvious candidate for the limit of a sequence, even though a preliminary analysis may suggest that convergence is likely. In this and the next two sections, we shall establish results that can be used to show a sequence is convergent even though the value of the limit is not known. The method we introduce in this section is more restricted in scope than the methods we give in the next two, but it is much easier to employ. It applies to sequences that are monotone in the following sense.

3.3.1 Definition Let $X = (x_n)$ be a sequence of real numbers. We say that X is **increasing** if it satisfies the inequalities

$$x_1 \leq x_2 \leq \cdots \leq x_n \leq x_{n+1} \leq \cdots .$$

We say that X is **decreasing** if it satisfies the inequalities

$$x_1 \geq x_2 \geq \cdots \geq x_n \geq x_{n+1} \geq \cdots .$$

We say that X is **monotone** if it is either increasing or decreasing.

The following sequences are increasing:

$$(1, 2, 3, 4, \cdots, n, \cdots), \qquad (1, 2, 2, 3, 3, 3, \cdots),$$
$$(a, a^2, a^3, \cdots, a^n, \cdots) \quad \text{if} \quad a > 1.$$

The following sequences are decreasing:

$$(1, 1/2, 1/3, \cdots, 1/n, \cdots), \qquad (1, 1/2, 1/2^2, \cdots, 1/2^{n-1}, \cdots),$$
$$(b, b^2, b^3, \cdots, b^n, \cdots) \quad \text{if} \quad 0 < b < 1.$$

The following sequences are not monotone:

$$\left(+1, -1, +1, \cdots, (-1)^{n+1}, \cdots\right), \qquad \left(-1, +2, -3, \cdots, (-1)^n n \cdots\right)$$

The following sequences are not monotone, but they are "ultimately" monotone:

$$(7, 6, 2, 1, 2, 3, 4, \cdots), \qquad (-2, 0, 1, 1/2, 1/3, 1/4, \cdots).$$

3.3.2 Monotone Convergence Theorem *A monotone sequence of real numbers is convergent if and only if it is bounded. Further:*

(a) *If $X = (x_n)$ is a bounded increasing sequence, then*

$$\lim(x_n) = \sup\{x_n : n \in \mathbb{N}\}.$$

(b) *If $Y = (y_n)$ is a bounded decreasing sequence, then*

$$\lim(y_n) = \inf\{y_n : n \in \mathbb{N}\}.$$

Proof. It was seen in Theorem 3.2.2 that a convergent sequence must be bounded.

Conversely, let X be a bounded monotone sequence. Then X is either increasing or decreasing.

(a) We first treat the case where $X = (x_n)$ is a bounded, increasing sequence. Since X is bounded, there exists a real number M such that $x_n \leq M$ for all $n \in \mathbb{N}$. According to the Completeness Property 2.3.6, the supremum $x^* = \sup\{x_n : n \in \mathbb{N}\}$ exists in $\mathbb{R}$; we will show that $x^* = \lim(x_n)$.

If $\varepsilon > 0$ is given, then $x^* - \varepsilon$ is not an upper bound of the set $\{x_n : n \in \mathbb{N}\}$, and hence there exists a member of set x_K such that $x^* - \varepsilon < x_K$. The fact that X is an increasing sequence implies that $x_K \leq x_n$ whenever $n \geq K$, so that

$$x^* - \varepsilon < x_K \leq x_n \leq x^* < x^* + \varepsilon \qquad \text{for all} \quad n \geq K.$$

Therefore we have

$$|x_n - x^*| < \varepsilon \qquad \text{for all} \quad n \geq K.$$

Since $\varepsilon > 0$ is arbitrary, we conclude that (x_n) converges to x^*.

(b) If $Y = (y_n)$ is a bounded decreasing sequence, then it is clear that $X := -Y = (-y_n)$ is a bounded increasing sequence. It was shown in part (a) that $\lim X = \sup\{-y_n : n \in \mathbb{N}\}$. Now $\lim X = -\lim Y$ and also, by Exercise 2.4.4(b), we have

$$\sup\{-y_n : n \in \mathbb{N}\} = -\inf\{y_n : n \in \mathbb{N}\}.$$

Therefore $\lim Y = -\lim X = \inf\{y_n : n \in \mathbb{N}\}$.   Q.E.D.

The Monotone Convergence Theorem establishes the existence of the limit of a bounded monotone sequence. It also gives us a way of calculating the limit of the sequence *provided* we can evaluate the supremum in case (a), or the infimum in case (b). Sometimes it is difficult to evaluate this supremum (or infimum), but once we know that it exists, it is often possible to evaluate the limit by other methods.

3.3.3 Examples   (a) $\lim(1/\sqrt{n}) = 0$.

It is possible to handle this sequence by using Theorem 3.2.10; however, we shall use the Monotone Convergence Theorem. Clearly 0 is a lower bound for the set $\{1/\sqrt{n}: n \in \mathbb{N}\}$, and it is not difficult to show that 0 is the infimum of the set $\{1/\sqrt{n}: n \in \mathbb{N}\}$; hence $0 = \lim(1/\sqrt{n})$.

On the other hand, once we know that $X := (1/\sqrt{n})$ is bounded and decreasing, we know that it converges to some real number x. Since $X = (1/\sqrt{n})$ converges to x, it follows from Theorem 3.2.3 that $X \cdot X = (1/n)$ converges to x^2. Therefore $x^2 = 0$, whence $x = 0$.

(b)   Let $x_n := 1 + 1/2 + 1/3 + \cdots + 1/n$ for $n \in \mathbb{N}$.

Since $x_{n+1} = x_n + 1/(n+1) > x_n$, we see that (x_n) is an increasing sequence. By the Monotone Convergence Theorem 3.3.2, the question of whether the sequence is convergent or not is reduced to the question of whether the sequence is bounded or not. Attempts to use direct numerical calculations to arrive at a conjecture concerning the possible boundedness of the sequence (x_n) lead to inconclusive frustration. A computer run will reveal the approximate values $x_n \approx 11.4$ for $n = 50,000$, and $x_n \approx 12.1$ for $n = 100,000$. Such numerical facts may lead the casual observer to conclude that the sequence is bounded. However, the sequence is in fact divergent, which is established by noting that

$$x_{2^n} = 1 + \frac{1}{2} + \left(\frac{1}{3} + \frac{1}{4}\right) + \cdots + \left(\frac{1}{2^{n-1}+1} + \cdots + \frac{1}{2^n}\right)$$

$$> 1 + \frac{1}{2} + \left(\frac{1}{4} + \frac{1}{4}\right) + \cdots + \left(\frac{1}{2^n} + \cdots + \frac{1}{2^n}\right)$$

$$= 1 + \frac{1}{2} + \frac{1}{2} + \cdots + \frac{1}{2}$$

$$= 1 + \frac{n}{2}.$$

Since (x_n) is unbounded, Theorem 3.2.2 implies that it is divergent.

The terms x_n increase extremely slowly. For example, it can be shown that to achieve $x_n > 50$ would entail approximately 5.2×10^{21} additions, and a normal computer performing 400 million additions a second would require more than 400,000 years to perform the calculation (there are 31,536,000 seconds in a year). Even a supercomputer that can perform more than a trillion additions a second, would take more than 164 years to reach that modest goal.   □

Sequences that are defined inductively must be treated differently. If such a sequence is known to converge, then the value of the limit can sometimes be determined by using the inductive relation.

For example, suppose that convergence has been established for the sequence (x_n) defined by

$$x_1 = 2, \qquad x_{n+1} = 2 + \frac{1}{x_n}, \qquad n \in \mathbb{N}.$$

If we let $x = \lim(x_n)$, then we also have $x = \lim(x_{n+1})$ since the 1-tail (x_{n+1}) converges to the same limit. Further, we see that $x_n \geq 2$, so that $x \neq 0$ and $x_n \neq 0$ for all $n \in \mathbb{N}$. Therefore, we may apply the limit theorems for sequences to obtain

$$x = \lim(x_{n+1}) = 2 + \frac{1}{\lim(x_n)} = 2 + \frac{1}{x}.$$

Thus, the limit x is a solution of the quadratic equation $x^2 - 2x - 1 = 0$, and since x must be positive, we find that the limit of the sequence is $x = 1 + \sqrt{2}$.

Of course, the issue of convergence must not be ignored or casually assumed. For example, if we assumed the sequence (y_n) defined by $y_1 := 1$, $y_{n+1} := 2y_n + 1$ is convergent with limit y, then we would obtain $y = 2y + 1$, so that $y = -1$. Of course, this is absurd.

In the following examples, we employ this method of evaluating limits, but only after carefully establishing convergence using the Monotone Convergence Theorem. Additional examples of this type will be given in Section 3.5.

3.3.4 Examples (a) Let $Y = (y_n)$ be defined inductively by $y_1 := 1$, $y_{n+1} := \frac{1}{4}(2y_n + 3)$ for $n \geq 1$. We shall show that $\lim Y = 3/2$.

Direct calculation shows that $y_2 = 5/4$. Hence we have $y_1 < y_2 < 2$. We show, by Induction, that $y_n < 2$ for all $n \in \mathbb{N}$. Indeed, this is true for $n = 1, 2$. If $y_k < 2$ holds for some $k \in \mathbb{N}$, then

$$y_{k+1} = \tfrac{1}{4}(2y_k + 3) < \tfrac{1}{4}(4 + 3) = \tfrac{7}{4} < 2,$$

so that $y_{k+1} < 2$. Therefore $y_n < 2$ for all $n \in \mathbb{N}$.

We now show, by Induction, that $y_n < y_{n+1}$ for all $n \in \mathbb{N}$. The truth of this assertion has been verified for $n = 1$. Now suppose that $y_k < y_{k+1}$ for some k; then $2y_k + 3 < 2y_{k+1} + 3$, whence it follows that

$$y_{k+1} = \tfrac{1}{4}(2y_k + 3) < \tfrac{1}{4}(2y_{k+1} + 3) = y_{k+2}.$$

Thus $y_k < y_{k+1}$ implies that $y_{k+1} < y_{k+2}$. Therefore $y_n < y_{n+1}$ for all $n \in \mathbb{N}$.

We have shown that the sequence $Y = (y_n)$ is increasing and bounded above by 2. It follows from the Monotone Convergence Theorem that Y converges to a limit that is at most 2. In this case it is not so easy to evaluate $\lim(y_n)$ by calculating $\sup\{y_n : n \in \mathbb{N}\}$. However, there is another way to evaluate its limit. Since $y_{n+1} = \frac{1}{4}(2y_n + 3)$ for all $n \in \mathbb{N}$, the nth term in the 1-tail Y_1 of Y has a simple algebraic relation to the nth term of Y. Since, by Theorem 3.1.9, we have $y := \lim Y_1 = \lim Y$, it therefore follows from Theorem 3.2.3 (why?) that

$$y = \tfrac{1}{4}(2y + 3),$$

from which it follows that $y = 3/2$.

(b) Let $Z = (z_n)$ be the sequence of real numbers defined by $z_1 := 1$, $z_{n+1} := \sqrt{2z_n}$ for $n \in \mathbb{N}$. We will show that $\lim(z_n) = 2$.

Note that $z_1 = 1$ and $z_2 = \sqrt{2}$; hence $1 \leq z_1 < z_2 < 2$. We claim that the sequence Z is increasing and bounded above by 2. To show this we will show, by Induction, that

$1 \leq z_n < z_{n+1} < 2$ for all $n \in \mathbb{N}$. This fact has been verified for $n = 1$. Suppose that it is true for $n = k$; then $2 \leq 2z_k < 2z_{k+1} < 4$, whence it follows (why?) that

$$1 < \sqrt{2} \leq z_{k+1} = \sqrt{2z_k} < z_{k+2} = \sqrt{2z_{k+1}} < \sqrt{4} = 2.$$

[In this last step we have used Example 2.1.13(a).] Hence the validity of the inequality $1 \leq z_k < z_{k+1} < 2$ implies the validity of $1 \leq z_{k+1} < z_{k+2} < 2$. Therefore $1 \leq z_n < z_{n+1} < 2$ for all $n \in \mathbb{N}$.

   Since $Z = (z_n)$ is a bounded increasing sequence, it follows from the Monotone Convergence Theorem that it converges to a number $z := \sup\{z_n\}$. It may be shown directly that $\sup\{z_n\} = 2$, so that $z = 2$. Alternatively we may use the method employed in part (a). The relation $z_{n+1} = \sqrt{2z_n}$ gives a relation between the nth term of the 1-tail Z_1 of Z and the nth term of Z. By Theorem 3.1.9, we have $\lim Z_1 = z = \lim Z$. Moreover, by Theorems 3.2.3 and 3.2.10, it follows that the limit z must satisfy the relation

$$z = \sqrt{2z}.$$

Hence z must satisfy the equation $z^2 = 2z$ which has the roots $z = 0, 2$. Since the terms of $z = (z_n)$ all satisfy $1 \leq z_n \leq 2$, it follows from Theorem 3.2.6 that we must have $1 \leq z \leq 2$. Therefore $z = 2$.   □

The Calculation of Square Roots

We now give an application of the Monotone Convergence Theorem to the calculation of square roots of positive numbers.

3.3.5 Example   Let $a > 0$; we will construct a sequence (s_n) of real numbers that converges to $\sqrt{a}$.

   Let $s_1 > 0$ be arbitrary and define $s_{n+1} := \frac{1}{2}(s_n + a/s_n)$ for $n \in \mathbb{N}$. We now show that the sequence (s_n) converges to $\sqrt{a}$. (This process for calculating square roots was known in Mesopotamia before 1500 B.C.)

   We first show that $s_n^2 \geq a$ for $n \geq 2$. Since s_n satisfies the quadratic equation $s_n^2 - 2s_{n+1}s_n + a = 0$, this equation has a real root. Hence the discriminant $4s_{n+1}^2 - 4a$ must be nonnegative; that is, $s_{n+1}^2 \geq a$ for $n \geq 1$.

   To see that (s_n) is ultimately decreasing, we note that for $n \geq 2$ we have

$$s_n - s_{n+1} = s_n - \frac{1}{2}\left(s_n + \frac{a}{s_n}\right) = \frac{1}{2} \cdot \frac{(s_n^2 - a)}{s_n} \geq 0.$$

Hence, $s_{n+1} \leq s_n$ for all $n \geq 2$. The Monotone Convergence Theorem implies that $s := \lim(s_n)$ exists. Moreover, from Theorem 3.2.3, the limit s must satisfy the relation

$$s = \frac{1}{2}\left(s + \frac{a}{s}\right),$$

whence it follows (why?) that $s = a/s$ or $s^2 = a$. Thus $s = \sqrt{a}$.

   For the purposes of calculation, it is often important to have an estimate of *how rapidly* the sequence (s_n) converges to $\sqrt{a}$. As above, we have $\sqrt{a} \leq s_n$ for all $n \geq 2$, whence it follows that $a/s_n \leq \sqrt{a} \leq s_n$. Thus we have

$$0 \leq s_n - \sqrt{a} \leq s_n - a/s_n = (s_n^2 - a)/s_n \qquad \text{for} \quad n \geq 2.$$

Using this inequality we can calculate $\sqrt{a}$ to any desired degree of accuracy.   □

Euler's Number _____

We conclude this section by introducing a sequence that converges to one of the most important "transcendental" numbers in mathematics, second in importance only to π.

3.3.6 Example Let $e_n := (1 + 1/n)^n$ for $n \in \mathbb{N}$. We will now show that the sequence $E = (e_n)$ is bounded and increasing; hence it is convergent. The limit of this sequence is the famous *Euler number* e, whose approximate value is $2.718\,281\,828\,459\,045\cdots$, which is taken as the base of the "natural" logarithm.

If we apply the Binomial Theorem, we have

$$e_n = \left(1 + \frac{1}{n}\right)^n = 1 + \frac{n}{1}\cdot\frac{1}{n} + \frac{n(n-1)}{2!}\cdot\frac{1}{n^2} + \frac{n(n-1)(n-2)}{3!}\cdot\frac{1}{n^3}$$
$$+\cdots+\frac{n(n-1)\cdots 2\cdot 1}{n!}\cdot\frac{1}{n^n}.$$

If we divide the powers of n into the terms in the numerators of the binomial coefficients, we get

$$e_n = 1 + 1 + \frac{1}{2!}\left(1 - \frac{1}{n}\right) + \frac{1}{3!}\left(1 - \frac{1}{n}\right)\left(1 - \frac{2}{n}\right)$$
$$+\cdots+\frac{1}{n!}\left(1 - \frac{1}{n}\right)\left(1 - \frac{2}{n}\right)\cdots\left(1 - \frac{n-1}{n}\right).$$

Similarly we have

$$e_{n+1} = 1 + 1 + \frac{1}{2!}\left(1 - \frac{1}{n+1}\right) + \frac{1}{3!}\left(1 - \frac{1}{n+1}\right)\left(1 - \frac{2}{n+1}\right)$$
$$+\cdots+\frac{1}{n!}\left(1 - \frac{1}{n+1}\right)\left(1 - \frac{2}{n+1}\right)\cdots\left(1 - \frac{n-1}{n+1}\right)$$
$$+\frac{1}{(n+1)!}\left(1 - \frac{1}{n+1}\right)\left(1 - \frac{2}{n+1}\right)\cdots\left(1 - \frac{n}{n+1}\right).$$

Note that the expression for e_n contains $n + 1$ terms, while that for e_{n+1} contains $n + 2$ terms. Moreover, each term appearing in e_n is less than or equal to the corresponding term in e_{n+1}, and e_{n+1} has one more positive term. Therefore we have $2 \le e_1 < e_2 < \cdots < e_n < e_{n+1} < \cdots$, so that the terms of E are increasing.

To show that the terms of E are bounded above, we note that if $p = 1, 2, \cdots, n$, then $(1 - p/n) < 1$. Moreover $2^{p-1} \le p!$ [see 1.2.4(e)] so that $1/p! \le 1/2^{p-1}$. Therefore, if $n > 1$, then we have

$$2 < e_n < 1 + 1 + \frac{1}{2} + \frac{1}{2^2} + \cdots + \frac{1}{2^{n-1}}.$$

Since it can be verified that [see 1.2.4(f)]

$$\frac{1}{2} + \frac{1}{2^2} + \cdots + \frac{1}{2^{n-1}} = 1 - \frac{1}{2^{n-1}} < 1,$$

we deduce that $2 < e_n < 3$ for all $n \in \mathbb{N}$. The Monotone Convergence Theorem implies that the sequence E converges to a real number that is between 2 and 3. We define the number e to be the limit of this sequence.

By refining our estimates we can find closer rational approximations to e, but we cannot evaluate it *exactly*, since e is an irrational number. However, it is possible to calculate e to as many decimal places as desired. The reader should use a calculator (or a computer) to evaluate e_n for "large" values of n. □

Leonhard Euler

Leonhard Euler (1707–1783) was born near Basel, Switzerland. His clergy-man father hoped that his son would follow him into the ministry, but when Euler entered the University of Basel at age 14, his mathematical talent was noted by Johann Bernoulli, who became his mentor. In 1727, Euler went to Russia to join Johann's son, Daniel, at the new St. Petersburg Academy. There he met and married Katharina Gsell, the daughter of a Swiss artist. During their long marriage they had 13 children, but only five survived childhood.

In 1741, Euler accepted an offer from Frederick the Great to join the Berlin Academy, where he stayed for 25 years. During this period he wrote landmark books on calculus and a steady stream of papers. In response to a request for instruction in science from the Princess of Anhalt-Dessau, he wrote a multi-volume work on science that became famous under the title *Letters to a German Princess*.

In 1766, he returned to Russia at the invitation of Catherine the Great. His eyesight had deteriorated over the years, and soon after his return to Russia he became totally blind. Incredibly, his blindness made little impact on his mathematical output, for he wrote several books and over 400 papers while blind. He remained busy and active until the day of his death.

Euler's productivity was remarkable: he wrote textbooks on physics, algebra, calculus, real and complex analysis, analytic and differential geometry, and the calculus of variations. He also wrote hundreds of original papers, many of which won prizes. A current edition of his collected works consists of 74 volumes.

Exercises for Section 3.3

1. Let $x_1 := 8$ and $x_{n+1} := \frac{1}{2}x_n + 2$ for $n \in \mathbb{N}$. Show that (x_n) is bounded and monotone. Find the limit.

2. Let $x_1 > 1$ and $x_{n+1} := 2 - 1/x_n$ for $n \in \mathbb{N}$. Show that (x_n) is bounded and monotone. Find the limit.

3. Let $x_1 \geq 2$ and $x_{n+1} := 1 + \sqrt{x_n - 1}$ for $n \in \mathbb{N}$. Show that (x_n) is decreasing and bounded below by 2. Find the limit.

4. Let $x_1 := 1$ and $x_{n+1} := \sqrt{2 + x_n}$ for $n \in \mathbb{N}$. Show that (x_n) converges and find the limit.

5. Let $y_1 := \sqrt{p}$, where $p > 0$, and $y_{n+1} := \sqrt{p + y_n}$ for $n \in \mathbb{N}$. Show that (y_n) converges and find the limit. [*Hint:* One upper bound is $1 + 2\sqrt{p}$.]

6. Let $a > 0$ and let $z_1 > 0$. Define $z_{n+1} := \sqrt{a + z_n}$ for $n \in \mathbb{N}$. Show that (z_n) converges and find the limit.

7. Let $x_1 := a > 0$ and $x_{n+1} := x_n + 1/x_n$ for $n \in \mathbb{N}$. Determine if (x_n) converges or diverges.

8. Let (a_n) be an increasing sequence, (b_n) a decreasing sequence, and assume that $a_n \leq b_n$ for all $n \in \mathbb{N}$. Show that $\lim(a_n) \leq \lim(b_n)$, and thereby deduce the Nested Intervals Property 2.5.2 from the Monotone Convergence Theorem 3.3.2.

9. Let A be an infinite subset of $\mathbb{R}$ that is bounded above and let $u := \sup A$. Show there exists an increasing sequence (x_n) with $x_n \in A$ for all $n \in \mathbb{N}$ such that $u = \lim(x_n)$.

10. Let (x_n) be a bounded sequence, and for each $n \in \mathbb{N}$ let $s_n := \sup\{x_k : k \geq n\}$ and $t_n := \inf\{x_k : k \geq n\}$. Prove that (s_n) and (t_n) are monotone and convergent. Also prove that if $\lim(s_n) = \lim(t_n)$, then (x_n) is convergent. [One calls $\lim(s_n)$ the **limit superior** of (x_n), and $\lim(t_n)$ the **limit inferior** of (x_n).]

11. Establish the convergence or the divergence of the sequence (y_n), where

$$y_n := \frac{1}{n+1} + \frac{1}{n+2} + \cdots + \frac{1}{2n} \qquad \text{for} \quad n \in \mathbb{N}.$$

12. Let $x_n := 1/1^2 + 1/2^2 + \cdots + 1/n^2$ for each $n \in \mathbb{N}$. Prove that (x_n) is increasing and bounded, and hence converges. [*Hint*: Note that if $k \geq 2$, then $1/k^2 \leq 1/k(k-1) = 1/(k-1) - 1/k$.]

13. Establish the convergence and find the limits of the following sequences.
     (a) $\left((1 + 1/n)^{n+1}\right)$,   (b) $\left((1 + 1/n)^{2n}\right)$,

     (c) $\left(\left(1 + \dfrac{1}{n+1}\right)^n\right)$,   (d) $\left((1 - 1/n)^n\right)$.

14. Use the method in Example 3.3.5 to calculate $\sqrt{2}$, correct to within 4 decimals.

15. Use the method in Example 3.3.5 to calculate $\sqrt{5}$, correct to within 5 decimals.

16. Calculate the number e_n in Example 3.3.6 for $n = 2, 4, 8, 16$.

17. Use a calculator to compute e_n for $n = 50$, $n = 100$, and $n = 1,000$.

Section 3.4   Subsequences and the Bolzano-Weierstrass Theorem

---

In this section we will introduce the notion of a subsequence of a sequence of real numbers. Informally, a subsequence of a sequence is a selection of terms from the given sequence such that the selected terms form a new sequence. Usually the selection is made for a definite purpose. For example, subsequences are often useful in establishing the convergence or the divergence of the sequence. We will also prove the important existence theorem known as the Bolzano-Weierstrass Theorem, which will be used to establish a number of significant results.

3.4.1 Definition   Let $X = (x_n)$ be a sequence of real numbers and let $n_1 < n_2 < \cdots < n_k < \cdots$ be a strictly increasing sequence of natural numbers. Then the sequence $X' = (x_{n_k})$ given by

$$\left(x_{n_1}, x_{n_2}, \cdots, x_{n_k}, \cdots\right)$$

is called a **subsequence** of X.

For example, if $X := \left(\frac{1}{1}, \frac{1}{2}, \frac{1}{3}, \cdots\right)$, then the selection of even indexed terms produces the subsequence

$$X' = \left(\frac{1}{2}, \frac{1}{4}, \frac{1}{6}, \cdots, \frac{1}{2k}, \cdots\right),$$

where $n_1 = 2, n_2 = 4, \cdots, n_k = 2k, \cdots$. Other subsequences of $X = (1/n)$ are the following:

$$\left(\frac{1}{1}, \frac{1}{3}, \frac{1}{5}, \cdots, \frac{1}{2k-1}, \cdots\right), \qquad \left(\frac{1}{2!}, \frac{1}{4!}, \frac{1}{6!}, \cdots, \frac{1}{(2k)!}, \cdots\right).$$

The following sequences are *not* subsequences of $X = (1/n)$:

$$\left(\frac{1}{2}, \frac{1}{1}, \frac{1}{4}, \frac{1}{3}, \frac{1}{6}, \frac{1}{5}, \cdots\right), \qquad \left(\frac{1}{1}, 0, \frac{1}{3}, 0, \frac{1}{5}, 0, \cdots\right).$$

A tail of a sequence (see 3.1.8) is a special type of subsequence. In fact, the m-tail corresponds to the sequence of indices

$$n_1 = m+1, n_2 = m+2, \cdots, n_k = m+k, \cdots.$$

But, clearly, not every subsequence of a given sequence need be a tail of the sequence.

Subsequences of convergent sequences also converge to the same limit, as we now show.

3.4.2 Theorem   *If a sequence $X = (x_n)$ of real numbers converges to a real number x, then any subsequence $X' = (x_{n_k})$ of X also converges to x.*

Proof.   Let $\varepsilon > 0$ be given and let $K(\varepsilon)$ be such that if $n \geq K(\varepsilon)$, then $|x_n - x| < \varepsilon$. Since $n_1 < n_2 < \cdots < n_k < \cdots$ is an increasing sequence of natural numbers, it is easily proved (by Induction) that $n_k \geq k$. Hence, if $k \geq K(\varepsilon)$, we also have $n_k \geq k \geq K(\varepsilon)$ so that $|x_{n_k} - x| < \varepsilon$. Therefore the subsequence (x_{n_k}) also converges to x.   Q.E.D.

3.4.3 Example   **(a)** $\lim(b^n) = 0$ if $0 < b < 1$.

We have already seen, in Example 3.1.11(b), that if $0 < b < 1$ and if $x_n := b^n$, then it follows from Bernoulli's Inequality that $\lim(x_n) = 0$. Alternatively, we see that since $0 < b < 1$, then $x_{n+1} = b^{n+1} < b^n = x_n$ so that the sequence (x_n) is decreasing. It is also clear that $0 \leq x_n \leq 1$, so it follows from the Monotone Convergence Theorem 3.3.2 that the sequence is convergent. Let $x := \lim x_n$. Since (x_{2n}) is a subsequence of (x_n) it follows from Theorem 3.4.2 that $x = \lim(x_{2n})$. Moreover, it follows from the relation $x_{2n} = b^{2n} = (b^n)^2 = x_n^2$ and Theorem 3.2.3 that

$$x = \lim(x_{2n}) = \left(\lim(x_n)\right)^2 = x^2.$$

Therefore we must either have $x = 0$ or $x = 1$. Since the sequence (x_n) is decreasing and bounded above by $b < 1$, we deduce that $x = 0$.

(b) $\lim(c^{1/n}) = 1$ for $c > 1$.

This limit has been obtained in Example 3.1.11(c) for $c > 0$, using a rather ingenious argument. We give here an alternative approach for the case $c > 1$. Note that if $z_n := c^{1/n}$, then $z_n > 1$ and $z_{n+1} < z_n$ for all $n \in \mathbb{N}$. (Why?) Thus by the Monotone Convergence Theorem, the limit $z := \lim(z_n)$ exists. By Theorem 3.4.2, it follows that $z = \lim(z_{2n})$. In addition, it follows from the relation

$$z_{2n} = c^{1/2n} = (c^{1/n})^{1/2} = z_n^{1/2}$$

and Theorem 3.2.10 that

$$z = \lim(z_{2n}) = \left(\lim(z_n)\right)^{1/2} = z^{1/2}.$$

Therefore we have $z^2 = z$ whence it follows that either $z = 0$ or $z = 1$. Since $z_n > 1$ for all $n \in \mathbb{N}$, we deduce that $z = 1$.

We leave it as an exercise to the reader to consider the case $0 < c < 1$.   □

The following result is based on a careful negation of the definition of $\lim(x_n) = x$. It leads to a convenient way to establish the divergence of a sequence.

3.4.4 Theorem   *Let $X = (x_n)$ be a sequence of real numbers. Then the following are equivalent:*

(i)   *The sequence $X = (x_n)$ does not converge to $x \in \mathbb{R}$.*

(ii)   There exists an $\varepsilon_0 > 0$ such that for any $k \in \mathbb{N}$, there exists $n_k \in \mathbb{N}$ such that $n_k \geq k$ and $|x_{n_k} - x| \geq \varepsilon_0$.

(iii)   There exists an $\varepsilon_0 > 0$ and a subsequence $X' = (x_{n_k})$ of X such that $|x_{n_k} - x| \geq \varepsilon_0$ for all $k \in \mathbb{N}$.

Proof.   (i) $\Rightarrow$ (ii)   If (x_n) does not converge to x, then for some $\varepsilon_0 > 0$ it is impossible to find a natural number k such that for all $n \geq k$ the terms x_n satisfy $|x_n - x| < \varepsilon_0$. That is, for each $k \in \mathbb{N}$ it is *not true* that for *all* $n \geq k$ the inequality $|x_n - x| < \varepsilon_0$ holds. In other words, for each $k \in \mathbb{N}$ there exists a natural number $n_k \geq k$ such that $|x_{n_k} - x| \geq \varepsilon_0$.

(ii) $\Rightarrow$ (iii)   Let ε_0 be as in (ii) and let $n_1 \in \mathbb{N}$ be such that $n_1 \geq 1$ and $|x_{n_1} - x| \geq \varepsilon_0$. Now let $n_2 \in \mathbb{N}$ be such that $n_2 > n_1$ and $|x_{n_2} - x| \geq \varepsilon_0$; let $n_3 \in \mathbb{N}$ be such that $n_3 > n_2$ and $|x_{n_3} - x| \geq \varepsilon_0$. Continue in this way to obtain a subsequence $X' = (x_{n_k})$ of X such that $|x_{n_k} - x| \geq \varepsilon_0$ for all $k \in \mathbb{N}$.

(iii) $\Rightarrow$ (i)   Suppose $X = (x_n)$ has a subsequence $X' = (x_{n_k})$ satisfying the condition in (iii). Then X cannot converge to x; for if it did, then, by Theorem 3.4.2, the subsequence X' would also converge to x. But this is impossible, since none of the terms of X' belongs to the ε_0-neighborhood of x.
$$\text{Q.E.D.}$$

Since all subsequences of a convergent sequence must converge to the same limit, we have part (i) in the following result. Part (ii) follows from the fact that a convergent sequence is bounded.

3.4.5 Divergence Criteria   If a sequence $X = (x_n)$ of real numbers has either of the following properties, then X is divergent.

(i) X has two convergent subsequences $X' = (x_{n_k})$ and $X'' = (x_{r_k})$ whose limits are not equal.

(ii) X is unbounded.

3.4.6 Examples   (a)   The sequence $X := ((-1^n))$ is divergent.

The subsequence $X' := ((-1)^{2n}) = (1, 1, \cdots)$ converges to 1, and the subsequence $X'' := ((-1)^{2n-1}) = (-1, -1, \cdots)$ converges to -1. Therefore, we conclude from Theorem 3.4.5(i) that X is divergent.

(b)   The sequence $\left(1, \frac{1}{2}, 3, \frac{1}{4}, \cdots\right)$ is divergent.

This is the sequence $Y = (y_n)$, where $y_n = n$ if n is odd, and $y_n = 1/n$ if n is even. It can easily be seen that Y is not bounded. Hence, by Theorem 3.4.5(ii), the sequence is divergent.

(c)   The sequence $S := (\sin n)$ is divergent.

This sequence is not so easy to handle. In discussing it we must, of course, make use of elementary properties of the sine function. We recall that $\sin(\pi/6) = \frac{1}{2} = \sin(5\pi/6)$ and that $\sin x > \frac{1}{2}$ for x in the interval $I_1 := (\pi/6, 5\pi/6)$. Since the length of I_1 is $5\pi/6 - \pi/6 = 2\pi/3 > 2$, there are at least two natural numbers lying inside I_1; we let n_1 be the first such number. Similarly, for each $k \in \mathbb{N}$, $\sin x > \frac{1}{2}$ for x in the interval

$$I_k := \left(\pi/6 + 2\pi(k-1), 5\pi/6 + 2\pi(k-1)\right).$$

Since the length of I_k is greater than 2, there are at least two natural numbers lying inside I_k; we let n_k be the first one. The subsequence $S' := (\sin n_k)$ of S obtained in this way has the property that all of its values lie in the interval $\left[\frac{1}{2}, 1\right]$.

Similarly, if $k \in \mathbb{N}$ and J_k is the interval

$$J_k := \left(7\pi/6 + 2\pi(k-1),\, 11\pi/6 + 2\pi(k-1)\right),$$

then it is seen that $\sin x < -\frac{1}{2}$ for all $x \in J_k$ and the length of J_k is greater than 2. Let m_k be the first natural number lying in J_k. Then the subsequence $S'' := (\sin m_k)$ of S has the property that all of its values lie in the interval $\left[-1, -\frac{1}{2}\right]$.

Given any real number c, it is readily seen that at least one of the subsequences S' and S'' lies entirely outside of the $\frac{1}{2}$-neighborhood of c. Therefore c cannot be a limit of S. Since $c \in \mathbb{R}$ is arbitrary, we deduce that S is divergent.   □

The Existence of Monotone Subsequences

While not every sequence is a monotone sequence, we will now show that every sequence has a monotone subsequence.

3.4.7 Monotone Subsequence Theorem   *If $X = (x_n)$ is a sequence of real numbers, then there is a subsequence of X that is monotone.*

Proof.   For the purpose of this proof, we will say that the mth term x_m is a "peak" if $x_m \geq x_n$ for all n such that $n \geq m$. (That is, x_m is never exceeded by any term that follows it in the sequence.) Note that, in a decreasing sequence, every term is a peak, while in an increasing sequence, no term is a peak.

We will consider two cases, depending on whether X has infinitely many, or finitely many, peaks.

Case 1: X has infinitely many peaks. In this case, we list the peaks by increasing subscripts: $x_{m_1}, x_{m_2}, \cdots, x_{m_k}, \cdots$. Since each term is a peak, we have

$$x_{m_1} \geq x_{m_2} \geq \cdots \geq x_{m_k} \geq \cdots.$$

Therefore, the subsequence (x_{m_k}) of peaks is a decreasing subsequence of X.

Case 2: X has a finite number (possibly zero) of peaks. Let these peaks be listed by increasing subscripts: $x_{m_1}, x_{m_2}, \cdots, x_{m_r}$. Let $s_1 := m_r + 1$ be the first index beyond the last peak. Since x_{s_1} is not a peak, there exists $s_2 > s_1$ such that $x_{s_1} < x_{s_2}$. Since x_{s_2} is not a peak, there exists $s_3 > s_2$ such that $x_{s_2} < x_{s_3}$. Continuing in this way, we obtain an increasing subsequence (x_{s_k}) of X.   Q.E.D.

It is not difficult to see that a given sequence may have one subsequence that is increasing, and another subsequence that is decreasing.

The Bolzano-Weierstrass Theorem

We will now use the Monotone Subsequence Theorem to prove the Bolzano-Weierstrass Theorem, which states that every bounded sequence has a convergent subsequence. Because of the importance of this theorem we will also give a second proof of it based on the Nested Interval Property.

3.4.8 The Bolzano-Weierstrass Theorem   *A bounded sequence of real numbers has a convergent subsequence.*

First Proof.   It follows from the Monotone Subsequence Theorem that if $X = (x_n)$ is a bounded sequence, then it has a subsequence $X' = (x_{n_k})$ that is monotone. Since this

3.4 SUBSEQUENCES AND THE BOLZANO-WEIERSTRASS THEOREM

subsequence is also bounded, it follows from the Monotone Convergence Theorem 3.3.2 that the subsequence is convergent.   Q.E.D.

Second Proof.   Since the set of values $\{x_n : n \in \mathbb{N}\}$ is bounded, this set is contained in an interval $I_1 := [a, b]$. We take $n_1 := 1$.

We now bisect I_1 into two equal subintervals I_1' and I_1'', and divide the set of indices $\{n \in \mathbb{N} : n > 1\}$ into two parts:

$$A_1 := \{n \in \mathbb{N} : n > n_1, x_n \in I_1'\}, \qquad B_1 = \{n \in \mathbb{N} : n > n_1, x_n \in I_1''\}.$$

If A_1 is infinite, we take $I_2 := I_1'$ and let n_2 be the smallest natural number in A_1. (See 1.2.1.) If A_1 is a finite set, then B_1 must be infinite, and we take $I_2 := I_1''$ and let n_2 be the smallest natural number in B_1.

We now bisect I_2 into two equal subintervals I_2' and I_2'', and divide the set $\{n \in \mathbb{N} : n > n_2\}$ into two parts:

$$A_2 = \{n \in \mathbb{N} : n > n_2, x_n \in I_2'\}, \qquad B_2 := \{n \in \mathbb{N} : n > n_2, x_n \in I_2''\}$$

If A_2 is infinite, we take $I_3 := I_2'$ and let n_3 be the smallest natural number in A_2. If A_2 is a finite set, then B_2 must be infinite, and we take $I_3 := I_2''$ and let n_3 be the smallest natural number in B_2.

We continue in this way to obtain a sequence of nested intervals $I_1 \supseteq I_2 \supseteq \cdots \supseteq I_k \supseteq \cdots$ and a subsequence (x_{n_k}) of X such that $x_{n_k} \in I_k$ for $k \in \mathbb{N}$. Since the length of I_k is equal to $(b - a)/2^{k-1}$, it follows from Theorem 2.5.3 that there is a (unique) common point $\xi \in I_k$ for all $k \in \mathbb{N}$. Moreover, since x_{n_k} and ξ both belong to I_k, we have

$$|x_{n_k} - \xi| \le (b - a)/2^{k-1},$$

whence it follows that the subsequence (x_{n_k}) of X converges to ξ.   Q.E.D.

Theorem 3.4.8 is sometimes called the Bolzano-Weierstrass Theorem *for sequences*, because there is another version of it that deals with bounded sets in $\mathbb{R}$ (see Exercise 11.2.6).

It is readily seen that a bounded sequence can have various subsequences that converge to different limits or even diverge. For example, the sequence $((-1)^n)$ has subsequences that converge to -1, other subsequences that converge to $+1$, and it has subsequences that diverge.

Let X be a sequence of real numbers and let X' be a subsequence of X. Then X' is a sequence in its own right, and so it has subsequences. We note that if X'' is a subsequence of X', then it is also a subsequence of X.

3.4.9 Theorem   *Let $X = (x_n)$ be a bounded sequence of real numbers and let $x \in \mathbb{R}$ have the property that every convergent subsequence of X converges to x. Then the sequence X converges to x.*

Proof.   Suppose $M > 0$ is a bound for the sequence X so that $|x_n| \le M$ for all $n \in \mathbb{N}$. If X does not converge to x, then Theorem 3.4.4 implies that there exist $\varepsilon_0 > 0$ and a subsequence $X' = (x_{n_k})$ of X such that

(1) $|x_{n_k} - x| \ge \varepsilon_0$ for all $k \in \mathbb{N}$.

Since X' is a subsequence of X, the number M is also a bound for X'. Hence the Bolzano-Weierstrass Theorem implies that X' has a convergent subsequence X''. Since X'' is also a subsequence of X, it converges to x by hypothesis. Thus, its terms ultimately belong to the ε_0-neighborhood of x, contradicting (1).   Q.E.D.

Exercises for Section 3.4

1. Give an example of an unbounded sequence that has a convergent subsequence.

2. Use the method of Example 3.4.3(b) to show that if $0 < c < 1$, then $\lim(c^{1/n}) = 1$.

3. Let (f_n) be the Fibonacci sequence of Example 3.1.2(d), and let $x_n := f_{n+1}/f_n$. Given that $\lim(x_n) = L$ exists, determine the value of L.

4. Show that the following sequences are divergent.
   (a) $(1 - (-1)^n + 1/n)$, (b) $(\sin n\pi/4)$.

5. Let $X = (x_n)$ and $Y = (y_n)$ be given sequences, and let the "shuffled" sequence $Z = (z_n)$ be defined by $z_1 := x_1, z_2 := y_1, \cdots, z_{2n-1} := x_n, z_{2n} := y_n, \cdots$. Show that Z is convergent if and only if both X and Y are convergent and $\lim X = \lim Y$.

6. Let $x_n := n^{1/n}$ for $n \in \mathbb{N}$.
   (a) Show that $x_{n+1} < x_n$ if and only if $(1 + 1/n)^n < n$, and infer that the inequality is valid for $n \geq 3$. (See Example 3.3.6.) Conclude that (x_n) is ultimately decreasing and that $x := \lim(x_n)$ exists.
   (b) Use the fact that the subsequence (x_{2n}) also converges to x to conclude that $x = 1$.

7. Establish the convergence and find the limits of the following sequences:
   (a) $\left((1 + 1/n^2)^{n^2}\right)$, (b) $\left((1 + 1/2n)^n\right)$,
   (c) $\left((1 + 1/n^2)^{2n^2}\right)$, (d) $\left((1 + 2/n)^n\right)$.

8. Determine the limits of the following.
   (a) $\left((3n)^{1/2n}\right)$, (b) $\left((1 + 1/2n)^{3n}\right)$.

9. Suppose that every subsequence of $X = (x_n)$ has a subsequence that converges to 0. Show that $\lim X = 0$.

10. Let (x_n) be a bounded sequence and for each $n \in \mathbb{N}$ let $s_n := \sup\{x_k : k \geq n\}$ and $S := \inf\{s_n\}$. Show that there exists a subsequence of (x_n) that converges to S.

11. Suppose that $x_n \geq 0$ for all $n \in \mathbb{N}$ and that $\lim\left((-1)^n x_n\right)$ exists. Show that (x_n) converges.

12. Show that if (x_n) is unbounded, then there exists a subsequence (x_{n_k}) such that $\lim(1/x_{n_k}) = 0$.

13. If $x_n := (-1)^n/n$, find the subsequence of (x_n) that is constructed in the second proof of the Bolzano-Weierstrass Theorem 3.4.8, when we take $I_1 := [-1, 1]$.

14. Let (x_n) be a bounded sequence and let $s := \sup\{x_n : n \in \mathbb{N}\}$. Show that if $s \notin \{x_n : n \in \mathbb{N}\}$, then there is a subsequence of (x_n) that converges to s.

15. Let (I_n) be a nested sequence of closed bounded intervals. For each $n \in \mathbb{N}$, let $x_n \in I_n$. Use the Bolzano-Weierstrass Theorem to give a proof of the Nested Intervals Property 2.5.2.

16. Give an example to show that Theorem 3.4.9 fails if the hypothesis that X is a bounded sequence is dropped.

Section 3.5   The Cauchy Criterion

The Monotone Convergence Theorem is extraordinarily useful and important, but it has the significant drawback that it applies only to sequences that are monotone. It is important for us to have a condition implying the convergence of a sequence that does not require us to know the value of the limit in advance, and is not restricted to monotone sequences. The Cauchy Criterion, which will be established in this section, is such a condition.

3.5.1 Definition   A sequence $X = (x_n)$ of real numbers is said to be a **Cauchy sequence** if for every $\varepsilon > 0$ there exists a natural number $H(\varepsilon)$ such that for all natural numbers $n, m \geq H(\varepsilon)$, the terms x_n, x_m satisfy $|x_n - x_m| < \varepsilon$.

The significance of the concept of Cauchy sequence lies in the main theorem of this section, which asserts that a sequence of real numbers is convergent if and only if it is a Cauchy sequence. This will give us a method of proving a sequence converges without knowing the limit of the sequence.

However, we will first highlight the definition of Cauchy sequence in the following examples.

3.5.2 Examples   (a)   The sequence $(1/n)$ is a Cauchy sequence.

If $\varepsilon > 0$ is given, we choose a natural number $H = H(\varepsilon)$ such that $H > 2/\varepsilon$. Then if $m, n \geq H$, we have $1/n \leq 1/H < \varepsilon/2$ and similarly $1/m < \varepsilon/2$. Therefore, it follows that if $m, n \geq H$, then

$$\left| \frac{1}{n} - \frac{1}{m} \right| \leq \frac{1}{n} + \frac{1}{m} < \frac{\varepsilon}{2} + \frac{\varepsilon}{2} = \varepsilon.$$

Since $\varepsilon > 0$ is arbitrary, we conclude that $(1/n)$ is a Cauchy sequence.

(b)   The sequence $(1 + (-1)^n)$ is *not* a Cauchy sequence.

The negation of the definition of Cauchy sequence is: There exists $\varepsilon_0 > 0$ such that for every H there exist at least one $n > H$ and at least one $m > H$ such that $|x_n - x_m| \geq \varepsilon_0$. For the terms $x_n := 1 + (-1)^n$, we observe that if n is even, then $x_n = 2$ and $x_{n+1} = 0$. If we take $\varepsilon_0 = 2$, then for any H we can choose an even number $n > H$ and let $m := n + 1$ to get

$$|x_n - x_{n+1}| = 2 = \varepsilon_0.$$

We conclude that (x_n) is not a Cauchy sequence.   □

Remark   We emphasize that to prove a sequence (x_n) is a Cauchy sequence, we may not assume a relationship between m and n, since the required inequality $|x_n - x_m| < \varepsilon$ must hold for *all* $n, m \geq H(\varepsilon)$. But to prove a sequence is *not* a Cauchy sequence, we may specify a relation between n and m as long as arbitrarily large values of n and m can be chosen so that $|x_n - x_m| \geq \varepsilon_0$.

Our goal is to show that the Cauchy sequences are precisely the convergent sequences. We first prove that a convergent sequence is a Cauchy sequence.

3.5.3 Lemma   *If $X = (x_n)$ is a convergent sequence of real numbers, then X is a Cauchy sequence.*

Proof.   If $x := \lim X$, then given $\varepsilon > 0$ there is a natural number $K(\varepsilon/2)$ such that if $n \geq K(\varepsilon/2)$ then $|x_n - x| < \varepsilon/2$. Thus, if $H(\varepsilon) := K(\varepsilon/2)$ and if $n, m \geq H(\varepsilon)$, then we have

$$|x_n - x_m| = |(x_n - x) + (x - x_m)|$$
$$\leq |x_n - x| + |x_m - x| < \varepsilon/2 + \varepsilon/2 = \varepsilon.$$

Since $\varepsilon > 0$ is arbitrary, it follows that (x_n) is a Cauchy sequence.   Q.E.D.

In order to establish that a Cauchy sequence is convergent, we will need the following result. (See Theorem 3.2.2.)

3.5.4 Lemma *A Cauchy sequence of real numbers is bounded.*

Proof. Let $X := (x_n)$ be a Cauchy sequence and let $\varepsilon := 1$. If $H := H(1)$ and $n \geq H$, then $|x_n - x_H| < 1$. Hence, by the Triangle Inequality, we have $|x_n| \leq |x_H| + 1$ for all $n \geq H$. If we set

$$M := \sup \{|x_1|, |x_2|, \cdots, |x_{H-1}|, |x_H| + 1\},$$

then it follows that $|x_n| \leq M$ for all $n \in \mathbb{N}$.   Q.E.D.

We now present the important Cauchy Convergence Criterion.

3.5.5 Cauchy Convergence Criterion *A sequence of real numbers is convergent if and only if it is a Cauchy sequence.*

Proof. We have seen, in Lemma 3.5.3, that a convergent sequence is a Cauchy sequence.

Conversely, let $X = (x_n)$ be a Cauchy sequence; we will show that X is convergent to some real number. First we observe from Lemma 3.5.4 that the sequence X is bounded. Therefore, by the Bolzano-Weierstrass Theorem 3.4.8, there is a subsequence $X' = (x_{n_k})$ of X that converges to some real number x^*. We shall complete the proof by showing that X converges to x^*.

Since $X = (x_n)$ is a Cauchy sequence, given $\varepsilon > 0$ there is a natural number $H(\varepsilon/2)$ such that if $n, m \geq H(\varepsilon/2)$ then

(1) $$|x_n - x_m| < \varepsilon/2.$$

Since the subsequence $X' = (x_{n_k})$ converges to x^*, there is a natural number $K \geq H(\varepsilon/2)$ belonging to the set $\{n_1, n_2, \cdots\}$ such that

$$|x_K - x^*| < \varepsilon/2.$$

Since $K \geq H(\varepsilon/2)$, it follows from (1) with $m = K$ that

$$|x_n - x_K| < \varepsilon/2 \qquad \text{for} \quad n \geq H(\varepsilon/2).$$

Therefore, if $n \geq H(\varepsilon/2)$, we have

$$\begin{aligned} |x_n - x^*| &= |(x_n - x_K) + (x_K - x^*)| \\ &\leq |x_n - x_K| + |x_K - x^*| \\ &< \varepsilon/2 + \varepsilon/2 = \varepsilon. \end{aligned}$$

Since $\varepsilon > 0$ is arbitrary, we infer that $\lim(x_n) = x^*$. Therefore the sequence X is convergent.   Q.E.D.

We will now give some examples of applications of the Cauchy Criterion.

3.5.6 Examples **(a)** Let $X = (x_n)$ be defined by

$$x_1 := 1, \qquad x_2 := 2, \qquad \text{and} \qquad x_n := \frac{1}{2}(x_{n-2} + x_{n-1}) \qquad \text{for} \quad n > 2.$$

It can be shown by Induction that $1 \leq x_n \leq 2$ for all $n \in \mathbb{N}$. (Do so.) Some calculation shows that the sequence X is not monotone. However, since the terms are formed by averaging, it is readily seen that

$$|x_n - x_{n+1}| = \frac{1}{2^{n-1}} \qquad \text{for} \quad n \in \mathbb{N}.$$

(Prove this by Induction.) Thus, if $m > n$, we may employ the Triangle Inequality to obtain

$$|x_n - x_m| \leq |x_n - x_{n+1}| + |x_{n+1} - x_{n+2}| + \cdots + |x_{m-1} - x_m|$$
$$= \frac{1}{2^{n-1}} + \frac{1}{2^n} + \cdots + \frac{1}{2^{m-2}}$$
$$= \frac{1}{2^{n-1}} \left(1 + \frac{1}{2} + \cdots + \frac{1}{2^{m-n-1}} \right) < \frac{1}{2^{n-2}}.$$

Therefore, given $\varepsilon > 0$, if n is chosen so large that $1/2^n < \varepsilon/4$ and if $m \geq n$, then it follows that $|x_n - x_m| < \varepsilon$. Therefore, X is a Cauchy sequence in $\mathbb{R}$. By the Cauchy Criterion 3.5.5 we infer that the sequence X converges to a number x.

To evaluate the limit x, we might first "pass to the limit" in the rule of definition $x_n = \frac{1}{2}(x_{n-1} + x_{n-2})$ to conclude that x must satisfy the relation $x = \frac{1}{2}(x + x)$, which is true, but not informative. Hence we must try something else.

Since X converges to x, so does the subsequence X' with odd indices. By Induction, the reader can establish that [see 1.2.4(f)]

$$x_{2n+1} = 1 + \frac{1}{2} + \frac{1}{2^3} + \cdots + \frac{1}{2^{2n-1}}$$
$$= 1 + \frac{2}{3} \left(1 - \frac{1}{4^n} \right).$$

It follows from this (how?) that $x = \lim X = \lim X' = 1 + \frac{2}{3} = \frac{5}{3}$.

(b)   Let $Y = (y_n)$ be the sequence of real numbers given by

$$y_1 := \frac{1}{1!}, \quad y_2 := \frac{1}{1!} - \frac{1}{2!}, \cdots, \quad y_n := \frac{1}{1!} - \frac{1}{2!} + \cdots + \frac{(-1)^{n+1}}{n!}, \cdots$$

Clearly, Y is not a monotone sequence. However, if $m > n$, then

$$y_m - y_n = \frac{(-1)^{n+2}}{(n+1)!} + \frac{(-1)^{n+3}}{(n+2)!} + \cdots + \frac{(-1)^{m+1}}{m!}.$$

Since $2^{r-1} \leq r!$ [see 1.2.4(e)], it follows that if $m > n$, then (why?)

$$|y_m - y_n| \leq \frac{1}{(n+1)!} + \frac{1}{(n+2)!} + \cdots + \frac{1}{m!}$$
$$\leq \frac{1}{2^n} + \frac{1}{2^{n+1}} + \cdots + \frac{1}{2^{m-1}} < \frac{1}{2^{n-1}}.$$

Therefore, it follows that (y_n) is a Cauchy sequence. Hence it converges to a limit y. At the present moment we cannot evaluate y directly; however, passing to the limit (with respect to m) in the above inequality, we obtain

$$|y_n - y| \leq 1/2^{n-1}.$$

Hence we can calculate y to any desired accuracy by calculating the terms y_n for sufficiently large n. The reader should do this and show that y is approximately equal to 0.632 120 559. (The exact value of y is $1 - 1/e$.)

(c)   The sequence $\left(\dfrac{1}{1} + \dfrac{1}{2} + \cdots + \dfrac{1}{n}\right)$ diverges.

Let $H := (h_n)$ be the sequence defined by

$$h_n := \frac{1}{1} + \frac{1}{2} + \cdots + \frac{1}{n} \qquad \text{for} \quad n \in \mathbb{N},$$

which was considered in 3.3.3(b). If $m > n$, then

$$h_m - h_n = \frac{1}{n+1} + \cdots + \frac{1}{m}.$$

Since each of these $m - n$ terms exceeds $1/m$, then $h_m - h_n > (m - n)/m = 1 - n/m$. In particular, if $m = 2n$ we have $h_{2n} - h_n > \frac{1}{2}$. This shows that H is not a Cauchy sequence (why?); therefore H is *not* a convergent sequence. (In terms that will be introduced in Section 3.7, we have just proved that the "harmonic series" $\sum_{n=1}^{\infty} 1/n$ is divergent.)   □

3.5.7 Definition   We say that a sequence $X = (x_n)$ of real numbers is **contractive** if there exists a constant $C, 0 < C < 1$, such that

$$|x_{n+2} - x_{n+1}| \le C|x_{n+1} - x_n|$$

for all $n \in \mathbb{N}$. The number C is called the **constant** of the contractive sequence.

3.5.8 Theorem   *Every contractive sequence is a Cauchy sequence, and therefore is convergent.*

Proof.   If we successively apply the defining condition for a contractive sequence, we can work our way back to the beginning of the sequence as follows:

$$|x_{n+2} - x_{n+1}| \le C|x_{n+1} - x_n| \le C^2|x_n - x_{n-1}|$$
$$\le C^3|x_{n-1} - x_{n-2}| \le \cdots \le C^n|x_2 - x_1|.$$

For $m > n$, we estimate $|x_m - x_n|$ by first applying the Triangle Inequality and then using the formula for the sum of a geometric progression (see 1.2.4(f)). This gives

$$|x_m - x_n| \le |x_m - x_{m-1}| + |x_{m-1} - x_{m-2}| + \cdots + |x_{n+1} - x_n|$$
$$\le \left(C^{m-2} + C^{m-3} + \cdots + C^{n-1}\right)|x_2 - x_1|$$
$$= C^{n-1}\left(\frac{1 - C^{m-n}}{1 - C}\right)|x_2 - x_1|$$
$$\le C^{n-1}\left(\frac{1}{1 - C}\right)|x_2 - x_1|.$$

Since $0 < C < 1$, we know $\lim(C^n) = 0$ [see 3.1.11(b)]. Therefore, we infer that (x_n) is a Cauchy sequence. It now follows from the Cauchy Convergence Criterion 3.5.5. that (x_n) is a convergent sequence.   Q.E.D.

In the process of calculating the limit of a contractive sequence, it is often very important to have an estimate of the error at the nth stage. In the next result we give two such estimates: the first one involves the first two terms in the sequence and n; the second one involves the difference $x_n - x_{n-1}$.

3.5.9 Corollary   *If $X := (x_n)$ is a contractive sequence with constant $C, 0 < C < 1$, and if $x^* := \lim X$, then*

(i) $\quad |x^* - x_n| \le \dfrac{C^{n-1}}{1 - C}|x_2 - x_1|,$

(ii) $\quad |x^* - x_n| \le \dfrac{C}{1 - C}|x_n - x_{n-1}|.$

Proof. From the preceding proof, if $m > n$, then $|x_m - x_n| \le (C^{n-1}/(1 - C))|x_2 - x_1|$. If we let $m \to \infty$ in this inequality, we obtain (i).

To prove (ii), recall that if $m > n$, then

$$|x_m - x_n| \le |x_m - x_{m-1}| + \cdots + |x_{n+1} - x_n|.$$

Since it is readily established, using Induction, that

$$|x_{n+k} - x_{n+k-1}| \le C^k |x_n - x_{n-1}|,$$

we infer that

$$|x_m - x_n| \le (C^{m-n} + \cdots + C^2 + C)|x_n - x_{n-1}|$$
$$\le \frac{C}{1 - C}|x_n - x_{n-1}|$$

We now let $m \to \infty$ in this inequality to obtain assertion (ii).   Q.E.D.

3.5.10 Example We are told that the cubic equation $x^3 - 7x + 2 = 0$ has a solution between 0 and 1 and we wish to approximate this solution. This can be accomplished by means of an iteration procedure as follows. We first rewrite the equation as $x = (x^3 + 2)/7$ and use this to define a sequence. We assign to x_1 an arbitrary value between 0 and 1, and then define

$$x_{n+1} := \tfrac{1}{7}(x_n^3 + 2) \qquad \text{for} \quad n \in \mathbb{N}.$$

Because $0 < x_1 < 1$, it follows that $0 < x_n < 1$ for all $n \in \mathbb{N}$. (Why?) Moreover, we have

$$|x_{n+2} - x_{n+1}| = \left|\tfrac{1}{7}(x_{n+1}^3 + 2) - \tfrac{1}{7}(x_n^3 + 2)\right| = \tfrac{1}{7}|x_{n+1}^3 - x_n^3|$$
$$= \tfrac{1}{7}|x_{n+1}^2 + x_{n+1}x_n + x_n^2||x_{n+1} - x_n| \le \tfrac{3}{7}|x_{n+1} - x_n|.$$

Therefore, (x_n) is a contractive sequence and hence there exists r such that $\lim(x_n) = r$. If we pass to the limit on both sides of the equality $x_{n+1} = (x_n^3 + 2)/7$, we obtain $r = (r^3 + 2)/7$ and hence $r^3 - 7r + 2 = 0$. Thus r is a solution of the equation.

We can approximate r by choosing x_1 and calculating $x_2, x_3, \cdots$ successively. For example, if we take $x_1 = 0.5$, we obtain (to nine decimal places):

$$x_2 = 0.303\,571\,429, \quad x_3 = 0.289\,710\,830,$$
$$x_4 = 0.289\,188\,016, \quad x_5 = 0.289\,169\,244,$$
$$x_6 = 0.289\,168\,571, \quad \text{etc.}$$

To estimate the accuracy, we note that $|x_2 - x_1| < 0.2$. Thus, after n steps it follows from Corollary 3.5.9(i) that we are sure that $|x^* - x_n| \le 3^{n-1}/(7^{n-2} \cdot 20)$. Thus, when $n = 6$, we are sure that

$$|x^* - x_6| \le 3^5/(7^4 \cdot 20) = 243/48\,020 < 0.0051.$$

Actually the approximation is substantially better than this. In fact, since $|x_6 - x_5| < 0.000\,0005$, it follows from 3.5.9(ii) that $|x^* - x_6| \le \tfrac{3}{4}|x_6 - x_5| < 0.000\,0004$. Hence the first five decimal places of x_6 are correct.   □

Exercises for Section 3.5

1. Give an example of a bounded sequence that is not a Cauchy sequence.

2. Show directly from the definition that the following are Cauchy sequences.

   (a) $\left(\dfrac{n+1}{n}\right)$,

   (b) $\left(1 + \dfrac{1}{2!} + \cdots + \dfrac{1}{n!}\right)$.

3. Show directly from the definition that the following are not Cauchy sequences.

   (a) $\left((-1)^n\right)$,

   (b) $\left(n + \dfrac{(-1)^n}{n}\right)$,

   (c) $(\ln n)$.

4. Show directly from the definition that if (x_n) and (y_n) are Cauchy sequences, then $(x_n + y_n)$ and $(x_n y_n)$ are Cauchy sequences.

5. If $x_n := \sqrt{n}$, show that (x_n) satisfies $\lim |x_{n+1} - x_n| = 0$, but that it is not a Cauchy sequence.

6. Let p be a given natural number. Give an example of a sequence (x_n) that is not a Cauchy sequence, but that satisfies $\lim |x_{n+p} - x_n| = 0$.

7. Let (x_n) be a Cauchy sequence such that x_n is an integer for every $n \in \mathbb{N}$. Show that (x_n) is ultimately constant.

8. Show directly that a bounded, monotone increasing sequence is a Cauchy sequence.

9. If $0 < r < 1$ and $|x_{n+1} - x_n| < r^n$ for all $n \in \mathbb{N}$, show that (x_n) is a Cauchy sequence.

10. If $x_1 < x_2$ are arbitrary real numbers and $x_n := \frac{1}{2}(x_{n-2} + x_{n-1})$ for $n > 2$, show that (x_n) is convergent. What is its limit?

11. If $y_1 < y_2$ are arbitrary real numbers and $y_n := \frac{1}{3}y_{n-1} + \frac{2}{3}y_{n-2}$ for $n > 2$, show that (y_n) is convergent. What is its limit?

12. If $x_1 > 0$ and $x_{n+1} := (2 + x_n)^{-1}$ for $n \geq 1$, show that (x_n) is a contractive sequence. Find the limit.

13. If $x_1 := 2$ and $x_{n+1} := 2 + 1/x_n$ for $n \geq 1$, show that (x_n) is a contractive sequence. What is its limit?

14. The polynomial equation $x^3 - 5x + 1 = 0$ has a root r with $0 < r < 1$. Use an appropriate contractive sequence to calculate r within 10^{-4}.

Section 3.6   Properly Divergent Sequences

For certain purposes it is convenient to define what is meant for a sequence (x_n) of real numbers to "tend to $\pm\infty$".

3.6.1 Definition   Let (x_n) be a sequence of real numbers.

(i)   We say that (x_n) **tends to** $+\infty$, and write $\lim(x_n) = +\infty$, if for every $\alpha \in \mathbb{R}$ there exists a natural number $K(\alpha)$ such that if $n \geq K(\alpha)$, then $x_n > \alpha$.

(ii)   We say that (x_n) **tends to** $-\infty$, and write $\lim(x_n) = -\infty$, if for every $\beta \in \mathbb{R}$ there exists a natural number $K(\beta)$ such that if $n \geq K(\beta)$, then $x_n < \beta$.

We say that (x_n) is **properly divergent** in case we have either $\lim(x_n) = +\infty$ or $\lim(x_n) = -\infty$.

The reader should realize that we are using the symbols $+\infty$ and $-\infty$ purely as a convenient *notation* in the above expressions. Results that have been proved in earlier sections for conventional limits $\lim(x_n) = L$ (for $L \in \mathbb{R}$) *may not* remain true when $\lim(x_n) = \pm\infty$.

3.6.2 Examples **(a)** $\lim(n) = +\infty$.
In fact, if $\alpha \in \mathbb{R}$ is given, let $K(\alpha)$ be any natural number such that $K(\alpha) > \alpha$.
(b) $\lim(n^2) = +\infty$.
If $K(\alpha)$ is a natural number such that $K(\alpha) > \alpha$, and if $n \geq K(\alpha)$ then we have $n^2 \geq n > \alpha$.
(c) If $c > 1$, then $\lim(c^n) = +\infty$.

Let $c = 1 + b$, where $b > 0$. If $\alpha \in \mathbb{R}$ is given, let $K(\alpha)$ be a natural number such that $K(\alpha) > \alpha/b$. If $n \geq K(\alpha)$ it follows from Bernoulli's Inequality that

$$c^n = (1 + b)^n \geq 1 + nb > 1 + \alpha > \alpha.$$

Therefore $\lim(c^n) = +\infty$. □

Monotone sequences are particularly simple in regard to their convergence. We have seen in the Monotone Convergence Theorem 3.3.2 that a monotone sequence is convergent if and only if it is bounded. The next result is a reformulation of that result.

3.6.3 Theorem *A monotone sequence of real numbers is properly divergent if and only if it is unbounded.*

(a) *If (x_n) is an unbounded increasing sequence, then $\lim(x_n) = +\infty$.*
(b) *If (x_n) is an unbounded decreasing sequence, then $\lim(x_n) = -\infty$.*

Proof. **(a)** Suppose that (x_n) is an increasing sequence. We know that if (x_n) is bounded, then it is convergent. If (x_n) is unbounded, then for any $\alpha \in \mathbb{R}$ there exists $n(\alpha) \in \mathbb{N}$ such that $\alpha < x_{n(\alpha)}$. But since (x_n) is increasing, we have $\alpha < x_n$ for all $n \geq n(\alpha)$. Since α is arbitrary, it follows that $\lim(x_n) = +\infty$.
Part (b) is proved in a similar fashion. Q.E.D.

The following "comparison theorem" is frequently used in showing that a sequence is properly divergent. [In fact, we implicitly used it in Example 3.6.2(c).]

3.6.4 Theorem *Let (x_n) and (y_n) be two sequences of real numbers and suppose that*

$$(1) \qquad\qquad x_n \leq y_n \qquad \text{for all} \quad n \in \mathbb{N}.$$

(a) *If $\lim(x_n) = +\infty$, then $\lim(y_n) = +\infty$.*
(b) *If $\lim(y_n) = -\infty$, then $\lim(x_n) = -\infty$.*

Proof. **(a)** If $\lim(x_n) = +\infty$, and if $\alpha \in \mathbb{R}$ is given, then there exists a natural number $K(\alpha)$ such that if $n \geq K(\alpha)$, then $\alpha < x_n$. In view of (1), it follows that $\alpha < y_n$ for all $n \geq K(\alpha)$. Since α is arbitrary, it follows that $\lim(y_n) = +\infty$.
The proof of (b) is similar. Q.E.D.

Remarks **(a)** Theorem 3.6.4 remains true if condition (1) is ultimately true; that is, if there exists $m \in \mathbb{N}$ such that $x_n \leq y_n$ for all $n \geq m$.

(b) If condition (1) of Theorem 3.6.4 holds and if $\lim(y_n) = +\infty$, it does *not* follow that $\lim(x_n) = +\infty$. Similarly, if (1) holds and if $\lim(x_n) = -\infty$, it does *not* follow that $\lim(y_n) = -\infty$. In using Theorem 3.6.4 to show that a sequence tends to $+\infty$ [respectively, $-\infty$] we need to show that the terms of the sequence are ultimately greater [respectively, less] than or equal to the corresponding terms of a sequence that is known to tend to $+\infty$ [respectively, $-\infty$].

Since it is sometimes difficult to establish an inequality such as (1), the following "limit comparison theorem" is often more convenient to use than Theorem 3.6.4.

3.6.5 Theorem   *Let* (x_n) *and* (y_n) *be two sequences of positive real numbers and suppose that for some* $L \in \mathbb{R}$, $L > 0$, *we have*

$$(2) \qquad\qquad\qquad \lim(x_n/y_n) = L.$$

Then $\lim(x_n) = +\infty$ *if and only if* $\lim(y_n) = +\infty$.

Proof.   If (2) holds, there exists $K \in \mathbb{N}$ such that

$$\tfrac{1}{2}L < x_n/y_n < \tfrac{3}{2}L \qquad \text{for all} \quad n \geq K.$$

Hence we have $\left(\tfrac{1}{2}L\right) y_n < x_n < \left(\tfrac{3}{2}L\right) y_n$ for all $n \geq K$. The conclusion now follows from a slight modification of Theorem 3.6.4. We leave the details to the reader.   Q.E.D.

The reader can show that the conclusion need not hold if either $L = 0$ or $L = +\infty$. However, there are some partial results that can be established in these cases, as will be seen in the exercises.

Exercises for Section 3.6

1. Show that if (x_n) is an unbounded sequence, then there exists a properly divergent subsequence.

2. Give examples of properly divergent sequences (x_n) and (y_n) with $y_n \neq 0$ for all $n \in \mathbb{N}$ such that:
    (a) (x_n/y_n) is convergent,   (b) (x_n/y_n) is properly divergent.

3. Show that if $x_n > 0$ for all $n \in \mathbb{N}$, then $\lim(x_n) = 0$ if and only if $\lim(1/x_n) = +\infty$.

4. Establish the proper divergence of the following sequences.
    (a) $\left(\sqrt{n}\right)$,   (b) $\left(\sqrt{n+1}\right)$,
    (c) $\left(\sqrt{n-1}\right)$,   (d) $\left(n/\sqrt{n+1}\right)$.

5. Is the sequence $(n \sin n)$ properly divergent?

6. Let (x_n) be properly divergent and let (y_n) be such that $\lim(x_n y_n)$ belongs to $\mathbb{R}$. Show that (y_n) converges to 0.

7. Let (x_n) and (y_n) be sequences of positive numbers such that $\lim(x_n/y_n) = 0$.
    (a)   Show that if $\lim(x_n) = +\infty$, then $\lim(y_n) = +\infty$.
    (b)   Show that if (y_n) is bounded, then $\lim(x_n) = 0$.

8. Investigate the convergence or the divergence of the following sequences:
    (a) $\left(\sqrt{n^2 + 2}\right)$,   (b) $\left(\sqrt{n}/(n^2 + 1)\right)$,
    (c) $\left(\sqrt{n^2 + 1}/\sqrt{n}\right)$,   (d) $(\sin \sqrt{n})$.

9. Let (x_n) and (y_n) be sequences of positive numbers such that $\lim(x_n/y_n) = +\infty$.
    (a)   Show that if $\lim(y_n) = +\infty$, then $\lim(x_n) = +\infty$.
    (b)   Show that if (x_n) is bounded, then $\lim(y_n) = 0$.

10. Show that if $\lim(a_n/n) = L$, where $L > 0$, then $\lim(a_n) = +\infty$.

Section 3.7   Introduction to Infinite Series

We will now give a brief introduction to infinite series of real numbers. This is a topic that will be discussed in more detail in Chapter 9, but because of its importance, we will establish a few results here. These results will be seen to be immediate consequences of theorems we have met in this chapter.

In elementary texts, an infinite series is sometimes "defined" to be "an expression of the form"

(1)
$$x_1 + x_2 + \cdots + x_n + \cdots.$$

However, this "definition" lacks clarity, since there is *a priori* no particular value that we can attach to this array of symbols, which calls for an *infinite* number of additions to be performed.

3.7.1 Definition   If $X := (x_n)$ is a sequence in $\mathbb{R}$, then the **infinite series** (or simply the **series**) **generated by** X is the sequence $S := (s_k)$ defined by

$$s_1 := x_1$$
$$s_2 := s_1 + x_2 \quad (= x_1 + x_2)$$
$$\cdots$$
$$s_k := s_{k-1} + x_k \quad (= x_1 + x_2 + \cdots + x_k)$$
$$\cdots$$

The numbers x_n are called the **terms** of the series and the numbers s_k are called the **partial sums** of this series. If $\lim S$ exists, we say that this series is **convergent** and call this limit the **sum** or the **value** of this series. If this limit does not exist, we say that the series S is **divergent**.

It is convenient to use symbols such as

(2)
$$\sum (x_n) \quad \text{or} \quad \sum x_n \quad \text{or} \quad \sum_{n=1}^{\infty} x_n$$

to denote both the infinite series S generated by the sequence $X = (x_n)$ and also to denote the value $\lim S$, in case this limit exists. Thus the symbols in (2) may be regarded merely as a way of exhibiting an infinite series whose convergence or divergence is to be investigated. In practice, this double use of these notations does not lead to any confusion, provided it is understood that the convergence (or divergence) of the series must be established.

Just as a sequence may be indexed such that its first element is not x_1, but is x_0, or x_5 or x_{99}, we will denote the series having these numbers as their first element by the symbols

$$\sum_{n=0}^{\infty} x_n \quad \text{or} \quad \sum_{n=5}^{\infty} x_n \quad \text{or} \quad \sum_{n=99}^{\infty} x_n.$$

It should be noted that when the first term in the series is x_N, then the first partial sum is denoted by s_N.

Warning   The reader should guard against confusing the words "sequence" and "series". In nonmathematical language, these words are interchangeable; however, in mathematics, these words are not synonyms. Indeed, a series is a sequence $S = (s_k)$ obtained from a given sequence $X = (x_n)$ according to the special procedure given in Definition 3.7.1.

3.7.2 Examples   (a)   Consider the sequence $X := (r^n)_{n=0}^{\infty}$ where $r \in \mathbb{R}$, which generates the **geometric series**:

$$(3) \qquad \sum_{n=0}^{\infty} r^n = 1 + r + r^2 + \cdots + r^n + \cdots .$$

We will show that if $|r| < 1$, then this series converges to $1/(1 - r)$. (See also Example 1.2.4(f).) Indeed, if $s_n := 1 + r + r^2 + \cdots + r^n$ for $n \geq 0$, and if we multiply s_n by r and subtract the result from s_n, we obtain (after some simplification):

$$s_n(1 - r) = 1 - r^{n+1}.$$

Therefore, we have

$$s_n - \frac{1}{1 - r} = -\frac{r^{n+1}}{1 - r},$$

from which it follows that

$$\left| s_n - \frac{1}{1 - r} \right| \leq \frac{|r|^{n+1}}{|1 - r|}.$$

Since $|r|^{n+1} \to 0$ when $|r| < 1$, it follows that the geometric series (3) converges to $1/(1 - r)$ when $|r| < 1$.

(b)   Consider the series generated by $((-1)^n)_{n=0}^{\infty}$; that is, the series:

$$(4) \qquad \sum_{n=0}^{\infty} (-1)^n = (+1) + (-1) + (+1) + (-1) + \cdots .$$

It is easily seen (by Mathematical Induction) that $s_n = 1$ if $n \geq 0$ is even and $s_n = 0$ if n is odd; therefore, the sequence of partial sums is $(1, 0, 1, 0, \cdots)$. Since this sequence is not convergent, the series (4) is divergent.

(c)   Consider the series

$$(5) \qquad \sum_{n=1}^{\infty} \frac{1}{n(n + 1)} = \frac{1}{1 \cdot 2} + \frac{1}{2 \cdot 3} + \frac{1}{3 \cdot 4} + \cdots .$$

By a stroke of insight, we note that

$$\frac{1}{k(k + 1)} = \frac{1}{k} - \frac{1}{k + 1}.$$

Hence, on adding these terms from $k = 1$ to $k = n$ and noting the telescoping that takes place, we obtain

$$s_n = \frac{1}{1} - \frac{1}{n + 1},$$

whence it follows that $s_n \to 1$. Therefore the series (5) converges to 1.   □

We now present a very useful and simple *necessary* condition for the convergence of a series. It is far from being sufficient, however.

3.7.3 The *n*th Term Test *If the series $\sum x_n$ converges, then $\lim(x_n) = 0$.*

Proof. By Definition 3.7.1, the convergence of $\sum x_n$ requires that $\lim(s_k)$ exists. Since $x_n = s_n - s_{n-1}$, then $\lim(x_n) = \lim(s_n) - \lim(s_{n-1}) = 0$.   Q.E.D.

Since the following Cauchy Criterion is precisely a reformulation of Theorem 3.5.5, we will omit its proof.

3.7.4 Cauchy Criterion for Series *The series $\sum x_n$ converges if and only if for every $\varepsilon > 0$ there exists $M(\varepsilon) \in \mathbb{N}$ such that if $m > n \geq M(\varepsilon)$, then*

(6)
$$|s_m - s_n| = |x_{n+1} + x_{n+2} + \cdots + x_m| < \varepsilon.$$

The next result, although limited in scope, is of great importance and utility.

3.7.5 Theorem *Let (x_n) be a sequence of nonnegative real numbers. Then the series $\sum x_n$ converges if and only if the sequence $S = (s_k)$ of partial sums is bounded. In this case,*

$$\sum_{n=1}^{\infty} x_n = \lim(s_k) = \sup\{s_k : k \in \mathbb{N}\}.$$

Proof. Since $x_n > 0$, the sequence S of partial sums is monotone increasing:

$$s_1 \leq s_2 \leq \cdots \leq s_k \leq \cdots.$$

By the Monotone Convergence Theorem 3.3.2, the sequence $S = (s_k)$ converges if and only if it is bounded, in which case its limit equals $\sup\{s_k\}$.   Q.E.D.

3.7.6 Examples **(a)** The geometric series (3) diverges if $|r| \geq 1$.
     This follows from the fact that the terms r^n do not approach 0 when $|r| \geq 1$.

(b) The **harmonic series** $\sum_{n=1}^{\infty} \dfrac{1}{n}$ diverges.

Since the terms $1/n \to 0$, we cannot use the *n*th Term Test 3.7.3 to establish this divergence. However, it was seen in Examples 3.3.3(b) and 3.5.6(c) that the sequence (s_n) of partial sums is not bounded. Therefore, it follows from Theorem 3.7.5 that the harmonic series is divergent.

(c) The **2-series** $\sum_{n=1}^{\infty} \dfrac{1}{n^2}$ is convergent.

Since the partial sums are monotone, it suffices (why?) to show that some subsequence of (s_k) is bounded. If $k_1 := 2^1 - 1 = 1$, then $s_{k_1} = 1$. If $k_2 := 2^2 - 1 = 3$, then

$$s_{k_2} = \frac{1}{1} + \left(\frac{1}{2^2} + \frac{1}{3^2}\right) < 1 + \frac{2}{2^2} = 1 + \frac{1}{2},$$

and if $k_3 := 2^3 - 1 = 7$, then we have

$$s_{k_3} = s_{k_2} + \left(\frac{1}{4^2} + \frac{1}{5^2} + \frac{1}{6^2} + \frac{1}{7^2}\right) < s_{k_2} + \frac{4}{4^2} < 1 + \frac{1}{2} + \frac{1}{2^2}.$$

By Mathematical Induction, we find that if $k_j := 2^j - 1$, then

$$0 < s_{k_j} < 1 + \tfrac{1}{2} + \left(\tfrac{1}{2}\right)^2 + \cdots + \left(\tfrac{1}{2}\right)^{j-1}.$$

Since the term on the right is a partial sum of a geometric series with $r = \frac{1}{2}$, it is dominated by $1/(1 - \frac{1}{2}) = 2$, and Theorem 3.7.5 implies that the 2-series converges.

(d) The p-**series** $\displaystyle\sum_{n=1}^{\infty} \frac{1}{n^p}$ converges when $p > 1$.

Since the argument is very similar to the special case considered in part (c), we will leave some of the details to the reader. As before, if $k_1 := 2^1 - 1 = 1$, then $s_{k_1} = 1$. If $k_2 := 2^2 - 1 = 3$, then since $2^p < 3^p$, we have

$$s_{k_2} = \frac{1}{1^p} + \left(\frac{1}{2^p} + \frac{1}{3^p}\right) < 1 + \frac{2}{2^p} = 1 + \frac{1}{2^{p-1}}.$$

Further, if $k_3 := 2^3 - 1$, then (how?) it is seen that

$$s_{k_3} < s_{k_2} + \frac{4}{4^p} < 1 + \frac{1}{2^{p-1}} + \frac{1}{4^{p-1}}.$$

Finally, we let $r := 1/2^{p-1}$; since $p > 1$, we have $0 < r < 1$. Using Mathematical Induction, we show that if $k_j := 2^j - 1$, then

$$0 < s_{k_j} < 1 + r + r^2 + \cdots + r^{j-1} < \frac{1}{1-r}.$$

Therefore, Theorem 3.7.5 implies that the p-series converges when $p > 1$.

(e) The p-**series** $\displaystyle\sum_{n=1}^{\infty} \frac{1}{n^p}$ diverges when $0 < p \leq 1$.

We will use the elementary inequality $n^p \leq n$ when $n \in \mathbb{N}$ and $0 < p \leq 1$. It follows that

$$\frac{1}{n} \leq \frac{1}{n^p} \qquad \text{for} \quad n \in \mathbb{N}.$$

Since the partial sums of the harmonic series are not bounded, this inequality shows that the partial sums of the p-series are not bounded when $0 < p \leq 1$. Hence the p-series diverges for these values of p.

(f) The **alternating harmonic series,** given by

(7)
$$\sum_{n=1}^{\infty} \frac{(-1)^{n+1}}{n} = \frac{1}{1} - \frac{1}{2} + \frac{1}{3} - \cdots + \frac{(-1)^{n+1}}{n} + \cdots$$

is convergent.

The reader should compare this series with the harmonic series in (b), which is divergent. Thus, the subtraction of some of the terms in (7) is essential if this series is to converge. Since we have

$$s_{2n} = \left(\frac{1}{1} - \frac{1}{2}\right) + \left(\frac{1}{3} - \frac{1}{4}\right) + \cdots + \left(\frac{1}{2n-1} - \frac{1}{2n}\right),$$

it is clear that the "even" subsequence (s_{2n}) is increasing. Similarly, the "odd" subsequence (s_{2n+1}) is decreasing since

$$s_{2n+1} = \frac{1}{1} - \left(\frac{1}{2} - \frac{1}{3}\right) - \left(\frac{1}{4} - \frac{1}{5}\right) - \cdots - \left(\frac{1}{2n} - \frac{1}{2n+1}\right).$$

Since $0 < s_{2n} < s_{2n} + 1/(2n+1) = s_{2n+1} \leq 1$, both of these subsequences are bounded below by 0 and above by 1. Therefore they are both convergent and to the same value. Thus the sequence (s_n) of partial sums converges, proving that the alternating harmonic series (7) converges. (It is far from obvious that the limit of this series is equal to $\ln 2$.)   □

Comparison Tests

Our first test shows that if the terms of a nonnegative series are dominated by the corresponding terms of a *convergent series,* then the first series is convergent.

3.7.7 Comparison Test   *Let $X := (x_n)$ and $Y := (y_n)$ be real sequences and suppose that for some $K \in \mathbb{N}$ we have*

(8) $0 \le x_n \le y_n$ *for $n \ge K$.*

(a)   *Then the convergence of $\sum y_n$ implies the convergence of $\sum x_n$.*
(b)   *The divergence of $\sum x_n$ imples the divergence of $\sum y_n$.*

Proof.   (a) Suppose that $\sum y_n$ converges and, given $\varepsilon > 0$, let $M(\varepsilon) \in \mathbb{N}$ be such that if $m > n \ge M(\varepsilon)$, then

$$y_{n+1} + \cdots + y_m < \varepsilon.$$

If $m > \sup\{K, M(\varepsilon)\}$, then it follows that

$$0 \le x_{n+1} + \cdots + x_m \le y_{n+1} + \cdots + y_m < \varepsilon,$$

from which the convergence of $\sum x_n$ follows.
(b) This statement is the contrapositive of (a).   Q.E.D.

Since it is sometimes difficult to establish the inequalities (8), the next result is frequently very useful.

3.7.8 Limit Comparison Test   *Suppose that $X := (x_n)$ and $Y := (y_n)$ are strictly positive sequences and suppose that the following limit exists in $\mathbb{R}$:*

(9) $$r := \lim\left(\frac{x_n}{y_n}\right).$$

(a)   *If $r \ne 0$ then $\sum x_n$ is convergent if and only if $\sum y_n$ is convergent.*
(b)   *If $r = 0$ and if $\sum y_n$ is convergent, then $\sum x_n$ is convergent.*

Proof.   (a) It follows from (9) and Exercise 3.1.17 that there exists $K \in \mathbb{N}$ such that $\frac{1}{2}r \le x_n/y_n \le 2r$ for $n \ge K$, whence

$$\left(\tfrac{1}{2}r\right) y_n \le x_n \le (2r)y_n \qquad \text{for}\quad n \ge K.$$

If we apply the Comparison Test 3.7.7 twice, we obtain the assertion in (a).
(b) If $r = 0$, then there exists $K \in \mathbb{N}$ such that

$$0 < x_n \le y_n \qquad \text{for}\quad n \ge K,$$

so that Theorem 3.7.7(a) applies.   Q.E.D.

Remark   The Comparison Tests 3.7.7 and 3.7.8 depend on having a stock of series that one knows to be convergent (or divergent). The reader will find that the p-series is often useful for this purpose.

3.7.9 Examples   **(a)**   The series $\displaystyle\sum_{n=1}^{\infty} \frac{1}{n^2 + n}$ converges.

It is clear that the inequality

$$0 < \frac{1}{n^2 + n} < \frac{1}{n^2} \qquad \text{for} \quad n \in \mathbb{N}$$

is valid. Since the series $\sum 1/n^2$ is convergent (by Example 3.7.6(c)), we can apply the Comparison Test 3.7.7 to obtain the convergence of the given series.

(b)   The series $\displaystyle\sum_{n=1}^{\infty} \frac{1}{n^2 - n + 1}$ is convergent.

If the inequality

(10)
$$\frac{1}{n^2 - n + 1} \leq \frac{1}{n^2}$$

were true, we could argue as in (a). However, (10) is *false* for all $n \in \mathbb{N}$. The reader can probably show that the inequality

$$0 < \frac{1}{n^2 - n + 1} \leq \frac{2}{n^2}$$

is valid for all $n \in \mathbb{N}$, and this inequality will work just as well. However, it might take some experimentation to think of such an inequality and then establish it.

Instead, if we take $x_n := 1/(n^2 - n + 1)$ and $y_n := 1/n^2$, then we have

$$\frac{x_n}{y_n} = \frac{n^2}{n^2 - n + 1} = \frac{1}{1 - (1/n) + (1/n^2)} \to 1.$$

Therefore, the convergence of the given series follows from the Limit Comparison Test 3.7.8(a).

(c)   The series $\displaystyle\sum_{n=1}^{\infty} \frac{1}{\sqrt{n + 1}}$ is divergent.

This series closely resembles the series $\sum 1/\sqrt{n}$ which is a p-series with $p = \frac{1}{2}$; by Example 3.7.6(e), it is divergent. If we let $x_n := 1/\sqrt{n + 1}$ and $y_n := 1/\sqrt{n}$, then we have

$$\frac{x_n}{y_n} = \frac{\sqrt{n}}{\sqrt{n + 1}} = \frac{1}{\sqrt{1 + 1/n}} \to 1.$$

Therefore the Limit Comparison Test 3.7.8(a) applies.

(d)   The series $\displaystyle\sum_{n=1}^{\infty} \frac{1}{n!}$ is convergent.

It would be possible to establish this convergence by showing (by Induction) that $n^2 < n!$ for $n \geq 4$, whence it follows that

$$0 < \frac{1}{n!} < \frac{1}{n^2} \qquad \text{for} \quad n \geq 4.$$

Alternatively, if we let $x := 1/n!$ and $y_n := 1/n^2$, then (when $n \geq 4$) we have

$$0 \leq \frac{x_n}{y_n} = \frac{n^2}{n!} = \frac{n}{1 \cdot 2 \cdots (n - 1)} < \frac{1}{n - 2} \to 0.$$

Therefore the Limit Comparison Test 3.7.8(b) applies. (Note that this test was a bit troublesome to apply since we do not presently know the convergence of any series for which the limit of x_n/y_n is really easy to determine.)   □

Exercises for Section 3.7

1. Let $\sum a_n$ be a given series and let $\sum b_n$ be the series in which the terms are the same and in the same order as in $\sum a_n$ except that the terms for which $a_n = 0$ have been omitted. Show that $\sum a_n$ converges to A if and only if $\sum b_n$ converges to A.

2. Show that the convergence of a series is not affected by changing a *finite* number of its terms. (Of course, the value of the sum may be changed.)

3. By using partial fractions, show that

   (a) $\displaystyle\sum_{n=0}^{\infty} \frac{1}{(n+1)(n+2)} = 1$,

   (b) $\displaystyle\sum_{n=0}^{\infty} \frac{1}{(\alpha+n)(\alpha+n+1)} = \frac{1}{\alpha} > 0$, if $\alpha > 0$.

   (c) $\displaystyle\sum_{n=1}^{\infty} \frac{1}{n(n+1)(n+2)} = \frac{1}{4}$.

4. If $\sum x_n$ and $\sum y_n$ are convergent, show that $\sum(x_n + y_n)$ is convergent.

5. Can you give an example of a convergent series $\sum x_n$ and a divergent series $\sum y_n$ such that $\sum(x_n + y_n)$ is convergent? Explain.

6. (a) Show that the series $\displaystyle\sum_{n=1}^{\infty} \cos n$ is divergent.

   (b) Show that the series $\displaystyle\sum_{n=1}^{\infty} (\cos n)/n^2$ is convergent.

7. Use an argument similar to that in Example 3.7.6(f) to show that the series $\displaystyle\sum_{n=1}^{\infty} \frac{(-1)^n}{\sqrt{n}}$ is convergent.

8. If $\sum a_n$ with $a_n > 0$ is convergent, then is $\sum a_n^2$ always convergent? Either prove it or give a counterexample.

9. If $\sum a_n$ with $a_n > 0$ is convergent, then is $\sum \sqrt{a_n}$ always convergent? Either prove it or give a counterexample.

10. If $\sum a_n$ with $a_n > 0$ is convergent, then is $\sum \sqrt{a_n a_{n+1}}$ always convergent? Either prove it or give a counterexample.

11. If $\sum a_n$ with $a_n > 0$ is convergent, and if $b_n := (a_1 + \cdots + a_n)/n$ for $n \in \mathbb{N}$, then show that $\sum b_n$ is always divergent.

12. Let $\displaystyle\sum_{n=1}^{\infty} a(n)$ be such that $(a(n))$ is a decreasing sequence of strictly positive numbers. If $s(n)$ denotes the nth partial sum, show (by grouping the terms in $s(2^n)$ in two different ways) that

$$\tfrac{1}{2}\left(a(1) + 2a(2) + \cdots + 2^n a(2^n)\right) \le s(2^n) \le \left(a(1) + 2a(2) + \cdots + 2^{n-1}a(2^{n-1})\right) + a(2^n).$$

Use these inequalities to show that $\displaystyle\sum_{n=1}^{\infty} a(n)$ converges if and only if $\displaystyle\sum_{n=1}^{\infty} 2^n a(2^n)$ converges. This result is often called the **Cauchy Condensation Test**; it is very powerful.

13. Use the Cauchy Condensation Test to discuss the *p*-series $\displaystyle\sum_{n=1}^{\infty}(1/n^p)$ for $p > 0$.

14. Use the Cauchy Condensation Test to establish the divergence of the series:

    (a) $\displaystyle\sum \frac{1}{n \ln n}$,

    (b) $\displaystyle\sum \frac{1}{n(\ln n)(\ln \ln n)}$,

    (c) $\displaystyle\sum \frac{1}{n(\ln n)(\ln \ln n)(\ln \ln \ln n)}$.

15. Show that if $c > 1$, then the following series are convergent:

    (a) $\displaystyle\sum \frac{1}{n(\ln n)^c}$,

    (b) $\displaystyle\sum \frac{1}{n(\ln n)(\ln \ln n)^c}$.

CHAPTER 4

LIMITS

"Mathematical analysis" is generally understood to refer to that area of mathematics in which systematic use is made of various limiting concepts. In the preceding chapter we studied one of these basic limiting concepts: the limit of a sequence of real numbers. In this chapter we will encounter the notion of the limit of a function.

The rudimentary notion of a limiting process emerged in the 1680s as Isaac Newton (1642–1727) and Gottfried Leibniz (1646–1716) struggled with the creation of the Calculus. Though each person's work was initially unknown to the other and their creative insights were quite different, both realized the need to formulate a notion of function and the idea of quantities being "close to" one another. Newton used the word "fluent" to denote a relationship between variables, and in his major work *Principia* in 1687 he discussed limits "to which they approach nearer than by any given difference, but never go beyond, nor in effect attain to, till the quantities are diminished *in infinitum*". Leibniz introduced the term "function" to indicate a quantity that depended on a variable, and he invented "infinitesimally small" numbers as a way of handling the concept of a limit. The term "function" soon became standard terminology, and Leibniz also introduced the term "calculus" for this new method of calculation.

In 1748, Leonhard Euler (1707–1783) published his two-volume treatise *Introductio in Analysin Infinitorum,* in which he discussed power series, the exponential and logarithmic functions, the trigonometric functions, and many related topics. This was followed by *Institutiones Calculi Differentialis* in 1755 and the three-volume *Institutiones Calculi Integralis* in 1768–70. These works remained the standard textbooks on calculus for many years. But the concept of limit was very intuitive and its looseness led to a number of problems. Verbal descriptions of the limit concept were proposed by other mathematicians of the era, but none was adequate to provide the basis for rigorous proofs.

In 1821, Augustin-Louis Cauchy (1789–1857) published his lectures on analysis in his *Cours d'Analyse,* which set the standard for mathematical exposition for many years. He was concerned with rigor and in many ways raised the level of precision in mathematical discourse. He formulated definitions and presented arguments with greater care than his predecessors, but the concept of limit still remained elusive. In an early chapter he gave the following definition:

> If the successive values attributed to the same variable approach indefinitely a fixed value, such that they finally differ from it by as little as one wishes, this latter is called the limit of all the others.

The final steps in formulating a precise definition of limit were taken by Karl Weierstrass (1815–1897). He insisted on precise language and rigorous proofs, and his definition of limit is the one we use today.

Gottfried Leibniz

Gottfried Wilhelm Leibniz (1646–1716) was born in Leipzig, Germany. He was six years old when his father, a professor of philosophy, died and left his son the key to his library and a life of books and learning. Leibniz entered the University of Leipzig at age 15, graduated at age 17, and received a Doctor of Law degree from the University of Altdorf four years later. He wrote on legal matters, but was more interested in philosophy. He also developed original theories about language and the nature of the universe. In 1672, he went to Paris as a diplomat for four years. While there he began to study  mathematics with the Dutch mathematician Christiaan Huygens. His travels to London to visit the Royal Academy further stimulated his interest in mathematics. His background in philosophy led him to very original, though not always rigorous, results.

Unaware of Newtons's unpublished work, Leibniz published papers in the 1680s that presented a method of finding areas that is known today as the Fundamental Theorem of Calculus. He coined the term "calculus" and invented the dy/dx and elongated S notations that are used today. Unfortunately, some followers of Newton accused Leibniz of plagiarism, resulting in a dispute that lasted until Leibniz's death. Their approaches to calculus were quite different and it is now evident that their discoveries were made independently. Leibniz is now renowned for his work in philosophy, but his mathematical fame rests on his creation of the calculus.

Section 4.1   Limits of Functions

In this section we will introduce the important notion of the limit of a function. The intuitive idea of the function f having a limit L at the point c is that the values $f(x)$ are close to L when x is close to (but different from) c. But it is necessary to have a technical way of working with the idea of "close to" and this is accomplished in the ε-δ definition given below.

In order for the idea of the limit of a function f at a point c to be meaningful, it is necessary that f be defined at points near c. It need not be defined at the point c, but it should be defined at enough points close to c to make the study interesting. This is the reason for the following definition.

4.1.1 Definition   Let $A \subseteq \mathbb{R}$. A point $c \in \mathbb{R}$ is a **cluster point** of A if for every $\delta > 0$ there exists at least one point $x \in A$, $x \neq c$ such that $|x - c| < \delta$.

This definition is rephrased in the language of neighborhoods as follows: A point c is a cluster point of the set A if every δ-neighborhood $V_\delta(c) = (c - \delta, c + \delta)$ of c contains at least one point of A distinct from c.

Note   The point c may or may not be a member of A, but even if it is in A, it is ignored when deciding whether it is a cluster point of A or not, since we explicitly require that there be points in $V_\delta(c) \cap A$ distinct from c in order for c to be a cluster point of A.

For example, if $A := \{1, 2\}$, then the point 1 is not a cluster point of A, since choosing $\delta := \frac{1}{2}$ gives a neighborhood of 1 that contains no points of A distinct from 1. The same is true for the point 2, so we see that A has no cluster points.

4.1.2 Theorem   *A number $c \in \mathbb{R}$ is a cluster point of a subset A of $\mathbb{R}$ if and only if there exists a sequence (a_n) in A such that $\lim(a_n) = c$ and $a_n \neq c$ for all $n \in \mathbb{N}$.*

Proof.   If c is a cluster point of A, then for any $n \in \mathbb{N}$ the $(1/n)$-neighborhood $V_{1/n}(c)$ contains at least one point a_n in A distinct from c. Then $a_n \in A$, $a_n \neq c$, and $|a_n - c| < 1/n$ implies $\lim(a_n) = c$.

Conversely, if there exists a sequence (a_n) in $A \setminus \{c\}$ with $\lim(a_n) = c$, then for any $\delta > 0$ there exists K such that if $n \geq K$, then $a_n \in V_\delta(c)$. Therefore the δ-neighborhood $V_\delta(c)$ of c contains the points a_n, for $n \geq K$, which belong to A and are distinct from c.

Q.E.D.

The next examples emphasize that a cluster point of a set may or may not belong to the set.

4.1.3 Examples   **(a)**   For the open interval $A_1 := (0, 1)$, every point of the closed interval $[0,1]$ is a cluster point of A_1. Note that the points $0,1$ are cluster points of A_1, but do not belong to A_1. All the points of A_1 are cluster points of A_1.

(b)   A finite set has no cluster points.

(c)   The infinite set $\mathbb{N}$ has no cluster points.

(d)   The set $A_4 := \{1/n : n \in \mathbb{N}\}$ has only the point 0 as a cluster point. None of the points in A_4 is a cluster point of A_4.

(e)   If $I := [0, 1]$, then the set $A_5 := I \cap \mathbb{Q}$ consists of all the rational numbers in I. It follows from the Density Theorem 2.4.8 that every point in I is a cluster point of A_5.   □

Having made this brief detour, we now return to the concept of the limit of a function at a cluster point of its domain.

The Definition of the Limit

We now state the precise definition of the limit of a function f at a point c. It is important to note that in this definition, it is immaterial whether f is defined at c or not. In any case, we exclude c from consideration in the determination of the limit.

4.1.4 Definition   Let $A \subseteq \mathbb{R}$, and let c be a cluster point of A. For a function $f : A \to \mathbb{R}$, a real number L is said to be a **limit of** f **at** c if, given any $\varepsilon > 0$ there exists a $\delta > 0$ such that if $x \in A$ and $0 < |x - c| < \delta$, then $|f(x) - L| < \varepsilon$.

Remarks   **(a)**   Since the value of δ usually depends on ε, we will sometimes write $\delta(\varepsilon)$ instead of δ to emphasize this dependence.

(b)   The inequality $0 < |x - c|$ is equivalent to saying $x \neq c$.

If L is a limit of f at c, then we also say that f **converges to** L **at** c. We often write

$$L = \lim_{x \to c} f(x) \quad \text{or} \quad L = \lim_{x \to c} f.$$

We also say that "$f(x)$ approaches L as x approaches c". (But it should be noted that the points do not actually move anywhere.) The symbolism

$$f(x) \to L \qquad \text{as} \quad x \to c$$

is also used sometimes to express the fact that f has limit L at c.

If the limit of f at c does not exist, we say that f **diverges** at c.

Our first result is that the value L of the limit is uniquely determined. This uniqueness is not part of the definition of limit, but must be deduced.

4.1.5 Theorem *If $f : A \to \mathbb{R}$ and if c is a cluster point of A, then f can have only one limit at c.*

Proof.   Suppose that numbers L and L' satisfy Definition 4.1.4. For any $\varepsilon > 0$, there exists $\delta(\varepsilon/2) > 0$ such that if $x \in A$ and $0 < |x - c| < \delta(\varepsilon/2)$, then $|f(x) - L| < \varepsilon/2$. Also there exists $\delta'(\varepsilon/2)$ such that if $x \in A$ and $0 < |x - c| < \delta'(\varepsilon/2)$, then $|f(x) - L'| < \varepsilon/2$. Now let $\delta := \inf\{\delta(\varepsilon/2), \delta'(\varepsilon/2)\}$. Then if $x \in A$ and $0 < |x - c| < \delta$, the Triangle Inequality implies that

$$|L - L'| \le |L - f(x)| + |f(x) - L'| < \varepsilon/2 + \varepsilon/2 = \varepsilon.$$

Since $\varepsilon > 0$ is arbitrary, we conclude that $L - L' = 0$, so that $L = L'$.   Q.E.D.

The definition of limit can be very nicely described in terms of neighborhoods. (See Figure 4.1.1.) We observe that because

$$V_\delta(c) = (c - \delta, c + \delta) = \{x : |x - c| < \delta\},$$

the inequality $0 < |x - c| < \delta$ is equivalent to saying that $x \ne c$ and x belongs to the δ-neighborhood $V_\delta(c)$ of c. Similarly, the inequality $|f(x) - L| < \varepsilon$ is equivalent to saying that $f(x)$ belongs to the ε-neighborhood $V_\varepsilon(L)$ of L. In this way, we obtain the following result. The reader should write out a detailed argument to establish the theorem.

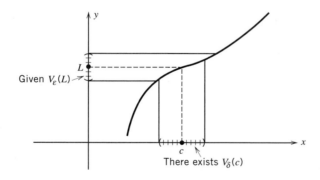

Figure 4.1.1   The limit of f at c is L.

4.1.6 Theorem *Let $f: A \to \mathbb{R}$ and let c be a cluster point of A. Then the following statements are equivalent.*

(i) $\displaystyle \lim_{x \to c} f(x) = L$.

(ii)   *Given any ε-neighborhood $V_\varepsilon(L)$ of L, there exists a δ-neighborhood $V_\delta(c)$ of c such that if $x \ne c$ is any point in $V_\delta(c) \cap A$, then $f(x)$ belongs $V_\varepsilon(L)$.*

We now give some examples that illustrate how the definition of limit is applied.

4.1.7 Examples   **(a)** $\displaystyle \lim_{x \to c} b = b$.

To be more explicit, let $f(x) := b$ for all $x \in \mathbb{R}$. We want to show that $\displaystyle \lim_{x \to c} f(x) = b$. If $\varepsilon > 0$ is given, we let $\delta := 1$. (In fact, any strictly positive δ will serve the purpose.) Then if $0 < |x - c| < 1$, we have $|f(x) - b| = |b - b| = 0 < \varepsilon$. Since $\varepsilon > 0$ is arbitrary, we conclude from Definition 4.1.4 that $\displaystyle \lim_{x \to c} f(x) = b$.

(b) $\lim_{x \to c} x = c$.

Let $g(x) := x$ for all $x \in \mathbb{R}$. If $\varepsilon > 0$, we choose $\delta(\varepsilon) := \varepsilon$. Then if $0 < |x - c| < \delta(\varepsilon)$, we have $|g(x) - c| = |x - c| < \varepsilon$. Since $\varepsilon > 0$ is arbitrary, we deduce that $\lim_{x \to c} g = c$.

(c) $\lim_{x \to c} x^2 = c^2$.

Let $h(x) := x^2$ for all $x \in \mathbb{R}$. We want to make the difference

$$\left| h(x) - c^2 \right| = \left| x^2 - c^2 \right|$$

less than a preassigned $\varepsilon > 0$ by taking x sufficiently close to c. To do so, we note that $x^2 - c^2 = (x + c)(x - c)$. Moreover, if $|x - c| < 1$, then

$$|x| \le |c| + 1 \qquad \text{so that} \qquad |x + c| \le |x| + |c| \le 2|c| + 1.$$

Therefore, if $|x - c| < 1$, we have

$$(1) \qquad \left| x^2 - c^2 \right| = |x + c| \, |x - c| \le (2|c| + 1) \, |x - c|.$$

Moreover this last term will be less than ε provided we take $|x - c| < \varepsilon/(2|c| + 1)$. Consequently, if we choose

$$\delta(\varepsilon) := \inf \left\{ 1, \frac{\varepsilon}{2|c| + 1} \right\},$$

then if $0 < |x - c| < \delta(\varepsilon)$, it will follow first that $|x - c| < 1$ so that (1) is valid, and therefore, since $|x - c| < \varepsilon/(2|c| + 1)$ that

$$\left| x^2 - c^2 \right| \le (2|c| + 1) \, |x - c| < \varepsilon.$$

Since we have a way of choosing $\delta(\varepsilon) > 0$ for an arbitrary choice of $\varepsilon > 0$, we infer that $\lim_{x \to c} h(x) = \lim_{x \to c} x^2 = c^2$.

(d) $\lim_{x \to c} \dfrac{1}{x} = \dfrac{1}{c}$ if $c > 0$.

Let $\varphi(x) := 1/x$ for $x > 0$ and let $c > 0$. To show that $\lim_{x \to c} \varphi = 1/c$ we wish to make the difference

$$\left| \varphi(x) - \frac{1}{c} \right| = \left| \frac{1}{x} - \frac{1}{c} \right|$$

less than a preassigned $\varepsilon > 0$ by taking x sufficiently close to $c > 0$. We first note that

$$\left| \frac{1}{x} - \frac{1}{c} \right| = \left| \frac{1}{cx}(c - x) \right| = \frac{1}{cx} |x - c|$$

for $x > 0$. It is useful to get an upper bound for the term $1/(cx)$ that holds in some neighborhood of c. In particular, if $|x - c| < \frac{1}{2}c$, then $\frac{1}{2}c < x < \frac{3}{2}c$ (why?), so that

$$0 < \frac{1}{cx} < \frac{2}{c^2} \qquad \text{for} \quad |x - c| < \tfrac{1}{2}c.$$

Therefore, for these values of x we have

$$(2) \qquad \left| \varphi(x) - \frac{1}{c} \right| \le \frac{2}{c^2} |x - c|.$$

In order to make this last term less than ε it suffices to take $|x - c| < \frac{1}{2}c^2\varepsilon$. Consequently, if we choose

$$\delta(\varepsilon) := \inf \left\{ \tfrac{1}{2}c, \tfrac{1}{2}c^2\varepsilon \right\},$$

then if $0 < |x - c| < \delta(\varepsilon)$, it will follow first that $|x - c| < \frac{1}{2}c$ so that (2) is valid, and therefore, since $|x - c| < \left(\frac{1}{2}c^2\right)\varepsilon$, that

$$\left|\varphi(x) - \frac{1}{c}\right| = \left|\frac{1}{x} - \frac{1}{c}\right| < \varepsilon.$$

Since we have a way of choosing $\delta(\varepsilon) > 0$ for an arbitrary choice of $\varepsilon > 0$, we infer that $\lim_{x \to c} \varphi = 1/c$.

(e) $\lim_{x \to 2} \dfrac{x^3 - 4}{x^2 + 1} = \dfrac{4}{5}.$

Let $\psi(x) := (x^3 - 4)/(x^2 + 1)$ for $x \in \mathbb{R}$. Then a little algebraic manipulation gives us

$$\left|\psi(x) - \frac{4}{5}\right| = \frac{|5x^3 - 4x^2 - 24|}{5(x^2 + 1)}$$
$$= \frac{|5x^2 + 6x + 12|}{5(x^2 + 1)} \cdot |x - 2|.$$

To get a bound on the coefficient of $|x - 2|$, we restrict x by the condition $1 < x < 3$. For x in this interval, we have $5x^2 + 6x + 12 \le 5 \cdot 3^2 + 6 \cdot 3 + 12 = 75$ and $5(x^2 + 1) \ge 5(1 + 1) = 10$, so that

$$\left|\psi(x) - \frac{4}{5}\right| \le \frac{75}{10}|x - 2| = \frac{15}{2}|x - 2|.$$

Now for given $\varepsilon > 0$, we choose

$$\delta(\varepsilon) := \inf\left\{1, \tfrac{2}{15}\varepsilon\right\}.$$

Then if $0 < |x - 2| < \delta(\varepsilon)$, we have $|\psi(x) - (4/5)| \le (15/2)|x - 2| < \varepsilon$. Since $\varepsilon > 0$ is arbitrary, the assertion is proved. $\square$

Sequential Criterion for Limits

The following important formulation of limit of a function is in terms of limits of sequences. This characterization permits the theory of Chapter 3 to be applied to the study of limits of functions.

4.1.8 Theorem (*Sequential Criterion*) *Let $f : A \to \mathbb{R}$ and let c be a cluster point of A. Then the following are equivalent.*

(i) $\lim_{x \to c} f = L.$

(ii) *For every sequence (x_n) in A that converges to c such that $x_n \ne c$ for all $n \in \mathbb{N}$, the sequence $\left(f(x_n)\right)$ converges to L.*

Proof. (i) $\Rightarrow$ (ii). Assume f has limit L at c, and suppose (x_n) is a sequence in A with $\lim(x_n) = c$ and $x_n \ne c$ for all n. We must prove that the sequence $\left(f(x_n)\right)$ converges to L. Let $\varepsilon > 0$ be given. Then by Definition 4.1.4, there exists $\delta > 0$ such that if $x \in A$ satisfies $0 < |x - c| < \delta$, then $f(x)$ satisfies $|f(x) - L| < \varepsilon$. We now apply the definition of convergent sequence for the given δ to obtain a natural number $K(\delta)$ such that if $n > K(\delta)$ then $|x_n - c| < \delta$. But for each such x_n we have $|f(x_n) - L| < \varepsilon$. Thus if $n > K(\delta)$, then $|f(x_n) - L| < \varepsilon$. Therefore, the sequence $\left(f(x_n)\right)$ converges to L.

(ii) $\Rightarrow$ (i). [The proof is a contrapositive argument.] If (i) is not true, then there exists an ε_0-neighborhood $V_{\varepsilon_0}(L)$ such that no matter what δ-neighborhood of c we pick, there will be at least one number x_δ in $A \cap V_\delta(c)$ with $x_\delta \neq c$ such that $f(x_\delta) \notin V_{\varepsilon_0}(L)$. Hence for every $n \in \mathbb{N}$, the $(1/n)$-neighborhood of c contains a number x_n such that

$$0 < |x_n - c| < 1/n \qquad \text{and} \qquad x_n \in A,$$

but such that

$$|f(x_n) - L| \geq \varepsilon_0 \qquad \text{for all} \quad n \in \mathbb{N}.$$

We conclude that the sequence (x_n) in $A\backslash\{c\}$ converges to c, but the sequence $(f(x_n))$ does not converge to L. Therefore we have shown that if (i) is not true, then (ii) is not true. We conclude that (ii) implies (i).   Q.E.D.

We shall see in the next section that many of the basic limit properties of functions can be established by using corresponding properties for convergent sequences. For example, we know from our work with sequences that if (x_n) is any sequence that converges to a number c, then (x_n^2) converges to c^2. Therefore, by the sequential criterion, we can conclude that the function $h(x) := x^2$ has limit $\lim_{x \to c} h(x) = c^2$.

Divergence Criteria

It is often important to be able to show (i) that a certain number is *not* the limit of a function at a point, or (ii) that the function *does not have* a limit at a point. The following result is a consequence of (the proof of) Theorem 4.1.8. We leave the details of its proof as an important exercise.

4.1.9 Divergence Criteria   *Let $A \subseteq \mathbb{R}$, let $f: A \to \mathbb{R}$ and let $c \in \mathbb{R}$ be a cluster point of A.*
(a)   *If $L \in \mathbb{R}$, then f does **not** have limit L at c if and only if there exists a sequence (x_n) in A with $x_n \neq c$ for all $n \in \mathbb{N}$ such that the sequence (x_n) converges to c but the sequence $(f(x_n))$ does **not** converge to L.*
(b)   *The function f does **not** have a limit at c if and only if there exists a sequence (x_n) in A with $x_n \neq c$ for all $n \in \mathbb{N}$ such that the sequence (x_n) converges to c but the sequence $(f(x_n))$ does **not** converge in $\mathbb{R}$.*

We now give some applications of this result to show how it can be used.

4.1.10 Example   **(a)** $\lim\limits_{x \to 0}(1/x)$ does not exist in $\mathbb{R}$.

As in Example 4.1.7(d), let $\varphi(x) := 1/x$ for $x > 0$. However, here we consider $c = 0$. The argument given in Example 4.1.7(d) breaks down if $c = 0$ since we cannot obtain a bound such as that in (2) of that example. Indeed, if we take the sequence (x_n) with $x_n := 1/n$ for $n \in \mathbb{N}$, then $\lim(x_n) = 0$, but $\varphi(x_n) = 1/(1/n) = n$. As we know, the sequence $(\varphi(x_n)) = (n)$ is not convergent in $\mathbb{R}$, since it is not bounded. Hence, by Theorem 4.1.9(b), $\lim\limits_{x \to 0}(1/x)$ does not exist in $\mathbb{R}$.
(b) $\lim\limits_{x \to 0} \text{sgn}(x)$ does not exist.

Let the **signum function** sgn be defined by

$$\text{sgn}(x) := \begin{cases} +1 & \text{for} \quad x > 0, \\ 0 & \text{for} \quad x = 0, \\ -1 & \text{for} \quad x < 0. \end{cases}$$

Note that $\text{sgn}(x) = x/|x|$ for $x \neq 0$. (See Figure 4.1.2.) We shall show that sgn does not have a limit at $x = 0$. We shall do this by showing that there is a sequence (x_n) such that $\lim(x_n) = 0$, but such that $(\text{sgn}(x_n))$ does not converge.

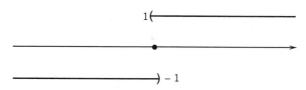

Figure 4.1.2   The signum function.

Indeed, let $x_n := (-1)^n/n$ for $n \in \mathbb{N}$ so that $\lim(x_n) = 0$. However, since

$$\text{sgn}(x_n) = (-1)^n \qquad \text{for} \quad n \in \mathbb{N},$$

it follows from Example 3.4.6(a) that $(\text{sgn}(x_n))$ does not converge. Therefore $\lim_{x \to 0} \text{sgn}(x)$ does not exist.

(c) † $\lim_{x \to 0} \sin(1/x)$ does not exist in $\mathbb{R}$.

Let $g(x) := \sin(1/x)$ for $x \neq 0$. (See Figure 4.1.3.) We shall show that g does not have a limit at $c = 0$, by exhibiting two sequences (x_n) and (y_n) with $x_n \neq 0$ and $y_n \neq 0$ for all $n \in \mathbb{N}$ and such that $\lim(x_n) = 0$ and $\lim(y_n) = 0$, but such that $\lim(g(x_n)) \neq \lim(g(y_n))$. In view of Theorem 4.1.9 this implies that $\lim_{x \to 0} g$ cannot exist. (Explain why.)

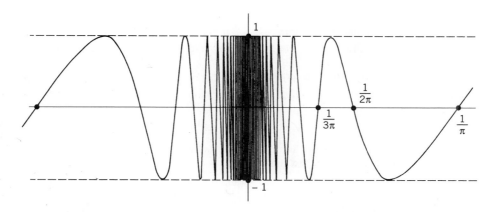

Figure 4.1.3   The function $g(x) = \sin(1/x)$ $(x \neq 0)$.

Indeed, we recall from calculus that $\sin t = 0$ if $t = n\pi$ for $n \in \mathbb{Z}$, and that $\sin t = +1$ if $t = \frac{1}{2}\pi + 2\pi n$ for $n \in \mathbb{Z}$. Now let $x_n := 1/n\pi$ for $n \in \mathbb{N}$; then $\lim(x_n) = 0$ and $g(x_n) = \sin n\pi = 0$ for all $n \in \mathbb{N}$, so that $\lim(g(x_n)) = 0$. On the other hand, let $y_n := \left(\frac{1}{2}\pi + 2\pi n\right)^{-1}$ for $n \in \mathbb{N}$; then $\lim(y_n) = 0$ and $g(y_n) = \sin\left(\frac{1}{2}\pi + 2\pi n\right) = 1$ for all $n \in \mathbb{N}$, so that $\lim(g(y_n)) = 1$. We conclude that $\lim_{x \to 0} \sin(1/x)$ does not exist. $\qquad \square$

---

†In order to have some interesting applications in this and later examples, we shall make use of well-known properties of trigonometric and exponential functions that will be established in Chapter 8.

$$\frac{2x+1/x}{2+2x} = \frac{3x+1}{2+2x}$$

Exercises for Section 4.1

1. Determine a condition on $|x - 1|$ that will assure that:
   (a) $|x^2 - 1| < \frac{1}{2}$,
   (b) $|x^2 - 1| < 1/10^{-3}$,
   (c) $|x^2 - 1| < 1/n$ for a given $n \in \mathbb{N}$,
   (d) $|x^3 - 1| < 1/n$ for a given $n \in \mathbb{N}$.

2. Determine a condition on $|x - 4|$ that will assure that:
   (a) $|\sqrt{x} - 2| < \frac{1}{2}$,
   (b) $|\sqrt{x} - 2| < 10^{-2}$.

3. Let c be a cluster point of $A \subseteq \mathbb{R}$ and let $f: A \to \mathbb{R}$. Prove that $\lim_{x \to c} f(x) = L$ if and only if $\lim_{x \to c} |f(x) - L| = 0$.

4. Let $f : \mathbb{R} \to \mathbb{R}$ and let $c \in \mathbb{R}$. Show that $\lim_{x \to c} f(x) = L$ if and only if $\lim_{x \to 0} f(x + c) = L$.

5. Let $I := (0, a)$ where $a > 0$, and let $g(x) := x^2$ for $x \in I$. For any points $x, c \in I$, show that $|g(x) - c^2| \le 2a|x - c|$. Use this inequality to prove that $\lim_{x \to c} x^2 = c^2$ for any $c \in I$.

6. Let I be an interval in $\mathbb{R}$, let $f : I \to \mathbb{R}$, and let $c \in I$. Suppose there exist constants K and L such that $|f(x) - L| \le K|x - c|$ for $x \in I$. Show that $\lim_{x \to c} f(x) = L$.

7. Show that $\lim_{x \to c} x^3 = c^3$ for any $c \in \mathbb{R}$.

8. Show that $\lim_{x \to c} \sqrt{x} = \sqrt{c}$ for any $c > 0$.

9. Use either the ε-δ definition of limit or the Sequential Criterion for limits, to establish the following limits.
   (a) $\lim_{x \to 2} \dfrac{1}{1 - x} = -1$,
   (b) $\lim_{x \to 1} \dfrac{x}{1 + x} = \dfrac{1}{2}$,
   (c) $\lim_{x \to 0} \dfrac{x^2}{|x|} = 0$,
   (d) $\lim_{x \to 1} \dfrac{x^2 - x + 1}{x + 1} = \dfrac{1}{2}$.

10. Use the definition of limit to show that
   (a) $\lim_{x \to 2} (x^2 + 4x) = 12$,
   (b) $\lim_{x \to -1} \dfrac{x + 5}{2x + 3} = 4$.

11. Show that the following limits do *not* exist.
   (a) $\lim_{x \to 0} \dfrac{1}{x^2}$ $(x > 0)$,
   (b) $\lim_{x \to 0} \dfrac{1}{\sqrt{x}}$ $(x > 0)$,
   (c) $\lim_{x \to 0} (x + \operatorname{sgn}(x))$,
   (d) $\lim_{x \to 0} \sin(1/x^2)$.

12. Suppose the function $f : \mathbb{R} \to \mathbb{R}$ has limit L at 0, and let $a > 0$. If $g : \mathbb{R} \to \mathbb{R}$ is defined by $g(x) := f(ax)$ for $x \in \mathbb{R}$, show that $\lim_{x \to 0} g(x) = L$.

13. Let $c \in \mathbb{R}$ and let $f : \mathbb{R} \to \mathbb{R}$ be such that $\lim_{x \to c} \left(f(x) \right)^2 = L$.
   (a) Show that if $L = 0$, then $\lim_{x \to c} f(x) = 0$.
   (b) Show by example that if $L \ne 0$, then f may not have a limit at c.

14. Let $f : \mathbb{R} \to \mathbb{R}$ be defined by setting $f(x) := x$ if x is rational, and $f(x) = 0$ if x is irrational.
   (a) Show that f has a limit at $x = 0$.
   (b) Use a sequential argument to show that if $c \ne 0$, then f does not have a limit at c.

15. Let $f : \mathbb{R} \to \mathbb{R}$, let I be an *open* interval in $\mathbb{R}$, and let $c \in I$. If f_1 is the restriction of f to I, show that f_1 has a limit at c if and only if f has a limit at c, and that the limits are equal.

16. Let $f : \mathbb{R} \to \mathbb{R}$, let J be a *closed* interval in $\mathbb{R}$, and let $c \in J$. If f_2 is the restriction of f to J, show that if f has a limit at c then f_2 has a limit at c. Show by example that it does *not* follow that if f_2 has a limit at c, then f has a limit at c.

Section 4.2   Limit Theorems

We shall now obtain results that are useful in calculating limits of functions. These results are parallel to the limit theorems established in Section 3.2 for sequences. In fact, in most cases these results can be proved by using Theorem 4.1.8 and results from Section 3.2. Alternatively, the results in this section can be proved by using ε-δ arguments that are very similar to the ones employed in Section 3.2.

4.2.1 Definition   Let $A \subseteq \mathbb{R}$, let $f: A \to \mathbb{R}$, and let $c \in \mathbb{R}$ be a cluster point of A. We say that f is **bounded on a neighborhood of** c if there exists a δ-neighborhood $V_\delta(c)$ of c and a constant $M > 0$ such that we have $|f(x)| \le M$ for all $x \in A \cap V_\delta(c)$.

4.2.2 Theorem   *If $A \subseteq \mathbb{R}$ and $f: A \to \mathbb{R}$ has a limit at $c \in \mathbb{R}$, then f is bounded on some neighborhood of c.*

Proof.   If $L := \lim_{x \to c} f$, then for $\varepsilon = 1$, there exists $\delta > 0$ such that if $0 < |x - c| < \delta$, then $|f(x) - L| < 1$; hence (by Corollary 2.2.4(a)),

$$|f(x)| - |L| \le |f(x) - L| < 1.$$

Therefore, if $x \in A \cap V_\delta(c), x \ne c$, then $|f(x)| \le |L| + 1$. If $c \notin A$, we take $M = |L| + 1$, while if $c \in A$ we take $M := \sup\{|f(c)|, |L| + 1\}\}$. It follows that if $x \in A \cap V_\delta(c)$, then $|f(x)| \le M$. This shows that f is bounded on the neighborhood $V_\delta(c)$ of c.   Q.E.D.

The next definition is similar to the definition for sums, differences, products, and quotients of sequences given in Section 3.2.

4.2.3 Definition   Let $A \subseteq \mathbb{R}$ and let f and g be functions defined on A to $\mathbb{R}$. We define the **sum** $f + g$, the **difference** $f - g$, and the **product** fg on A to $\mathbb{R}$ to be the functions given by

$$(f + g)(x) := f(x) + g(x), \qquad (f - g)(x) := f(x) - g(x),$$
$$(fg)(x) := f(x)g(x) \qquad\qquad ,$$

for all $x \in A$. Further, if $b \in \mathbb{R}$, we define the **multiple** bf to be the function given by

$$(bf)(x) := bf(x) \qquad \text{for all}\quad x \in A.$$

Finally, if $h(x) \ne 0$ for $x \in A$, we define the **quotient** f/h to be the function given by

$$\left(\frac{f}{h}\right)(x) := \frac{f(x)}{h(x)} \qquad \text{for all}\quad x \in A.$$

4.2.4 Theorem   *Let $A \subseteq \mathbb{R}$, let f and g be functions on A to $\mathbb{R}$, and let $c \in \mathbb{R}$ be a cluster point of A. Further, let $b \in \mathbb{R}$.*

(a)   *If $\lim_{x \to c} f = L$ and $\lim_{x \to c} g = M$, then:*

$$\lim_{x \to c}(f + g) = L + M, \qquad \lim_{x \to c}(f - g) = L - M,$$
$$\lim_{x \to c}(fg) = LM, \qquad \lim_{x \to c}(bf) = bL.$$

(b) *If* $h: A \to \mathbb{R}$, *if* $h(x) \neq 0$ *for all* $x \in A$, *and if* $\lim_{x \to c} h = H \neq 0$, *then*

$$\lim_{x \to c} \left(\frac{f}{h} \right) = \frac{L}{H}.$$

Proof.   One proof of this theorem is exactly similar to that of Theorem 3.2.3. Alternatively, it can be proved by making use of Theorems 3.2.3 and 4.1.8. For example, let (x_n) be any sequence in A such that $x_n \neq c$ for $n \in \mathbb{N}$, and $c = \lim(x_n)$. It follows from Theorem 4.1.8 that

$$\lim \left(f(x_n) \right) = L, \qquad \lim \left(g(x_n) \right) = M.$$

On the other hand, Definition 4.2.3 implies that

$$(fg)(x_n) = f(x_n)g(x_n) \qquad \text{for} \quad n \in \mathbb{N}.$$

Therefore an application of Theorem 3.2.3 yields

$$\lim \left((fg)(x_n) \right) = \lim \left(f(x_n)g(x_n) \right)$$
$$= \left[\lim \left(f(x_n) \right) \right] \left[\lim \left(g(x_n) \right) \right] = LM.$$

Consequently, it follows from Theorem 4.1.8 that

$$\lim_{x \to c}(fg) = \lim \left((fg)(x_n) \right) = LM.$$

The other parts of this theorem are proved in a similar manner. We leave the details to the reader.   Q.E.D.

Remarks   **(1)** We note that, in part (b), the additional assumption that $H = \lim_{x \to c} h \neq 0$ is made. If this assumption is not satisfied, then the limit

$$\lim_{x \to c} \frac{f(x)}{h(x)}$$

may or may not exist. But even if this limit does exist, we *cannot* use Theorem 4.2.4(b) to evaluate it.

(2) Let $A \subseteq \mathbb{R}$, and let $f_1, f_2, \cdots, f_n$ be functions on A to $\mathbb{R}$, and let c be a cluster point of A. If

$$L_k := \lim_{x \to c} f_k \qquad \text{for} \quad k = 1, \cdots, n,$$

then it follows from Theorem 4.2.4 by an Induction argument that

$$L_1 + L_2 + \cdots + L_n = \lim_{x \to c}(f_1 + f_2 + \cdots + f_n),$$

and

$$L_1 \cdot L_2 \cdots L_n = \lim(f_1 \cdot f_2 \cdots f_n).$$

In particular, we deduce that if $L = \lim_{x \to c} f$ and $n \in \mathbb{N}$, then

$$L^n = \lim_{x \to c} \left(f(x) \right)^n.$$

4.2.5 Examples (a) Some of the limits that were established in Section 4.1 can be proved by using Theorem 4.2.4. For example, it follows from this result that since $\lim_{x \to c} x = c$, then $\lim_{x \to c} x^2 = c^2$, and that if $c > 0$, then

$$\lim_{x \to c} \frac{1}{x} = \frac{1}{\lim_{x \to c} x} = \frac{1}{c}.$$

(b) $\lim_{x \to 2} (x^2 + 1)(x^3 - 4) = 20.$

It follows from Theorem 4.2.4 that

$$\lim_{x \to 2} (x^2 + 1)(x^3 - 4) = \left(\lim_{x \to 2} (x^2 + 1) \right) \left(\lim_{x \to 2} (x^3 - 4) \right)$$
$$= 5 \cdot 4 = 20.$$

(c) $\lim_{x \to 2} \left(\frac{x^3 - 4}{x^2 + 1} \right) = \frac{4}{5}.$

If we apply Theorem 4.2.4(b), we have

$$\lim_{x \to 2} \frac{x^3 - 4}{x^2 + 1} = \frac{\lim_{x \to 2} (x^3 - 4)}{\lim_{x \to 2} (x^2 + 1)} = \frac{4}{5}.$$

Note that since the limit in the denominator [i.e., $\lim_{x \to 2} (x^2 + 1) = 5$] is not equal to 0, then Theorem 4.2.4(b) is applicable.

(d) $\lim_{x \to 2} = \frac{x^2 - 4}{3x - 6} = \frac{4}{3}.$

If we let $f(x) := x^2 - 4$ and $h(x) := 3x - 6$ for $x \in \mathbb{R}$, then we *cannot* use Theorem 4.2.4(b) to evaluate $\lim_{x \to 2} (f(x)/h(x))$ because

$$H = \lim_{x \to 2} h(x) = \lim_{x \to 2} (3x - 6)$$
$$= 3 \lim_{x \to 2} x - 6 = 3 \cdot 2 - 6 = 0.$$

However, if $x \neq 2$, then it follows that

$$\frac{x^2 - 4}{3x - 6} = \frac{(x + 2)(x - 2)}{3(x - 2)} = \frac{1}{3}(x + 2).$$

Therefore we have

$$\lim_{x \to 2} \frac{x^2 - 4}{3x - 6} = \lim_{x \to 2} \frac{1}{3} (x + 2) = \frac{1}{3} \left(\lim_{x \to 2} x + 2 \right) = \frac{4}{3}.$$

Note that the function $g(x) = (x^2 - 4)/(3x - 6)$ has a limit at $x = 2$ *even though it is not defined there.*

(e) $\lim_{x \to 0} \frac{1}{x}$ does not exist in $\mathbb{R}$.

Of course $\lim_{x \to 0} 1 = 1$ and $H := \lim_{x \to 0} x = 0$. However, since $H = 0$, we *cannot* use Theorem 4.2.4(b) to evaluate $\lim_{x \to 0} (1/x)$. In fact, as was seen in Example 4.1.10(a), the function $\varphi(x) = 1/x$ does not have a limit at $x = 0$. This conclusion also follows from Theorem 4.2.2 since the function $\varphi(x) = 1/x$ is not bounded on a neighborhood of $x = 0$. (Why?)

(f)   If p is a polynomial function, then $\lim_{x \to c} p(x) = p(c)$.

Let p be a polynomial function on $\mathbb{R}$ so that $p(x) = a_n x^n + a_{n-1} x^{n-1} + \cdots + a_1 x + a_0$ for all $x \in \mathbb{R}$. It follows from Theorem 4.2.4 and the fact that $\lim_{x \to c} x^k = c^k$, that

$$
\begin{aligned}
\lim_{x \to c} p(x) &= \lim_{x \to c} \left[a_n x^n + a_{n-1} x^{n-1} + \cdots + a_1 x + a_0 \right] \\
&= \lim_{x \to c} (a_n x^n) + \lim_{x \to c} (a_{n-1} x^{n-1}) + \cdots + \lim_{x \to c} (a_1 x) + \lim_{x \to c} a_0 \\
&= a_n c^n + a_{n-1} c^{n-1} + \cdots + a_1 c + a_0 \\
&= p(c).
\end{aligned}
$$

Hence $\lim_{x \to c} p(x) = p(c)$ for any polynomial function p.

(g)   If p and q are polynomial functions on $\mathbb{R}$ and if $q(c) \neq 0$, then

$$
\lim_{x \to c} \frac{p(x)}{q(x)} = \frac{p(c)}{q(c)}.
$$

Since $q(x)$ is a polynomial function, it follows from a theorem in algebra that there are at most a finite number of real numbers $\alpha_1, \cdots, \alpha_m$ [the real zeroes of $q(x)$] such that $q(\alpha_j) = 0$ and such that if $x \notin \{\alpha_1, \cdots, \alpha_m\}$, then $q(x) \neq 0$. Hence, if $x \notin \{\alpha_1, \cdots, \alpha_m\}$, we can define

$$
r(x) := \frac{p(x)}{q(x)}.
$$

If c is not a zero of $q(x)$, then $q(c) \neq 0$, and it follows from part (f) that $\lim_{x \to c} q(x) = q(c) \neq 0$. Therefore we can apply Theorem 4.2.4(b) to conclude that

$$
\lim_{x \to c} \frac{p(x)}{q(x)} = \frac{\lim_{x \to c} p(x)}{\lim_{x \to c} q(x)} = \frac{p(c)}{q(c)}. \qquad \square
$$

The next result is a direct analogue of Theorem 3.2.6.

4.2.6 Theorem   *Let $A \subseteq \mathbb{R}$, let $f: A \to \mathbb{R}$ and let $c \in \mathbb{R}$ be a cluster point of A. If*

$$
a \le f(x) \le b \qquad \text{for all} \quad x \in A, x \neq c,
$$

and if $\lim_{x \to c} f$ exists, then $a \le \lim_{x \to c} f \le b$.

Proof.   Indeed, if $L = \lim_{x \to c} f$, then it follows from Theorem 4.1.8 that if (x_n) is any sequence of real numbers such that $c \neq x_n \in A$ for all $n \in \mathbb{N}$ and if the sequence (x_n) converges to c, then the sequence $(f(x_n))$ converges to L. Since $a \le f(x_n) \le b$ for all $n \in \mathbb{N}$, it follows from Theorem 3.2.6 that $a \le L \le b$.   Q.E.D.

We now state an analogue of the Squeeze Theorem 3.2.7. We leave its proof to the reader.

4.2.7 Squeeze Theorem   *Let $A \subseteq \mathbb{R}$, let $f, g, h: A \to \mathbb{R}$, and let $c \in \mathbb{R}$ be a cluster point of A. If*

$$
f(x) \le g(x) \le h(x) \qquad \text{for all} \quad x \in A, x \neq c,
$$

and if $\lim_{x \to c} f = L = \lim_{x \to c} h$, then $\lim_{x \to c} g = L$.

4.2.8 Examples   (a) $\lim\limits_{x \to 0} x^{3/2} = 0 \ (x > 0).$

Let $f(x) := x^{3/2}$ for $x > 0$. Since the inequality $x < x^{1/2} \le 1$ holds for $0 < x \le 1$ (why?), it follows that $x^2 \le f(x) = x^{3/2} \le x$ for $0 < x \le 1$. Since

$$\lim_{x \to 0} x^2 = 0 \qquad \text{and} \qquad \lim_{x \to 0} x = 0,$$

it follows from the Squeeze Theorem 4.2.7 that $\lim\limits_{x \to 0} x^{3/2} = 0$.

(b) $\lim\limits_{x \to 0} \sin x = 0.$

It will be proved later (see Theorem 8.4.8), that

$$-x \le \sin x \le x \qquad \text{for all} \quad x \ge 0.$$

Since $\lim\limits_{x \to 0} (\pm x) = 0$, it follows from the Squeeze Theorem that $\lim\limits_{x \to 0} \sin x = 0$.

(c) $\lim\limits_{x \to 0} \cos x = 1.$

It will be proved later (see Theorem 8.4.8) that

(1) $$1 - \tfrac{1}{2}x^2 \le \cos x \le 1 \qquad \text{for all} \quad x \in \mathbb{R}.$$

Since $\lim\limits_{x \to 0} \left(1 - \tfrac{1}{2}x^2\right) = 1$, it follows from the Squeeze Theorem that $\lim\limits_{x \to 0} \cos x = 1$.

(d) $\lim\limits_{x \to 0} \left(\dfrac{\cos x - 1}{x}\right) = 0.$

We cannot use Theorem 4.2.4(b) to evaluate this limit. (Why not?) However, it follows from the inequality (1) in part (c) that

$$-\tfrac{1}{2}x \le (\cos x - 1)/x \le 0 \qquad \text{for} \quad x > 0$$

and that

$$0 \le (\cos x - 1)/x \le -\tfrac{1}{2}x \qquad \text{for} \quad x < 0.$$

Now let $f(x) := -x/2$ for $x \ge 0$ and $f(x) := 0$ for $x < 0$, and let $h(x) := 0$ for $x \ge 0$ and $h(x) := -x/2$ for $x < 0$. Then we have

$$f(x) \le (\cos x - 1)/x \le h(x) \qquad \text{for} \quad x \ne 0.$$

Since it is readily seen that $\lim\limits_{x \to 0} f = 0 = \lim\limits_{x \to 0} h$, it follows from the Squeeze Theorem that $\lim\limits_{x \to 0} (\cos x - 1)/x = 0$.

(e) $\lim\limits_{x \to 0} \left(\dfrac{\sin x}{x}\right) = 1.$

Again we cannot use Theorem 4.2.4(b) to evaluate this limit. However, it will be proved later (see Theorem 8.4.8) that

$$x - \tfrac{1}{6}x^3 \le \sin x \le x \qquad \text{for} \quad x \ge 0$$

and that

$$x \le \sin x \le x - \tfrac{1}{6}x^3 \qquad \text{for} \quad x \le 0.$$

Therefore it follows (why?) that

$$1 - \tfrac{1}{6}x^2 \le (\sin x)/x \le 1 \qquad \text{for all} \quad x \ne 0.$$

But since $\lim\limits_{x \to 0} (1 - \tfrac{1}{6}x^2) = 1 - \tfrac{1}{6} \cdot \lim\limits_{x \to 0} x^2 = 1$, we infer from the Squeeze Theorem that $\lim\limits_{x \to 0} (\sin x)/x = 1$.

(f) $\lim_{x \to 0} (x \sin(1/x)) = 0$.

Let $f(x) = x \sin(1/x)$ for $x \neq 0$. Since $-1 \leq \sin z \leq 1$ for all $z \in \mathbb{R}$, we have the inequality

$$- |x| \leq f(x) = x \sin(1/x) \leq |x|$$

for all $x \in \mathbb{R}$, $x \neq 0$. Since $\lim_{x \to 0} |x| = 0$, it follows from the Squeeze Theorem that $\lim_{x \to 0} f = 0$. For a graph, see Figure 5.1.3 or the cover of this book.   □

There are results that are parallel to Theorems 3.2.9 and 3.2.10; however, we will leave them as exercises. We conclude this section with a result that is, in some sense, a partial converse to Theorem 4.2.6.

4.2.9 Theorem   Let $A \subseteq \mathbb{R}$, let $f : A \to \mathbb{R}$ and let $c \in \mathbb{R}$ be a cluster point of A. If

$$\lim_{x \to c} f > 0 \qquad \left[\textit{respectively, } \lim_{x \to c} f < 0 \right],$$

then there exists a neighborhood $V_\delta(c)$ of c such that $f(x) > 0$ [respectively, $f(x) < 0$] for all $x \in A \cap V_\delta(c)$, $x \neq c$.

Proof.   Let $L := \lim_{x \to c} f$ and suppose that $L > 0$. We take $\varepsilon = \frac{1}{2}L > 0$ in Definition 4.1.4, and obtain a number $\delta > 0$ such that if $0 < |x - c| < \delta$ and $x \in A$, then $|f(x) - L| < \frac{1}{2}L$. Therefore (why?) it follows that if $x \in A \cap V_\delta(c)$, $x \neq c$, then $f(x) > \frac{1}{2}L > 0$.

If $L < 0$, a similar argument applies.   Q.E.D.

Exercises for Section 4.2

1.   Apply Theorem 4.2.4 to determine the following limits:

(a) $\lim_{x \to 1} (x + 1) (2x + 3) \quad (x \in \mathbb{R})$,

(b) $\lim_{x \to 1} \dfrac{x^2 + 2}{x^2 - 2} \quad (x > 0)$,

(c) $\lim_{x \to 2} \left(\dfrac{1}{x + 1} - \dfrac{1}{2x} \right) \quad (x > 0)$,

(d) $\lim_{x \to 0} \dfrac{x + 1}{x^2 + 2} \quad (x \in \mathbb{R})$.

2.   Determine the following limits and state which theorems are used in each case. (You may wish to use Exercise 14 below.)

(a) $\lim_{x \to 2} \sqrt{\dfrac{2x + 1}{x + 3}} \quad (x > 0)$,

(b) $\lim_{x \to 2} \dfrac{x^2 - 4}{x - 2} \quad (x > 0)$,

(c) $\lim_{x \to 0} \dfrac{(x + 1)^2 - 1}{x} \quad (x > 0)$,

(d) $\lim_{x \to 1} \dfrac{\sqrt{x} - 1}{x - 1} \quad (x > 0)$.

3.   Find $\lim_{x \to 0} \dfrac{\sqrt{1 + 2x} - \sqrt{1 + 3x}}{x + 2x^2}$ where $x > 0$.

4.   Prove that $\lim_{x \to 0} \cos(1/x)$ does not exist but that $\lim_{x \to 0} x \cos(1/x) = 0$.

5.   Let f, g be defined on $A \subseteq \mathbb{R}$ to $\mathbb{R}$, and let c be a cluster point of A. Suppose that f is bounded on a neighborhood of c and that $\lim_{x \to c} g = 0$. Prove that $\lim_{x \to c} fg = 0$.

6.   Use the definition of the limit to prove the first assertion in Theorem 4.2.4(a).

7.   Use the sequential formulation of the limit to prove Theorem 4.2.4(b).

8. Let $n \in \mathbb{N}$ be such that $n \geq 3$. Derive the inequality $-x^2 \leq x^n \leq x^2$ for $-1 < x < 1$. Then use the fact that $\lim_{x \to 0} x^2 = 0$ to show that $\lim_{x \to 0} x^n = 0$.

9. Let f, g be defined on A to $\mathbb{R}$ and let c be a cluster point of A.
    (a) Show that if both $\lim_{x \to c} f$ and $\lim_{x \to c}(f + g)$ exist, then $\lim_{x \to c} g$ exists.

    (b) If $\lim_{x \to c} f$ and $\lim_{x \to c} fg$ exist, does it follow that $\lim_{x \to c} g$ exists?

10. Give examples of functions f and g such that f and g do not have limits at a point c, but such that both $f + g$ and fg have limits at c. $(c=0$

11. Determine whether the following limits exist in $\mathbb{R}$.
    (a) $\lim_{x \to 0} \sin(1/x^2)$ $(x \neq 0)$,
    (b) $\lim_{x \to 0} x \sin(1/x^2)$ $(x \neq 0)$,
    (c) $\lim_{x \to 0} \operatorname{sgn} \sin(1/x)$ $(x \neq 0)$,
    (d) $\lim_{x \to 0} \sqrt{x} \sin(1/x^2)$ $(x > 0)$.

12. Let $f : \mathbb{R} \to \mathbb{R}$ be such that $f(x + y) = f(x) + f(y)$ for all x, y in $\mathbb{R}$. Assume that $\lim_{x \to 0} f = L$ exists. Prove that $L = 0$, and then prove that f has a limit at every point $c \in \mathbb{R}$. [*Hint*: First note that $f(2x) = f(x) + f(x) = 2f(x)$ for $x \in \mathbb{R}$. Also note that $f(x) = f(x - c) + f(c)$ for x, c in $\mathbb{R}$.]

13. Let $A \subseteq \mathbb{R}$, let $f : A \to \mathbb{R}$ and let $c \in \mathbb{R}$ be a cluster point of A. If $\lim_{x \to c} f$ exists, and if $|f|$ denotes the function defined for $x \in A$ by $|f|(x) := |f(x)|$, prove that $\lim_{x \to c} |f| = |\lim_{x \to c} f|$.

14. Let $A \subseteq \mathbb{R}$, let $f : A \to \mathbb{R}$, and let $c \in \mathbb{R}$ be a cluster point of A. In addition, suppose that $f(x) \geq 0$ for all $x \in A$, and let $\sqrt{f}$ be the function defined for $x \in A$ by $\left(\sqrt{f} \right)(x) := \sqrt{f(x)}$. If $\lim_{x \to c} f$ exists, prove that $\lim_{x \to c} \sqrt{f} = \sqrt{\lim_{x \to c} f}$.

Section 4.3   Some Extensions of the Limit Concept[†]

In this section, we shall present three types of extensions of the notion of a limit of a function that often occur. Since all the ideas here are closely parallel to ones we have already encountered, this section can be read easily.

One-sided Limits

There are times when a function f may not possess a limit at a point c, yet a limit does exist when the function is restricted to an interval on one side of the cluster point c.

For example, the signum function considered in Example 4.1.10(b), and illustrated in Figure 4.1.2, has no limit at $c = 0$. However, if we restrict the signum function to the interval $(0, \infty)$, the resulting function has a limit of 1 at $c = 0$. Similarly, if we restrict the signum function to the interval $(-\infty, 0)$, the resulting function has a limit of -1 at $c = 0$. These are elementary examples of right-hand and left-hand limits at $c = 0$.

4.3.1 Definition   Let $A \in \mathbb{R}$ and let $f : A \to \mathbb{R}$.

(i)   If $c \in \mathbb{R}$ is a cluster point of the set $A \cap (c, \infty) = \{x \in A : x > c\}$, then we say that $L \in \mathbb{R}$ is a **right-hand limit of** f **at** c and we write

$$\lim_{x \to c+} f = L \quad \text{or} \quad \lim_{x \to c+} f(x) = L$$

[†]This section can be largely omitted on a first reading of this chapter.

if given any $\varepsilon > 0$ there exists a $\delta = \delta(\varepsilon) > 0$ such that for all $x \in A$ with $0 < x - c < \delta$, then $|f(x) - L| < \varepsilon$.

(ii)   If $c \in \mathbb{R}$ is a cluster point of the set $A \cap (-\infty, c) = \{x \in A : x < c\}$, then we say that $L \in \mathbb{R}$ is a **left-hand limit of** f **at** c and we write

$$\lim_{x \to c-} f = L \qquad \text{or} \qquad \lim_{x \to c-} f(x) = L$$

if given any $\varepsilon > 0$ there exists a $\delta > 0$ such that for all $x \in A$ with $0 < c - x < \delta$, then $|f(x) - L| < \varepsilon$.

Notes   (1)   The limits $\lim_{x \to c+} f$ and $\lim_{x \to c-} f$ are called **one-sided limits of** f **at** c. It is possible that neither one-sided limit may exist. Also, one of them may exist without the other existing. Similarly, as is the case for $f(x) := \text{sgn}(x)$ at $c = 0$, they may both exist and be different.

(2)   If A is an interval with left endpoint c, then it is readily seen that $f : A \to \mathbb{R}$ has a limit at c if and only if it has a right-hand limit at c. Moreover, in this case the limit $\lim_{x \to c} f$ and the right-hand limit $\lim_{x \to c+} f$ are equal. (A similar situation occurs for the left-hand limit when A is an interval with right endpoint c.)

The reader can show that f can have only one right-hand (respectively, left-hand) limit at a point. There are results analogous to those established in Sections 4.1 and 4.2 for two-sided limits. In particular, the existence of one-sided limits can be reduced to sequential considerations.

4.3.2 Theorem   Let $A \subseteq \mathbb{R}$, let $f : A \to \mathbb{R}$, and let $c \in \mathbb{R}$ be a cluster point of $A \cap (c, \infty)$. Then the following statements are equivalent:

(i) $\lim_{x \to c+} f = L$.

(ii)   For every sequence (x_n) that converges to c such that $x_n \in A$ and $x_n > c$ for all $n \in \mathbb{N}$, the sequence $(f(x_n))$ converges to L.

We leave the proof of this result (and the formulation and proof of the analogous result for left-hand limits) to the reader. We will not take the space to write out the formulations of the one-sided version of the other results in Sections 4.1 and 4.2.

The following result relates the notion of the limit of a function to one-sided limits. We leave its proof as an exercise.

4.3.3 Theorem   Let $A \subseteq \mathbb{R}$, let $f : A \to \mathbb{R}$, and let $c \in \mathbb{R}$ be a cluster point of both of the sets $A \cap (c, \infty)$ and $A \cap (-\infty, c)$. Then $\lim_{x \to c} f = L$ if and only if $\lim_{x \to c+} f = L = \lim_{x \to c-} f$.

4.3.4 Examples   **(a)**   Let $f(x) := \text{sgn}(x)$.

We have seen in Example 4.1.10(b) that sgn does not have a limit at 0. It is clear that $\lim_{x \to 0+} \text{sgn}(x) = +1$ and that $\lim_{x \to 0-} \text{sgn}(x) = -1$. Since these one-sided limits are different, it also follows from Theorem 4.3.3 that $\text{sgn}(x)$ does not have a limit at 0.

(b)   Let $g(x) := e^{1/x}$ for $x \neq 0$. (See Figure 4.3.1.)

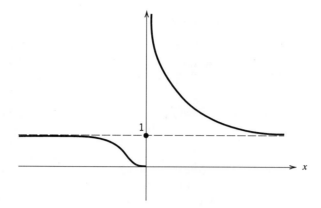

Figure 4.3.1   Graph of $g(x) = e^{1/x}$ $(x \neq 0)$.

We first show that g does not have a finite right-hand limit at $c = 0$ since it is not bounded on any right-hand neighborhood $(0, \delta)$ of 0. We shall make use of the inequality

$$(1) \qquad\qquad 0 < t < e^t \qquad \text{for} \quad t > 0,$$

which will be proved later (see Corollary 8.3.3). It follows from (1) that if $x > 0$, then $0 < 1/x < e^{1/x}$. Hence, if we take $x_n = 1/n$, then $g(x_n) > n$ for all $n \in \mathbb{N}$. Therefore $\lim_{x \to 0+} e^{1/x}$ does not exist in $\mathbb{R}$.

However, $\lim_{x \to 0-} e^{1/x} = 0$. Indeed, if $x < 0$ and we take $t = -1/x$ in (1) we obtain $0 < -1/x < e^{-1/x}$. Since $x < 0$, this implies that $0 < e^{1/x} < -x$ for all $x < 0$. It follows from this inequality that $\lim_{x \to 0-} e^{1/x} = 0$.

(c)   Let $h(x) := 1/(e^{1/x} + 1)$ for $x \neq 0$. (See Figure 4.3.2.)

We have seen in part (b) that $0 < 1/x < e^{1/x}$ for $x > 0$, whence

$$0 < \frac{1}{e^{1/x} + 1} < \frac{1}{e^{1/x}} < x,$$

which implies that $\lim_{x \to 0+} h = 0$.

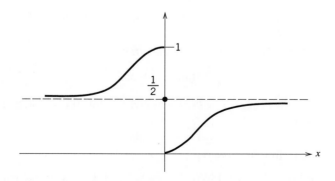

Figure 4.3.2   Graph of $h(x) = 1/(e^{1/x} + 1)$ $(x \neq 0)$.

Since we have seen in part (b) that $\lim\limits_{x \to 0-} e^{1/x} = 0$, it follows from the analogue of Theorem 4.2.4(b) for left-hand limits that

$$\lim_{x \to 0-} \left(\frac{1}{e^{1/x} + 1} \right) = \frac{1}{\lim\limits_{x \to 0-} e^{1/x} + 1} = \frac{1}{0 + 1} = 1.$$

Note that for this function, both one-sided limits exist in $\mathbb{R}$, but they are unequal. $\square$

Infinite Limits

The function $f(x) := 1/x^2$ for $x \neq 0$ (see Figure 4.3.3) is not bounded on a neighborhood of 0, so it cannot have a limit in the sense of Definition 4.1.4. While the symbols ∞ $(= +\infty)$ and $-\infty$ do not represent real numbers, it is sometimes useful to be able to say that "$f(x) = 1/x^2$ tends to ∞ as $x \to 0$". This use of $\pm\infty$ will not cause any difficulties, provided we exercise caution and *never* interpret ∞ or $-\infty$ as being real numbers.

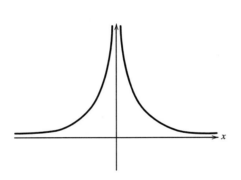

Figure 4.3.3   Graph of
$f(x) = 1/x^2$ $(x \neq 0)$

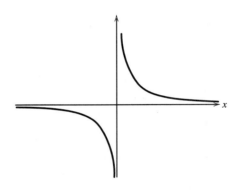

Figure 4.3.4   Graph of
$g(x) = 1/x$ $(x \neq 0)$

4.3.5 Definition   Let $A \subseteq \mathbb{R}$, let $f: A \to \mathbb{R}$, and let $c \in \mathbb{R}$ be a cluster point of A.

(i)   We say that f **tends to ∞ as** $x \to c$, and write

$$\lim_{x \to c} f = \infty,$$

if for every $\alpha \in \mathbb{R}$ there exists $\delta = \delta(\alpha) > 0$ such that for all $x \in A$ with $0 < |x - c| < \delta$, then $f(x) > \alpha$.

(ii)   We say that f **tends to $-\infty$ as** $x \to c$, and write

$$\lim_{x \to c} f = -\infty,$$

if for every $\beta \in \mathbb{R}$ there exists $\delta = \delta(\beta) > 0$ such that for all $x \in A$ with $0 < |x - c| < \delta$, then $f(x) < \beta$.

4.3.6 Examples   (a) $\lim\limits_{x \to 0}(1/x^2) = \infty$.

For, if $\alpha > 0$ is given, let $\delta := 1/\sqrt{\alpha}$. It follows that if $0 < |x| < \delta$, then $x^2 < 1/\alpha$ so that $1/x^2 > \alpha$.

(b)   Let $g(x) := 1/x$ for $x \neq 0$. (See Figure 4.3.4.)

The function g does *not* tend to either ∞ or $-\infty$ as $x \to 0$. For, if $\alpha > 0$ then $g(x) < \alpha$ for all $x < 0$, so that g does not tend to ∞ as $x \to 0$. Similarly, if $\beta < 0$ then $g(x) > \beta$ for all $x > 0$, so that g does not tend to $-\infty$ as $x \to 0$.   □

While many of the results in Sections 4.1 and 4.2 have extensions to this limiting notion, not all of them do since $\pm\infty$ are not real numbers. The following result is an analogue of the Squeeze Theorem 4.2.7. (See also Theorem 3.6.4.)

4.3.7 Theorem   Let $A \subseteq \mathbb{R}$, let $f, g: A \to \mathbb{R}$, and let $c \in \mathbb{R}$ be a cluster point of A. Suppose that $f(x) \leq g(x)$ for all $x \in A, x \neq c$.

(a)   If $\lim\limits_{x \to c} f = \infty$, then $\lim\limits_{x \to c} g = \infty$.

(b)   If $\lim\limits_{x \to c} g = -\infty$, then $\lim\limits_{x \to c} f = -\infty$.

Proof.   (a) If $\lim\limits_{x \to c} f = \infty$ and $\alpha \in \mathbb{R}$ is given, then there exists $\delta(\alpha) > 0$ such that if $0 < |x - c| < \delta(\alpha)$ and $x \in A$, then $f(x) > a$. But since $f(x) \leq g(x)$ for all $x \in A, x \neq c$, it follows that if $0 < |x - c| < \delta(\alpha)$ and $x \in A$, then $g(x) > \alpha$. Therefore $\lim\limits_{x \to c} g = \infty$.

The proof of (b) is similar.   Q.E.D.

The function $g(x) = 1/x$ considered in Example 4.3.6(b) suggests that it might be useful to consider one-sided infinite limits. We will define only right-hand infinite limits.

4.3.8 Definition   Let $A \subseteq \mathbb{R}$ and let $f : A \to \mathbb{R}$. If $c \in \mathbb{R}$ is a cluster point of the set $A \cap (c, \infty) = \{x \in A: x > c\}$, then we say that f **tends to** ∞ [respectively, $-\infty$] **as** $x \to c+$, and we write

$$\lim_{x \to c+} f = \infty \quad \left[\text{respectively, } \lim_{x \to c+} f = -\infty\right],$$

if for every $\alpha \in \mathbb{R}$ there is $\delta = \delta(\alpha) > 0$ such that for all $x \in A$ with $0 < x - c < \delta$, then $f(x) > \alpha$ [respectively, $f(x) < \alpha$].

4.3.9 Examples   **(a)**   Let $g(x) := 1/x$ for $x \neq 0$. We have noted in Example 4.3.6(b) that $\lim\limits_{x \to 0} g$ does not exist. However, it is an easy exercise to show that

$$\lim_{x \to 0+} (1/x) = \infty \quad \text{and} \quad \lim_{x \to 0-} (1/x) = -\infty.$$

(b)   It was seen in Example 4.3.4(b) that the function $g(x) := e^{1/x}$ for $x \neq 0$ is not bounded on any interval $(0, \delta)$, $\delta > 0$. Hence the right-hand limit of $e^{1/x}$ as $x \to 0+$ does not exist in the sense of Definition 4.3.1(i). However, since

$$1/x < e^{1/x} \quad \text{for} \quad x > 0,$$

it is readily seen that $\lim\limits_{x \to 0+} e^{1/x} = \infty$ in the sense of Definition 4.3.8.   □

Limits at Infinity

It is also desirable to define the notion of the limit of a function as $x \to \infty$. The definition as $x \to -\infty$ is similar.

4.3.10 Definition   Let $A \subseteq \mathbb{R}$ and let $f \colon A \to \mathbb{R}$. Suppose that $(a, \infty) \subseteq A$ for some $a \in \mathbb{R}$. We say that $L \in \mathbb{R}$ is **a limit of** f **as** $x \to \infty$, and write

$$\lim_{x \to \infty} f = L \qquad \text{or} \qquad \lim_{x \to \infty} f(x) = L,$$

if given any $\varepsilon > 0$ there exists $K = K(\varepsilon) > a$ such that for any $x > K$, then $|f(x) - L| < \varepsilon$.

The reader should note the close resemblance between 4.3.10 and the definition of a limit of a sequence.

We leave it to the reader to show that the limits of f as $x \to \pm\infty$ are unique whenever they exist. We also have sequential criteria for these limits; we shall only state the criterion as $x \to \infty$. This uses the notion of the limit of a properly divergent sequence (see Definition 3.6.1).

4.3.11 Theorem   Let $A \subseteq \mathbb{R}$, let $f : A \to \mathbb{R}$, and suppose that $(a, \infty) \subseteq A$ for some $a \in \mathbb{R}$. Then the following statements are equivalent:

(i) $L = \lim\limits_{x \to \infty} f$.

(ii)   For every sequence (x_n) in $A \cap (a, \infty)$ such that $\lim(x_n) = \infty$, the sequence $\big(f(x_n)\big)$ converges to L.

We leave it to the reader to prove this theorem and to formulate and prove the companion result concerning the limit as $x \to -\infty$.

4.3.12 Examples   **(a)** Let $g(x) := 1/x$ for $x \neq 0$.

It is an elementary exercise to show that $\lim\limits_{x \to \infty} (1/x) = 0 = \lim\limits_{x \to -\infty} (1/x)$. (See Figure 4.3.4.)

(b) Let $f(x) := 1/x^2$ for $x \neq 0$.

The reader may show that $\lim\limits_{x \to \infty} (1/x^2) = 0 = \lim\limits_{x \to -\infty} (1/x^2)$. (See Figure 4.3.3.) One way to do this is to show that if $x \geq 1$ then $0 \leq 1/x^2 \leq 1/x$. In view of part (a), this implies that $\lim\limits_{x \to \infty} (1/x^2) = 0$. $\square$

Just as it is convenient to be able to say that $f(x) \to \pm\infty$ as $x \to c$ for $c \in \mathbb{R}$, it is convenient to have the corresponding notion as $x \to \pm\infty$. We will treat the case where $x \to \infty$.

4.3.13 Definition   Let $A \subseteq \mathbb{R}$ and let $f \colon A \to \mathbb{R}$. Suppose that $(a, \infty) \subseteq A$ for some $a \in A$. We say that f **tends to** ∞ [respectively, $-\infty$] **as** $x \to \infty$, and write

$$\lim_{x \to \infty} f = \infty \qquad \left[\text{respectively, } \lim_{x \to \infty} f = -\infty\right]$$

if given any $\alpha \in \mathbb{R}$ there exists $K = K(\alpha) > a$ such that for any $x > K$, then $f(x) > \alpha$ [respectively, $f(x) < \alpha$].

As before there is a sequential criterion for this limit.

4.3.14 Theorem   Let $A \in \mathbb{R}$, let $f \colon A \to \mathbb{R}$, and suppose that $(a, \infty) \subseteq A$ for some $a \in \mathbb{R}$. Then the following statements are equivalent:

(i) $\lim\limits_{x \to \infty} f = \infty$ [respectively, $\lim\limits_{x \to \infty} f = -\infty$].

(ii)   For every sequence (x_n) in (a, ∞) such that $\lim(x_n) = \infty$, then $\lim\left(f(x_n)\right) = \infty$ [respectively, $\lim(f(x_n)) = -\infty$].

   The next result is an analogue of Theorem 3.6.5.

4.3.15 Theorem   Let $A \subseteq \mathbb{R}$, let f, $g: A \to \mathbb{R}$, and suppose that $(a, \infty) \subseteq A$ for some $a \in \mathbb{R}$. Suppose further that $g(x) > 0$ for all $x > a$ and that for some $L \in \mathbb{R}$, $L \neq 0$, we have

$$\lim_{x \to \infty} \frac{f(x)}{g(x)} = L.$$

(i)   If $L > 0$, then $\lim\limits_{x \to \infty} f = \infty$ if and only if $\lim\limits_{x \to \infty} g = \infty$.

(ii)   If $L < 0$, then $\lim\limits_{x \to \infty} f = -\infty$ if and only if $\lim\limits_{x \to \infty} g = \infty$.

Proof.   (i)   Since $L > 0$, the hypothesis implies that there exists $a_1 > a$ such that

$$0 < \tfrac{1}{2}L \leq \frac{f(x)}{g(x)} < \tfrac{3}{2}L \qquad \text{for} \quad x > a_1.$$

Therefore we have $\left(\tfrac{1}{2}L\right) g(x) < f(x) < \left(\tfrac{3}{2}L\right) g(x)$ for all $x > a_1$, from which the conclusion follows readily.

   The proof of (ii) is similar.   Q.E.D.

   We leave it to the reader to formulate the analogous result as $x \to -\infty$.

4.3.16 Examples   **(a)** $\lim\limits_{x \to \infty} x^n = \infty$ for $n \in \mathbb{N}$.

   Let $g(x) := x^n$ for $x \in (0, \infty)$. Given $\alpha \in \mathbb{R}$, let $K := \sup\{1, \alpha\}$. Then for all $x > K$, we have $g(x) = x^n \geq x > \alpha$. Since $\alpha \in \mathbb{R}$ is arbitrary, it follows that $\lim\limits_{x \to \infty} g = \infty$.

(b) $\lim\limits_{x \to -\infty} x^n = \infty$ for $n \in \mathbb{N}$, n even, and $\lim\limits_{x \to -\infty} x^n = -\infty$ for $n \in \mathbb{N}$, n odd.

   We will treat the case n odd, say $n = 2k + 1$ with $k = 0, 1, \cdots$. Given $\alpha \in \mathbb{R}$, let $K := \inf\{\alpha, -1\}$. For any $x < K$, then since $(x^2)^k \geq 1$, we have $x^n = (x^2)^k x \leq x < \alpha$. Since $\alpha \in \mathbb{R}$ is arbitrary, it follows that $\lim\limits_{x \to -\infty} x^n = -\infty$.

(c)   Let $p: \mathbb{R} \to \mathbb{R}$ be the polynomial function

$$p(x) := a_n x^n + a_{n-1} x^{n-1} + \cdots + a_1 x + a_0.$$

Then $\lim\limits_{x \to \infty} p = \infty$ if $a_n > 0$, and $\lim\limits_{x \to \infty} p = -\infty$ if $a_n < 0$.

   Indeed, let $g(x) := x^n$ and apply Theorem 4.3.15. Since

$$\frac{p(x)}{g(x)} = a_n + a_{n-1}\left(\frac{1}{x}\right) + \cdots + a_1\left(\frac{1}{x^{n-1}}\right) + a_0\left(\frac{1}{x^n}\right),$$

it follows that $\lim\limits_{x \to \infty} (p(x)/g(x)) = a_n$. Since $\lim\limits_{x \to \infty} g = \infty$, the assertion follows from Theorem 4.3.15.

(d)   Let p be the polynomial function in part (c). Then $\lim\limits_{x \to -\infty} p = \infty$ [respectively, $-\infty$] if n is even [respectively, odd] and $a_n > 0$.

   We leave the details to the reader.   □

Exercises for Section 4.3

1. Prove Theorem 4.3.2.

2. Give an example of a function that has a right-hand limit but not a left-hand limit at a point.

3. Let $f(x) := |x|^{-1/2}$ for $x \neq 0$. Show that $\lim\limits_{x \to 0+} f(x) = \lim\limits_{x \to 0-} f(x) = +\infty$.

4. Let $c \in \mathbb{R}$ and let f be defined for $x \in (c, \infty)$ and $f(x) > 0$ for all $x \in (c, \infty)$. Show that $\lim\limits_{x \to c} f = \infty$ if and only if $\lim\limits_{x \to c} 1/f = 0$.

5. Evaluate the following limits, or show that they do not exist.

   (a) $\lim\limits_{x \to 1+} \dfrac{x}{x-1} \quad (x \neq 1)$,

   (b) $\lim\limits_{x \to 1} \dfrac{x}{x-1} \quad (x \neq 1)$,

   (c) $\lim\limits_{x \to 0+} (x+2)/\sqrt{x} \quad (x > 0)$,

   (d) $\lim\limits_{x \to \infty} (x+2)/\sqrt{x} \quad (x > 0)$,

   (e) $\lim\limits_{x \to 0} \left(\sqrt{x+1}\right)/x \quad (x > -1)$,

   (f) $\lim\limits_{x \to \infty} \left(\sqrt{x+1}\right)/x \quad (x > 0)$,

   (g) $\lim\limits_{x \to \infty} \dfrac{\sqrt{x}-5}{\sqrt{x}+3} \quad (x > 0)$,

   (h) $\lim\limits_{x \to \infty} \dfrac{\sqrt{x}-x}{\sqrt{x}+x} \quad (x > 0)$.

6. Prove Theorem 4.3.11.

7. Suppose that f and g have limits in $\mathbb{R}$ as $x \to \infty$ and that $f(x) \leq g(x)$ for all $x \in (a, \infty)$. Prove that $\lim\limits_{x \to \infty} f \leq \lim\limits_{x \to \infty} g$.

8. Let f be defined on $(0, \infty)$ to $\mathbb{R}$. Prove that $\lim\limits_{x \to \infty} f(x) = L$ if and only if $\lim\limits_{x \to 0+} f(1/x) = L$.

9. Show that if $f : (a, \infty) \to \mathbb{R}$ is such that $\lim\limits_{x \to \infty} x f(x) = L$ where $L \in \mathbb{R}$, then $\lim\limits_{x \to \infty} f(x) = 0$.

10. Prove Theorem 4.3.14.

11. Suppose that $\lim\limits_{x \to c} f(x) = L$ where $L > 0$, and that $\lim\limits_{x \to c} g(x) = \infty$. Show that $\lim\limits_{x \to c} f(x)g(x) = \infty$. If $L = 0$, show by example that this conclusion may fail.

12. Find functions f and g defined on $(0, \infty)$ such that $\lim\limits_{x \to \infty} f = \infty$ and $\lim\limits_{x \to \infty} g = \infty$, and $\lim\limits_{x \to \infty} (f - g) = 0$. Can you find such functions, with $g(x) > 0$ for all $x \in (0, \infty)$, such that $\lim\limits_{x \to \infty} f/g = 0$?

13. Let f and g be defined on (a, ∞) and suppose $\lim\limits_{x \to \infty} f = L$ and $\lim\limits_{x \to \infty} g = \infty$. Prove that $\lim\limits_{x \to \infty} f \circ g = L$.

CHAPTER 5

CONTINUOUS FUNCTIONS

We now begin the study of the most important class of functions that arises in real analysis: the class of continuous functions. The term "continuous" has been used since the time of Newton to refer to the motion of bodies or to describe an unbroken curve, but it was not made precise until the nineteenth century. Work of Bernhard Bolzano in 1817 and Augustin-Louis Cauchy in 1821 identified continuity as a very significant property of functions and proposed definitions, but since the concept is tied to that of limit, it was the careful work of Karl Weierstrass in the 1870s that brought proper understanding to the idea of continuity.

We will first define the notions of continuity at a point and continuity on a set, and then show that various combinations of continuous functions give rise to continuous functions. Then in Section 5.3 we establish the fundamental properties that make continuous functions so important. For instance, we will prove that a continuous function on a closed bounded interval must attain a maximum and a minimum value. We also prove that a continuous function must take on every value intermediate to any two values it attains. These properties and others are not possessed by general functions, as various examples illustrate, and thus they distinguish continuous functions as a very special class of functions.

In Section 5.4 we introduce the very important notion of uniform continuity. The distinction between continuity and uniform continuity is somewhat subtle and was not fully appreciated until the work of Weierstrass and the mathematicians of his era, but it proved to

---

Karl Weierstrass

Karl Weierstrass (=Weierstraß) (1815–1897) was born in Westphalia, Germany. His father, a customs officer in a salt works, insisted that he study law and public finance at the University of Bonn, but he had more interest in drinking and fencing, and left Bonn without receiving a diploma. He then enrolled in the Academy of Münster where he studied mathematics with Christoph Gudermann. From 1841–1854 he taught at various *gymnasia* in Prussia. Despite the fact that he had no contact with the mathematical world during this time, he worked hard on mathematical research and was able  to publish a few papers, one of which attracted considerable attention. Indeed, the University of Königsberg gave him an honorary doctoral degree for this work in 1855. The next year, he secured positions at the Industrial Institute of Berlin and the University of Berlin. He remained at Berlin until his death.

A methodical and painstaking scholar, Weierstrass distrusted intuition and worked to put everything on a firm and logical foundation. He did fundamental work on the foundations of arithmetic and analysis, on complex analysis, the calculus of variations, and algebraic geometry. Due to his meticulous preparation, he was an extremely popular lecturer; it was not unusual for him to speak about advanced mathematical topics to audiences of more than 250. Among his auditors are counted Georg Cantor, Sonya Kovalevsky, Gösta Mittag-Leffler, Max Planck, Otto Hölder, David Hilbert, and Oskar Bolza (who had many American doctoral students). Through his writings and his lectures, Weierstrass had a profound influence on contemporary mathematics.

be very significant in applications. We present one application to the idea of approximating continuous functions by more elementary functions (such as polynomials).

The notion of a "gauge" is introduced in Section 5.5 and is used to provide an alternative method of proving the fundamental properties of continuous functions. The main significance of this concept, however, is in the area of integration theory where gauges are essential in defining the generalized Riemann integral. This will be discussed in Chapter 10.

Monotone functions are an important class of functions with strong continuity properties and they are discussed in Section 5.6.

Section 5.1   Continuous Functions

In this section, which is very similar to Section 4.1, we will define what it means to say that a function is continuous at a point, or on a set. This notion of continuity is one of the central concepts of mathematical analysis, and it will be used in almost all of the following material in this book. Consequently, it is essential that the reader master it.

5.1.1 Definition   Let $A \subseteq \mathbb{R}$, let $f : A \to \mathbb{R}$, and let $c \in A$. We say that f is **continuous at c** if, given any number $\varepsilon > 0$ there exists $\delta > 0$ such that if x is any point of A satisfying $|x - c| < \delta$, then $|f(x) - f(c)| < \varepsilon$.

If f fails to be continuous at c, then we say that f is **discontinuous at c**.

As with the definition of limit, the definition of continuity at a point can be formulated very nicely in terms of neighborhoods. This is done in the next result. We leave the verification as an important exercise for the reader. See Figure 5.1.1.

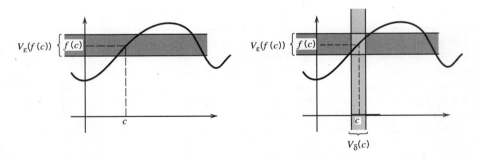

Figure 5.1.1   Given $V_\varepsilon(f(c))$, a neighborhood $V_\delta(c)$ is to be determined.

5.1.2 Theorem   *A function $f : A \to \mathbb{R}$ is continuous at a point $c \in A$ if and only if given any ε-neighborhood $V_\varepsilon(f(c))$ of $f(c)$ there exists a δ-neighborhood $V_\delta(c)$ of c such that if x is any point of $A \cap V_\delta(c)$, then $f(x)$ belongs to $V_\varepsilon(f(c))$, that is,*

$$f(A \cap V_\delta(c)) \subseteq V_\varepsilon(f(c)).$$

Remark   **(1)**   If $c \in A$ is a cluster point of A, then a comparison of Definitions 4.1.4 and 5.1.1 show that f is continuous at c if and only if

(1)
$$f(c) = \lim_{x \to c} f(x).$$

Thus, if c is a cluster point of A, then three conditions must hold for f to be continuous at c:

    (i) f must be defined at c (so that $f(c)$ makes sense),

    (ii)   the limit of f at c must exist in $\mathbb{R}$ (so that $\lim_{x \to c} f(x)$ makes sense), and

    (iii)   these two values must be equal.

(2)   If $c \in A$ is not a cluster point of A, then there exists a neighborhood $V_\delta(c)$ of c such that $A \cap V_\delta(c) = \{c\}$. Thus we conclude that a function f is automatically continuous at a point $c \in A$ that is not a cluster point of A. Such points are often called "isolated points" of A. They are of little practical interest to us, since they have no relation to a limiting process. Since continuity is automatic for such points, we generally test for continuity only at cluster points. Thus we regard condition (1) as being characteristic for continuity at c.

A slight modification of the proof of Theorem 4.1.8 for limits yields the following sequential version of continuity at a point.

5.1.3 Sequential Criterion for Continuity   *A function $f : A \to \mathbb{R}$ is continuous at the point $c \in A$ if and only if for every sequence (x_n) in A that converges to c, the sequence $(f(x_n))$ converges to $f(c)$.*

The following Discontinuity Criterion is a consequence of the last theorem. It should be compared with the Divergence Criterion 4.1.9(a) with $L = f(c)$. Its proof should be written out in detail by the reader.

5.1.4 Discontinuity Criterion   *Let $A \subseteq \mathbb{R}$, let $f : A \to \mathbb{R}$, and let $c \in A$. Then f is discontinuous at c if and only if there exists a sequence (x_n) in A such that (x_n) converges to c, but the sequence $(f(x_n))$ does not converge to $f(c)$.*

So far we have discussed continuity at a *point*. To talk about the continuity of a function on a *set*, we will simply require that the function be continuous at each point of the set. We state this formally in the next definition.

5.1.5 Definition   Let $A \subseteq \mathbb{R}$ and let $f : A \to \mathbb{R}$. If B is a subset of A, we say that f is **continuous on the set** B if f is continuous at every point of B.

5.1.6 Examples **(a)** The constant function $f(x) := b$ is continuous on $\mathbb{R}$.

It was seen in Example 4.1.7(a) that if $c \in \mathbb{R}$, then $\lim_{x \to c} f(x) = b$. Since $f(c) = b$, we have $\lim_{x \to c} f(x) = f(c)$, and thus f is continuous at every point $c \in \mathbb{R}$. Therefore f is continuous on $\mathbb{R}$.

(b) $g(x) := x$ is continuous on $\mathbb{R}$.

It was seen in Example 4.1.7(b) that if $c \in \mathbb{R}$, then we have $\lim_{x \to c} g = c$. Since $g(c) = c$, then g is continuous at every point $c \in \mathbb{R}$. Thus g is continuous on $\mathbb{R}$.

(c) $h(x) := x^2$ is continuous on $\mathbb{R}$.

It was seen in Example 4.1.7(c) that if $c \in \mathbb{R}$, then we have $\lim_{x \to c} h = c^2$. Since $h(c) = c^2$, then h is continuous at every point $c \in \mathbb{R}$. Thus h is continuous on $\mathbb{R}$.

(d) $\varphi(x) := 1/x$ is continuous on $A := \{x \in \mathbb{R} : x > 0\}$.

It was seen in Example 4.1.7(d) that if $c \in A$, then we have $\lim_{x \to c} \varphi = 1/c$. Since $\varphi(c) := 1/c$, this shows that φ is continuous at every point $c \in A$. Thus φ is continuous on A.

(e) $\varphi(x) := 1/x$ is not continuous at $x = 0$.

Indeed, if $\varphi(x) = 1/x$ for $x > 0$, then φ is not defined for $x = 0$, so it cannot be continuous there. Alternatively, it was seen in Example 4.1.10(a) that $\lim_{x \to 0} \varphi$ does not exist in $\mathbb{R}$, so φ cannot be continuous at $x = 0$.

(f)   The signum function sgn is not continuous at 0.

The signum function was defined in Example 4.1.10(b), where it was also shown that $\lim_{x \to 0} \text{sgn}(x)$ does not exist in $\mathbb{R}$. Therefore sgn is not continuous at $x = 0$ (even though sgn 0 is defined).

It is an exercise to show that sgn is continuous at every point $c \neq 0$.

(g)   Let $A := \mathbb{R}$ and let f be Dirichlet's "discontinuous function" defined by

$$f(x) := \begin{cases} 1 & \text{if} \quad x \text{ is rational}, \\ 0 & \text{if} \quad x \text{ is irrational}. \end{cases}$$

We claim that f is *not continuous at any point of* $\mathbb{R}$. (This function was introduced in 1829 by P. G. L. Dirichlet.)

Indeed, if c is a rational number, let (x_n) be a sequence of irrational numbers that converges to c. (Corollary 2.4.9 to the Density Theorem 2.4.8 assures us that such a sequence does exist.) Since $f(x_n) = 0$ for all $n \in \mathbb{N}$, we have $\lim(f(x_n)) = 0$, while $f(c) = 1$. Therefore f is not continuous at the rational number c.

On the other hand, if b is an irrational number, let (y_n) be a sequence of rational numbers that converge to b. (The Density Theorem 2.4.8 assures us that such a sequence does exist.) Since $f(y_n) = 1$ for all $n \in \mathbb{N}$, we have $\lim(f(y_n)) = 1$, while $f(b) = 0$. Therefore f is not continuous at the irrational number b.

Since every real number is either rational or irrational, we deduce that f is not continuous at any point in $\mathbb{R}$.

(h)   Let $A := \{x \in \mathbb{R}: x > 0\}$. For any irrational number $x > 0$ we define $h(x) = 0$. For a rational number in A of the form m/n, with natural numbers m, n having no common factors except 1, we define $h(m/n) := 1/n$. (Sometimes we also define $h(0) := 1$.)

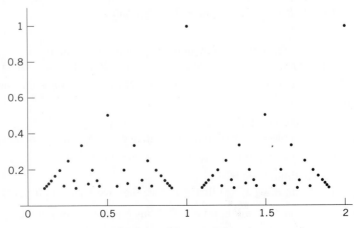

Figure 5.1.2   Thomae's function.

We claim that *h is continuous at every irrational number in A, and is discontinuous at every rational number in A.* (This function was introduced in 1875 by K. J. Thomae.)

Indeed, if $a > 0$ is rational, let (x_n) be a sequence of irrational numbers in A that converges to a. Then $\lim \left(h(x_n) \right) = 0$, while $h(a) > 0$. Hence h is discontinuous at a.

On the other hand, if b is an irrational number and $\varepsilon > 0$, then (by the Archimedean Property) there is a natural number n_0 such that $1/n_0 < \varepsilon$. There are only a finite number of rationals with denominator less than n_0 in the interval $(b - 1, b + 1)$. (Why?) Hence $\delta > 0$ can be chosen so small that the neighborhood $(b - \delta, b + \delta)$ contains no rational numbers with denominator less than n_0. It then follows that for $|x - b| < \delta, x \in A$, we have $|h(x) - h(b)| = |h(x)| \leq 1/n_0 < \varepsilon$. Thus h is continuous at the irrational number b.

Consequently, we deduce that Thomae's function h is continuous precisely at the irrational points in A. $\qquad\qquad\square$

5.1.7 Remarks **(a)** Sometimes a function $f: A \to \mathbb{R}$ is not continuous at a point c because it is not defined at this point. However, if the function f has a limit L at the point c and if we define F on $A \cup \{c\} \to \mathbb{R}$ by

$$F(x) := \begin{cases} L & \text{for} \quad x = c, \\ f(x) & \text{for} \quad x \in A, \end{cases}$$

then F is continuous at c. To see this, one needs to check that $\lim_{x \to c} F = L$, but this follows (why?), since $\lim_{x \to c} f = L$.

(b) If a function $g : A \to \mathbb{R}$ does not have a limit at c, then there is no way that we can obtain a function $G : A \cup \{c\} \to \mathbb{R}$ that is continuous at c by defining

$$G(x) := \begin{cases} C & \text{for} \quad x = c, \\ g(x) & \text{for} \quad x \in A. \end{cases}$$

To see this, observe that if $\lim_{x \to c} G$ exists and equals C, then $\lim_{x \to c} g$ must also exist and equal C.

5.1.8 Examples **(a)** The function $g(x) := \sin(1/x)$ for $x \neq 0$ (see Figure 4.1.3) does not have a limit at $x = 0$ (see Example 4.1.10(c)). Thus there is no value that we can assign at $x = 0$ to obtain a continuous extension of g at $x = 0$.

(b) Let $f(x) = x \sin(1/x)$ for $x \neq 0$. (See Figure 5.1.3.) Since f is not defined at $x = 0$, the function f cannot be continuous at this point. However, it was seen in Example 4.2.8(f) that $\lim_{x \to 0} \left(x \sin(1/x) \right) = 0$. Therefore it follows from Remark 5.1.7(a) that if we define $F : \mathbb{R} \to \mathbb{R}$ by

$$F(x) := \begin{cases} 0 & \text{for} \quad x = 0, \\ x \sin(1/x) & \text{for} \quad x \neq 0, \end{cases}$$

then F is continuous at $x = 0$. $\qquad\qquad\square$

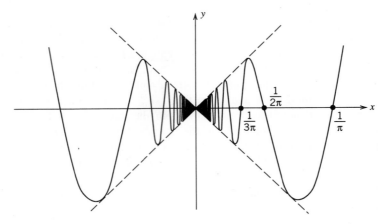

Figure 5.1.3   Graph of $f(x) = x \sin(1/x)$ $(x \neq 0)$.

Exercises for Section 5.1

1. Prove the Sequential Criterion 5.1.3.

2. Establish the Discontinuity Criterion 5.1.4.

3. Let $a < b < c$. Suppose that f is continuous on $[a, b]$, that g is continuous on $[b, c]$, and that $f(b) = g(b)$. Define h on $[a, c]$ by $h(x) := f(x)$ for $x \in [a, b]$ and $h(x) := g(x)$ for $x \in (b, c]$. Prove that h is continuous on $[a, c]$.

4. If $x \in \mathbb{R}$, we define $[\![x]\!]$ to be the greatest integer $n \in \mathbb{Z}$ such that $n \leq x$. (Thus, for example, $[\![8.3]\!] = 8$, $[\![\pi]\!] = 3$, $[\![-\pi]\!] = -4$.) The function $x \mapsto [\![x]\!]$ is called the **greatest integer function**. Determine the points of continuity of the following functions:
    (a) $f(x) := [\![x]\!]$
    (b) $g(x) := x[\![x]\!]$,
    (c) $h(x) := [\![\sin x]\!]$,
    (d) $k(x) := [\![1/x]\!]$ $(x \neq 0)$.

5. Let f be defined for all $x \in \mathbb{R}$, $x \neq 2$, by $f(x) = (x^2 + x - 6)/(x - 2)$. Can f be defined at $x = 2$ in such a way that f is continuous at this point?

6. Let $A \subseteq \mathbb{R}$ and let $f : A \to \mathbb{R}$ be continuous at a point $c \in A$. Show that for any $\varepsilon > 0$, there exists a neighborhood $V_\delta(c)$ of c such that if $x, y \in A \cap V_\delta(c)$, then $|f(x) - f(y)| < \varepsilon$.

7. Let $f : \mathbb{R} \to \mathbb{R}$ be continuous at c and let $f(c) > 0$. Show that there exists a neighborhood $V_\delta(c)$ of c such that if $x \in V_\delta(c)$, then $f(x) > 0$.

8. Let $f : \mathbb{R} \to \mathbb{R}$ be continuous on $\mathbb{R}$ and let $S := \{x \in \mathbb{R} : f(x) = 0\}$ be the "zero set" of f. If (x_n) is in S and $x = \lim(x_n)$, show that $x \in S$.

9. Let $A \subseteq B \subseteq \mathbb{R}$, let $f: B \to \mathbb{R}$ and let g be the restriction of f to A (that is, $g(x) = f(x)$ for $x \in A$).
    (a) If f is continuous at $c \in A$, show that g is continuous at c.
    (b) Show by example that if g is continuous at c, it need not follow that f is continuous at c.

10. Show that the absolute value function $f(x) := |x|$ is continuous at every point $c \in \mathbb{R}$.

11. Let $K > 0$ and let $f : \mathbb{R} \to \mathbb{R}$ satisfy the condition $|f(x) - f(y)| \leq K|x - y|$ for all $x, y \in \mathbb{R}$. Show that f is continuous at every point $c \in \mathbb{R}$.

12. Suppose that $f : \mathbb{R} \to \mathbb{R}$ is continuous on $\mathbb{R}$ and that $f(r) = 0$ for every rational number r. Prove that $f(x) = 0$ for all $x \in \mathbb{R}$.

13. Define $g : \mathbb{R} \to \mathbb{R}$ by $g(x) := 2x$ for x rational, and $g(x) := x + 3$ for x irrational. Find all points at which g is continuous.

14. Let $A := (0, \infty)$ and let $k : A \to \mathbb{R}$ be defined as follows. For $x \in A$, x irrational, we define $k(x) = 0$; for $x \in A$ rational and of the form $x = m/n$ with natural numbers m, n having no common factors except 1, we define $k(x) := n$. Prove that k is unbounded on every open interval in A. Conclude that k is not continuous at any point of A. (See Example 5.1.6(h).)

15. Let $f : (0, 1) \to \mathbb{R}$ be bounded but such that $\lim\limits_{x \to 0} f$ does not exist. Show that there are two sequences (x_n) and (y_n) in $(0, 1)$ with $\lim(x_n) = 0 = \lim(y_n)$, but such that $\lim\left(f(x_n)\right)$ and $\lim\left(f(y_n)\right)$ exist but are not equal.

Section 5.2   Combinations of Continuous Functions

Let $A \subseteq \mathbb{R}$ and let f and g be functions that are defined on A to $\mathbb{R}$ and let $b \in \mathbb{R}$. In Definition 4.2.3 we defined the sum, difference, product, and multiple functions denoted by $f + g, f - g, fg, bf$. In addition, if $h : A \to \mathbb{R}$ is such that $h(x) \neq 0$ for all $x \in A$, then we defined the quotient function denoted by f/h.

The next result is similar to Theorem 4.2.4, from which it follows.

5.2.1 Theorem   Let $A \subseteq \mathbb{R}$, let f and g be functions on A to $\mathbb{R}$, and let $b \in \mathbb{R}$. Suppose that $c \in A$ and that f and g are continuous at c.

(a)   Then $f + g, f - g, fg$, and bf are continuous at c.

(b)   If $h : A \to \mathbb{R}$ is continuous at $c \in A$ and if $h(x) \neq 0$ for all $x \in A$, then the quotient f/h is continuous at c.

Proof.   If $c \in A$ is not a cluster point of A, then the conclusion is automatic. Hence we assume that c is a cluster point of A.

(a) Since f and g are continuous at c, then

$$f(c) = \lim_{x \to c} f \quad \text{and} \quad g(c) = \lim_{x \to c} g.$$

Hence it follows from Theorem 4.2.4(a) that

$$(f + g)(c) = f(c) + g(c) = \lim_{x \to c}(f + g).$$

Therefore $f + g$ is continuous at c. The remaining assertions in part (a) are proved in a similar fashion.

(b) Since $c \in A$, then $h(c) \neq 0$. But since $h(c) = \lim\limits_{x \to c} h$, it follows from Theorem 4.2.4(b) that

$$\frac{f}{h}(c) = \frac{f(c)}{h(c)} = \frac{\lim\limits_{x \to c} f}{\lim\limits_{x \to c} h} = \lim_{x \to c}\left(\frac{f}{h}\right).$$

Therefore f/h is continuous at c.   Q.E.D.

The next result is an immediate consequence of Theorem 5.2.1, applied to every point of A. However, since it is an extremely important result, we shall state it formally.

5.2.2 Theorem   *Let $A \subseteq \mathbb{R}$, let f and g be continuous on A to $\mathbb{R}$, and let $b \in \mathbb{R}$.*

(a)   *The functions $f + g$, $f - g$, fg, and bf are continuous on A.*

(b)   *If $h : A \to \mathbb{R}$ is continuous on A and $h(x) \neq 0$ for $x \in A$, then the quotient f/h is continuous on A.*

Remark   To define quotients, it is sometimes more convenient to proceed as follows. If $\varphi : A \to \mathbb{R}$, let $A_1 := \{x \in A : \varphi(x) \neq 0\}$. We can define the quotient f/φ on the set A_1 by

$$(1) \qquad\qquad \left(\frac{f}{\varphi}\right)(x) := \frac{f(x)}{\varphi(x)} \qquad \text{for} \quad x \in A_1.$$

If φ is continuous at a point $c \in A_1$, it is clear that the restriction φ_1 of φ to A_1 is also continuous at c. Therefore it follows from Theorem 5.2.1(b) applied to φ_1 that f/φ_1 is continuous at $c \in A$. Since $(f/\varphi)(x) = (f/\varphi_1)(x)$ for $x \in A_1$ it follows that f/φ is continuous at $c \in A_1$. Similarly, if f and φ are continuous on A, then the function f/φ, defined on A_1 by (1), is continuous on A_1.

5.2.3 Examples   **(a)**   Polynomial functions.
    If p is a polynomial function, so that $p(x) = a_n x^n + a_{n-1} x^{n-1} + \cdots + a_1 x + a_0$ for all $x \in \mathbb{R}$, then it follows from Example 4.2.5(f) that $p(c) = \lim_{x \to c} p$ for any $c \in \mathbb{R}$. Thus *a polynomial function is continuous on $\mathbb{R}$.*

(b)   Rational functions.
    If p and q are polynomial functions on $\mathbb{R}$, then there are at most a finite number $\alpha_1, \cdots, \alpha_m$ of real roots of q. If $x \notin \{\alpha_1, \cdots, \alpha_m\}$ then $q(x) \neq 0$ so that we can define the rational function r by

$$r(x) := \frac{p(x)}{q(x)} \qquad \text{for} \quad x \notin \{\alpha_1, \cdots, \alpha_m\}.$$

It was seen in Example 4.2.5(g) that if $q(c) \neq 0$, then

$$r(c) = \frac{p(c)}{q(c)} = \lim_{x \to c} \frac{p(x)}{q(x)} = \lim_{x \to c} r(x).$$

In other words, r is continuous at c. Since c is any real number that is not a root of q, we infer that *a rational function is continuous at every real number for which it is defined.*

(c)   We shall show that the sine function sin is continuous on $\mathbb{R}$.
    To do so we make use of the following properties of the sine and cosine functions. (See Section 8.4.) For all $x, y, z \in \mathbb{R}$ we have:

$$|\sin z| \leq |z|, \qquad |\cos z| \leq 1,$$

$$\sin x - \sin y = 2 \sin\left[\tfrac{1}{2}(x - y)\right] \cos\left[\tfrac{1}{2}(x + y)\right].$$

Hence if $c \in \mathbb{R}$, then we have

$$|\sin x - \sin c| \leq 2 \cdot \tfrac{1}{2}|x - c| \cdot 1 = |x - c|.$$

Therefore sin is continuous at c. Since $c \in \mathbb{R}$ is arbitrary, it follows that sin is continuous on $\mathbb{R}$.

(d)   The cosine function is continuous on $\mathbb{R}$.

We make use of the following properties of the sine and cosine functions. For all $x, y, z \in \mathbb{R}$ we have:

$$|\sin z| \leq |z|, \qquad |\sin z| \leq 1,$$
$$\cos x - \cos y = -2 \sin[\tfrac{1}{2}(x + y)] \sin[\tfrac{1}{2}(x - y)].$$

Hence if $c \in \mathbb{R}$, then we have

$$|\cos x - \cos c| \leq 2 \cdot 1 \cdot \tfrac{1}{2}|c - x| = |x - c|.$$

Therefore cos is continuous at c. Since $c \in \mathbb{R}$ is arbitrary, it follows that cos is continuous on $\mathbb{R}$. (Alternatively, we could use the relation $\cos x = \sin(x + \pi/2)$.)

(e)   The functions tan, cot, sec, csc are continuous where they are defined.
For example, the cotangent function is defined by

$$\cot x := \frac{\cos x}{\sin x}$$

provided $\sin x \neq 0$ (that is, provided $x \neq n\pi, n \in \mathbb{Z}$). Since sin and cos are continuous on $\mathbb{R}$, it follows (see the Remark before Example 5.2.3) that the function cot is continuous on its domain. The other trigonometric functions are treated similarly.   □

5.2.4 Theorem   Let $A \subseteq \mathbb{R}$, let $f : A \to \mathbb{R}$, and let $|f|$ be defined by $|f|(x) := |f(x)|$ for $x \in A$.

(a)   If f is continuous at a point $c \in A$, then $|f|$ is continuous at c.

(b)   If f is continuous on A, then $|f|$ is continuous on A.

Proof.   This is an immediate consequence of Exercise 4.2.13.   Q.E.D.

5.2.5 Theorem   Let $A \subseteq \mathbb{R}$, let $f : A \to \mathbb{R}$, and let $f(x) \geq 0$ for all $x \in A$. We let $\sqrt{f}$ be defined for $x \in A$ by $\left(\sqrt{f}\right)(x) := \sqrt{f(x)}$.

(a)   If f is continuous at a point $c \in A$, then $\sqrt{f}$ is continuous at c.

(b)   If f is continuous on A, then $\sqrt{f}$ is continuous on A.

Proof.   This is an immediate consequence of Exercise 4.2.14.   Q.E.D.

Composition of Continuous Functions ────────────────────────────

We now show that if the function $f : A \to \mathbb{R}$ is continuous at a point c and if $g : B \to \mathbb{R}$ is continuous at $b = f(c)$, then the composition $g \circ f$ is continuous at c. In order to assure that $g \circ f$ is defined on all of A, we also need to assume that $f(A) \subseteq B$.

5.2.6 Theorem   Let $A, B \subseteq \mathbb{R}$ and let $f : A \to \mathbb{R}$ and $g: B \to \mathbb{R}$ be functions such that $f(A) \subseteq B$. If f is continuous at a point $c \in A$ and g is continuous at $b = f(c) \in B$, then the composition $g \circ f : A \to \mathbb{R}$ is continuous at c.

Proof.   Let W be an ε-neighborhood of $g(b)$. Since g is continuous at b, there is a δ-neighborhood V of $b = f(c)$ such that if $y \in B \cap V$ then $g(y) \in W$. Since f is continuous at c, there is a γ-neighborhood U of c such that if $x \in A \cap U$, then $f(x) \in V$. (See Figure 5.2.1.) Since $f(A) \subseteq B$, it follows that if $x \in A \cap U$, then $f(x) \in B \cap V$ so that $g \circ f(x) = g(f(x)) \in W$. But since W is an arbitrary ε-neighborhood of $g(b)$, this implies that $g \circ f$ is continuous at c.   Q.E.D.

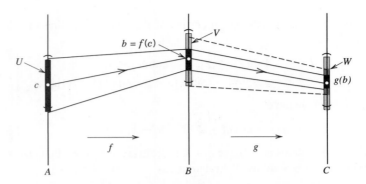

Figure 5.2.1   The composition of f and g.

5.2.7 Theorem   *Let $A, B \subseteq \mathbb{R}$, let $f : A \to \mathbb{R}$ be continuous on A, and let $g : B \to \mathbb{R}$ be continuous on B. If $f(A) \subseteq B$, then the composite function $g \circ f : A \to \mathbb{R}$ is continuous on A.*

Proof.   The theorem follows immediately from the preceding result, if f and g are continuous at every point of A and B, respectively.   Q.E.D.

Theorems 5.2.6 and 5.2.7 are very useful in establishing that certain functions are continuous. They can be used in many situations where it would be difficult to apply the definition of continuity directly.

5.2.8 Examples   (a)   Let $g_1(x) := |x|$ for $x \in \mathbb{R}$. It follows from the Triangle Inequality that

$$\left| g_1(x) - g_1(c) \right| \le |x - c|$$

for all $x, c \in \mathbb{R}$. Hence g_1 is continuous at $c \in \mathbb{R}$. If $f : A \to \mathbb{R}$ is any function that is continuous on A, then Theorem 5.2.7 implies that $g_1 \circ f = |f|$ is continuous on A. This gives another proof of Theorem 5.2.4.

(b)   Let $g_2(x) := \sqrt{x}$ for $x \ge 0$. It follows from Theorems 3.2.10 and 5.1.3 that g_2 is continuous at any number $c \ge 0$. If $f : A \to \mathbb{R}$ is continuous on A and if $f(x) \ge 0$ for all $x \in A$, then it follows from Theorem 5.2.7 that $g_2 \circ f = \sqrt{f}$ is continuous on A. This gives another proof of Theorem 5.2.5.

(c)   Let $g_3(x) := \sin x$ for $x \in \mathbb{R}$. We have seen in Example 5.2.3(c) that g_3 is continuous on $\mathbb{R}$. If $f : A \to \mathbb{R}$ is continuous on A, then it follows from Theorem 5.2.7 that $g_3 \circ f$ is continuous on A.

In particular, if $f(x) := 1/x$ for $x \ne 0$, then the function $g(x) := \sin(1/x)$ is continuous at every point $c \ne 0$. [We have seen, in Example 5.1.8(a), that g cannot be defined at 0 in order to become continuous at that point.]   □

Exercises for Section 5.2

1.   Determine the points of continuity of the following functions and state which theorems are used in each case.

(a) $f(x) := \dfrac{x^2 + 2x + 1}{x^2 + 1}$ $(x \in \mathbb{R})$,   (b) $g(x) := \sqrt{x + \sqrt{x}}$ $(x \ge 0)$,

(c) $h(x) := \dfrac{\sqrt{1 + |\sin x|}}{x}$ $(x \ne 0)$,   (d) $k(x) := \cos\sqrt{1 + x^2}$ $(x \in \mathbb{R})$.

2. Show that if $f : A \to \mathbb{R}$ is continuous on $A \subseteq \mathbb{R}$ and if $n \in \mathbb{N}$, then the function f^n defined by $f^n(x) = (f(x))^n$ for $x \in A$, is continuous on A.

3. Give an example of functions f and g that are both discontinuous at a point c in $\mathbb{R}$ such that (a) the sum $f + g$ is continuous at c, (b) the product fg is continuous at c.

4. Let $x \mapsto [\![x]\!]$ denote the greatest integer function (see Exercise 5.1.4). Determine the points of continuity of the function $f(x) := x - [\![x]\!]$, $x \in \mathbb{R}$.

5. Let g be defined on $\mathbb{R}$ by $g(1) := 0$, and $g(x) := 2$ if $x \neq 1$, and let $f(x) := x + 1$ for all $x \in \mathbb{R}$. Show that $\lim_{x \to 0} g \circ f \neq (g \circ f)(0)$. Why doesn't this contradict Theorem 5.2.6?

6. Let f, g be defined on $\mathbb{R}$ and let $c \in \mathbb{R}$. Suppose that $\lim_{x \to c} f = b$ and that g is continuous at b. Show that $\lim_{x \to c} g \circ f = g(b)$. (Compare this result with Theorem 5.2.7 and the preceding exercise.)

7. Give an example of a function $f : [0, 1] \to \mathbb{R}$ that is discontinuous at every point of $[0, 1]$ but such that $|f|$ is continuous on $[0, 1]$.

8. Let f, g be continuous from $\mathbb{R}$ to $\mathbb{R}$, and suppose that $f(r) = g(r)$ for all rational numbers r. Is it true that $f(x) = g(x)$ for all $x \in \mathbb{R}$?

9. Let $h: \mathbb{R} \to \mathbb{R}$ be continuous on $\mathbb{R}$ satisfying $h(m/2^n) = 0$ for all $m \in \mathbb{Z}$, $n \in \mathbb{N}$. Show that $h(x) = 0$ for all $x \in \mathbb{R}$.

10. Let $f: \mathbb{R} \to \mathbb{R}$ be continuous on $\mathbb{R}$, and let $P := \{x \in \mathbb{R} : f(x) > 0\}$. If $c \in P$, show that there exists a neighborhood $V_\delta(c) \subseteq P$.

11. If f and g are continuous on $\mathbb{R}$, let $S := \{x \in \mathbb{R} : f(x) \geq g(x)\}$. If $(s_n) \subseteq S$ and $\lim(s_n) = s$, show that $s \in S$.

12. A function $f : \mathbb{R} \to \mathbb{R}$ is said to be **additive** if $f(x + y) = f(x) + f(y)$ for all x, y in $\mathbb{R}$. Prove that if f is continuous at some point x_0, then it is continuous at every point of $\mathbb{R}$. (See Exercise 4.2.12.)

13. Suppose that f is a continuous additive function on $\mathbb{R}$. If $c := f(1)$, show that we have $f(x) = cx$ for all $x \in \mathbb{R}$. [*Hint:* First show that if r is a rational number, then $f(r) = cr$.]

14. Let $g : \mathbb{R} \to \mathbb{R}$ satisfy the relation $g(x + y) = g(x)g(y)$ for all x, y in $\mathbb{R}$. Show that if g is continuous at $x = 0$, then g is continuous at every point of $\mathbb{R}$. Also if we have $g(a) = 0$ for some $a \in \mathbb{R}$, then $g(x) = 0$ for all $x \in \mathbb{R}$.

15. Let $f, g : \mathbb{R} \to \mathbb{R}$ be continuous at a point c, and let $h(x) := \sup\{f(x), g(x)\}$ for $x \in \mathbb{R}$. Show that $h(x) = \frac{1}{2}\big(f(x) + g(x)\big) + \frac{1}{2}|f(x) - g(x)|$ for all $x \in \mathbb{R}$. Use this to show that h is continuous at c.

Section 5.3   Continuous Functions on Intervals

Functions that are continuous on intervals have a number of very important properties that are not possessed by general continuous functions. In this section, we will establish some deep results that are of considerable importance and that will be applied later. Alternative proofs of these results will be given in Section 5.5.

5.3.1 Definition   A function $f : A \to \mathbb{R}$ is said to be **bounded on** A if there exists a constant $M > 0$ such that $|f(x)| \leq M$ for all $x \in A$.

In other words, a function is bounded on a set if its range is a bounded set in $\mathbb{R}$. To say that a function is *not* bounded on a given set is to say that no particular number can serve as a bound for its range. In exact language, a function f is not bounded on the set A if given any $M > 0$, there exists a point $x_M \in A$ such that $|f(x_M)| > M$. We often say that f is **unbounded on** A in this case.

For example, the function f defined on the interval $A := (0, \infty)$ by $f(x) := 1/x$ is not bounded on A because for any $M > 0$ we can take the point $x_M := 1/(M + 1)$ in A to get $f(x_M) = 1/x_M = M + 1 > M$. This example shows that continuous functions need not be bounded. In the next theorem, however, we show that continuous functions on a certain type of interval are necessarily bounded.

5.3.2 Boundedness Theorem[†]   Let $I := [a, b]$ be a closed bounded interval and let $f : I \to \mathbb{R}$ be continuous on I. Then f is bounded on I.

Proof.   Suppose that f is not bounded on I. Then, for any $n \in \mathbb{N}$ there is a number $x_n \in I$ such that $|f(x_n)| > n$. Since I is bounded, the sequence $X := (x_n)$ is bounded. Therefore, the Bolzano-Weierstrass Theorem 3.4.8 implies that there is a subsequence $X' = (x_{n_r})$ of X that converges to a number x. Since I is closed and the elements of X' belong to I, it follows from Theorem 3.2.6 that $x \in I$. Then f is continuous at x, so that $(f(x_{n_r}))$ converges to $f(x)$. We then conclude from Theorem 3.2.2 that the convergent sequence $(f(x_{n_r}))$ must be bounded. But this is a contradiction since

$$\left| f(x_{n_r}) \right| > n_r \geq r \qquad \text{for} \quad r \in \mathbb{N}.$$

Therefore the supposition that the continuous function f is not bounded on the closed bounded interval I leads to a contradiction.   Q.E.D.

To show that each hypothesis of the Boundedness Theorem is needed, we can construct examples that show the conclusion fails if any one of the hypotheses is relaxed.

(i)   The interval must be bounded. The function $f(x) := x$ for x in the unbounded, closed interval $A := [0, \infty)$ is continuous but not bounded on A.

(ii)   The interval must be closed. The function $g(x) := 1/x$ for x in the half-open interval $B := (0, 1]$ is continuous but not bounded on B.

(iii)   The function must be continuous. The function h defined on the closed interval $C := [0, 1]$ by $h(x) := 1/x$ for $x \in (0, 1]$ and $h(0) := 1$ is discontinuous and unbounded on C.

The Maximum-Minimum Theorem _____

5.3.3 Definition   Let $A \subseteq \mathbb{R}$ and let $f : A \to \mathbb{R}$. We say that f **has an absolute maximum** on A if there is a point $x^* \in A$ such that

$$f(x^*) \geq f(x) \qquad \text{for all} \quad x \in A.$$

We say that f **has an absolute minimum** on A if there is a point $x_* \in A$ such that

$$f(x_*) \leq f(x) \qquad \text{for all} \quad x \in A.$$

We say that x^* is an **absolute maximum point** for f on A, and that x_* is an **absolute minimum point** for f on A, if they exist.

[†]This theorem, as well as 5.3.4, is true for an arbitrary closed bounded set. For these developments, see Sections 11.2 and 11.3.

We note that a continuous function on a set A does not necessarily have an absolute maximum or an absolute minimum on the set. For example, $f(x) := 1/x$ has neither an absolute maximum nor an absolute minimum on the set $A := (0, \infty)$. (See Figure 5.3.1). There can be no absolute maximum for f on A since f is not bounded above on A, and there is no point at which f attains the value $0 = \inf\{f(x) : x \in A\}$. The same function has neither an absolute maximum nor an absolute minimum when it is restricted to the set $(0, 1)$, while it has *both* an absolute maximum and an absolute minimum when it is restricted to the set $[1, 2]$. In addition, $f(x) = 1/x$ has an absolute maximum but no absolute minimum when restricted to the set $[1, \infty)$, but no absolute maximum and no absolute minimum when restricted to the set $(1, \infty)$.

It is readily seen that if a function has an absolute maximum point, then this point is not necessarily uniquely determined. For example, the function $g(x) := x^2$ defined for $x \in A := [-1, +1]$ has the two points $x = \pm 1$ giving the absolute maximum on A, and the single point $x = 0$ yielding its absolute minimum on A. (See Figure 5.3.2.) To pick an extreme example, the constant function $h(x) := 1$ for $x \in \mathbb{R}$ is such that *every point* of $\mathbb{R}$ is both an absolute maximum and an absolute minimum point for h.

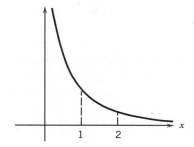

Figure 5.3.1   The function
$f(x) = 1/x$ $(x > 0)$.

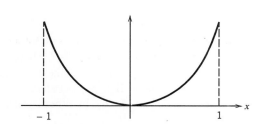

Figure 5.3.2   The function
$g(x) = x^2$ $(|x| \le 1)$.

5.3.4 Maximum-Minimum Theorem   *Let* $I := [a, b]$ *be a closed bounded interval and let* $f : I \to \mathbb{R}$ *be continuous on* I. *Then* f *has an absolute maximum and an absolute minimum on* I.

Proof.   Consider the nonempty set $f(I) := \{f(x) : x \in I\}$ of values of f on I. In Theorem 5.3.2 it was established that $f(I)$ is a bounded subset of $\mathbb{R}$. Let $s^* := \sup f(I)$ and $s_* := \inf f(I)$. We claim that there exist points x^* and x_* in I such that $s^* = f(x^*)$ and $s_* = f(x_*)$. We will establish the existence of the point x^*, leaving the proof of the existence of x_* to the reader.

Since $s^* = \sup f(I)$, if $n \in \mathbb{N}$, then the number $s^* - 1/n$ is not an upper bound of the set $f(I)$. Consequently there exists a number $x_n \in I$ such that

(1) $$s^* - \frac{1}{n} < f(x_n) \le s^* \qquad \text{for all} \quad n \in \mathbb{N}.$$

Since I is bounded, the sequence $X := (x_n)$ is bounded. Therefore, by the Bolzano–Weierstrass Theorem 3.4.8, there is a subsequence $X' = (x_{n_r})$ of X that converges to some number x^*. Since the elements of X' belong to $I = [a, b]$, it follows from Theorem 3.2.6

that $x^* \in I$. Therefore f is continuous at x^* so that $\left(\lim f(x_{n_r})\right) = f(x^*)$. Since it follows from (1) that

$$s^* - \frac{1}{n_r} < f(x_{n_r}) \leq s^* \qquad \text{for all} \quad r \in \mathbb{N},$$

we conclude from the Squeeze Theorem 3.2.7 that $\lim(f(x_{n_r})) = s^*$. Therefore we have

$$f(x^*) = \lim\left(f(x_{n_r})\right) = s^* = \sup f(I).$$

We conclude that x^* is an absolute maximum point of f on I. Q.E.D.

The next result is the theoretical basis for locating roots of a continuous function by means of sign changes of the function. The proof also provides an algorithm, known as the **Bisection Method,** for the calculation of roots to a specified degree of accuracy and can be readily programmed for a computer. It is a standard tool for finding solutions of equations of the form $f(x) = 0$, where f is a continuous function. An alternative proof of the theorem is indicated in Exercise 11.

5.3.5 Location of Roots Theorem   *Let $I = [a, b]$ and let $f : I \to \mathbb{R}$ be continuous on I. If $f(a) < 0 < f(b)$, or if $f(a) > 0 > f(b)$, then there exists a number $c \in (a, b)$ such that $f(c) = 0$.*

Proof.   We assume that $f(a) < 0 < f(b)$. We will generate a sequence of intervals by successive bisections. Let $I_1 := [a_1, b_1]$, where $a_1 := a$, $b_1 := b$, and let p_1 be the midpoint $p_1 := \frac{1}{2}(a_1 + b_1)$. If $f(p_1) = 0$, we take $c := p_1$ and we are done. If $f(p_1) \neq 0$, then either $f(p_1) > 0$ or $f(p_1) < 0$. If $f(p_1) > 0$, then we set $a_2 := a_1$, $b_2 := p_1$, while if $f(p_1) < 0$, then we set $a_2 := p_1$, $b_2 := b_1$. In either case, we let $I_2 := [a_2, b_2]$; then we have $I_2 \subset I_1$ and $f(a_2) < 0$, $f(b_2) > 0$.

We continue the bisection process. Suppose that the intervals $I_1, I_2, \cdots, I_k$ have been obtained by successive bisection in the same manner. Then we have $f(a_k) < 0$ and $f(b_k) > 0$, and we set $p_k := \frac{1}{2}(a_k + b_k)$. If $f(p_k) = 0$, we take $c := p_k$ and we are done. If $f(p_k) > 0$, we set $a_{k+1} := a_k$, $b_{k+1} := p_k$, while if $f(p_k) < 0$, we set $a_{k+1} := p_k$, $b_{k+1} := b_k$. In either case, we let $I_{k+1} := [a_{k+1}, b_{k+1}]$; then $I_{k+1} \subset I_k$ and $f(a_{k+1}) < 0$, $f(b_{k+1}) > 0$.

If the process terminates by locating a point p_n such that $f(p_n) = 0$, then we are done. If the process does not terminate, then we obtain a nested sequence of closed bounded intervals $I_n := [a_n, b_n]$ such that for every $n \in \mathbb{N}$ we have

$$f(a_n) < 0 \qquad \text{and} \qquad f(b_n) > 0.$$

Furthermore, since the intervals are obtained by repeated bisection, the length of I_n is equal to $b_n - a_n = (b - a)/2^{n-1}$. It follows from the Nested Intervals Property 2.5.2 that there exists a point c that belongs to I_n for all $n \in \mathbb{N}$. Since $a_n \leq c \leq b_n$ for all $n \in \mathbb{N}$, we have $0 \leq c - a_n \leq b_n - a_n = (b - a)/2^{n-1}$, and $0 \leq b_n - c \leq b_n - a_n = (b - a)/2^{n-1}$. Hence, it follows that $\lim(a_n) = c = \lim(b_n)$. Since f is continuous at c, we have

$$\lim\left(f(a_n)\right) = f(c) = \lim\left(f(b_n)\right).$$

The fact that $f(a_n) < 0$ for all $n \in \mathbb{N}$ implies that $f(c) = \lim\left(f(a_n)\right) \leq 0$. Also, the fact that $f(b_n) \geq 0$ for all $n \in \mathbb{N}$ implies that $f(c) = \lim\left(f(b_n)\right) \geq 0$. Thus, we conclude that $f(c) = 0$. Consequently, c is a root of f. Q.E.D.

The following example illustrates how the Bisection Method for finding roots is applied in a systematic fashion.

5.3.6 Example   The equation $f(x) = xe^x - 2 = 0$ has a root c in the interval $[0, 1]$, because f is continuous on this interval and $f(0) = -2 < 0$ and $f(1) = e - 2 > 0$. We construct the following table, where the sign of $f(p_n)$ determines the interval at the next step. The far right column is an upper bound on the error when p_n is used to approximate the root c, because we have

$$|p_n - c| \leq \tfrac{1}{2}(b_n - a_n) = 1/2^n.$$

We will find an approximation p_n with error less than 10^{-2}.

n	a_n	b_n	p_n	$f(p_n)$	$\tfrac{1}{2}(b_n - a_n)$
1	0	1	.5	−1.176	.5
2	.5	1	.75	−.412	.25
3	.75	1	.875	+.099	.125
4	.75	.875	.8125	−.169	.0625
5	.8125	.875	.84375	−.0382	.03125
6	.84375	.875	.859375	+.0296	.015625
7	.84375	.859375	.8515625	—	.0078125

We have stopped at $n = 7$, obtaining $c \approx p_7 = .8515625$ with error less than .0078125. This is the first step in which the error is less than 10^{-2}. The decimal place values of p_7 past the second place cannot be taken seriously, but we can conclude that $.843 < c < .860$.   □

Bolzano's Theorem

The next result is a generalization of the Location of Roots Theorem. It assures us that a continuous function on an interval takes on (at least once) any number that lies between two of its values.

5.3.7 Bolzano's Intermediate Value Theorem   *Let I be an interval and let $f : I \to \mathbb{R}$ be continuous on I. If $a, b \in I$ and if $k \in \mathbb{R}$ satisfies $f(a) < k < f(b)$, then there exists a point $c \in I$ between a and b such that $f(c) = k$.*

Proof.   Suppose that $a < b$ and let $g(x) := f(x) - k$; then $g(a) < 0 < g(b)$. By the Location of Roots Theorem 5.3.5 there exists a point c with $a < c < b$ such that $0 = g(c) = f(c) - k$. Therefore $f(c) = k$.

If $b < a$, let $h(x) := k - f(x)$ so that $h(b) < 0 < h(a)$. Therefore there exists a point c with $b < c < a$ such that $0 = h(c) = k - f(c)$, whence $f(c) = k$.   Q.E.D.

5.3.8 Corollary   *Let $I = [a, b]$ be a closed, bounded interval and let $f : I \to \mathbb{R}$ be continuous on I. If $k \in \mathbb{R}$ is any number satisfying*

$$\inf f(I) \leq k \leq \sup f(I),$$

then there exists a number $c \in I$ such that $f(c) = k$.

Proof.   It follows from the Maximum-Minimum Theorem 5.3.4 that there are points c_* and c^* in I such that

$$\inf f(I) = f(c_*) \leq k \leq f(c^*) = \sup f(I).$$

The conclusion now follows from Bolzano's Theorem 5.3.7.   Q.E.D.

The next theorem summarizes the main results of this section. It states that the image of a closed bounded interval under a continuous function is also a closed bounded interval. The endpoints of the **image interval** are the absolute minimum and absolute maximum values of the function, and the statement that all values between the absolute minimum and the absolute maximum values belong to the image is a way of describing Bolzano's Intermediate Value Theorem.

5.3.9 Theorem   *Let I be a closed bounded interval and let $f : I \to \mathbb{R}$ be continuous on I. Then the set $f(I) := \{f(x) : x \in I\}$ is a closed bounded interval.*

Proof.   If we let $m := \inf f(I)$ and $M := \sup f(I)$, then we know from the Maximum-Minimum Theorem 5.3.4 that m and M belong to $f(I)$. Moreover, we have $f(I) \subseteq [m, M]$. If k is any element of $[m, M]$, then it follows from the preceding corollary that there exists a point $c \in I$ such that $k = f(c)$. Hence, $k \in f(I)$ and we conclude that $[m, M] \subseteq f(I)$. Therefore, $f(I)$ is the interval $[m, M]$.   Q.E.D.

Warning   If $I := [a, b]$ is an interval and $f : I \to \mathbb{R}$ is continuous on I, we have proved that $f(I)$ is the interval $[m, M]$. We have *not* proved (and it is not always true) that $f(I)$ is the interval $[f(a), f(b)]$. (See Figure 5.3.3.)

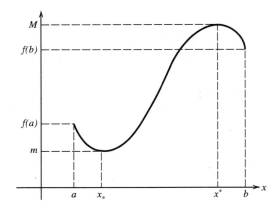

Figure 5.3.3 $f(I) = [m, M]$.

The preceding theorem is a "preservation" theorem in the sense that it states that the continuous image of a closed bounded interval is a set of the same type. The next theorem extends this result to general intervals. However, it should be noted that although the continuous image of an interval is shown to be an interval, it is *not* true that the image interval necessarily has the *same form* as the domain interval. For example, the continuous image of an open interval need not be an open interval, and the continuous image of an unbounded closed interval need not be a closed interval. Indeed, if $f(x) := 1/(x^2 + 1)$ for $x \in \mathbb{R}$, then f is continuous on $\mathbb{R}$ [see Example 5.2.3(b)]. It is easy to see that if $I_1 := (-1, 1)$, then $f(I_1) = (\frac{1}{2}, 1]$, which is not an open interval. Also, if $I_2 := [0, \infty)$, then $f(I_2) = (0, 1]$, which is not a closed interval. (See Figure 5.3.4.)

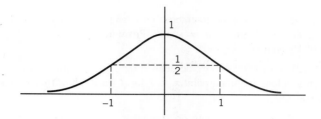

Figure 5.3.4 Graph of $f(x) = 1/(x^2 + 1)$ $(x \in \mathbb{R})$.

To prove the Preservation of Intervals Theorem 5.3.10, we will use Theorem 2.5.1 characterizing intervals.

5.3.10 Preservation of Intervals Theorem *Let I be an interval and let $f : I \to \mathbb{R}$ be continuous on I. Then the set $f(I)$ is an interval.*

Proof. Let $\alpha, \beta \in f(I)$ with $\alpha < \beta$; then there exist points $a, b \in I$ such that $\alpha = f(a)$ and $\beta = f(b)$. Further, it follows from Bolzano's Intermediate Value Theorem 5.3.7 that if $k \in (\alpha, \beta)$ then there exists a number $c \in I$ with $k = f(c) \in f(I)$. Therefore $[\alpha, \beta] \subseteq f(I)$, showing that $f(I)$ possesses property (1) of Theorem 2.5.1. Therefore $f(I)$ is an interval.   Q.E.D.

Exercises for Section 5.3

1. Let $I := [a, b]$ and let $f : I \to \mathbb{R}$ be a continuous function such that $f(x) > 0$ for each x in I. Prove that there exists a number $\alpha > 0$ such that $f(x) \geq \alpha$ for all $x \in I$.

2. Let $I := [a, b]$ and let $f : I \to \mathbb{R}$ and $g : I \to \mathbb{R}$ be continuous functions on I. Show that the set $E := \{x \in I : f(x) = g(x)\}$ has the property that if $(x_n) \subseteq E$ and $x_n \to x_0$, then $x_0 \in E$.

3. Let $I := [a, b]$ and let $f : I \to \mathbb{R}$ be a continuous function on I such that for each x in I there exists y in I such that $|f(y)| \leq \frac{1}{2}|f(x)|$. Prove there exists a point c in I such that $f(c) = 0$.

4. Show that every polynomial of odd degree with real coefficients has at least one real root.

5. Show that the polynomial $p(x) := x^4 + 7x^3 - 9$ has at least two real roots. Use a calculator to locate these roots to within two decimal places.

6. Let f be continuous on the interval $[0, 1]$ to $\mathbb{R}$ and such that $f(0) = f(1)$. Prove that there exists a point c in $[0, \frac{1}{2}]$ such that $f(c) = f(c + \frac{1}{2})$. [*Hint*: Consider $g(x) = f(x) - f(x + \frac{1}{2})$.] Conclude that there are, at any time, antipodal points on the earth's equator that have the same temperature.

7. Show that the equation $x = \cos x$ has a solution in the interval $[0, \pi/2]$. Use the Bisection Method and a calculator to find an approximate solution of this equation, with error less than 10^{-3}.

8. Show that the function $f(x) := 2 \ln x + \sqrt{x} - 2$ has root in the interval $[1, 2]$. Use the Bisection Method and a calculator to find the root with error less than 10^{-2}.

9. (a) The function $f(x) := (x - 1)(x - 2)(x - 3)(x - 4)(x - 5)$ has five roots in the interval $[0, 7]$. If the Bisection Method is applied on this interval, which of the roots is located?
   (b) Same question for $g(x) := (x - 2)(x - 3)(x - 4)(x - 5)(x - 6)$ on the interval $[0, 7]$.

10. If the Bisection Method is used on an interval of length 1 to find p_n with error $|p_n - c| < 10^{-5}$, determine the least value of n that will assure this accuracy.

11. Let $I := [a, b]$, let $f : I \to \mathbb{R}$ be continuous on I, and assume that $f(a) < 0$, $f(b) > 0$. Let $W := \{x \in I : f(x) < 0\}$, and let $w := \sup W$. Prove that $f(w) = 0$. (This provides an alternative proof of Theorem 5.3.5.)

12. Let $I := [0, \pi/2]$ and let $f : I \to \mathbb{R}$ be defined by $f(x) := \sup\{x^2, \cos x\}$ for $x \in I$. Show there exists an absolute minimum point $x_0 \in I$ for f on I. Show that x_0 is a solution to the equation $\cos x = x^2$.

13. Suppose that $f : \mathbb{R} \to \mathbb{R}$ is continuous on $\mathbb{R}$ and that $\lim_{x \to -\infty} f = 0$ and $\lim_{x \to \infty} f = 0$. Prove that f is bounded on $\mathbb{R}$ and attains either a maximum or minimum on $\mathbb{R}$. Give an example to show that both a maximum and a minimum need not be attained.

14. Let $f : \mathbb{R} \to \mathbb{R}$ be continuous on $\mathbb{R}$ and let $\beta \in \mathbb{R}$. Show that if $x_0 \in \mathbb{R}$ is such that $f(x_0) < \beta$, then there exists a δ-neighborhood U of x_0 such that $f(x) < \beta$ for all $x \in U$.

15. Examine which open [respectively, closed] intervals are mapped by $f(x) := x^2$ for $x \in \mathbb{R}$ onto open [respectively, closed] intervals.

16. Examine the mapping of open [respectively, closed] intervals under the functions $g(x) := 1/(x^2 + 1)$ and $h(x) := x^3$ for $x \in \mathbb{R}$.

17. If $f : [0, 1] \to \mathbb{R}$ is continuous and has only rational [respectively, irrational] values, must f be constant? Prove your assertion.

18. Let $I := [a, b]$ and let $f : I \to \mathbb{R}$ be a (not necessarily continuous) function with the property that for every $x \in I$, the function f is bounded on a neighborhood $V_{\delta_x}(x)$ of x (in the sense of Definition 4.2.1). Prove that f is bounded on I.

19. Let $J := (a, b)$ and let $g : J \to \mathbb{R}$ be a continuous function with the property that for every $x \in J$, the function g is bounded on a neighborhood $V_{\delta_x}(x)$ of x. Show by example that g is not necessarily bounded on J.

Section 5.4   Uniform Continuity

Let $A \subseteq \mathbb{R}$ and let $f : A \to \mathbb{R}$. Definition 5.1.1 states that the following statements are equivalent:

(i) f is continuous at every point $u \in A$;

(ii)   given $\varepsilon > 0$ and $u \in A$, there is a $\delta(\varepsilon, u) > 0$ such that for all x such that $x \in A$ and $|x - u| < \delta(\varepsilon, u)$, then $|f(x) - f(u)| < \varepsilon$.

The point we wish to emphasize here is that δ depends, in general, on *both* $\varepsilon > 0$ and $u \in A$. The fact that δ depends on u is a reflection of the fact that the function f may change its values rapidly near certain points and slowly near other points. [For example, consider $f(x) := \sin(1/x)$ for $x > 0$; see Figure 4.1.3.]

Now it often happens that the function f is such that the number δ can be chosen to be independent of the point $u \in A$ and to depend only on ε. For example, if $f(x) := 2x$ for all $x \in \mathbb{R}$, then

$$|f(x) - f(u)| = 2|x - u|,$$

and so we can choose $\delta(\varepsilon, u) := \varepsilon/2$ for all $\varepsilon > 0$, $u \in \mathbb{R}$. (Why?)

On the other hand if $g(x) := 1/x$ for $x \in A := \{x \in \mathbb{R} : x > 0\}$, then

(1)
$$g(x) - g(u) = \frac{u - x}{ux}.$$

If $u \in A$ is given and if we take

(2) $$\delta(\varepsilon, u) := \inf\left\{\tfrac{1}{2}u, \tfrac{1}{2}u^2\varepsilon\right\},$$

then if $|x - u| < \delta(\varepsilon, u)$, we have $|x - u| < \tfrac{1}{2}u$ so that $\tfrac{1}{2}u < x < \tfrac{3}{2}u$, whence it follows that $1/x < 2/u$. Thus, if $|x - u| < \tfrac{1}{2}u$, the equality (1) yields the inequality

(3) $$|g(x) - g(u)| \leq (2/u^2)\,|x - u|.$$

Consequently, if $|x - u| < \delta(\varepsilon, u)$, then (2) and (3) imply that

$$|g(x) - g(u)| < (2/u^2)\left(\tfrac{1}{2}u^2\varepsilon\right) = \varepsilon.$$

We have seen that the selection of $\delta(\varepsilon, u)$ by the formula (2) "works" in the sense that it enables us to give a value of δ that will ensure that $|g(x) - g(u)| < \varepsilon$ when $|x - u| < \delta$ and $x, u \in A$. We note that the value of $\delta(\varepsilon, u)$ given in (2) certainly depends on the point $u \in A$. If we wish to consider *all* $u \in A$, formula (2) does not lead to one value $\delta(\varepsilon) > 0$ that will "work" simultaneously for all $u > 0$, since $\inf\{\delta(\varepsilon, u) : u > 0\} = 0$.

An alert reader will have observed that there are other selections that can be made for δ. (For example we could also take $\delta_1(\varepsilon, u) := \inf\left\{\tfrac{1}{3}u, \tfrac{2}{3}u^2\varepsilon\right\}$, as the reader can show; however, we still have $\inf\left\{\delta_1(\varepsilon, u): u > 0\right\} = 0$.) In fact, there is no way of choosing one value of δ that will "work" for all $u > 0$ for the function $g(x) = 1/x$, as we shall see.

The situation is exhibited graphically in Figures 5.4.1 and 5.4.2 where, for a given ε-neighborhood $V_\varepsilon\left(\tfrac{1}{2}\right)$ about $\tfrac{1}{2} = f(2)$ and $V_\varepsilon(2)$ about $2 = f\left(\tfrac{1}{2}\right)$, the corresponding maximum values of δ are seen to be considerably different. As u tends to 0, the permissible values of δ tend to 0.

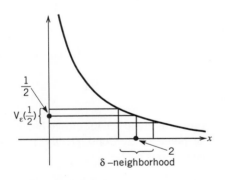

Figure 5.4.1 $g(x) = 1/x$ $(x > 0)$.

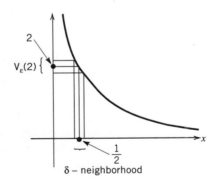

Figure 5.4.2 $g(x) = 1/x$ $(x > 0)$.

5.4.1 Definition   Let $A \subseteq \mathbb{R}$ and let $f : A \to \mathbb{R}$. We say that f is **uniformly continuous** on A if for each $\varepsilon > 0$ there is a $\delta(\varepsilon) > 0$ such that if $x, u \in A$ are any numbers satisfying $|x - u| < \delta(\varepsilon)$, then $|f(x) - f(u)| < \varepsilon$.

It is clear that if f is uniformly continuous on A, then it is continuous at every point of A. In general, however, the converse does not hold, as is shown by the function $g(x) = 1/x$ on the set $A := \{x \in \mathbb{R} : x > 0\}$.

It is useful to formulate a condition equivalent to saying that f is *not* uniformly continuous on A. We give such criteria in the next result, leaving the proof to the reader as an exercise.

5.4.2 Nonuniform Continuity Criteria   *Let $A \subseteq \mathbb{R}$ and let $f : A \to \mathbb{R}$. Then the following statements are equivalent:*

(i)   *f is not uniformly continuous on A.*

(ii)   *There exists an $\varepsilon_0 > 0$ such that for every $\delta > 0$ there are points x_δ, u_δ in A such that $|x_\delta - u_\delta| < \delta$ and $|f(x_\delta) - f(u_\delta)| \geq \varepsilon_0$.*

(iii)   *There exists an $\varepsilon_0 > 0$ and two sequences (x_n) and (u_n) in A such that $\lim(x_n - u_n) = 0$ and $|f(x_n) - f(u_n)| \geq \varepsilon_0$ for all $n \in \mathbb{N}$.*

We can apply this result to show that $g(x) := 1/x$ is not uniformly continuous on $A := \{x \in \mathbb{R} : x > 0\}$. For, if $x_n := 1/n$ and $u_n := 1/(n+1)$, then we have $\lim(x_n - u_n) = 0$, but $|g(x_n) - g(u_n)| = 1$ for all $n \in \mathbb{N}$.

We now present an important result that assures that a continuous function on a closed bounded interval I is uniformly continuous on I. Other proofs of this theorem are given in Sections 5.5 and 11.3.

5.4.3 Uniform Continuity Theorem   *Let I be a closed bounded interval and let $f : I \to \mathbb{R}$ be continuous on I. Then f is uniformly continuous on I.*

Proof.   If f is not uniformly continuous on I then, by the preceding result, there exists $\varepsilon_0 > 0$ and two sequences (x_n) and (u_n) in I such that $|x_n - u_n| < 1/n$ and $|f(x_n) - f(u_n)| \geq \varepsilon_0$ for all $n \in \mathbb{N}$. Since I is bounded, the sequence (x_n) is bounded; by the Bolzano-Weierstrass Theorem 3.4.8 there is a subsequence (x_{n_k}) of (x_n) that converges to an element z. Since I is closed, the limit z belongs to I, by Theorem 3.2.6. It is clear that the corresponding subsequence (u_{n_k}) also converges to z, since

$$|u_{n_k} - z| \leq |u_{n_k} - x_{n_k}| + |x_{n_k} - z|.$$

Now if f is continuous at the point z, then both of the sequences $\left(f(x_{n_k})\right)$ and $\left(f(u_{n_k})\right)$ must converge to $f(z)$. But this is not possible since

$$|f(x_n) - f(u_n)| \geq \varepsilon_0$$

for all $n \in \mathbb{N}$. Thus the hypothesis that f is not uniformly continuous on the closed bounded interval I implies that f is not continuous at some point $z \in I$. Consequently, if f is continuous at every point of I, then f is uniformly continuous on I.   Q.E.D.

Lipschitz Functions

If a uniformly continuous function is given on a set that is not a closed bounded interval, then it is sometimes difficult to establish its uniform continuity. However, there is a condition that frequently occurs that is sufficient to guarantee uniform continuity.

5.4.4 Definition   Let $A \subseteq \mathbb{R}$ and let $f : A \to \mathbb{R}$. If there exists a constant $K > 0$ such that

(4) $$|f(x) - f(u)| \leq K|x - u|$$

for all $x, u \in A$, then f is said to be a **Lipschitz function** (or to satisfy a **Lipschitz condition**) on A.

The condition (4) that a function $f : I \to \mathbb{R}$ on an interval I is a Lipschitz function can be interpreted geometrically as follows. If we write the condition as

$$\left| \frac{f(x) - f(u)}{x - u} \right| \leq K, \qquad x, u \in I, x \neq u,$$

then the quantity inside the absolute values is the slope of a line segment joining the points $(x, f(x))$ and $(u, f(u))$. Thus a function f satisfies a Lipschitz condition if and only if the slopes of all line segments joining two points on the graph of $y = f(x)$ over I are bounded by some number K.

5.4.5 Theorem   *If $f : A \to \mathbb{R}$ is a Lipschitz function, then f is uniformly continuous on A.*

Proof.   If condition (4) is satisfied, then given $\varepsilon > 0$, we can take $\delta := \varepsilon / K$. If $x, u \in A$ satisfy $|x - u| < \delta$, then

$$|f(x) - f(u)| < K \cdot \frac{\varepsilon}{K} = \varepsilon.$$

Therefore f is uniformly continuous on A.   Q.E.D.

5.4.6 Examples   **(a)**   If $f(x) := x^2$ on $A := [0, b]$, where $b > 0$, then

$$|f(x) - f(u)| = |x + u| \, |x - u| \leq 2b \, |x - u|$$

for all x, u in $[0, b]$. Thus f satisfies (4) with $K := 2b$ on A, and therefore f is uniformly continuous on A. Of course, since f is continuous and A is a closed bounded interval, this can also be deduced from the Uniform Continuity Theorem. (Note that f does *not* satisfy a Lipschitz condition on the interval $[0, \infty)$.)

(b)   Not every uniformly continuous function is a Lipschitz function.

Let $g(x) := \sqrt{x}$ for x in the closed bounded interval $I := [0, 2]$. Since g is continuous on I, it follows from the Uniform Continuity Theorem 5.4.3 that g is uniformly continuous on I. However, there is no number $K > 0$ such that $|g(x)| \leq K|x|$ for all $x \in I$. (Why not?) Therefore, g is not a Lipschitz function on I.

(c)   The Uniform Continuity Theorem and Theorem 5.4.5 can sometimes be combined to establish the uniform continuity of a function on a set.

We consider $g(x) := \sqrt{x}$ on the set $A := [0, \infty)$. The uniform continuity of g on the interval $I := [0, 2]$ follows from the Uniform Continuity Theorem as noted in (b). If $J := [1, \infty)$, then if both x, u are in J, we have

$$|g(x) - g(u)| = \left| \sqrt{x} - \sqrt{u} \right| = \frac{|x - u|}{\sqrt{x} + \sqrt{u}} \leq \tfrac{1}{2} |x - u|.$$

Thus g is a Lipschitz function on J with constant $K = \tfrac{1}{2}$, and hence by Theorem 5.4.5, g is uniformly continuous on $[1, \infty)$. Since $A = I \cup J$, it follows $\left[$by taking $\delta(\varepsilon) := \inf \left\{ 1, \delta_I(\varepsilon), \delta_J(\varepsilon) \right\} \right]$ that g is uniformly continuous on A. We leave the details to the reader.   □

The Continuous Extension Theorem _____

We have seen examples of functions that are continuous but not uniformly continuous on open intervals; for example, the function $f(x) = 1/x$ on the interval $(0, 1)$. On the other hand, by the Uniform Continuity Theorem, a function that is continuous on a closed bounded interval is always uniformly continuous. So the question arises: Under what conditions is a function uniformly continuous on a bounded *open* interval? The answer reveals the strength of uniform continuity, for it will be shown that a function on (a, b) is uniformly continuous if and only if it can be defined at the endpoints to produce a function that is continuous on the closed interval. We first establish a result that is of interest in itself.

5.4.7 Theorem   *If $f : A \to \mathbb{R}$ is uniformly continuous on a subset A of $\mathbb{R}$ and if (x_n) is a Cauchy sequence in A, then $\big(f(x_n)\big)$ is a Cauchy sequence in $\mathbb{R}$.*

Proof.   Let (x_n) be a Cauchy sequence in A, and let $\varepsilon > 0$ be given. First choose $\delta > 0$ such that if x, u in A satisfy $|x - u| < \delta$, then $|f(x) - f(u)| < \varepsilon$. Since (x_n) is a Cauchy sequence, there exists $H(\delta)$ such that $|x_n - x_m| < \delta$ for all $n, m > H(\delta)$. By the choice of δ, this implies that for $n, m > H(\delta)$, we have $|f(x_n) - f(x_m)| < \varepsilon$. Therefore the sequence $(f(x_n))$ is a Cauchy sequence.   Q.E.D.

The preceding result gives us an alternative way of seeing that $f(x) := 1/x$ is not uniformly continuous on $(0, 1)$. We note that the sequence given by $x_n := 1/n$ in $(0, 1)$ is a Cauchy sequence, but the image sequence, where $f(x_n) = n$, is not a Cauchy sequence.

5.4.8 Continuous Extension Theorem   *A function f is uniformly continuous on the interval (a, b) if and only if it can be defined at the endpoints a and b such that the extended function is continuous on $[a, b]$.*

Proof.   ($\Leftarrow$) This direction is trivial.
($\Rightarrow$) Suppose f is uniformly continuous on (a, b). We shall show how to extend f to a; the argument for b is similar. This is done by showing that $\lim_{x \to c} f(x) = L$ exists, and this is accomplished by using the sequential criterion for limits. If (x_n) is a sequence in (a, b) with $\lim(x_n) = a$, then it is a Cauchy sequence, and by the preceding theorem, the sequence $\big(f(x_n)\big)$ is also a Cauchy sequence, and so is convergent by Theorem 3.5.5. Thus the limit $\lim\big(f(x_n)\big) = L$ exists. If (u_n) is any other sequence in (a, b) that converges to a, then $\lim(u_n - x_n) = a - a = 0$, so by the uniform continuity of f we have

$$\lim\big(f(u_n)\big) = \lim\big(f(u_n) - f(x_n)\big) + \lim\big(f(x_n)\big)$$
$$= 0 + L = L.$$

Since we get the same value L for every sequence converging to a, we infer from the sequential criterion for limits that f has limit L at a. If we define $f(a) := L$, then f is continuous at a. The same argument applies to b, so we conclude that f has a continuous extension to the interval $[a, b]$.   Q.E.D.

Since the limit of $f(x) := \sin(1/x)$ at 0 does not exist, we infer from the Continuous Extension Theorem that the function is not uniformly continuous on $(0, b]$ for any $b > 0$. On the other hand, since $\lim_{x \to 0} x \sin(1/x) = 0$ exists, the function $g(x) := x \sin(1/x)$ is uniformly continuous on $(0, b]$ for all $b > 0$.

Approximation[†]

In many applications it is important to be able to approximate continuous functions by functions of an elementary nature. Although there are a variety of definitions that can be used to make the word "approximate" more precise, one of the most natural (as well as one of

---

[†]The rest of this section can be omitted on a first reading of this chapter.

the most important) is to require that, at every point of the given domain, the approximating function shall not differ from the given function by more than the preassigned error.

5.4.9 Definition　Let $I \subseteq \mathbb{R}$ be an interval and let $s : I \to \mathbb{R}$. Then s is called a **step function** if it has only a finite number of distinct values, each value being assumed on one or more intervals in I.

For example, the function $s : [-2, 4] \to \mathbb{R}$ defined by

$$
s(x) := \begin{cases}
0, & -2 \leq x < -1, \\
1, & -1 \leq x \leq 0, \\
\frac{1}{2}, & 0 < x < \frac{1}{2}, \\
3, & \frac{1}{2} \leq x < 1, \\
-2, & 1 \leq x \leq 3, \\
2, & 3 < x \leq 4,
\end{cases}
$$

is a step function. (See Figure 5.4.3.)

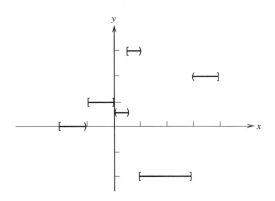

Figure 5.4.3　Graph of $y = s(x)$.

We will now show that a continuous function on a closed bounded interval I can be approximated arbitrarily closely by step functions.

5.4.10 Theorem　*Let I be a closed bounded interval and let $f : I \to \mathbb{R}$ be continuous on I. If $\varepsilon > 0$, then there exists a step function $s_\varepsilon : I \to \mathbb{R}$ such that $|f(x) - s_\varepsilon(x)| < \varepsilon$ for all $x \in I$.*

Proof.　Since (by the Uniform Continuity Theorem 5.4.3) the function f is uniformly continuous, it follows that given $\varepsilon > 0$ there is a number $\delta(\varepsilon) > 0$ such that if $x, y \in I$ and $|x - y| < \delta(\varepsilon)$, then $|f(x) - f(y)| < \varepsilon$. Let $I := [a, b]$ and let $m \in \mathbb{N}$ be sufficiently large so that $h := (b - a)/m < \delta(\varepsilon)$. We now divide $I = [a, b]$ into m disjoint intervals of length h; namely, $I_1 := [a, a + h]$, and $I_k := \left(a + (k - 1)h, a + kh\right]$ for $k = 2, \cdots, m$. Since the length of each subinterval I_k is $h < \delta(\varepsilon)$, the difference between any two values of f in I_k is less than ε. We now define

$$(5) \qquad\qquad s_\varepsilon(x) := f(a + kh) \qquad \text{for} \quad x \in I_k, \quad k = 1, \cdots, m,$$

so that s_ε is constant on each interval I_k. (In fact the value of s_ε on I_k is the value of f at the right endpoint of I_k. See Figure 5.4.4.) Consequently if $x \in I_k$, then

$$|f(x) - s_\varepsilon(x)| = |f(x) - f(a + kh)| < \varepsilon.$$

Therefore we have $|f(x) - s_\varepsilon(x)| < \varepsilon$ for all $x \in I$.   Q.E.D.

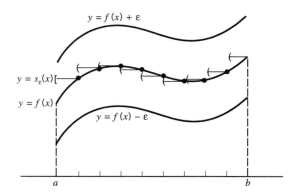

Figure 5.4.4   Approximation by step functions.

Note that the proof of the preceding theorem establishes somewhat more than was announced in the statement of the theorem. In fact, we have proved the following, more precise, assertion.

5.4.11 Corollary   Let $I := [a, b]$ be a closed bounded interval and let $f : I \to \mathbb{R}$ be continuous on I. If $\varepsilon > 0$, there exists a natural number m such that if we divide I into m disjoint intervals I_k having length $h := (b - a)/m$, then the step function s_ε defined in equation (5) satisfies $|f(x) - s_\varepsilon(x)| < \varepsilon$ for all $x \in I$.

Step functions are extremely elementary in character, but they are not continuous (except in trivial cases). Since it is often desirable to approximate continuous functions by elementary continuous functions, we now shall show that we can approximate continuous functions by continuous piecewise linear functions.

5.4.12 Definition   Let $I := [a, b]$ be an interval. Then a function $g : I \to \mathbb{R}$ is said to be **piecewise linear** on I if I is the union of a finite number of disjoint intervals $I_1, \cdots, I_m$, such that the restriction of g to each interval I_k is a linear function.

Remark   It is evident that in order for a piecewise linear function g to be continuous on I, the line segments that form the graph of g must meet at the endpoints of adjacent subintervals I_k, I_{k+1} $(k = 1, \cdots, m - 1)$.

5.4.13 Theorem   Let I be a closed bounded interval and let $f : I \to \mathbb{R}$ be continuous on I. If $\varepsilon > 0$, then there exists a continuous piecewise linear function $g_\varepsilon : I \to \mathbb{R}$ such that $|f(x) - g_\varepsilon(x)| < \varepsilon$ for all $x \in I$.

Proof.   Since f is uniformly continuous on $I := [a, b]$, there is a number $\delta(\varepsilon) > 0$ such that if $x, y \in I$ and $|x - y| < \delta(\varepsilon)$, then $|f(x) - f(y)| < \varepsilon$. Let $m \in \mathbb{N}$ be sufficiently large so that $h := (b - a)/m < \delta(\varepsilon)$. Divide $I = [a, b]$ into m disjoint intervals of length h; namely let $I_1 = [a, a + h]$, and let $I_k = (a + (k - 1)h, a + kh]$ for $k = 2, \cdots, m$. On each interval I_k we define g_ε to be the linear function joining the points

$$\big(a + (k - 1)h, f(a + (k - 1)h)\big) \qquad \text{and} \qquad \big(a + kh, f(a + kh)\big).$$

Then g_ε is a continuous piecewise linear function on I. Since, for $x \in I_k$ the value $f(x)$ is within ε of $f(a + (k - 1)h)$ and $f(a + kh)$, it is an exercise to show that $|f(x) - g_\varepsilon(x)| < \varepsilon$ for all $x \in I_k$; therefore this inequality holds for all $x \in I$. (See Figure 5.4.5.)   Q.E.D.

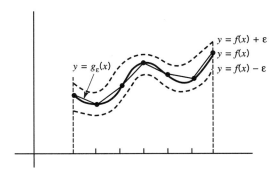

Figure 5.4.5   Approximation by piecewise linear function.

We shall close this section by stating the important theorem of Weierstrass concerning the approximation of continuous functions by polynomial functions. As would be expected, in order to obtain an approximation within an arbitrarily preassigned $\varepsilon > 0$, we must be prepared to use polynomials of arbitrarily high degree.

5.4.14 Weierstrass Approximation Theorem   *Let $I = [a, b]$ and let $f : I \to \mathbb{R}$ be a continuous function. If $\varepsilon > 0$ is given, then there exists a polynomial function p_ε such that $|f(x) - p_\varepsilon(x)| < \varepsilon$ for all $x \in I$.*

There are a number of proofs of this result. Unfortunately, all of them are rather intricate, or employ results that are not yet at our disposal. One of the most elementary proofs is based on the following theorem, due to Serge Bernstein, for continuous functions on $[0, 1]$. Given $f : [0, 1] \to \mathbb{R}$, Bernstein defined the sequence of polynomials:

(6) $$B_n(x) = \sum_{k=0}^{n} f\left(\frac{k}{n}\right) \binom{n}{k} x^k (1 - x)^{n-k}.$$

The polynomial function B_n is called the nth **Bernstein polynomial for** f; it is a polynomial of degree at most n and its coefficients depend on the values of the function f at the $n + 1$ equally spaced points $0, 1/n, 2/n, \cdots, k/n, \cdots, 1$, and on the binomial coefficients

$$\binom{n}{k} = \frac{n!}{k!(n - k)} = \frac{n(n - 1) \cdots (n - k + 1)}{1 \cdot 2 \cdots k}.$$

5.4.15 Bernstein's Approximation Theorem   *Let $f : [0, 1] \to \mathbb{R}$ be continuous and let $\varepsilon > 0$. There exists an $n_\varepsilon \in \mathbb{N}$ such that if $n \geq n_\varepsilon$, then we have $|f(x) - B_n(x)| < \varepsilon$ for all $x \in [0, 1]$.*

The proof of Bernstein's Approximation Theorem is given in [ERA, pp. 169–172].

The Weierstrass Approximation Theorem 5.4.14 can be derived from the Bernstein Approximation Theorem 5.4.15 by a change of variable. Specifically, we replace $f : [a, b] \to \mathbb{R}$ by a function $F : [0, 1] \to \mathbb{R}$, defined by

$$F(t) := f\big(a + (b - a)t\big) \qquad \text{for} \quad t \in [0, 1].$$

The function F can be approximated by Bernstein polynomials for F on the interval $[0, 1]$, which can then yield polynomials on $[a, b]$ that approximate f.

Exercises for Section 5.4

1. Show that the function $f(x) := 1/x$ is uniformly continuous on the set $A := [a, \infty)$, where a is a positive constant.

2. Show that the function $f(x) := 1/x^2$ is uniformly continuous on $A := [1, \infty)$, but that it is not uniformly continuous on $B := (0, \infty)$.

3. Use the Nonuniform Continuity Criterion 5.4.2 to show that the following functions are not uniformly continuous on the given sets.
   (a) $f(x) := x^2$, $A := [0, \infty)$.
   (b) $g(x) := \sin(1/x)$, $B := (0, \infty)$.

4. Show that the function $f(x) := 1/(1 + x^2)$ for $x \in \mathbb{R}$ is uniformly continuous on $\mathbb{R}$.

5. Show that if f and g are uniformly continuous on a subset A of $\mathbb{R}$, then $f + g$ is uniformly continuous on A.

6. Show that if f and g are uniformly continuous on $A \subseteq \mathbb{R}$ and if they are *both* bounded on A, then their product fg is uniformly continuous on A.

7. If $f(x) := x$ and $g(x) := \sin x$, show that both f and g are uniformly continuous on $\mathbb{R}$, but that their product fg is not uniformly continuous on $\mathbb{R}$.

8. Prove that if f and g are each uniformly continuous on $\mathbb{R}$, then the composite function $f \circ g$ is uniformly continuous on $\mathbb{R}$.

9. If f is uniformly continuous on $A \subseteq \mathbb{R}$, and $|f(x)| \geq k > 0$ for all $x \in A$, show that $1/f$ is uniformly continuous on A.

10. Prove that if f is uniformly continuous on a bounded subset A of $\mathbb{R}$, then f is bounded on A.

11. If $g(x) := \sqrt{x}$ for $x \in [0, 1]$, show that there does not exist a constant K such that $|g(x)| \leq K|x|$ for all $x \in [0, 1]$. Conclude that the uniformly continuous g is not a Lipschitz function on $[0, 1]$.

12. Show that if f is continuous on $[0, \infty)$ and uniformly continuous on $[a, \infty)$ for some positive constant a, then f is uniformly continuous on $[0, \infty)$.

13. Let $A \subseteq \mathbb{R}$ and suppose that $f : A \to \mathbb{R}$ has the following property: for each $\varepsilon > 0$ there exists a function $g_\varepsilon : A \to \mathbb{R}$ such that g_ε is uniformly continuous on A and $|f(x) - g_\varepsilon(x)| < \varepsilon$ for all $x \in A$. Prove that f is uniformly continuous on A.

14. A function $f: \mathbb{R} \to \mathbb{R}$ is said to be **periodic** on $\mathbb{R}$ if there exists a number $p > 0$ such that $f(x + p) = f(x)$ for all $x \in \mathbb{R}$. Prove that a continuous periodic function on $\mathbb{R}$ is bounded and uniformly continuous on $\mathbb{R}$.

15. If $f_0(x) := 1$ for $x \in [0, 1]$, calculate the first few Bernstein polynomials for f_0. Show that they coincide with f_0. [*Hint*: The Binomial Theorem asserts that

$$(a + b)^n = \sum_{k=0}^{n} \binom{n}{k} a^k b^{n-k}.]$$

16. If $f_1(x) := x$ for $x \in [0, 1]$, calculate the first few Bernstein polynomials for f_1. Show that they coincide with f_1.

17. If $f_2(x) := x^2$ for $x \in [0, 1]$, calculate the first few Bernstein polynomials for f_2. Show that $B_n(x) = (1 - 1/n)x^2 + (1/n)x$.

Section 5.5 Continuity and Gauges

We will now introduce some concepts that will be used later—especially in Chapters 7 and 10 on integration theory. However, we wish to introduce the notion of a "gauge" now because of its connection with the study of continuous functions. We first define the notion of a tagged partition of an interval.

5.5.1 Definition A **partition** of an interval $I := [a, b]$ is a collection $\mathcal{P} = \{I_1, \cdots, I_n\}$ of non-overlapping closed intervals whose union is $[a, b]$. We ordinarily denote the intervals by $I_i := [x_{i-1}, x_i]$, where

$$a = x_0 < \cdots < x_{i-1} < x_i < \cdots < x_n = b.$$

The points x_i $(i = 0, \cdots, n)$ are called the **partition points** of $\mathcal{P}$. If a point t_i has been chosen from each interval I_i, for $i = 1, \cdots, n$, then the points t_i are called the **tags** and the set of ordered pairs

$$\dot{\mathcal{P}} = \{(I_1, t_1), \cdots, (I_n, t_n)\}$$

is called a **tagged partition** of I. (The dot signifies that the partition is tagged.)

The "fineness" of a partition $\mathcal{P}$ refers to the lengths of the subintervals in $\mathcal{P}$. Instead of requiring that all subintervals have length less than some specific quantity, it is often useful to allow varying degrees of fineness for different subintervals I_i in $\mathcal{P}$. This is accomplished by the use of a "gauge", which we now define.

5.5.2 Definition A **gauge** on I is a strictly positive function defined on I. If δ is a gauge on I, then a (tagged) partition $\dot{\mathcal{P}}$ is said to be δ-**fine** if

(1) $$t_i \in I_i \subseteq [t_i - \delta(t_i), t_i + \delta(t_i)] \qquad \text{for} \quad i = 1, \cdots, n.$$

We note that the notion of δ-fineness requires that the partition be tagged, so we do not need to say "tagged partition" in this case.

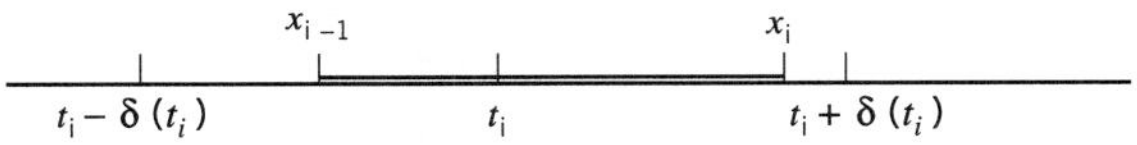

Figure 5.5.1 Inclusion (1).

A gauge δ on an interval I assigns an interval $[t - \delta(t), t + \delta(t)]$ to each point $t \in I$. The δ-fineness of a partition $\dot{\mathcal{P}}$ requires that each subinterval I_i of $\dot{\mathcal{P}}$ is contained in the interval determined by the gauge δ and the tag t_i for that subinterval. This is indicated by the inclusions in (1); see Figure 5.5.1. Note that the length of the subintervals is also controlled by the gauge and the tags; the next lemma reflects that control.

5.5.3 Lemma   *If a partition $\dot{\mathcal{P}}$ of $I := [a, b]$ is δ-fine and $x \in I$, then there exists a tag t_i in $\dot{\mathcal{P}}$ such that $|x - t_i| \leq \delta(t_i)$.*

Proof.   If $x \in I$, there exists a subinterval $[x_{i-1}, x_i]$ from $\dot{\mathcal{P}}$ that contains x. Since $\dot{\mathcal{P}}$ is δ-fine, then

$$(2) \qquad t_i - \delta(t_i) \leq x_{i-1} \leq x \leq x_i \leq t_i + \delta(t_i),$$

whence it follows that $|x - t_i| \leq \delta(t_i)$.   Q.E.D.

In the theory of Riemann integration, we will use gauges δ that are constant functions to control the fineness of the partition; in the theory of the *generalized* Riemann integral, the use of nonconstant gauges is essential. But nonconstant gauge functions arise quite naturally in connection with continuous functions. For, let $f : I \to \mathbb{R}$ be continuous on I and let $\varepsilon > 0$ be given. Then, for each point $t \in I$ there exists $\delta_\varepsilon(t) > 0$ such that if $|x - t| < \delta_\varepsilon(t)$ and $x \in I$, then $|f(x) - f(t)| < \varepsilon$. Since δ_ε is defined and is strictly positive on I, the function δ_ε is a gauge on I. Later in this section, we will use the relations between gauges and continuity to give alternative proofs of the fundamental properties of continuous functions discussed in Sections 5.3 and 5.4.

5.5.4 Examples   (a)   If δ and γ are gauges on $I := [a, b]$ and if $0 < \delta(x) \leq \gamma(x)$ for all $x \in I$, then every partition $\dot{\mathcal{P}}$ that is δ-fine is also γ-fine. This follows immediately from the inequalities

$$t_i - \gamma(t_i) \leq t_i - \delta(t_i) \qquad \text{and} \qquad t_i + \delta(t_i) \leq t_i + \gamma(t_i)$$

which imply that

$$t_i \in \left[t_i - \delta(t_i), t_i + \delta(t_i) \right] \subseteq \left[t_i - \gamma(t_i), t_i + \gamma(t_i) \right] \qquad \text{for} \quad i = 1, \cdots, n.$$

(b)   If δ_1 and δ_2 are gauges on $I := [a, b]$ and if

$$\delta(x) := \min\{\delta_1(x), \delta_2(x)\} \qquad \text{for all} \quad x \in I,$$

then δ is also a gauge on I. Moreover, since $\delta(x) \leq \delta_1(x)$, then every δ-fine partition is δ_1-fine. Similarly, every δ-fine partition is also δ_2-fine.

(c)   Suppose that δ is defined on $I := [0, 1]$ by

$$\delta(x) := \begin{cases} \frac{1}{10} & \text{if} \quad x = 0, \\ \frac{1}{2}x & \text{if} \quad 0 < x \leq 1. \end{cases}$$

Then δ is a gauge on $[0, 1]$. If $0 < t \leq 1$, then $[t - \delta(t), t + \delta(t)] = \left[\frac{1}{2}t, \frac{3}{2}t\right]$, which does not contain the point 0. Thus, if $\dot{\mathcal{P}}$ is a δ-fine partition of I, then the only subinterval in $\dot{\mathcal{P}}$ that contains 0 must have the point 0 as its tag.

(d)   Let γ be defined on $I := [0, 1]$ by

$$\gamma(x) := \begin{cases} \frac{1}{10} & \text{if} \quad x = 0 \text{ or } x = 1, \\ \frac{1}{2}x & \text{if} \quad 0 < x \leq \frac{1}{2}, \\ \frac{1}{2}(1 - x) & \text{if} \quad \frac{1}{2} < x < 1. \end{cases}$$

Then γ is a gauge on I, and it is an exercise to show that the subintervals in any γ-fine partition that contain the points 0 or 1 must have these points as tags.   □

Existence of δ-Fine Partitions _____

In view of the above examples, it is not obvious that an arbitrary gauge δ admits a δ-fine partition. We now use the Supremum Property of $\mathbb{R}$ to establish the existence of δ-fine partitions. In the Exercises, we will sketch a proof based on the Nested Intervals Theorem 2.5.2.

5.5.5 Theorem *If δ is a gauge defined on the interval $[a, b]$, then there exists a δ-fine partition of $[a, b]$.*

Proof. Let E denote the set of all points $x \in [a, b]$ such that there exists a δ-fine partition of the subinterval $[a, x]$. The set E is not empty, since the pair $([a, x], a)$ is a δ-fine partition of the interval $[a, x]$ when $x \in [a, a + \delta(a)]$ and $x \leq b$. Since $E \subseteq [a, b]$, the set E is also bounded. Let $u := \sup E$ so that $a < u \leq b$. We will show that $u \in E$ and that $u = b$.

We claim that $u \in E$. Since $u - \delta(u) < u = \sup E$, there exists $v \in E$ such that $u - \delta(u) < v < u$. Let $\dot{\mathcal{P}}_1$ be a δ-fine partition of $[a, v]$ and let $\dot{\mathcal{P}}_2 := \dot{\mathcal{P}}_1 \cup ([v, u], u)$. Then $\dot{\mathcal{P}}_2$ is a δ-fine partition of $[a, u]$, so that $u \in E$.

If $u < b$, let $w \in [a, b]$ be such that $u < w < u + \delta(u)$. If $\dot{\mathcal{Q}}_1$ is a δ-fine partition of $[a, u]$, we let $\dot{\mathcal{Q}}_2 := \dot{\mathcal{Q}}_1 \cup ([u, w], u)$. Then $\dot{\mathcal{Q}}_2$ is a δ-fine partition of $[a, w]$, whence $w \in E$. But this contradicts the supposition that u is an upper bound of E. Therefore $u = b$.
Q.E.D.

Some Applications _____

Following R. A. Gordon (see his *Monthly* article), we will now show that some of the major theorems in the two preceding sections can be proved by using gauges.

Alternate Proof of Theorem 5.3.2: Boundedness Theorem. Since f is continuous on I, then for each $t \in I$ there exists $\delta(t) > 0$ such that if $x \in I$ and $|x - t| \leq \delta(t)$, then $|f(x) - f(t)| \leq 1$. Thus δ is a gauge on I. Let $\{(I_i, t_i)\}_{i=1}^n$ be a δ-fine partition of I and let $K := \max\{|f(t_i)| : i = 1, \cdots, n\}$. By Lemma 5.5.3, given any $x \in I$ there exists i with $|x - t_i| \leq \delta(t_i)$, whence

$$|f(x)| \leq |f(x) - f(t_i)| + |f(t_i)| \leq 1 + K.$$

Since $x \in I$ is arbitrary, then f is bounded by $1 + K$ on I.
Q.E.D.

Alternate Proof of Theorem 5.3.4: Maximum-Minimum Theorem. We will prove the existence of x^*. Let $M := \sup\{f(x) : x \in I\}$ and suppose that $f(x) < M$ for all $x \in I$. Since f is continuous on I, for each $t \in I$ there exists $\delta(t) > 0$ such that if $x \in I$ and $|x - t| \leq \delta(t)$, then $f(x) < \frac{1}{2}(M + f(t))$. Thus δ is a gauge on I, and if $\{(I_i, t_i)\}_{i=1}^n$ is a δ-fine partition of I, we let

$$\tilde{M} := \tfrac{1}{2} \max\{M + f(t_1), \cdots, M + f(t_n)\}.$$

By Lemma 5.5.3, given any $x \in I$, there exists i with $|x - t_i| \leq \delta(t_i)$, whence

$$f(x) < \tfrac{1}{2}\left(M + f(t_i)\right) \leq \tilde{M}.$$

Since $x \in I$ is arbitrary, then $\tilde{M}$ ($< M$) is an upper bound for f on I, contrary to the definition of M as the supremum of f.
Q.E.D.

Alternate Proof of Theorem 5.3.5: Location of Roots Theorem. We assume that $f(t) \neq 0$ for all $t \in I$. Since f is continuous at t, Exercise 5.1.7 implies that there exists $\delta(t) > 0$ such that if $x \in I$ and $|x - t| \leq \delta(t)$, then $f(x) < 0$ if $f(t) < 0$, and $f(x) > 0$ if $f(t) > 0$. Then δ is a gauge on I and we let $\{(I_i, t_i)\}_{i=1}^n$ be a δ-fine partition. Note that for each i, either $f(x) < 0$ for all $x \in [x_{i-1}, x_i]$ or $f(x) > 0$ for all such x. Since $f(x_0) = f(a) < 0$, this implies that $f(x_1) < 0$, which in turn implies that $f(x_2) < 0$. Continuing in this way, we have $f(b) = f(x_n) < 0$, contrary to the hypothesis that $f(b) > 0$.   Q.E.D.

Alternate Proof of Theorem 5.4.3: Uniform Continuity Theorem. Let $\varepsilon > 0$ be given. Since f is continuous at $t \in I$, there exists $\delta(t) > 0$ such that if $x \in I$ and $|x - t| \leq 2\delta(t)$, then $|f(x) - f(t)| \leq \frac{1}{2}\varepsilon$. Thus δ is a gauge on I. If $\{(I_i, t_i)\}_{i=1}^n$ is a δ-fine partition of I, let $\delta_\varepsilon := \min\{\delta(t_1), \cdots, \delta(t_n)\}$. Now suppose that $x, u \in I$ and $|x - u| \leq \delta_\varepsilon$, and choose i with $|x - t_i| \leq \delta(t_i)$. Since

$$|u - t_i| \leq |u - x| + |x - t_i| \leq \delta_\varepsilon + \delta(t_i) \leq 2\delta(t_i),$$

then it follows that

$$|f(x) - f(u)| \leq |f(x) - f(t_i)| + |f(t_i) - f(u)| \leq \tfrac{1}{2}\varepsilon + \tfrac{1}{2}\varepsilon = \varepsilon.$$

Therefore, f is uniformly continuous on I.   Q.E.D.

Exercises for Section 5.5

1. Let δ be the gauge on $[0, 1]$ defined by $\delta(0) := \frac{1}{4}$ and $\delta(t) := \frac{1}{2}t$ for $t \in (0, 1]$.
   (a) Show that $\dot{P}_1 := \{([0, \frac{1}{4}], 0), ([\frac{1}{4}, \frac{1}{2}], \frac{1}{2}), ([\frac{1}{2}, 1], \frac{3}{4})\}$ is δ-fine.
   (b) Show that $\dot{P}_2 := \{([0, \frac{1}{4}], 0), ([\frac{1}{4}, \frac{1}{2}], \frac{1}{2}), ([\frac{1}{2}, 1], \frac{3}{5})\}$ is not δ-fine.

2. Suppose that δ_1 is the gauge defined by $\delta_1(0) := \frac{1}{4}, \delta_1(t) := \frac{3}{4}t$ for $t \in (0, 1]$. Are the partitions given in Exercise 1 δ_1-fine? Note that $\delta(t) \leq \delta_1(t)$ for all $t \in [0, 1]$.

3. Suppose that δ_2 is the gauge defined by $\delta_2(0) := \frac{1}{10}$ and $\delta_2(t) := \frac{9}{10}t$ for $t \in (0, 1]$. Are the partitions given in Exercise 1 δ_2-fine?

4. Let γ be the gauge in Example 5.5.4(d).
   (a) If $t \in (0, \frac{1}{2}]$ show that $[t - \gamma(t), t + \gamma(t)] = [\frac{1}{2}t, \frac{3}{2}t] \subseteq (0, \frac{3}{4}]$.
   (b) If $t \in (\frac{1}{2}, 1)$ show that $[t - \gamma(t), t + \gamma(t)] \subseteq (\frac{1}{4}, 1)$.

5. Let $a < c < b$ and let δ be a gauge on $[a, b]$. If $\dot{P}'$ is a δ-fine partition of $[a, c]$ and if $\dot{P}''$ is a δ-fine partition of $[c, b]$, show that $\dot{P}' \cup \dot{P}''$ is δ-fine partition of $[a, b]$ having c as a partition point.

6. Let $a < c < b$ and let δ' and δ'' be gauges on $[a, c]$ and $[c, b]$, respectively. If δ is defined on $[a, b]$ by

$$\delta(t) := \begin{cases} \delta'(t) & \text{if } t \in [a, c), \\ \min\{\delta'(c), \delta''(c)\} & \text{if } t = c, \\ \delta''(t) & \text{if } t \in (c, b], \end{cases}$$

then δ is a gauge on $[a, b]$. Moreover, if $\dot{P}'$ is a δ'-fine partition of $[a, c]$ and $\dot{P}''$ is a δ''-fine partition of $[c, b]$, then $\dot{P}' \cup \dot{P}''$ is a tagged partition of $[a, b]$ having c as a partition point. Explain why $\dot{P}' \cup \dot{P}''$ may not be δ-fine. Give an example.

7. Let δ' and δ'' be as in the preceding exercise and let δ^* be defined by

$$\delta^*(t) := \begin{cases} \min\{\delta'(t), \frac{1}{2}(c - t)\} & \text{if } t \in [a, c), \\ \min\{\delta'(c), \delta''(c)\} & \text{if } t = c, \\ \min\{\delta''(t), \frac{1}{2}(t - c)\} & \text{if } t \in (c, b]. \end{cases}$$

Show that δ^* is a gauge on $[a, b]$ and that every δ^*-fine partition $\dot{P}$ of $[a, b]$ having c as a partition point gives rise to a δ'-fine partition $\dot{P}'$ of $[a, c]$ and a δ''-fine partition $\dot{P}''$ of $[c, b]$ such that $\dot{P} = \dot{P}' \cup \dot{P}''$.

8. Let δ be a gauge on $I := [a, b]$ and suppose that I *does not* have a δ-fine partition.
   (a) Let $c := \frac{1}{2}(a + b)$. Show that at least one of the intervals $[a, c]$ and $[c, b]$ does not have a δ-fine partition.
   (b) Construct a nested sequence (I_n) of subintervals with the length of I_n equal to $(b - a)/2^n$ such that I_n does not have a δ-fine partition.
   (c) Let $\xi \in \cap_{n=1}^{\infty} I_n$ and let $p \in \mathbb{N}$ be such that $(b - a)/2^p < \delta(\xi)$. Show that $I_p \subseteq [\xi - \delta(\xi), \xi + \delta(\xi)]$, so the pair (I_p, ξ) is a δ-fine partition of I_p.

9. Let $I := [a, b]$ and let $f : I \to \mathbb{R}$ be a (not necessarily continuous) function. We say that f is "locally bounded" at $c \in I$ if there exists $\delta(c) > 0$ such that f is bounded on $I \cap [c - \delta(c), c + \delta(c)]$. Prove that if f is locally bounded at every point of I, then f is bounded on I.

10. Let $I := [a, b]$ and $f : I \to \mathbb{R}$. We say that f is "locally increasing" at $c \in I$ if there exists $\delta(c) > 0$ such that f is increasing on $I \cap [c - \delta(c), c + \delta(c)]$. Prove that if f is locally increasing at every point of I, then f is increasing on I.

Section 5.6   Monotone and Inverse Functions

Recall that if $A \subseteq \mathbb{R}$, then a function $f : A \to \mathbb{R}$ is said to be **increasing on** A if whenever $x_1, x_2 \in A$ and $x_1 \leq x_2$, then $f(x_1) \leq f(x_2)$. The function f is said to be **strictly increasing on** A if whenever $x_1, x_2 \in A$ and $x_1 < x_2$, then $f(x_1) < f(x_2)$. Similarly, $g : A \to \mathbb{R}$ is said to be **decreasing on** A if whenever $x_1, x_2 \in A$ and $x_1 \leq x_2$ then $g(x_1) \geq g(x_2)$. The function g is said to be **strictly decreasing on** A if whenever $x_1, x_2 \in A$ and $x_1 < x_2$ then $g(x_1) > g(x_2)$.

If a function is either increasing or decreasing on A, we say that it is **monotone** on A. If f is either strictly increasing or strictly decreasing on A, we say that f is **strictly monotone** on A.

We note that if $f : A \to \mathbb{R}$ is increasing on A then $g := -f$ is decreasing on A; similarly if $\varphi : A \to \mathbb{R}$ is decreasing on A then $\psi := -\varphi$ is increasing on A.

In this section, we will be concerned with monotone functions that are defined on an interval $I \subseteq \mathbb{R}$. We will discuss increasing functions explicitly, but it is clear that there are corresponding results for decreasing functions. These results can either be obtained directly from the results for increasing functions or proved by similar arguments.

Monotone functions are not necessarily continuous. For example, if $f(x) := 0$ for $x \in [0, 1]$ and $f(x) := 1$ for $x \in (1, 2]$, then f is increasing on $[0, 2]$, but fails to be continuous at $x = 1$. However, the next result shows that a monotone function always has both one-sided limits (see Definition 4.3.1) in $\mathbb{R}$ at every point that is not an endpoint of its domain.

5.6.1 Theorem   *Let $I \subseteq \mathbb{R}$ be an interval and let $f : I \to \mathbb{R}$ be increasing on I. Suppose that $c \in I$ is not an endpoint of I. Then*

(i) $\displaystyle\lim_{x \to c-} f = \sup\{f(x) : x \in I, x < c\}$,

(ii) $\displaystyle\lim_{x \to c+} f = \inf\{f(x) : x \in I, x > c\}$.

Proof.   (i) First note that if $x \in I$ and $x < c$, then $f(x) \leq f(c)$. Hence the set $\{f(x) : x \in I, x < c\}$, which is nonvoid since c is not an endpoint of I, is bounded above by $f(c)$. Thus the indicated supremum exists; we denote it by L. If $\varepsilon > 0$ is given, then $L - \varepsilon$ is not an upper bound of this set. Hence there exists $y_\varepsilon \in I$, $y_\varepsilon < c$ such that $L - \varepsilon < f(y_\varepsilon) \leq L$.

Since f is increasing, we deduce that if $\delta_\varepsilon := c - y_\varepsilon$ and if $0 < c - y < \delta_\varepsilon$, then $y_\varepsilon < y < c$ so that

$$L - \varepsilon < f(y_\varepsilon) \le f(y) \le L.$$

Therefore $|f(y) - L| < \varepsilon$ when $0 < c - y < \delta_\varepsilon$. Since $\varepsilon > 0$ is arbitrary we infer that (i) holds.

The proof of (ii) is similar.   Q.E.D.

The next result gives criteria for the continuity of an increasing function f at a point c that is not an endpoint of the interval on which f is defined.

5.6.2 Corollary   *Let $I \subseteq \mathbb{R}$ be an interval and let $f : I \to \mathbb{R}$ be increasing on I. Suppose that $c \in I$ is not an endpoint of I. Then the following statements are equivalent.*

(a)   *f is continuous at c.*

(b) $\displaystyle\lim_{x \to c-} f = f(c) = \lim_{x \to c+} f$.

(c) $\sup\{f(x) : x \in I, x < c\} = f(c) = \inf\{f(x) : x \in I, x > c\}$.

This follows easily from Theorems 5.6.1 and 4.3.3. We leave the details to the reader.

Let I be an interval and let $f : I \to \mathbb{R}$ be an increasing function. If a is the left endpoint of I, it is an exercise to show that f is continuous at a if and only if

$$f(a) = \inf\{f(x) : x \in I, a < x\}$$

or if and only if $f(a) = \displaystyle\lim_{x \to a+} f$. Similar conditions apply at a right endpoint, and for decreasing functions.

If $f : I \to \mathbb{R}$ is increasing on I and if c is not an endpoint of I, we define the **jump of f at c** to be $j_f(c) := \displaystyle\lim_{x \to c+} f - \lim_{x \to c-} f$. (See Figure 5.6.1.) It follows from Theorem 5.5.1 that

$$j_f(c) = \inf\{f(x) : x \in I, x > c\} - \sup\{f(x) : x \in I, x < c\}$$

for an increasing function. If the left endpoint a of I belongs to I, we define the **jump of f at a** to be $j_f(a) := \displaystyle\lim_{x \to a+} f - f(a)$. If the right endpoint b of I belongs to I, we define the **jump of f at b** to be $j_f(b) := f(b) - \displaystyle\lim_{x \to b-} f$.

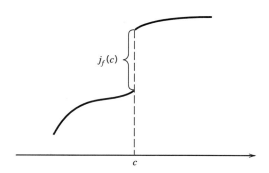

Figure 5.6.1   The jump of f at c.

5.6.3 Theorem   *Let $I \subseteq \mathbb{R}$ be an interval and let $f : I \to \mathbb{R}$ be increasing on I. If $c \in I$, then f is continuous at c if and only if $j_f(c) = 0$.*

Proof. If c is not an endpoint, this follows immediately from Corollary 5.6.2. If $c \in I$ is the left endpoint of I, then f is continuous at c if and only if $f(c) = \lim_{x \to c+} f$, which is equivalent to $j_f(c) = 0$. Similar remarks apply to the case of a right endpoint. Q.E.D.

We now show that there can be at most a countable set of points at which a monotone function is discontinuous.

5.6.4 Theorem *Let $I \subseteq \mathbb{R}$ be an interval and let $f : I \to \mathbb{R}$ be monotone on I. Then the set of points $D \subseteq I$ at which f is discontinuous is a countable set.*

Proof. We shall suppose that f is increasing on I. It follows from Theorem 5.6.3 that $D = \{x \in I : j_f(x) \neq 0\}$. We shall consider the case that $I := [a, b]$ is a closed bounded interval, leaving the case of an arbitrary interval to the reader.

We first note that since f is increasing, then $j_f(c) \geq 0$ for all $c \in I$. Moreover, if $a \leq x_1 < \cdots < x_n \leq b$, then (why?) we have

(1) $$f(a) \leq f(a) + j_f(x_1) + \cdots + j_f(x_n) \leq f(b),$$

whence it follows that

$$j_f(x_1) + \cdots + j_f(x_n) \leq f(b) - f(a).$$

(See Figure 5.6.2.) Consequently there can be at most k points in $I = [a, b]$ where $j_f(x) \geq (f(b) - f(a))/k$. We conclude that there is at most one point $x \in I$ where $j_f(x) = f(b) - f(a)$; there are at most two points in I where $j_f(x) \geq (f(b) - f(a))/2$; at most three points in I where $j_f(x) \geq (f(b) - f(a))/3$, and so on. Therefore there is at most a countable set of points x where $j_f(x) > 0$. But since every point in D must be included in this set, we deduce that D is a countable set. Q.E.D.

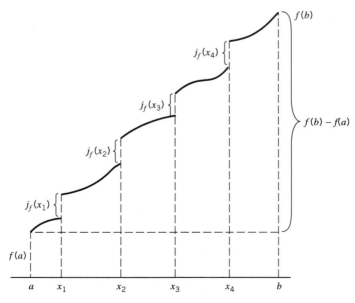

Figure 5.6.2 $j_f(x_1) + \cdots + j_f(x_n) \leq f(b) - f(a)$.

Theorem 5.6.4 has some useful applications. For example, it was seen in Exercise 5.2.12 that if $h : \mathbb{R} \to \mathbb{R}$ satisfies the identity

$$(2) \qquad\qquad h(x+y) = h(x) + h(y) \qquad \text{for all} \quad x, y \in \mathbb{R},$$

and if h is continuous at a single point x_0, then h is continuous at *every* point of $\mathbb{R}$. Thus, if h is a monotone function satisfying (2), then h must be continuous on $\mathbb{R}$. [It follows from this that $h(x) = Cx$ for all $x \in \mathbb{R}$, where $C := h(1)$.]

Inverse Functions

We shall now consider the existence of inverses for functions that are continuous on an interval $I \subseteq \mathbb{R}$. We recall (see Section 1.1) that a function $f : I \to \mathbb{R}$ has an inverse function if and only if f is injective (= one-one); that is, $x, y \in I$ and $x \neq y$ imply that $f(x) \neq f(y)$. We note that a strictly monotone function is injective and so has an inverse. In the next theorem, we show that if $f : I \to \mathbb{R}$ is a strictly monotone *continuous* function, then f has an inverse function g on $J := f(I)$ that is strictly monotone and continuous on J. In particular, if f is strictly increasing then so is g, and if f is strictly decreasing then so is g.

5.6.5 Continuous Inverse Theorem   Let $I \subseteq \mathbb{R}$ be an interval and let $f : I \to \mathbb{R}$ be *strictly monotone and continuous on I. Then the function g inverse to f is strictly monotone and continuous on $J := f(I)$.*

Proof.   We consider the case that f is strictly increasing, leaving the case that f is strictly decreasing to the reader.

Since f is continuous and I is an interval, it follows from the Preservation of Intervals Theorem 5.3.10 that $J := f(I)$ is an interval. Moreover, since f is strictly increasing on I, it is injective on I; therefore the function $g : J \to \mathbb{R}$ inverse to f exists. We claim that g is strictly increasing. Indeed, if $y_1, y_2 \in J$ with $y_1 < y_2$, then $y_1 = f(x_1)$ and $y_2 = f(x_2)$ for some $x_1, x_2 \in I$. We must have $x_1 < x_2$; otherwise $x_1 \geq x_2$, which implies that $y_1 = f(x_1) \geq f(x_2) = y_2$, contrary to the hypothesis that $y_1 < y_2$. Therefore we have $g(y_1) = x_1 < x_2 = g(y_2)$. Since y_1 and y_2 are arbitrary elements of J with $y_1 < y_2$, we conclude that g is strictly increasing on J.

It remains to show that g is continuous on J. However, this is a consequence of the fact that $g(J) = I$ is an interval. Indeed, if g is discontinuous at a point $c \in J$, then the jump of g at c is nonzero so that $\lim_{y \to c-} g < \lim_{y \to c+} g$. If we choose any number $x \neq g(c)$ satisfying $\lim_{x \to c-} g < x < \lim_{x \to c+} g$, then x has the property that $x \neq g(y)$ for any $y \in J$. (See Figure 5.6.3.) Hence $x \notin I$, which contradicts the fact that I is an interval. Therefore we conclude that g is continuous on J.   Q.E.D.

The nth Root Function

We will apply the Continuous Inverse Theorem 5.6.5 to the nth power function. We need to distinguish two cases: (i) n even, and (ii) n odd.

(i) n **even.** In order to obtain a function that is strictly monotone, we restrict our attention to the interval $I := [0, \infty)$. Thus, let $f(x) := x^n$ for $x \in I$. (See Figure 5.6.4.) We have seen (in Exercise 2.1.23) that if $0 \leq x < y$, then $f(x) = x^n < y^n = f(y)$; therefore f is strictly increasing on I. Moreover, it follows from Example 5.2.3(a) that f is continuous on I. Therefore, by the Preservation of Intervals Theorem 5.3.10, $J := f(I)$ is an interval.

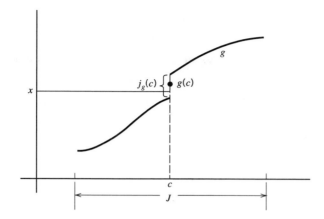

Figure 5.6.3 $g(y) \neq x$ for $y \in J$.

We will show that $J = [0, \infty)$. Let $y \geq 0$ be arbitrary; by the Archimedean Property, there exists $k \in \mathbb{N}$ such that $0 \leq y < k$. Since

$$f(0) = 0 \leq y < k \leq k^n = f(k),$$

it follows from Bolzano's Intermediate Value Theorem 5.3.7 that $y \in J$. Since $y \geq 0$ is arbitrary, we deduce that $J = [0, \infty)$.

We conclude from the Continuous Inverse Theorem 5.6.5 that the function g that is inverse to $f(x) = x^n$ on $I = [0, \infty)$ is strictly increasing and continuous on $J = [0, \infty)$. We usually write

$$g(x) = x^{1/n} \qquad \text{or} \qquad g(x) = \sqrt[n]{x}$$

for $x \geq 0$ (n even), and call $x^{1/n} = \sqrt[n]{x}$ the **nth root** of $x \geq 0$ (n even). The function g is called the **nth root function** (n even). (See Figure 5.6.5.)

Since g is inverse to f we have

$$g\big(f(x)\big) = x \qquad \text{and} \qquad f\big(g(x)\big) = x \qquad \text{for all} \quad x \in [0, \infty).$$

We can write these equations in the following form:

$$\left(x^n\right)^{1/n} = x \qquad \text{and} \qquad \left(x^{1/n}\right)^n = x$$

for all $x \in [0, \infty)$ and n even.

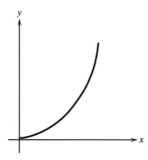

Figure 5.6.4   Graph of $f(x) = x^n$ ($x \geq 0$, n even).

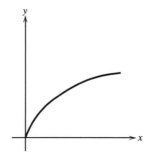

Figure 5.6.5   Graph of $g(x) = x^{1/n}$ ($x \geq 0$, n even).

(ii) *n* **odd**. In this case we let $F(x) := x^n$ for all $x \in \mathbb{R}$; by 5.2.3(a), F is continuous on $\mathbb{R}$. We leave it to the reader to show that F is strictly increasing on $\mathbb{R}$ and that $F(\mathbb{R}) = \mathbb{R}$. (See Figure 5.6.6.)

It follows from the Continuous Inverse Theorem 5.6.5 that the function G that is inverse to $F(x) = x^n$ for $x \in \mathbb{R}$, is strictly increasing and continuous on $\mathbb{R}$. We usually write

$$G(x) = x^{1/n} \quad \text{or} \quad G(x) = \sqrt[n]{x} \quad \text{for } x \in \mathbb{R}, n \text{ odd},$$

and call $x^{1/n}$ the *n*th **root of** $x \in \mathbb{R}$. The function G is called the *n*th **root function** (*n* odd). (See Figure 5.6.7.) Here we have

$$\left(x^n\right)^{1/n} = x \qquad \text{and} \qquad \left(x^{1/n}\right)^n = x$$

for all $x \in \mathbb{R}$ and *n* odd.

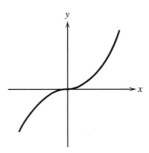

Figure 5.6.6   Graph of $F(x) = x^n$ ($x \in \mathbb{R}$, *n* odd).

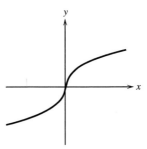

Figure 5.6.7   Graph of $G(x) = x^{1/n}$ ($x \in \mathbb{R}$, *n* odd).

Rational Powers

Now that the *n*th root functions have been defined for $n \in \mathbb{N}$, it is easy to define rational powers.

5.6.6 Definition   (i)   If $m, n \in \mathbb{N}$ and $x \geq 0$, we define $x^{m/n} := (x^{1/n})^m$.
(ii)   If $m, n \in \mathbb{N}$ and $x > 0$, we define $x^{-m/n} := (x^{1/n})^{-m}$.

Hence we have defined x^r when r is a rational number and $x > 0$. The graphs of $x \mapsto x^r$ depend on whether $r > 1, r = 1, 0 < r < 1, r = 0$, or $r < 0$. (See Figure 5.6.8.) Since a rational number $r \in \mathbb{Q}$ can be written in the form $r = m/n$ with $m \in \mathbb{Z}, n \in \mathbb{N}$, in many ways, it should be shown that Definition 5.6.6 is not ambiguous. That is if $r = m/n = p/q$ with $m, p \in \mathbb{Z}$ and $n, q \in \mathbb{N}$ and if $x > 0$, then $(x^{1/n})^m = (x^{1/q})^p$. We leave it as an exercise to the reader to establish this relation.

5.6.7 Theorem   If $m \in \mathbb{Z}, n \in \mathbb{N}$, and $x > 0$, then $x^{m/n} = (x^m)^{1/n}$.

Proof.   If $x > 0$ and $m, n \in \mathbb{Z}$, then $(x^m)^n = x^{mn} = (x^n)^m$. Now let $y := x^{m/n} = (x^{1/n})^m > 0$ so that $y^n = ((x^{1/n})^m)^n = ((x^{1/n})^n)^m = x^m$. Therefore it follows that $y = (x^m)^{1/n}$.   Q.E.D.

The reader should also show, as an exercise, that if $x > 0$ and $r, s \in \mathbb{Q}$, then

$$x^r x^s = x^{r+s} = x^s x^r \quad \text{and} \quad (x^r)^s = x^{rs} = (x^s)^r.$$

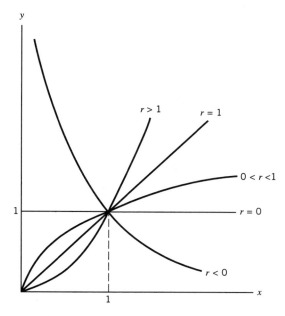

Figure 5.6.8   Graphs of $x \to x^r$ $(x \ge 0)$.

Exercises for Section 5.6

1. If $I := [a, b]$ is an interval and $f : I \to \mathbb{R}$ is an increasing function, then the point a [respectively, b] is an absolute minimum [respectively, maximum] point for f on I. If f is strictly increasing, then a is the only absolute minimum point for f on I.

2. If f and g are increasing functions on an interval $I \subseteq \mathbb{R}$, show that $f + g$ is an increasing function on I. If f is also strictly increasing on I, then $f + g$ is strictly increasing on I.

3. Show that both $f(x) := x$ and $g(x) := x - 1$ are strictly increasing on $I := [0, 1]$, but that their product fg is not increasing on I.

4. Show that if f and g are positive increasing functions on an interval I, then their product fg is increasing on I.

5. Show that if $I := [a, b]$ and $f : I \to \mathbb{R}$ is increasing on I, then f is continuous at a if and only if $f(a) = \inf\{f(x) : x \in (a, b]\}$.

6. Let $I \subseteq \mathbb{R}$ be an interval and let $f : I \to \mathbb{R}$ be increasing on I. Suppose that $c \in I$ is not an endpoint of I. Show that f is continuous at c if and only if there exists a sequence (x_n) in I such that $x_n < c$ for $n = 1, 3, 5, \cdots$; $x_n > c$ for $n = 2, 4, 6, \cdots$; and such that $c = \lim(x_n)$ and $f(c) = \lim\left(f(x_n)\right)$.

7. Let $I \subseteq \mathbb{R}$ be an interval and let $f : I \to \mathbb{R}$ be increasing on I. If c is not an endpoint of I, show that the jump $j_f(c)$ of f at c is given by $\inf\{f(y) - f(x) : x < c < y, x, y \in I\}$.

8. Let f, g be increasing on an interval $I \subseteq \mathbb{R}$ and let $f(x) > g(x)$ for all $x \in I$. If $y \in f(I) \cap g(I)$, show that $f^{-1}(y) < g^{-1}(y)$. [*Hint*: First interpret this statement geometrically.]

9. Let $I := [0, 1]$ and let $f : I \to \mathbb{R}$ be defined by $f(x) := x$ for x rational, and $f(x) := 1 - x$ for x irrational. Show that f is injective on I and that $f\left(f(x)\right) = x$ for all $x \in I$. (Hence f is its own inverse function!) Show that f is continuous only at the point $x = \frac{1}{2}$.

10. Let $I := [a, b]$ and let $f : I \to \mathbb{R}$ be continuous on I. If f has an absolute maximum [respectively, minimum] at an interior point c of I, show that f is not injective on I.

11. Let $f(x) := x$ for $x \in [0, 1]$, and $f(x) := 1 + x$ for $x \in (1, 2]$. Show that f and f^{-1} are strictly increasing. Are f and f^{-1} continuous at every point?

12. Let $f : [0, 1] \to \mathbb{R}$ be a continuous function that does not take on any of its values twice and with $f(0) < f(1)$. Show that f is strictly increasing on $[0, 1]$.

13. Let $h : [0, 1] \to \mathbb{R}$ be a function that takes on each of its values exactly twice. Show that h cannot be continuous at every point. [*Hint*: If $c_1 < c_2$ are the points where h attains its supremum, show that $c_1 = 0, c_2 = 1$. Now examine the points where h attains its infimum.]

14. Let $x \in \mathbb{R}, x > 0$. Show that if $m, p \in \mathbb{Z}, n, q \in \mathbb{N}$, and $mq = np$, then $(x^{1/n})^m = (x^{1/q})^p$.

15. If $x \in \mathbb{R}, x > 0$, and if $r, s \in \mathbb{Q}$, show that $x^r x^s = x^{r+s} = x^s x^r$ and $(x^r)^s = x^{rs} = (x^s)^r$.

CHAPTER 6

DIFFERENTIATION

Prior to the seventeenth century, a curve was generally described as a locus of points satisfying some geometric condition, and tangent lines were obtained through geometric construction. This viewpoint changed dramatically with the creation of analytic geometry in the 1630s by René Descartes (1596–1650) and Pierre de Fermat (1601–1665). In this new setting geometric problems were recast in terms of algebraic expressions, and new classes of curves were defined by algebraic rather than geometric conditions. The concept of derivative evolved in this new context. The problem of finding tangent lines and the seemingly unrelated problem of finding maximum or minimum values were first seen to have a connection by Fermat in the 1630s. And the relation between tangent lines to curves and the velocity of a moving particle was discovered in the late 1660s by Isaac Newton. Newton's theory of "fluxions", which was based on an intuitive idea of limit, would be familiar to any modern student of differential calculus once some changes in terminology and notation were made. But the vital observation, made by Newton and, independently, by Gottfried Leibniz in the 1680s, was that areas under curves could be calculated by reversing the differentiation process. This exciting technique, one that solved previously difficult area problems with ease, sparked enormous interest among the mathematicians of the era and led to a coherent theory that became known as the differential and integral calculus.

Isaac Newton

Isaac Newton (1642–1727) was born in Woolsthorpe, in Lincolnshire, England, on Christmas Day; his father, a farmer, had died three months earlier. His mother remarried when he was three years old and he was sent to live with his grandmother. He returned to his mother at age eleven, only to be sent to boarding school in Grantham the next year. Fortunately, a perceptive teacher noticed his mathematical talent and, in 1661, Newton entered Trinity College at Cambridge University, where he studied with Isaac Barrow.

When the bubonic plague struck in 1665–1666, leaving dead nearly 70,000 persons in London, the university closed and Newton spent two years back in Woolsthorpe. It was during this period that he formulated his basic ideas concerning optics, gravitation, and his method of "fluxions", later called "calculus". He returned to Cambridge in 1667 and was appointed Lucasian Professor in 1669. His theories of universal gravitation and planetary motion were published to world acclaim in 1687 under the title *Philosophiæ Naturalis Principia Mathematica*. However, he neglected to publish his method of inverse tangents for finding areas and other work in calculus, and this led to a controversy over priority with Leibniz.

Following an illness, he retired from Cambridge University and in 1696 was appointed Warden of the British mint. However, he maintained contact with advances in science and mathematics and served as President of the Royal Society from 1703 until his death in 1727. At his funeral, Newton was eulogized as "the greatest genius that ever existed". His place of burial in Westminster Abbey is a popular tourist site.

In this chapter we will develop the theory of differentiation. Integration theory, including the fundamental theorem that relates differentiation and integration, will be the subject of the next chapter. We will assume that the reader is already familiar with the geometrical and physical interpretations of the derivative of a function as described in introductory calculus courses. Consequently, we will concentrate on the mathematical aspects of the derivative and not go into its applications in geometry, physics, economics, and so on.

The first section is devoted to a presentation of the basic results concerning the differentiation of functions. In Section 6.2 we discuss the fundamental Mean Value Theorem and some of its applications. In Section 6.3 the important L'Hospital Rules are presented for the calculation of certain types of "indeterminate" limits.

In Section 6.4 we give a brief discussion of Taylor's Theorem and a few of its applications—for example, to convex functions and to Newton's Method for the location of roots.

Section 6.1   The Derivative

In this section we will present some of the elementary properties of the derivative. We begin with the definition of the derivative of a function.

6.1.1 Definition   Let $I \subseteq \mathbb{R}$ be an interval, let $f : I \to \mathbb{R}$, and let $c \in I$. We say that a real number L is the **derivative of f at c** if given any $\varepsilon > 0$ there exists $\delta(\varepsilon) > 0$ such that if $x \in I$ satisfies $0 < |x - c| < \delta(\varepsilon)$, then

(1)
$$\left| \frac{f(x) - f(c)}{x - c} - L \right| < \varepsilon.$$

In this case we say that f is **differentiable** at c, and we write $f'(c)$ for L.

In other words, the derivative of f at c is given by the limit

(2)
$$f'(c) = \lim_{x \to c} \frac{f(x) - f(c)}{x - c}$$

provided this limit exists. (We allow the possibility that c may be the endpoint of the interval.)

Note   It is possible to define the derivative of a function having a domain more general than an interval (since the point c need only be an element of the domain and also a cluster point of the domain) but the significance of the concept is most naturally apparent for functions defined on intervals. Consequently we shall limit our attention to such functions.

Whenever the derivative of $f : I \to \mathbb{R}$ exists at a point $c \in I$, its value is denoted by $f'(c)$. In this way we obtain a function f' whose domain is a subset of the domain of f. In working with the function f', it is convenient to regard it also as a function of x. For example, if $f(x) := x^2$ for $x \in \mathbb{R}$, then at any c in $\mathbb{R}$ we have

$$f'(c) = \lim_{x \to c} \frac{f(x) - f(c)}{x - c} = \lim_{x \to c} \frac{x^2 - c^2}{x - c} = \lim_{x \to c} (x + c) = 2c.$$

Thus, in this case, the function f' is defined on all of $\mathbb{R}$ and $f'(x) = 2x$ for $x \in \mathbb{R}$.

We now show that continuity of f at a point c is a necessary (but not sufficient) condition for the existence of the derivative at c.

6.1.2 Theorem   *If $f: I \to \mathbb{R}$ has a derivative at $c \in I$, then f is continuous at c.*

Proof.   For all $x \in I$, $x \neq c$, we have

$$f(x) - f(c) = \left(\frac{f(x) - f(c)}{x - c}\right)(x - c).$$

Since $f'(c)$ exists, we may apply Theorem 4.2.4 concerning the limit of a product to conclude that

$$\lim_{x \to c}(f(x) - f(c)) = \lim_{x \to c}\left(\frac{f(x) - f(c)}{x - c}\right)\left(\lim_{x \to c}(x - c)\right)$$
$$= f'(c) \cdot 0 = 0.$$

Therefore, $\lim_{x \to c} f(x) = f(c)$ so that f is continuous at c.   Q.E.D.

The continuity of $f: I \to \mathbb{R}$ at a point does not assure the existence of the derivative at that point. For example, if $f(x) := |x|$ for $x \in \mathbb{R}$, then for $x \neq 0$ we have $(f(x) - f(0))/(x - 0) = |x|/x$ which is equal to 1 if $x > 0$, and equal to -1 if $x < 0$. Thus the limit at 0 does not exist [see Example 4.1.10(b)], and therefore the function is not differentiable at 0. Hence, continuity at a point c is *not* a sufficient condition for the derivative to exist at c.

Remark   By taking simple algebraic combinations of functions of the form $x \mapsto |x - c|$, it is not difficult to construct continuous functions that do not have a derivative at a finite (or even a countable) number of points. In 1872, Karl Weierstrass astounded the mathematical world by giving an example of a function that is *continuous at every point but whose derivative does not exist anywhere.* Such a function defied geometric intuition about curves and tangent lines, and consequently spurred much deeper investigations into the concepts of real analysis. It can be shown that the function f defined by the series

$$f(x) := \sum_{n=0}^{\infty} \frac{1}{2^n} \cos(3^n x)$$

has the stated property. A very interesting historical discussion of this and other examples of continuous, nondifferentiable functions is given in Kline, p. 955–966, and also in Hawkins, p. 44–46. A detailed proof for a slightly different example can be found in Appendix E.

There are a number of basic properties of the derivative that are very useful in the calculation of the derivatives of various combinations of functions. We now provide the justification of some of these properties, which will be familiar to the reader from earlier courses.

6.1.3 Theorem   *Let $I \subseteq \mathbb{R}$ be an interval, let $c \in I$, and let $f: I \to \mathbb{R}$ and $g: I \to \mathbb{R}$ be functions that are differentiable at c. Then:*

(a)   *If $\alpha \in \mathbb{R}$, then the function αf is differentiable at c, and*

(3) $$(\alpha f)'(c) = \alpha f'(c).$$

(b) *The function $f + g$ is differentiable at c, and*

(4)
$$(f + g)'(c) = f'(c) + g'(c).$$

(c) *(Product Rule) The function fg is differentiable at c, and*

(5)
$$(fg)'(c) = f'(c)g(c) + f(c)g'(c).$$

(d) *(Quotient Rule) If $g(c) \neq 0$, then the function f/g is differentiable at c, and*

(6)
$$\left(\frac{f}{g}\right)'(c) = \frac{f'(c)g(c) - f(c)g'(c)}{\left(g(c)\right)^2}.$$

Proof.   We shall prove (c) and (d), leaving (a) and (b) as exercises for the reader.

(c) Let $p := fg$; then for $x \in I$, $x \neq c$, we have

$$\frac{p(x) - p(c)}{x - c} = \frac{f(x)g(x) - f(c)g(c)}{x - c}$$

$$= \frac{f(x)g(x) - f(c)g(x) + f(c)g(x) - f(c)g(c)}{x - c}$$

$$= \frac{f(x) - f(c)}{x - c} \cdot g(x) + f(c) \cdot \frac{g(x) - g(c)}{x - c}.$$

Since g is continuous at c, by Theorem 6.1.2, then $\lim_{x \to c} g(x) = g(c)$. Since f and g are differentiable at c, we deduce from Theorem 4.2.4 on properties of limits that

$$\lim_{x \to c} \frac{p(x) - p(c)}{x - c} = f'(c)g(c) + f(c)g'(c).$$

Hence $p := fg$ is differentiable at c and (5) holds.

(d) Let $q := f/g$. Since g is differentiable at c, it is continuous at that point (by Theorem 6.1.2). Therefore, since $g(c) \neq 0$, we know from Theorem 4.2.9 that there exists an interval $J \subseteq I$ with $c \in J$ such that $g(x) \neq 0$ for all $x \in J$. For $x \in J$, $x \neq c$, we have

$$\frac{q(x) - q(c)}{x - c} = \frac{f(x)/g(x) - f(c)/g(c)}{x - c} = \frac{f(x)g(c) - f(c)g(x)}{g(x)g(c)(x - c)}$$

$$= \frac{f(x)g(c) - f(c)g(c) + f(c)g(c) - f(c)g(x)}{g(x)g(c)(x - c)}$$

$$= \frac{1}{g(x)g(c)}\left[\frac{f(x) - f(c)}{x - c} \cdot g(c) - f(c) \cdot \frac{g(x) - g(c)}{x - c}\right].$$

Using the continuity of g at c and the differentiability of f and g at c, we get

$$q'(c) = \lim_{x \to c} \frac{q(x) - q(c)}{x - c} = \frac{f'(c)g(c) - f(c)g'(c)}{\left(g(c)\right)^2}.$$

Thus, $q = f/g$ is differentiable at c and equation (6) holds.   Q.E.D.

Mathematical Induction may be used to obtain the following extensions of the differentiation rules.

6.1.4 Corollary   *If $f_1, f_2, \cdots, f_n$ are functions on an interval I to $\mathbb{R}$ that are differentiable at $c \in I$, then:*

(a) *The function $f_1 + f_2 + \cdots + f_n$ is differentiable at c and*

(7)
$$(f_1 + f_2 + \cdots + f_n)'(c) = f_1'(c) + f_2'(c) + \cdots + f_n'(c).$$

(b) The function $f_1 f_2 \cdots f_n$ is differentiable at c, and

(8)
$$(f_1 f_2 \cdots f_n)'(c) = f_1'(c) f_2(c) \cdots f_n(c) + f_1(c) f_2'(c) \cdots f_n(c)$$
$$+ \cdots + f_1(c) f_2(c) \cdots f_n'(c).$$

An important special case of the extended product rule (8) occurs if the functions are equal, that is, $f_1 = f_2 = \cdots = f_n = f$. Then (8) becomes

(9)
$$(f^n)'(c) = n(f(c))^{n-1} f'(c).$$

In particular, if we take $f(x) := x$, then we find the derivative of $g(x) := x^n$ to be $g'(x) = nx^{n-1}$, $n \in \mathbb{N}$. The formula is extended to include negative integers by applying the Quotient Rule 6.1.3(d).

Notation If $I \subseteq \mathbb{R}$ is an interval and $f : I \to \mathbb{R}$, we have introduced the notation f' to denote the function whose domain is a subset of I and whose value at a point c is the derivative $f'(c)$ of f at c. There are other notations that are sometimes used for f'; for example, one sometimes writes Df for f'. Thus one can write formulas (4) and (5) in the form:

$$D(f + g) = Df + Dg, \qquad D(fg) = (Df) \cdot g + f \cdot (Dg).$$

When x is the "independent variable", it is common practice in elementary courses to write df/dx for f'. Thus formula (5) is sometimes written in the form

$$\frac{d}{dx}(f(x)g(x)) = \left(\frac{df}{dx}(x)\right) g(x) + f(x) \left(\frac{dg}{dx}(x)\right).$$

This last notation, due to Leibniz, has certain advantages. However, it also has certain disadvantages and must be used with some care.

The Chain Rule

We now turn to the theorem on the differentiation of composite functions known as the "Chain Rule". It provides a formula for finding the derivative of a composite function $g \circ f$ in terms of the derivatives of g and f.

We first establish the following theorem concerning the derivative of a function at a point that gives us a very nice method for proving the Chain Rule. It will also be used to derive the formula for differentiating inverse functions.

6.1.5 Carathéodory's Theorem *Let f be defined on an interval I containing the point c. Then f is differentiable at c if and only if there exists a function φ on I that is continuous at c and satisfies*

(10)
$$f(x) - f(c) = \varphi(x)(x - c) \qquad for \quad x \in I.$$

In this case, we have $\varphi(c) = f'(c)$.

Proof. ($\Rightarrow$) If $f'(c)$ exists, we can define φ by

$$\varphi(x) := \begin{cases} \dfrac{f(x) - f(c)}{x - c} & for \quad x \neq c, x \in I, \\ f'(c) & for \quad x = c. \end{cases}$$

The continuity of φ follows from the fact that $\lim_{x \to c} \varphi(x) = f'(c)$. If $x = c$, then both sides of (10) equal 0, while if $x \neq c$, then multiplication of $\varphi(x)$ by $x - c$ gives (10) for all other $x \in I$.

($\Leftarrow$) Now assume that a function φ that is continuous at c and satisfying (10) exists. If we divide (10) by $x - c \neq 0$, then the continuity of φ implies that

$$\varphi(c) = \lim_{x \to c} \varphi(x) = \lim_{x \to c} \frac{f(x) - f(c)}{x - c}$$

exists. Therefore f is differentiable at c and $f'(c) = \varphi(c)$.   Q.E.D.

To illustrate Carathéodory's Theorem, we consider the function f defined by $f(x) := x^3$ for $x \in \mathbb{R}$. For $c \in \mathbb{R}$, we see from the factorization

$$x^3 - c^3 = (x^2 + cx + c^2)(x - c)$$

that $\varphi(x) := x^2 + cx + c^2$ satisfies the conditions of the theorem. Therefore, we conclude that f is differentiable at $c \in \mathbb{R}$ and that $f'(c) = \varphi(c) = 3c^2$.

We will now establish the Chain Rule. If f is differentiable at c and g is differentiable at $f(c)$, then the Chain Rule states that the derivative of the composite function $g \circ f$ at c is the product $(g \circ f)'(c) = g'(f(c)) \cdot f'(c)$. Note this can be written

$$(g \circ f)' = (g' \circ f) \cdot f'.$$

One approach to the Chain Rule is the observation that the difference quotient can be written, when $f(x) \neq f(c)$, as the product

$$\frac{g(f(x)) - g(f(c))}{x - c} = \frac{g(f(x)) - g(f(c))}{f(x) - f(c)} \cdot \frac{f(x) - f(c)}{x - c}.$$

This suggests the correct limiting value. Unfortunately, the first factor in the product on the right is undefined if the denominator $f(x) - f(c)$ equals 0 for values of x near c, and this presents a problem. However, the use of Carathéodory's Theorem neatly avoids this difficulty.

6.1.6 Chain Rule   Let I, J be intervals in $\mathbb{R}$, let $g : I \to \mathbb{R}$ and $f : J \to \mathbb{R}$ be functions such that $f(J) \subseteq I$, and let $c \in J$. If f is differentiable at c and if g is differentiable at $f(c)$, then the composite function $g \circ f$ is differentiable at c and

(11) $$(g \circ f)'(c) = g'(f(c)) \cdot f'(c).$$

Proof.   Since $f'(c)$ exists, Carathéodory's Theorem 6.1.5 implies that there exists a function φ on J such that φ is continuous at c and $f(x) - f(c) = \varphi(x)(x - c)$ for $x \in J$, and where $\varphi(c) = f'(c)$. Also, since $g'(f(c))$ exists, there is a function ψ defined on I such that ψ is continuous at $d := f(c)$ and $g(y) - g(d) = \psi(y)(y - d)$ for $y \in I$, where $\psi(d) = g'(d)$. Substitution of $y = f(x)$ and $d = f(c)$ then produces

$$g(f(x)) - g(f(c)) = \psi(f(x))(f(x) - f(c)) = [(\psi \circ f(x)) \cdot \varphi(x)](x - c)$$

for all $x \in J$ such that $f(x) \in I$. Since the function $(\psi \circ f) \cdot \varphi$ is continuous at c and its value at c is $g'(f(c)) \cdot f'(c)$, Carathéodory's Theorem gives (11).   Q.E.D.

If g is differentiable on I, if f is differentiable on J and if $f(J) \subseteq I$, then it follows from the Chain Rule that $(g \circ f)' = (g' \circ f) \cdot f'$ which can also be written in the form $D(g \circ f) = (Dg \circ f) \cdot Df$.

6.1.7 Examples   **(a)** If $f : I \to \mathbb{R}$ is differentiable on I and $g(y) := y^n$ for $y \in \mathbb{R}$ and $n \in \mathbb{N}$, then since $g'(y) = ny^{n-1}$, it follows from the Chain Rule 6.1.6 that

$$(g \circ f)'(x) = g'(f(x)) \cdot f'(x) \qquad \text{for} \quad x \in I.$$

Therefore we have $(f^n)'(x) = n(f(x))^{n-1} f'(x)$ for all $x \in I$ as was seen in (9).

(b) Suppose that $f : I \to \mathbb{R}$ is differentiable on I and that $f(x) \neq 0$ and $f'(x) \neq 0$ for $x \in I$. If $h(y) := 1/y$ for $y \neq 0$, then it is an exercise to show that $h'(y) = -1/y^2$ for $y \in \mathbb{R}$, $y \neq 0$. Therefore we have

$$\left(\frac{1}{f}\right)'(x) = (h \circ f)'(x) = h'(f(x)) f'(x) = -\frac{f'(x)}{(f(x))^2} \qquad \text{for} \quad x \in I.$$

(c) The absolute value function $g(x) := |x|$ is differentiable at all $x \neq 0$ and has derivative $g'(x) = \text{sgn}(x)$ for $x \neq 0$. (The signum function is defined in Example 4.1.10(b).) Though sgn is defined everywhere, it is not equal to g' at $x = 0$ since $g'(0)$ does not exist.

Now if f is a differentiable function, then the Chain Rule implies that the function $g \circ f = |f|$ is also differentiable at all points x where $f(x) \neq 0$, and its derivative is given by

$$|f|'(x) = \text{sgn}(f(x)) \cdot f'(x) = \begin{cases} f'(x) & \text{if} \quad f(x) > 0, \\ -f'(x) & \text{if} \quad f(x) < 0. \end{cases}$$

If f is differentiable at a point c with $f(c) = 0$, then it is an exercise to show that $|f|$ is differentiable at c if and only if $f'(c) = 0$. (See Exercise 7.)

For example, if $f(x) := x^2 - 1$ for $x \in \mathbb{R}$, then the derivative of its absolute value $|f|(x) = |x^2 - 1|$ is equal to $|f|'(x) = \text{sgn}(x^2 - 1) \cdot (2x)$ for $x \neq 1, -1$. See Figure 6.1.1 for a graph of $|f|$.

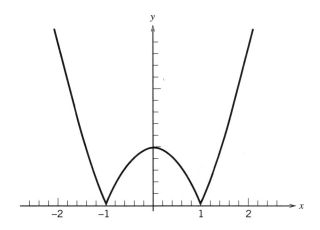

Figure 6.1.1   The function $|f|(x) = |x^2 - 1|$.

(d) It will be proved later that if $S(x) := \sin x$ and $C(x) := \cos x$ for all $x \in \mathbb{R}$, then

$$S'(x) = \cos x = C(x) \qquad \text{and} \qquad C'(x) = -\sin x = -S(x)$$

for all $x \in \mathbb{R}$. If we use these facts together with the definitions

$$\tan x := \frac{\sin x}{\cos x}, \qquad \sec x := \frac{1}{\cos x},$$

for $x \neq (2k+1)\pi/2$, $k \in \mathbb{Z}$, and apply the Quotient Rule 6.1.3(d), we obtain

$$D \tan x = \frac{(\cos x)(\cos x) - (\sin x)(-\sin x)}{(\cos x)^2} = (\sec x)^2,$$

$$D \sec x = \frac{0 - 1(-\sin x)}{(\cos x)^2} = \frac{\sin x}{(\cos x)^2} = (\sec x)(\tan x)$$

for $x \neq (2k+1)\pi/2$, $k \in \mathbb{Z}$.
    Similarly, since

$$\cot x := \frac{\cos x}{\sin x}, \qquad \csc x := \frac{1}{\sin x}$$

for $x \neq k\pi$, $k \in \mathbb{Z}$, then we obtain

$$D \cot x = -(\csc x)^2 \qquad \text{and} \qquad D \csc x = -(\csc x)(\cot x)$$

for $x \neq k\pi$, $k \in \mathbb{Z}$.

(e)   Suppose that f is defined by

$$f(x) := \begin{cases} x^2 \sin(1/x) & \text{for} \quad x \neq 0, \\ 0 & \text{for} \quad x = 0. \end{cases}$$

If we use the fact that $D \sin x = \cos x$ for all $x \in \mathbb{R}$ and apply the Product Rule 6.1.3(c) and the Chain Rule 6.1.6, we obtain (why?)

$$f'(x) = 2x \sin(1/x) - \cos(1/x) \qquad \text{for} \quad x \neq 0.$$

If $x = 0$, none of the calculational rules may be applied. (Why?) Consequently, the derivative of f at $x = 0$ must be found by applying the definition of derivative. We find that

$$f'(0) = \lim_{x \to 0} \frac{f(x) - f(0)}{x - 0} = \lim_{x \to 0} \frac{x^2 \sin(1/x)}{x} = \lim_{x \to 0} x \sin(1/x) = 0.$$

Hence, the derivative f' of f exists at all $x \in \mathbb{R}$. However, the function f' does not have a limit at $x = 0$ (why?), and consequently f' is discontinuous at $x = 0$. Thus, a function f that is differentiable at every point of $\mathbb{R}$ need not have a continuous derivative f'. $\square$

Inverse Functions

We will now relate the derivative of a function to the derivative of its inverse function, when this inverse function exists. We will limit our attention to a continuous strictly monotone function and use the Continuous Inverse Theorem 5.6.5 to ensure the existence of a continuous inverse function.
    If f is a continuous strictly monotone function on an interval I, then its inverse function $g = f^{-1}$ is defined on the interval $J := f(I)$ and satisfies the relation

$$g(f(x)) = x \qquad \text{for} \quad x \in I.$$

If $c \in I$ and $d := f(c)$, and if we knew that both $f'(c)$ and $g'(d)$ exist, then we could differentiate both sides of the equation and apply the Chain Rule to the left side to get $g'(f(c)) \cdot f'(c) = 1$. Thus, if $f'(c) \neq 0$, we would obtain

$$g'(d) = \frac{1}{f'(c)}.$$

However, it is necessary to deduce the differentiability of the inverse function g from the assumed differentiability of f before such a calculation can be performed. This is nicely accomplished by using Carathéodory's Theorem.

6.1.8 Theorem   *Let I be an interval in $\mathbb{R}$ and let $f : I \to \mathbb{R}$ be strictly monotone and continuous on I. Let $J := f(I)$ and let $g : J \to \mathbb{R}$ be the strictly monotone and continuous function inverse to f. If f is differentiable at $c \in I$ and $f'(c) \neq 0$, then g is differentiable at $d := f(c)$ and*

$$(12) \qquad g'(d) = \frac{1}{f'(c)} = \frac{1}{f'(g(d))}.$$

Proof.   Given $c \in \mathbb{R}$, we obtain from Carathéodory's Theorem 6.1.5 a function φ on I with properties that φ is continuous at c, $f(x) - f(c) = \varphi(x)(x - c)$ for $x \in I$, and $\varphi(c) = f'(c)$. Since $\varphi(c) \neq 0$ by hypothesis, there exists a neighborhood $V := (c - \delta, c + \delta)$ such that $\varphi(x) \neq 0$ for all $x \in V \cap I$. (See Theorem 4.2.9.) If $U := f(V \cap I)$, then the inverse function g satisfies $f(g(y)) = y$ for all $y \in U$, so that

$$y - d = f(g(y)) - f(c) = \varphi(g(y)) \cdot (g(y) - g(d)).$$

Since $\varphi(g(y)) \neq 0$ for $y \in U$, we can divide to get

$$g(y) - g(d) = \frac{1}{\varphi(g(y))} \cdot (y - d).$$

Since the function $1/(\varphi \circ g)$ is continuous at d, we apply Theorem 6.1.5 to conclude that $g'(d)$ exists and $g'(d) = 1/\varphi(g(d)) = 1/\varphi(c) = 1/f'(c)$. Q.E.D.

Note   The hypothesis, made in Theorem 6.1.8, that $f'(c) \neq 0$ is essential. In fact, if $f'(c) = 0$, then the inverse function g is *never* differentiable at $d = f(c)$, since the assumed existence of $g'(d)$ would lead to $1 = f'(c)g'(d) = 0$, which is impossible. The function $f(x) := x^3$ with $c = 0$ is such an example.

6.1.9 Theorem   *Let I be an interval and let $f : I \to \mathbb{R}$ be strictly monotone on I. Let $J := f(I)$ and let $g : J \to \mathbb{R}$ be the function inverse to f. If f is differentiable on I and $f'(x) \neq 0$ for $x \in I$, then g is differentiable on J and*

$$(13) \qquad g' = \frac{1}{f' \circ g}.$$

Proof.   If f is differentiable on I, then Theorem 6.1.2 implies that f is continuous on I, and by the Continuous Inverse Theorem 5.6.5, the inverse function g is continuous on J. Equation (13) now follows from Theorem 6.1.8. Q.E.D.

Remark   If f and g are the functions of Theorem 6.1.9, and if $x \in I$ and $y \in J$ are related by $y = f(x)$ and $x = g(y)$, then equation (13) can be written in the form

$$g'(y) = \frac{1}{(f' \circ g)(y)}, \quad y \in J, \qquad \text{or} \qquad (g' \circ f)(x) = \frac{1}{f'(x)}, \quad x \in I.$$

It can also be written in the form $g'(y) = 1/f'(x)$, *provided* that it is kept in mind that x and y are related by $y = f(x)$ and $x = g(y)$.

6.1.10 Examples **(a)** The function $f : \mathbb{R} \to \mathbb{R}$ defined by $f(x) := x^5 + 4x + 3$ is continuous and strictly monotone increasing (since it is the sum of two strictly increasing functions). Moreover, $f'(x) = 5x^4 + 4$ is never zero. Therefore, by Theorem 6.1.8, the inverse function $g = f^{-1}$ is differentiable at every point. If we take $c = 1$, then since $f(1) = 8$, we obtain $g'(8) = g'(f(1)) = 1/f'(1) = 1/9$.

(b) Let $n \in \mathbb{N}$ be even, let $I := [0, \infty)$, and let $f(x) := x^n$ for $x \in I$. It was seen at the end of Section 5.6 that f is strictly increasing and continuous on I, so that its inverse function $g(y) := y^{1/n}$ for $y \in J := [0, \infty)$ is also strictly increasing and continuous on J. Moreover, we have $f'(x) = nx^{n-1}$ for all $x \in I$. Hence it follows that if $y > 0$, then $g'(y)$ exists and

$$g'(y) = \frac{1}{f'(g(y))} = \frac{1}{n(g(y))^{n-1}} = \frac{1}{ny^{(n-1)/n}}.$$

Hence we deduce that

$$g'(y) = \frac{1}{n} y^{(1/n)-1} \qquad \text{for} \quad y > 0.$$

However, g is *not* differentiable at 0. (For a graph of f and g, see Figures 5.6.4 and 5.6.5.)

(c) Let $n \in \mathbb{N}, n \neq 1$, be odd, let $F(x) := x^n$ for $x \in \mathbb{R}$, and let $G(y) := y^{1/n}$ be its inverse function defined for all $y \in \mathbb{R}$. As in part (b) we find that G is differentiable for $y \neq 0$ and that $G'(y) = (1/n)y^{(1/n)-1}$ for $y \neq 0$. However, G is not differentiable at 0, even though G is differentiable for all $y \neq 0$. (For a graph of F and G, see Figures 5.6.6 and 5.6.7.)

(d) Let $r := m/n$ be a positive rational number, let $I := [0, \infty)$, and let $R(x) := x^r$ for $x \in I$. (Recall Definition 5.6.6.) Then R is the composition of the functions $f(x) := x^m$ and $g(x) := x^{1/n}$, $x \in I$. That is, $R(x) = f(g(x))$ for $x \in I$. If we apply the Chain Rule 6.1.6 and the results of (b) [or (c), depending on whether n is even or odd], then we obtain

$$R'(x) = f'(g(x))g'(x) = m(x^{1/n})^{m-1} \cdot \frac{1}{n} x^{(1/n)-1}$$

$$= \frac{m}{n} x^{(m/n)-1} = rx^{r-1}$$

for all $x > 0$. If $r > 1$, then it is an exercise to show that the derivative also exists at $x = 0$ and $R'(0) = 0$. (For a graph of R see Figure 5.6.8.)

(e) The sine function is strictly increasing on the interval $I := [-\pi/2, \pi/2]$; therefore its inverse function, which we will denote by Arcsin, exists on $J := [-1, 1]$. That is, if $x \in [-\pi/2, \pi/2]$ and $y \in [-1, 1]$ then $y = \sin x$ if and only if Arcsin $y = x$. It was asserted (without proof) in Example 6.1.7(d) that sin is differentiable on I and that $D \sin x = \cos x$ for $x \in I$. Since $\cos x \neq 0$ for x in $(-\pi/2, \pi/2)$ it follows from Theorem 6.1.8 that

$$D \text{ Arcsin } y = \frac{1}{D \sin x} = \frac{1}{\cos x}$$

$$= \frac{1}{\sqrt{1 - (\sin x)^2}} = \frac{1}{\sqrt{1 - y^2}}$$

for all $y \in (-1, 1)$. The derivative of Arcsin does *not* exist at the points -1 and 1.   □

Exercises for Section 6.1

1. Use the definition to find the derivative of each of the following functions:
   (a) $f(x) := x^3$ for $x \in \mathbb{R}$,
   (b) $g(x) := 1/x$ for $x \in \mathbb{R}, x \neq 0$,
   (c) $h(x) := \sqrt{x}$ for $x > 0$,
   (d) $k(x) := 1/\sqrt{x}$ for $x > 0$.

2. Show that $f(x) := x^{1/3}$, $x \in \mathbb{R}$, is not differentiable at $x = 0$. *use limit quotient lim → ∞*

3. Prove Theorem 6.1.3(a), (b).

4. Let $f : \mathbb{R} \to \mathbb{R}$ be defined by $f(x) := x^2$ for x rational, $f(x) := 0$ for x irrational. Show that f is differentiable at $x = 0$, and find $f'(0)$.

5. Differentiate and simplify:
    (a) $f(x) := \dfrac{x}{1+x^2}$,
    (b) $g(x) := \sqrt{5 - 2x + x^2}$,
    (c) $h(x) := (\sin x^k)^m$ for $m, k \in \mathbb{N}$,
    (d) $k(x) := \tan(x^2)$ for $|x| < \sqrt{\pi/2}$.

6. Let $n \in \mathbb{N}$ and let $f : \mathbb{R} \to \mathbb{R}$ be defined by $f(x) := x^n$ for $x \geq 0$ and $f(x) := 0$ for $x < 0$. For which values of n is f' continuous at 0? For which values of n is f' differentiable at 0?

7. Suppose that $f : \mathbb{R} \to \mathbb{R}$ is differentiable at c and that $f(c) = 0$. Show that $g(x) := |f(x)|$ is differentiable at c if and only if $f'(c) = 0$.

8. Determine where each of the following functions from $\mathbb{R}$ to $\mathbb{R}$ is differentiable and find the derivative:
    (a) $f(x) := |x| + |x + 1|$,
    (b) $g(x) := 2x + |x|$,
    (c) $h(x) := x|x|$,
    (d) $k(x) := |\sin x|$,

9. Prove that if $f : \mathbb{R} \to \mathbb{R}$ is an **even function** [that is, $f(-x) = f(x)$ for all $x \in \mathbb{R}$] and has a derivative at every point, then the derivative f' is an **odd function** [that is, $f'(-x) = -f'(x)$ for all $x \in \mathbb{R}$]. Also prove that if $g : \mathbb{R} \to \mathbb{R}$ is a differentiable odd function, then g' is an even function.

10. Let $g : \mathbb{R} \to \mathbb{R}$ be defined by $g(x) := x^2 \sin(1/x^2)$ for $x \neq 0$, and $g(0) := 0$. Show that g is differentiable for all $x \in \mathbb{R}$. Also show that the derivative g' is not bounded on the interval $[-1, 1]$.

11. Assume that there exists a function $L : (0, \infty) \to \mathbb{R}$ such that $L'(x) = 1/x$ for $x > 0$. Calculate the derivatives of the following functions:
    (a) $f(x) := L(2x + 3)$ for $x > 0$,
    (b) $g(x) := (L(x^2))^3$ for $x > 0$,
    (c) $h(x) := L(ax)$ for $a > 0, x > 0$,
    (d) $k(x) := L(L(x))$ when $L(x) > 0, x > 0$.

12. If $r > 0$ is a rational number, let $f : \mathbb{R} \to \mathbb{R}$ be defined by $f(x) := x^r \sin(1/x)$ for $x \neq 0$, and $f(0) := 0$. Determine those values of r for which $f'(0)$ exists.

13. If $f : \mathbb{R} \to \mathbb{R}$ is differentiable at $c \in \mathbb{R}$, show that
$$f'(c) = \lim \left(n\{f(c + 1/n) - f(c)\} \right).$$
However, show by example that the existence of the limit of this sequence does not imply the existence of $f'(c)$.

14. Given that the function $h(x) := x^3 + 2x + 1$ for $x \in \mathbb{R}$ has an inverse h^{-1} on $\mathbb{R}$, find the value of $(h^{-1})'(y)$ at the points corresponding to $x = 0, 1, -1$.

15. Given that the restriction of the cosine function cos to $I := [0, \pi]$ is strictly decreasing and that $\cos 0 = 1$, $\cos \pi = -1$, let $J := [-1, 1]$, and let Arccos: $J \to \mathbb{R}$ be the function inverse to the restriction of cos to I. Show that Arccos is differentiable on $(-1, 1)$ and $D\text{Arccos } y = (-1)/(1 - y^2)^{1/2}$ for $y \in (-1, 1)$. Show that Arccos is not differentiable at -1 and 1.

16. Given that the restriction of the tangent function tan to $I := (-\pi/2, \pi/2)$ is strictly increasing and that $\tan(I) = \mathbb{R}$, let Arctan: $\mathbb{R} \to \mathbb{R}$ be the function inverse to the restriction of tan to I. Show that Arctan is differentiable on $\mathbb{R}$ and that $D\text{Arctan}(y) = (1 + y^2)^{-1}$ for $y \in \mathbb{R}$.

17. Let $f : I \to \mathbb{R}$ be differentiable at $c \in I$. Establish the **Straddle Lemma:** Given $\varepsilon > 0$ there exists $\delta(\varepsilon) > 0$ such that if $u, v \in I$ satisfy $c - \delta(\varepsilon) < u \leq c \leq v < c + \delta(\varepsilon)$, then we have $|f(v) - f(u) - (v - u)f'(c)| \leq \varepsilon(v - u)$. [*Hint:* The $\delta(\varepsilon)$ is given by Definition 6.1.1. Subtract and add the term $f(c) - cf'(c)$ on the left side and use the Triangle Inequality.]

Section 6.2   The Mean Value Theorem

The Mean Value Theorem, which relates the values of a function to values of its derivative, is one of the most useful results in real analysis. In this section we will establish this important theorem and sample some of its many consequences.

We begin by looking at the relationship between the relative extrema of a function and the values of its derivative. Recall that the function $f : I \to \mathbb{R}$ is said to have a **relative maximum** [respectively, **relative minimum**] at $c \in I$ if there exists a neighborhood $V := V_\delta(c)$ of c such that $f(x) \le f(c)$ [respectively, $f(c) \le f(x)$] for all x in $V \cap I$. We say that f has a **relative extremum** at $c \in I$ if it has either a relative maximum or a relative minimum at c.

The next result provides the theoretical justification for the familiar process of finding points at which f has relative extrema by examining the zeros of the derivative. However, it must be realized that this procedure applies only to *interior* points of the interval. For example, if $f(x) := x$ on the interval $I := [0, 1]$, then the endpoint $x = 0$ yields the unique relative minimum and the endpoint $x = 1$ yields the unique maximum of f on I, but neither point is a zero of the derivative of f.

6.2.1 Interior Extremum Theorem   *Let c be an interior point of the interval I at which $f : I \to \mathbb{R}$ has a relative extremum. If the derivative of f at c exists, then $f'(c) = 0$.*

Proof.   We will prove the result only for the case that f has a relative maximum at c; the proof for the case of a relative minimum is similar.

If $f'(c) > 0$, then by Theorem 4.2.9 there exists a neighborhood $V \subseteq I$ of c such that

$$\frac{f(x) - f(c)}{x - c} > 0 \qquad \text{for} \quad x \in V, \, x \ne c.$$

If $x \in V$ and $x > c$, then we have

$$f(x) - f(c) = (x - c) \cdot \frac{f(x) - f(c)}{x - c} > 0.$$

But this contradicts the hypothesis that f has a relative maximum at c. Thus we cannot have $f'(c) > 0$. Similarly (how?), we cannot have $f'(c) < 0$. Therefore we must have $f'(c) = 0$.   Q.E.D.

6.2.2 Corollary   *Let $f : I \to \mathbb{R}$ be continuous on an interval I and suppose that f has a relative extremum at an interior point c of I. Then either the derivative of f at c does not exist, or it is equal to zero.*

We note that if $f(x) := |x|$ on $I := [-1, 1]$, then f has an interior minimum at $x = 0$; however, the derivative of f fails to exist at $x = 0$.

6.2.3 Rolle's Theorem   *Suppose that f is continuous on a closed interval $I := [a, b]$, that the derivative f' exists at every point of the open interval (a, b), and that $f(a) = f(b) = 0$. Then there exists at least one point c in (a, b) such that $f'(c) = 0$.*

Proof.   If f vanishes identically on I, then any c in (a, b) will satisfy the conclusion of the theorem. Hence we suppose that f does not vanish identically; replacing f by $-f$ if necessary, we may suppose that f assumes some positive values. By the Maximum–Minimum Theorem 5.3.4, the function f attains the value $\sup\{f(x) : x \in I\} > 0$ at some point c in I. Since $f(a) = f(b) = 0$, the point c must lie in (a, b); therefore $f'(c)$ exists.

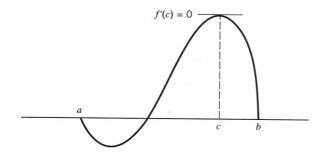

Figure 6.2.1   Rolle's Theorem

Since f has a relative maximum at c, we conclude from the Interior Extremum Theorem 6.2.1 that $f'(c) = 0$. (See Figure 6.2.1.)

Q.E.D.

As a consequence of Rolle's Theorem, we obtain the fundamental Mean Value Theorem.

6.2.4 Mean Value Theorem   *Suppose that f is continuous on a closed interval $I :=$ $[a, b]$, and that f has a derivative in the open interval (a, b). Then there exists at least one point c in (a, b) such that*

$$f(b) - f(a) = f'(c)(b - a).$$

Proof.   Consider the function φ defined on I by

$$\varphi(x) := f(x) - f(a) - \frac{f(b) - f(a)}{b - a}(x - a).$$

[The function φ is simply the difference of f and the function whose graph is the line segment joining the points $(a, f(a))$ and $(b, f(b))$; see Figure 6.2.2.] The hypotheses of

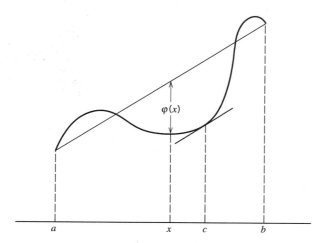

Figure 6.2.2   The Mean Value Theorem

Rolle's Theorem are satisfied by φ since φ is continuous on $[a, b]$, differentiable on (a, b), and $\varphi(a) = \varphi(b) = 0$. Therefore, there exists a point c in (a, b) such that

$$0 = \varphi'(c) = f'(c) - \frac{f(b) - f(a)}{b - a}.$$

Hence, $f(b) - f(a) = f'(c)(b - a)$.   Q.E.D.

Remark   The geometric view of the Mean Value Theorem is that there is some point on the curve $y = f(x)$ at which the tangent line is parallel to the line segment through the points $(a, f(a))$ and $(b, f(b))$. Thus it is easy to remember the statement of the Mean Value Theorem by drawing appropriate diagrams. While this should not be discouraged, it tends to suggest that its importance is geometrical in nature, which is quite misleading. In fact the Mean Value Theorem is a wolf in sheep's clothing and is *the* Fundamental Theorem of Differential Calculus. In the remainder of this section, we will present some of the consequences of this result. Other applications will be given later.

The Mean Value Theorem permits one to draw conclusions about the nature of a function f from information about its derivative f'. The following results are obtained in this manner.

6.2.5 Theorem   *Suppose that f is continuous on the closed interval $I := [a, b]$, that f is differentiable on the open interval (a, b), and that $f'(x) = 0$ for $x \in (a, b)$. Then f is constant on I.*

Proof.   We will show that $f(x) = f(a)$ for all $x \in I$. Indeed, if $x \in I$, $x > a$, is given, we apply the Mean Value Theorem to f on the closed interval $[a, x]$. We obtain a point c (depending on x) between a and x such that $f(x) - f(a) = f'(c)(x - a)$. Since $f'(c) = 0$ (by hypothesis), we deduce that $f(x) - f(a) = 0$. Hence, $f(x) = f(a)$ for any $x \in I$.
                                                                                Q.E.D.

6.2.6 Corollary   *Suppose that f and g are continuous on $I := [a, b]$, that they are differentiable on (a, b), and that $f'(x) = g'(x)$ for all $x \in (a, b)$. Then there exists a constant C such that $f = g + C$ on I.*

Recall that a function $f : I \to \mathbb{R}$ is said to be **increasing** on the interval I if whenever x_1, x_2 in I satisfy $x_1 < x_2$, then $f(x_1) \le f(x_2)$. Also recall that f is **decreasing** on I if the function $-f$ is increasing on I.

6.2.7 Theorem   *Let $f : I \to \mathbb{R}$ be differentiable on the interval I. Then:*

(a)   *f is increasing on I if and only if $f'(x) \ge 0$ for all $x \in I$.*
(b)   *f is decreasing on I if and only if $f'(x) \le 0$ for all $x \in I$.*

Proof.   (a) Suppose that $f'(x) \ge 0$ for all $x \in I$. If x_1, x_2 in I satisfy $x_1 < x_2$, then we apply the Mean Value Theorem to f on the closed interval $J := [x_1, x_2]$ to obtain a point c in (x_1, x_2) such that

$$f(x_2) - f(x_1) = f'(c)(x_2 - x_1).$$

Since $f'(c) \geq 0$ and $x_2 - x_1 > 0$, it follows that $f(x_2) - f(x_1) \geq 0$. (Why?) Hence, $f(x_1) \leq f(x_2)$ and, since $x_1 < x_2$ are arbitrary points in I, we conclude that f is increasing on I.

For the converse assertion, we suppose that f is differentiable and increasing on I. Thus, for any point $x \neq c$ in I, we have $(f(x) - f(c))/(x - c) \geq 0$. (Why?) Hence, by Theorem 4.2.6 we conclude that

$$f'(c) = \lim_{x \to c} \frac{f(x) - f(c)}{x - c} \geq 0.$$

(b) The proof of part (b) is similar and will be omitted.   Q.E.D.

A function f is said to be **strictly increasing** on an interval I if for any points x_1, x_2 in I such that $x_1 < x_2$, we have $f(x_1) < f(x_2)$. An argument along the same lines of the proof of Theorem 6.2.7 can be made to show that a function having a strictly positive derivative on an interval is strictly increasing there. (See Exercise 13.) However, the converse assertion is not true, since a strictly increasing differentiable function may have a derivative that vanishes at certain points. For example, the function $f : \mathbb{R} \to \mathbb{R}$ defined by $f(x) := x^3$ is strictly increasing on $\mathbb{R}$, but $f'(0) = 0$. The situation for strictly decreasing functions is similar.

Remark   It is reasonable to define a function to be **increasing at a point** if there is a neighborhood of the point on which the function is increasing. One might suppose that, if the derivative is strictly positive at a point, then the function is increasing at this point. However, this supposition is false; indeed, the differentiable function defined by

$$g(x) := \begin{cases} x + 2x^2 \sin(1/x) & \text{if } x \neq 0, \\ 0 & \text{if } x = 0, \end{cases}$$

is such that $g'(0) = 1$, yet it can be shown that g is not increasing in any neighborhood of $x = 0$. (See Exercise 10.)

We next obtain a sufficient condition for a function to have a relative extremum at an interior point of an interval.

6.2.8 First Derivative Test for Extrema   *Let f be continuous on the interval $I := [a, b]$ and let c be an interior point of I. Assume that f is differentiable on (a, c) and (c, b). Then:*

(a) *If there is a neighborhood $(c - \delta, c + \delta) \subseteq I$ such that $f'(x) \geq 0$ for $c - \delta < x < c$ and $f'(x) \leq 0$ for $c < x < c + \delta$, then f has a relative maximum at c.*
(b) *If there is a neighborhood $(c - \delta, c + \delta) \subseteq I$ such that $f'(x) \leq 0$ for $c - \delta < x < c$ and $f'(x) \geq 0$ for $c < x < c + \delta$, then f has a relative minimum at c.*

Proof.   (a) If $x \in (c - \delta, c)$, then it follows from the Mean Value Theorem that there exists a point $c_x \in (x, c)$ such that $f(c) - f(x) = (c - x)f'(c_x)$. Since $f'(c_x) \geq 0$ we infer that $f(x) \leq f(c)$ for $x \in (c - \delta, c)$. Similarly, it follows (how?) that $f(x) \leq f(c)$ for $x \in (c, c + \delta)$. Therefore $f(x) \leq f(c)$ for all $x \in (c - \delta, c + \delta)$ so that f has a relative maximum at c.

(b) The proof is similar.   Q.E.D.

Remark   The converse of the First Derivative Test 6.2.8 is *not* true. For example, there exists a differentiable function $f : \mathbb{R} \to \mathbb{R}$ with absolute minimum at $x = 0$ but such that

f' takes on both positive and negative values on both sides of (and arbitrarily close to) $x = 0$. (See Exercise 9.)

Further Applications of the Mean Value Theorem

We will continue giving other types of applications of the Mean Value Theorem; in doing so we will draw more freely than before on the past experience of the reader and his or her knowledge concerning the derivatives of certain well-known functions.

6.2.9 Examples (a) Rolle's Theorem can be used for the location of roots of a function. For, if a function g can be identified as the derivative of a function f, then between any two roots of f there is at least one root of g. For example, let $g(x) := \cos x$, then g is known to be the derivative of $f(x) := \sin x$. Hence, between any two roots of $\sin x$ there is at least one root of $\cos x$. On the other hand, $g'(x) = -\sin x = -f(x)$, so another application of Rolle's Theorem tells us that between any two roots of $\cos$ there is at least one root of $\sin$. Therefore, we conclude that the roots of $\sin$ and $\cos$ *interlace each other*. This conclusion is probably not news to the reader; however, the same type of argument can be applied to the *Bessel functions* J_n of order $n = 0, 1, 2, \cdots$ by using the relations

$$[x^n J_n(x)]' = x^n J_{n-1}(x), \quad [x^{n-1} J_n(x)]' = -x^{-n} J_{n+1}(x) \quad \text{for} \quad x > 0.$$

The details of this argument should be supplied by the reader.

(b) We can apply the Mean Value Theorem for approximate calculations and to obtain error estimates. For example, suppose it is desired to evaluate $\sqrt{105}$. We employ the Mean Value Theorem with $f(x) := \sqrt{x}, a = 100, b = 105$, to obtain

$$\sqrt{105} - \sqrt{100} = \frac{5}{2\sqrt{c}}$$

for some number c with $100 < c < 105$. Since $10 < \sqrt{c} < \sqrt{105} < \sqrt{121} = 11$, we can assert that

$$\frac{5}{2(11)} < \sqrt{105} - 10 < \frac{5}{2(10)},$$

whence it follows that $10.2272 < \sqrt{105} < 10.2500$. This estimate may not be as sharp as desired. It is clear that the estimate $\sqrt{c} < \sqrt{105} < \sqrt{121}$ was wasteful and can be improved by making use of our conclusion that $\sqrt{105} < 10.2500$. Thus, $\sqrt{c} < 10.2500$ and we easily determine that

$$0.2439 < \frac{5}{2(10.2500)} < \sqrt{105} - 10.$$

Our improved estimate is $10.2439 < \sqrt{105} < 10.2500$. $\qquad\square$

Inequalities

One very important use of the Mean Value Theorem is to obtain certain inequalities. Whenever information concerning the range of the derivative of a function is available, this information can be used to deduce certain properties of the function itself. The following examples illustrate the valuable role that the Mean Value Theorem plays in this respect.

6.2.10 Examples   (a)   The exponential function $f(x) := e^x$ has the derivative $f'(x) = e^x$ for all $x \in \mathbb{R}$. Thus $f'(x) > 1$ for $x > 0$, and $f'(x) < 1$ for $x < 0$. From these relationships, we will derive the inequality

(1) $$e^x \geq 1 + x \qquad \text{for} \quad x \in \mathbb{R},$$

with equality occurring if and only if $x = 0$.

If $x = 0$, we have equality with both sides equal to 1. If $x > 0$, we apply the Mean Value Theorem to the function f on the interval $[0, x]$. Then for some c with $0 < c < x$ we have

$$e^x - e^0 = e^c(x - 0).$$

Since $e^0 = 1$ and $e^c > 1$, this becomes $e^x - 1 > x$ so that we have $e^x > 1 + x$ for $x > 0$. A similar argument establishes the same strict inequality for $x < 0$. Thus the inequality (1) holds for all x, and equality occurs only if $x = 0$.

(b)   The function $g(x) := \sin x$ has the derivative $g'(x) = \cos x$ for all $x \in \mathbb{R}$. On the basis of the fact that $-1 \leq \cos x \leq 1$ for all $x \in \mathbb{R}$, we will show that

(2) $$-x \leq \sin x \leq x \quad \text{for all} \quad x \geq 0.$$

Indeed, if we apply the Mean Value Theorem to g on the interval $[0, x]$, where $x > 0$, we obtain

$$\sin x - \sin 0 = (\cos c)(x - 0)$$

for some c between 0 and x. Since $\sin 0 = 0$ and $-1 \leq \cos c \leq 1$, we have $-x \leq \sin x \leq x$. Since equality holds at $x = 0$, the inequality (2) is established.

(c)   (Bernoulli's inequality) If $\alpha > 1$, then

(3) $$(1 + x)^\alpha \geq 1 + \alpha x \qquad \text{for all} \quad x > -1,$$

with equality if and only if $x = 0$.

This inequality was established earlier, in Example 2.1.13(c), for positive integer values of α by using Mathematical Induction. We now derive the more general version by employing the Mean Value Theorem.

If $h(x) := (1 + x)^\alpha$ then $h'(x) = \alpha(1 + x)^{\alpha-1}$ for all $x > -1$. [For rational α this derivative was established in Example 6.1.10(c). The extension to irrational will be discussed in Section 8.3.] If $x > 0$, we infer from the Mean Value Theorem applied to h on the interval $[0, x]$ that there exists c with $0 < c < x$ such that $h(x) - h(0) = h'(c)(x - 0)$. Thus, we have

$$(1 + x)^\alpha - 1 = \alpha(1 + c)^{\alpha-1}x.$$

Since $c > 0$ and $\alpha - 1 > 0$, it follows that $(1 + c)^{\alpha-1} > 1$ and hence that $(1 + x)^\alpha > 1 + \alpha x$. If $-1 < x < 0$, a similar use of the Mean Value Theorem on the interval $[x, 0]$ leads to the same strict inequality. Since the case $x = 0$ results in equality, we conclude that (3) is valid for all $x > -1$ with equality if and only if $x = 0$.

(d)   Let α be a real number satisfying $0 < \alpha < 1$ and let $g(x) = \alpha x - x^\alpha$ for $x \geq 0$. Then $g'(x) = \alpha(1 - x^{\alpha-1})$, so that $g'(x) < 0$ for $0 < x < 1$ and $g'(x) > 0$ for $x > 1$. Consequently, if $x \geq 0$, then $g(x) \geq g(1)$ and $g(x) = g(1)$ if and only if $x = 1$. Therefore, if $x \geq 0$ and $0 < \alpha < 1$, then we have

$$x^\alpha \leq \alpha x + (1 - \alpha).$$

If $a > 0$ and $b > 0$ and if we let $x = a/b$ and multiply by b, we obtain the inequality

$$a^\alpha b^{1-\alpha} \leq \alpha a + (1 - \alpha)b,$$

where equality holds if and only if $a = b$. $\qquad\qquad\qquad\qquad\qquad\qquad\qquad\qquad\square$

The Intermediate Value Property of Derivatives

We conclude this section with an interesting result, often referred to as Darboux's Theorem. It states that if a function f is differentiable at every point of an interval I, then the function f' has the Intermediate Value Property. This means that if f' takes on values A and B, then it also takes on all values between A and B. The reader will recognize this property as one of the important consequences of continuity as established in Theorem 5.3.7. It is remarkable that derivatives, which need not be continuous functions, also possess this property.

6.2.11 Lemma Let $I \subseteq \mathbb{R}$ be an interval, let $f : I \to \mathbb{R}$, let $c \in I$, and assume that f has a derivative at c. Then:

(a) If $f'(c) > 0$, then there is a number $\delta > 0$ such that $f(x) > f(c)$ for $x \in I$ such that $c < x < c + \delta$.

(b) If $f'(c) < 0$, then there is a number $\delta > 0$ such that $f(x) > f(c)$ for $x \in I$ such that $c - \delta < x < c$.

Proof. (a) Since

$$\lim_{x \to c} \frac{f(x) - f(c)}{x - c} = f'(c) > 0,$$

it follows from Theorem 4.2.9 that there is a number $\delta > 0$ such that if $x \in I$ and $0 < |x - c| < \delta$, then

$$\frac{f(x) - f(c)}{x - c} > 0.$$

If $x \in I$ also satisfies $x > c$, then we have

$$f(x) - f(c) = (x - c) \cdot \frac{f(x) - f(c)}{x - c} > 0.$$

Hence, if $x \in I$ and $c < x < c + \delta$, then $f(x) > f(c)$.

The proof of (b) is similar. $\qquad\qquad\qquad\qquad\qquad\qquad\qquad\qquad\qquad$ Q.E.D.

6.2.12 Darboux's Theorem If f is differentiable on $I = [a, b]$ and if k is a number between $f'(a)$ and $f'(b)$, then there is at least one point c in (a, b) such that $f'(c) = k$.

Proof. Suppose that $f'(a) < k < f'(b)$. We define g on I by $g(x) := kx - f(x)$ for $x \in I$. Since g is continuous, it attains a maximum value on I. Since $g'(a) = k - f'(a) > 0$, it follows from Lemma 6.2.11(a) that the maximum of g does not occur at $x = a$. Similarly, since $g'(b) = k - f'(b) < 0$, it follows from Lemma 6.2.11(b) that the maximum does not occur at $x = b$. Therefore, g attains its maximum at some c in (a, b). Then from Theorem 6.2.1 we have $0 = g'(c) = k - f'(c)$. Hence, $f'(c) = k$. $\qquad\qquad\qquad$ Q.E.D.

6.2.13 Example The function $g: [-1, 1] \to \mathbb{R}$ defined by

$$g(x) := \begin{cases} 1 & \text{for} \quad 0 < x \leq 1, \\ 0 & \text{for} \quad x = 0, \\ -1 & \text{for} \quad -1 \leq x < 0, \end{cases}$$

(which is a restriction of the signum function) clearly fails to satisfy the intermediate value property on the interval $[-1, 1]$. Therefore, by Darboux's Theorem, there does not exist a function f such that $f'(x) = g(x)$ for all $x \in [-1, 1]$. In other words, g is *not* the derivative on $[-1, 1]$ of any function. $\square$

Exercises for Section 6.2

1. For each of the following functions on $\mathbb{R}$ to $\mathbb{R}$, find points of relative extrema, the intervals on which the function is increasing, and those on which it is decreasing:
    (a) $f(x) := x^2 - 3x + 5$,
    (b) $g(x) := 3x - 4x^2$,
    (c) $h(x) := x^3 - 3x - 4$,
    (d) $k(x) := x^4 + 2x^2 - 4$.

2. Find the points of relative extrema, the intervals on which the following functions are increasing, and those on which they are decreasing:
    (a) $f(x) := x + 1/x$ for $x \neq 0$,
    (b) $g(x) := x/(x^2 + 1)$ for $x \in \mathbb{R}$,
    (c) $h(x) := \sqrt{x} - 2\sqrt{x+2}$ for $x > 0$,
    (d) $k(x) := 2x + 1/x^2$ for $x \neq 0$.

3. Find the points of relative extrema of the following functions on the specified domain:
    (a) $f(x) := |x^2 - 1|$ for $-4 \leq x \leq 4$,
    (b) $g(x) := 1 - (x - 1)^{2/3}$ for $0 \leq x \leq 2$,
    (c) $h(x) := x|x^2 - 12|$ for $-2 \leq x \leq 3$,
    (d) $k(x) := x(x - 8)^{1/3}$ for $0 \leq x \leq 9$.

4. Let $a_1, a_2, \cdots, a_n$ be real numbers and let f be defined on $\mathbb{R}$ by
    $$f(x) := \sum_{i=1}^{n}(a_i - x)^2 \quad \text{for} \quad x \in \mathbb{R}.$$
    Find the unique point of relative minimum for f.

5. Let $a > b > 0$ and let $n \in \mathbb{N}$ satisfy $n \geq 2$. Prove that $a^{1/n} - b^{1/n} < (a - b)^{1/n}$. [*Hint*: Show that $f(x) := x^{1/n} - (x - 1)^{1/n}$ is decreasing for $x \geq 1$, and evaluate f at 1 and a/b.]

6. Use the Mean Value Theorem to prove that $|\sin x - \sin y| \leq |x - y|$ for all x, y in $\mathbb{R}$.

7. Use the Mean Value Theorem to prove that $(x - 1)/x < \ln x < x - 1$ for $x > 1$. [*Hint*: Use the fact that $D \ln x = 1/x$ for $x > 0$.]

8. Let $f: [a, b] \to \mathbb{R}$ be continuous on $[a, b]$ and differentiable in (a, b). Show that if $\lim_{x \to a} f'(x) = A$, then $f'(a)$ exists and equals A. [*Hint*: Use the definition of $f'(a)$ and the Mean Value Theorem.]

9. Let $f : \mathbb{R} \to \mathbb{R}$ be defined by $f(x) := 2x^4 + x^4 \sin(1/x)$ for $x \neq 0$ and $f(0) := 0$. Show that f has an absolute minimum at $x = 0$, but that its derivative has both positive and negative values in every neighborhood of 0.

10. Let $g : \mathbb{R} \to \mathbb{R}$ be defined by $g(x) := x + 2x^2 \sin(1/x)$ for $x \neq 0$ and $g(0) := 0$. Show that $g'(0) = 1$, but in every neighborhood of 0 the derivative $g'(x)$ takes on both positive and negative values. Thus g is not monotonic in any neighborhood of 0.

11. Give an example of a uniformly continuous function on $[0, 1]$ that is differentiable on $(0, 1)$ but whose derivative is not bounded on $(0, 1)$.

12. If $h(x) := 0$ for $x < 0$ and $h(x) := 1$ for $x \geq 0$, prove there does not exist a function $f : \mathbb{R} \to \mathbb{R}$ such that $f'(x) = h(x)$ for all $x \in \mathbb{R}$. Give examples of two functions, not differing by a constant, whose derivatives equal $h(x)$ for all $x \neq 0$.

13. Let I be an interval and let $f : I \to \mathbb{R}$ be differentiable on I. Show that if f' is positive on I, then f is strictly increasing on I.

14. Let I be an interval and let $f : I \to \mathbb{R}$ be differentiable on I. Show that if the derivative f' is never 0 on I, then either $f'(x) > 0$ for all $x \in I$ or $f'(x) < 0$ for all $x \in I$.

15. Let I be an interval. Prove that if f is differentiable on I and if the derivative f' is bounded on I, then f satisfies a Lipschitz condition on I. (See Definition 5.4.4.)

16. Let $f: [0, \infty) \to \mathbb{R}$ be differentiable on $(0, \infty)$ and assume that $f'(x) \to b$ as $x \to \infty$.
    (a) Show that for any $h > 0$, we have $\lim_{x \to \infty} \left(f(x + h) - f(x) \right)/h = b$.
    (b) Show that if $f(x) \to a$ as $x \to \infty$, then $b = 0$.
    (c) Show that $\lim_{x \to \infty} \left(f(x)/x \right) = b$.

17. Let f, g be differentiable on $\mathbb{R}$ and suppose that $f(0) = g(0)$ and $f'(x) \le g'(x)$ for all $x \ge 0$. Show that $f(x) \le g(x)$ for all $x \ge 0$.

18. Let $I := [a, b]$ and let $f : I \to \mathbb{R}$ be differentiable at $c \in I$. Show that for every $\varepsilon > 0$ there exists $\delta > 0$ such that if $0 < |x - y| < \delta$ and $a \le x \le c \le y \le b$, then
$$\left| \frac{f(x) - f(y)}{x - y} - f'(c) \right| < \varepsilon.$$

19. A differentiable function $f : I \to \mathbb{R}$ is said to be **uniformly differentiable** on $I := [a, b]$ if for every $\varepsilon > 0$ there exists $\delta > 0$ such that if $0 < |x - y| < \delta$ and $x, y \in I$, then
$$\left| \frac{f(x) - f(y)}{x - y} - f'(x) \right| < \varepsilon.$$
Show that if f is uniformly differentiable on I, then f' is continuous on I.

20. Suppose that $f : [0, 2] \to \mathbb{R}$ is continuous on $[0, 2]$ and differentiable on $(0, 2)$, and that $f(0) = 0$, $f(1) = 1$, $f(2) = 1$.
    (a) Show that there exists $c_1 \in (0, 1)$ such that $f'(c_1) = 1$.
    (b) Show that there exists $c_2 \in (1, 2)$ such that $f'(c_2) = 0$.
    (c) Show that there exists $c \in (0, 2)$ such that $f'(c) = 1/3$.

Section 6.3   L'Hospital's Rules

The Marquis Guillame François L'Hospital (1661–1704) was the author of the first calculus book, *L'Analyse des infiniment petits*, published in 1696. He studied the then new differential calculus from Johann Bernoulli (1667–1748), first when Bernoulli visited L'Hospital's country estate and subsequently through a series of letters. The book was the result of L'Hospital's studies. The limit theorem that became known as L'Hospital's Rule first appeared in this book, though in fact it was discovered by Bernoulli.

The initial theorem was refined and extended, and the various results are collectively referred to as L'Hospital's (or L'Hôpital's) Rules. In this section we establish the most basic of these results and indicate how others can be derived.

Indeterminate Forms

In the preceding chapters we have often been concerned with methods of evaluating limits. It was shown in Theorem 4.2.4(b) that if $A := \lim_{x \to c} f(x)$ and $B := \lim_{x \to c} g(x)$, and if $B \ne 0$, then
$$\lim_{x \to c} \frac{f(x)}{g(x)} = \frac{A}{B}.$$
However, if $B = 0$, then no conclusion was deduced. It will be seen in Exercise 2 that if $B = 0$ and $A \ne 0$, then the limit is infinite (when it exists).

The case $A = 0$, $B = 0$ has not been covered previously. In this case, the limit of the quotient f/g is said to be "indeterminate". We will see that in this case the limit may not exist or may be any real value, depending on the particular functions f and g. The symbolism $0/0$ is used to refer to this situation. For example, if α is any real number, and if we define $f(x) := \alpha x$ and $g(x) := x$, then

$$\lim_{x \to 0} \frac{f(x)}{g(x)} = \lim_{x \to 0} \frac{\alpha x}{x} = \lim_{x \to 0} \alpha = \alpha.$$

Thus the indeterminate form $0/0$ can lead to any real number α as a limit.

Other indeterminate forms are represented by the symbols ∞/∞, $0 \cdot \infty$, 0^0, 1^∞, ∞^0, and $\infty - \infty$. These notations correspond to the indicated limiting behavior and juxtaposition of the functions f and g. Our attention will be focused on the indeterminate forms $0/0$ and ∞/∞. The other indeterminate cases are usually reduced to the form $0/0$ or ∞/∞ by taking logarithms, exponentials, or algebraic manipulations.

A Preliminary Result _____

To show that the use of differentiation in this context is a natural and not surprising development, we first establish an elementary result that is based simply on the definition of the derivative.

6.3.1 Theorem   Let f and g be defined on $[a, b]$, let $f(a) = g(a) = 0$, and let $g(x) \neq 0$ for $a < x < b$. If f and g are differentiable at a and if $g'(a) \neq 0$, then the limit of f/g at a exists and is equal to $f'(a)/g'(a)$. Thus

$$\lim_{x \to a+} \frac{f(x)}{g(x)} = \frac{f'(a)}{g'(a)}.$$

Proof.   Since $f(a) = g(a) = 0$, we can write the quotient $f(x)/g(x)$ for $a < x < b$ as follows:

$$\frac{f(x)}{g(x)} = \frac{f(x) - f(a)}{g(x) - g(a)} = \frac{\dfrac{f(x) - f(a)}{x - a}}{\dfrac{g(x) - g(a)}{x - a}}.$$

Applying Theorem 4.2.4(b), we obtain

$$\lim_{x \to a+} \frac{f(x)}{g(x)} = \frac{\displaystyle\lim_{x \to a+} \frac{f(x) - f(a)}{x - a}}{\displaystyle\lim_{x \to a+} \frac{g(x) - g(a)}{x - a}} = \frac{f'(a)}{g'(a)}. \qquad \text{Q.E.D.}$$

Warning   The hypothesis that $f(a) = g(a) = 0$ is essential here. For example, if $f(x) := x + 17$ and $g(x) := 2x + 3$ for $x \in \mathbb{R}$, then

$$\lim_{x \to 0} \frac{f(x)}{g(x)} = \frac{17}{3}, \quad \text{while} \quad \frac{f'(0)}{g'(0)} = \frac{1}{2}.$$

The preceding result enables us to deal with limits such as

$$\lim_{x \to 0} \frac{x^2 + x}{\sin 2x} = \frac{2 \cdot 0 + 1}{2 \cos 0} = \frac{1}{2}.$$

To handle limits where f and g are not differentiable at the point a, we need a more general version of the Mean Value Theorem due to Cauchy.

6.3.2 Cauchy Mean Value Theorem   *Let f and g be continuous on [a, b] and differentiable on (a, b), and assume that $g'(x) \neq 0$ for all x in (a, b). Then there exists c in (a, b) such that*

$$\frac{f(b) - f(a)}{g(b) - g(a)} = \frac{f'(c)}{g'(c)}.$$

Proof.   As in the proof of the Mean Value Theorem, we introduce a function to which Rolle's Theorem will apply. First we note that since $g'(x) \neq 0$ for all x in (a, b), it follows from Rolle's Theorem that $g(a) \neq g(b)$. For x in [a, b], we now define

$$h(x) := \frac{f(b) - f(a)}{g(b) - g(a)} \big(g(x) - g(a)\big) - \big(f(x) - f(a)\big).$$

Then h is continuous on [a, b], differentiable on (a, b), and $h(a) = h(b) = 0$. Therefore, it follows from Rolle's Theorem 6.2.3 that there exists a point c in (a, b) such that

$$0 = h'(c) = \frac{f(b) - f(a)}{g(b) - g(a)} g'(c) - f'(c).$$

Since $g'(c) \neq 0$, we obtain the desired result by dividing by $g'(c)$.   Q.E.D.

Remarks   The preceding theorem has a geometric interpretation that is similar to that of the Mean Value Theorem 6.2.4. The functions f and g can be viewed as determining a curve in the plane by means of the parametric equations $x = f(t)$, $y = g(t)$ where $a \leq t \leq b$. Then the conclusion of the theorem is that there exists a point $(f(c), g(c))$ on the curve for some c in (a, b) such that the slope $g'(c)/f'(c)$ of the line tangent to the curve at that point is equal to the slope of the line segment joining the endpoints of the curve.

Note that if $g(x) = x$, then the Cauchy Mean Value Theorem reduces to the Mean Value Theorem 6.2.4.

L'Hospital's Rule, I

We will now establish the first of L'Hospital's Rules. For convenience, we will consider right-hand limits at a point a; left-hand limits, and two-sided limits are treated in exactly the same way. In fact, the theorem even allows the possibility that $a = -\infty$. The reader should observe that, in contrast with Theorem 6.3.1, the following result does not assume the differentiability of the functions at the point a. The result asserts that the limiting behavior of $f(x)/g(x)$ as $x \to a+$ is the same as the limiting behavior of $f'(x)/g'(x)$ as $x \to a+$, including the case where this limit is infinite. An important hypothesis here is that both f and g approach 0 as $x \to a+$.

6.3.3 L'Hospital's Rule, I   *Let $-\infty \leq a < b \leq \infty$ and let f, g be differentiable on (a, b) such that $g'(x) \neq 0$ for all $x \in (a, b)$. Suppose that*

(1) $$\lim_{x \to a+} f(x) = 0 = \lim_{x \to a+} g(x).$$

(a)   *If $\displaystyle\lim_{x \to a+} \frac{f'(x)}{g'(x)} = L \in \mathbb{R}$, then $\displaystyle\lim_{x \to a+} \frac{f(x)}{g(x)} = L$.*

(b)   *If $\displaystyle\lim_{x \to a+} \frac{f'(x)}{g'(x)} = L \in \{-\infty, \infty\}$, then $\displaystyle\lim_{x \to a+} \frac{f(x)}{g(x)} = L$.*

Proof. If $a < \alpha < \beta < b$, then Rolle's Theorem implies that $g(\beta) \neq g(\alpha)$. Further, by the Cauchy Mean Value Theorem 6.3.2, there exists $u \in (\alpha, \beta)$ such that

$$(2) \qquad \frac{f(\beta) - f(\alpha)}{g(\beta) - g(\alpha)} = \frac{f'(u)}{g'(u)}.$$

Case (a): If $L \in \mathbb{R}$ and if $\varepsilon > 0$ is given, there exists $c \in (a, b)$ such that

$$L - \varepsilon < \frac{f'(u)}{g'(u)} < L + \varepsilon \qquad \text{for} \quad u \in (a, c),$$

whence it follows from (2) that

$$(3) \qquad L - \varepsilon < \frac{f(\beta) - f(\alpha)}{g(\beta) - g(\alpha)} < L + \varepsilon \qquad \text{for} \quad a < \alpha < \beta \leq c.$$

If we take the limit in (3) as $\alpha \to a+$, we have

$$L - \varepsilon \leq \frac{f(\beta)}{g(\beta)} \leq L + \varepsilon \qquad \text{for} \quad \beta \in (a, c].$$

Since $\varepsilon > 0$ is arbitrary, the assertion follows.

Case (b): If $L = +\infty$ and if $M > 0$ is given, there exists $c \in (a, b)$ such that

$$\frac{f'(u)}{g'(u)} > M \qquad \text{for} \quad u \in (a, c),$$

whence it follows from (2) that

$$(4) \qquad \frac{f(\beta) - f(\alpha)}{g(\beta) - g(\alpha)} > M \qquad \text{for} \quad a < \alpha < \beta < c.$$

If we take the limit in (4) as $\alpha \to a+$, we have

$$\frac{f(\beta)}{g(\beta)} \geq M \qquad \text{for} \quad \beta \in (a, c).$$

Since $M > 0$ is arbitrary, the assertion follows.

If $L = -\infty$, the argument is similar.   Q.E.D.

6.3.4 Examples   (a) We have

$$\lim_{x \to 0+} \frac{\sin x}{\sqrt{x}} = \lim_{x \to 0+} \left[\frac{\cos x}{1/(2\sqrt{x})} \right] = \lim_{x \to 0+} 2\sqrt{x} \cos x = 0.$$

Observe that the denominator is not differentiable at $x = 0$ so that Theorem 6.3.1 cannot be applied. However $f(x) := \sin x$ and $g(x) := \sqrt{x}$ are differentiable on $(0, \infty)$ and both approach 0 as $x \to 0+$. Moreover, $g'(x) \neq 0$ on $(0, \infty)$, so that 6.3.3 is applicable.

(b) We have $\displaystyle\lim_{x \to 0} \left[\frac{1 - \cos x}{x^2} \right] = \lim_{x \to 0} \frac{\sin x}{2x}$.

We need to consider both left and right hand limits here. The quotient in the second limit is again indeterminate in the form 0/0. However, the hypotheses of 6.3.3 are again satisfied so that a second application of L'Hospital's Rule is permissible. Hence, we obtain

$$\lim_{x \to 0} \left[\frac{1 - \cos x}{x^2} \right] = \lim_{x \to 0} \frac{\sin x}{2x} = \lim_{x \to 0} \frac{\cos x}{2} = \frac{1}{2}.$$

(c)   We have $\lim\limits_{x \to 0} \dfrac{e^x - 1}{x} = \lim\limits_{x \to 0} \dfrac{e^x}{1} = 1$.

Again, both left- and right-hand limits need to be considered. Similarly, we have

$$\lim_{x \to 0} \left[\frac{e^x - 1 - x}{x^2} \right] = \lim_{x \to 0} \frac{e^x - 1}{2x} = \lim_{x \to 0} \frac{e^x}{2} = \frac{1}{2}.$$

(d)   We have $\lim\limits_{x \to 1} \left[\dfrac{\ln x}{x - 1} \right] = \lim\limits_{x \to 1} \dfrac{(1/x)}{1} = 1$.   □

L'Hospital's Rule, II

This Rule is very similar to the first one, except that it treats the case where the denominator becomes infinite as $x \to a+$. Again we will consider only right-hand limits, but it is possible that $a = -\infty$. Left-hand limits and two-sided limits are handled similarly.

6.3.5 L'Hospital's Rule, II   *Let* $-\infty \le a < b \le \infty$ *and let* f, g *be differentiable on* (a, b) *such that* $g'(x) \ne 0$ *for all* $x \in (a, b)$. *Suppose that*

$$\lim_{x \to a+} g(x) = \pm\infty. \tag{5}$$

(a)   *If* $\lim\limits_{x \to a+} \dfrac{f'(x)}{g'(x)} = L \in \mathbb{R}$, *then* $\lim\limits_{x \to a+} \dfrac{f(x)}{g(x)} = L$.

(b)   *If* $\lim\limits_{x \to a+} \dfrac{f'(x)}{g'(x)} = L \in \{-\infty, \infty\}$, *then* $\lim\limits_{x \to a+} \dfrac{f(x)}{g(x)} = L$.

Proof.   We will suppose that (5) holds with limit ∞.

As before, we have $g(\beta) \ne g(\alpha)$ for $\alpha, \beta \in (a, b), \alpha < \beta$. Further, equation (2) in the proof of 6.3.3 holds for some $u \in (\alpha, \beta)$.

Case (a): If $L \in \mathbb{R}$ with $L > 0$ and $\varepsilon > 0$ is given, there is $c \in (a, b)$ such that (3) in the proof of 6.3.3 holds when $a < \alpha < \beta \le c$. Since $g(x) \to \infty$, we may also assume that $g(c) > 0$. Taking $\beta = c$ in (3), we have

$$L - \varepsilon < \frac{f(c) - f(\alpha)}{g(c) - g(\alpha)} < L + \varepsilon \qquad \text{for} \quad \alpha \in (a, c). \tag{6}$$

Since $g(c)/g(\alpha) \to 0$ as $\alpha \to a+$, we may assume that $0 < g(c)/g(\alpha) < 1$ for all $\alpha \in (a, c)$, whence it follows that

$$\frac{g(\alpha) - g(c)}{g(\alpha)} = 1 - \frac{g(c)}{g(\alpha)} > 0 \qquad \text{for} \quad \alpha \in (a, c).$$

If we multiply (6) by $(g(\alpha) - g(c))/g(\alpha) > 0$, we have

$$(L - \varepsilon)\left(1 - \frac{g(c)}{g(\alpha)} \right) < \frac{f(\alpha)}{g(\alpha)} - \frac{f(c)}{g(\alpha)} < (L + \varepsilon)\left(1 - \frac{g(c)}{g(\alpha)} \right). \tag{7}$$

Now, since $g(c)/g(\alpha) \to 0$ and $f(c)/g(\alpha) \to 0$ as $\alpha \to a+$, then for any δ with $0 < \delta < 1$ there exists $d \in (a, c)$ such that $0 < g(c)/g(\alpha) < \delta$ and $|f(c)|/g(\alpha) < \delta$ for all $\alpha \in (a, d)$, whence (7) gives

$$(L - \varepsilon)(1 - \delta) - \delta < \frac{f(\alpha)}{g(\alpha)} < (L + \varepsilon) + \delta. \tag{8}$$

If we take $\delta := \min\{1, \varepsilon, \varepsilon/(|L| + 1)\}$, it is an exercise to show that

$$L - 2\varepsilon \le \frac{f(\alpha)}{g(\alpha)} \le L + 2\varepsilon.$$

Since $\varepsilon > 0$ is arbitrary, this yields the assertion. The cases $L = 0$ and $L < 0$ are handled similarly.

Case (b): If $L = +\infty$, let $M > 1$ be given and $c \in (a, b)$ be such that $f'(u)/g'(u) > M$ for all $u \in (a, c)$. Then it follows as before that

(9)
$$\frac{f(\beta) - f(\alpha)}{g(\beta) - g(\alpha)} > M \qquad \text{for} \quad a < \alpha < \beta \le c.$$

Since $g(x) \to \infty$ as $x \to a+$, we may suppose that c also satisfies $g(c) > 0$, that $|f(c)|/g(\alpha) < \frac{1}{2}$, and that $0 < g(c)/g(\alpha) < \frac{1}{2}$ for all $\alpha \in (a, c)$. If we take $\beta = c$ in (9) and multiply by $1 - g(c)/g(\alpha) > \frac{1}{2}$, we get

$$\frac{f(\alpha) - f(c)}{g(\alpha)} > M\left(1 - \frac{g(c)}{g(\alpha)}\right) > \tfrac{1}{2}M,$$

so that

$$\frac{f(\alpha)}{g(\alpha)} > \tfrac{1}{2}M + \frac{f(c)}{g(\alpha)} > \tfrac{1}{2}(M - 1) \qquad \text{for} \quad \alpha \in (a, c).$$

Since $M > 1$ is arbitrary, it follows that $\lim_{\alpha \to a+} f(\alpha)/g(\alpha) = \infty$.

If $L = -\infty$, the argument is similar.   Q.E.D.

6.3.6 Examples (a) We consider $\lim_{x \to \infty} \dfrac{\ln x}{x}$.

Here $f(x) := \ln x$ and $g(x) := x$ on the interval $(0, \infty)$. If we apply the left-hand version of 6.3.5, we obtain $\lim_{x \to \infty} \dfrac{\ln x}{x} = \lim_{x \to \infty} \dfrac{1/x}{1} = 0.$

(b) We consider $\lim_{x \to \infty} e^{-x}x^2$.

Here we take $f(x) := x^2$ and $g(x) := e^x$ on $\mathbb{R}$. We obtain

$$\lim_{x \to \infty} \frac{x^2}{e^x} = \lim_{x \to \infty} \frac{2x}{e^x} = \lim_{x \to \infty} \frac{2}{e^x} = 0.$$

(c) We consider $\lim_{x \to 0+} \dfrac{\ln \sin x}{\ln x}$.

Here we take $f(x) := \ln \sin x$ and $g(x) := \ln x$ on $(0, \pi)$. If we apply 6.3.5, we obtain

$$\lim_{x \to 0+} \frac{\ln \sin x}{\ln x} = \lim_{x \to 0+} \frac{\cos x / \sin x}{1/x} = \lim_{x \to 0+} \left[\frac{x}{\sin x}\right] \cdot [\cos x].$$

Since $\lim_{x \to 0+} [x/\sin x] = 1$ and $\lim_{x \to 0+} \cos x = 1$, we conclude that the limit under consideration equals 1.   □

Other Indeterminate Forms

Indeterminate forms such as $\infty - \infty$, $0 \cdot \infty$, 1^∞, 0^0, ∞^0 can be reduced to the previously considered cases by algebraic manipulations and the use of the logarithmic and exponential functions. Instead of formulating these variations as theorems, we illustrate the pertinent techniques by means of examples.

6.3.7 Examples (a) Let $I := (0, \pi/2)$ and consider

$$\lim_{x \to 0+} \left(\frac{1}{x} - \frac{1}{\sin x}\right),$$

which has the indeterminate form $\infty - \infty$. We have

$$\lim_{x\to 0+} \left(\frac{1}{x} - \frac{1}{\sin x}\right) = \lim_{x\to 0+} \frac{\sin x - x}{x \sin x} = \lim_{x\to 0+} \frac{\cos x - 1}{\sin x + x \cos x}$$

$$= \lim_{x\to 0+} \frac{-\sin x}{2\cos x - x \sin x} = \frac{0}{2} = 0.$$

(b) Let $I := (0, \infty)$ and consider $\lim_{x\to 0+} x \ln x$, which has the indeterminate form $0 \cdot (-\infty)$. We have

$$\lim_{x\to 0+} x \ln x = \lim_{x\to 0+} \frac{\ln x}{1/x} = \lim_{x\to 0+} \frac{1/x}{-1/x^2} = \lim_{x\to 0+} (-x) = 0.$$

(c) Let $I := (0, \infty)$ and consider $\lim_{x\to 0+} x^x$, which has the indeterminate form 0^0.

We recall from calculus (see also Section 8.3) that $x^x = e^{x \ln x}$. It follows from part (b) and the continuity of the function $y \mapsto e^y$ at $y = 0$ that $\lim_{x\to 0+} x^x = e^0 = 1$.

(d) Let $I := (1, \infty)$ and consider $\lim_{x\to\infty} (1 + 1/x)^x$, which has the indeterminate form 1^∞.

We note that

$$(10) \qquad\qquad (1 + 1/x)^x = e^{x \ln(1+1/x)}.$$

Moreover, we have

$$\lim_{x\to\infty} x \ln(1 + 1/x) = \lim_{x\to\infty} \frac{\ln(1 + 1/x)}{1/x}$$

$$= \lim_{x\to\infty} \frac{(1 + 1/x)^{-1}(-x^{-2})}{-x^{-2}} = \lim_{x\to\infty} \frac{1}{1 + 1/x} = 1.$$

Since $y \mapsto e^y$ is continuous at $y = 1$, we infer that $\lim_{x\to\infty} (1 + 1/x)^x = e$.

(e) Let $I := (0, \infty)$ and consider $\lim_{x\to 0+} (1 + 1/x)^x$, which has the indeterminate form ∞^0.

In view of formula (10), we consider

$$\lim_{x\to 0+} x \ln(1 + 1/x) = \lim_{x\to 0+} \frac{\ln(1 + 1/x)}{1/x} = \lim_{x\to 0+} \frac{1}{1 + 1/x} = 0.$$

Therefore we have $\lim_{x\to 0+} (1 + 1/x)^x = e^0 = 1.$ □

Exercises for Section 6.3

1. Suppose that f and g are continuous on $[a, b]$, differentiable on (a, b), that $c \in [a, b]$ and that $g(x) \neq 0$ for $x \in [a, b]$, $x \neq c$. Let $A := \lim_{x\to c} f$ and $B := \lim_{x\to c} g$. If $B = 0$, and if $\lim_{x\to c} f(x)/g(x)$ exists in $\mathbb{R}$, show that we must have $A = 0$. [Hint: $f(x) = \{f(x)/g(x)\}g(x)$.]

2. In addition to the suppositions of the preceding exercise, let $g(x) > 0$ for $x \in [a, b]$, $x \neq c$. If $A > 0$ and $B = 0$, prove that we must have $\lim_{x\to c} f(x)/g(x) = \infty$. If $A < 0$ and $B = 0$, prove that we must have $\lim_{x\to c} f(x)/g(x) = -\infty$.

3. Let $f(x) := x^2 \sin(1/x)$ for $0 < x \leq 1$ and $f(0) := 0$, and let $g(x) := x^2$ for $x \in [0, 1]$. Then both f and g are differentiable on $[0, 1]$ and $g(x) > 0$ for $x \neq 0$. Show that $\lim_{x\to 0} f(x) = 0 = \lim_{x\to 0} g(x)$ and that $\lim_{x\to 0} f(x)/g(x)$ does not exist.

4. Let $f(x) := x^2$ for x rational, let $f(x) := 0$ for x irrational, and let $g(x) := \sin x$ for $x \in \mathbb{R}$. Use Theorem 6.3.1 to show that $\lim_{x \to 0} f(x)/g(x) = 0$. Explain why Theorem 6.3.3 cannot be used.

5. Let $f(x) := x^2 \sin(1/x)$ for $x \neq 0$, let $f(0) := 0$, and let $g(x) := \sin x$ for $x \in \mathbb{R}$. Show that $\lim_{x \to 0} f(x)/g(x) = 0$ but that $\lim_{x \to 0} f'(x)/g'(x)$ does not exist.

6. Evaluate the following limits, where the domain of the quotient is as indicated.
   (a) $\lim_{x \to 0+} \dfrac{\ln(x+1)}{\sin x}$ $(0, \pi/2)$,
   (b) $\lim_{x \to 0+} \dfrac{\tan x}{x}$ $(0, \pi/2)$,
   (c) $\lim_{x \to 0+} \dfrac{\ln \cos x}{x}$ $(0, \pi/2)$,
   (d) $\lim_{x \to 0+} \dfrac{\tan x - x}{x^3}$ $(0, \pi/2)$.

7. Evaluate the following limits:
   (a) $\lim_{x \to 0} \dfrac{\text{Arctan} x}{x}$ $(-\infty, \infty)$,
   (b) $\lim_{x \to 0} \dfrac{1}{x(\ln x)^2}$ $(0, 1)$,
   (c) $\lim_{x \to 0+} x^3 \ln x$ $(0, \infty)$,
   (d) $\lim_{x \to \infty} \dfrac{x^3}{e^x}$ $(0, \infty)$.

8. Evaluate the following limits:
   (a) $\lim_{x \to \infty} \dfrac{\ln x}{x^2}$ $(0, \infty)$,
   (b) $\lim_{x \to \infty} \dfrac{\ln x}{\sqrt{x}}$ $(0, \infty)$,
   (c) $\lim_{x \to 0} x \ln \sin x$ $(0, \pi)$,
   (d) $\lim_{x \to \infty} \dfrac{x + \ln x}{x \ln x}$ $(0, \infty)$.

9. Evaluate the following limits:
   (a) $\lim_{x \to 0+} x^{2x}$ $(0, \infty)$,
   (b) $\lim_{x \to 0} (1 + 3/x)^x$ $(0, \infty)$,
   (c) $\lim_{x \to \infty} (1 + 3/x)^x$ $(0, \infty)$,
   (d) $\lim_{x \to 0+} \left(\dfrac{1}{x} - \dfrac{1}{\text{Arctan } x} \right)$ $(0, \infty)$.

10. Evaluate the following limits:
    (a) $\lim_{x \to \infty} x^{1/x}$ $(0, \infty)$,
    (b) $\lim_{x \to 0+} (\sin x)^x$ $(0, \pi)$,
    (c) $\lim_{x \to 0+} x^{\sin x}$ $(0, \infty)$,
    (d) $\lim_{x \to \pi/2-} (\sec x - \tan x)$ $(0, \pi/2)$.

11. Let f be differentiable on $(0, \infty)$ and suppose that $\lim_{x \to \infty} \left(f(x) + f'(x) \right) = L$. Show that $\lim_{x \to \infty} f(x) = L$ and $\lim_{x \to \infty} f'(x) = 0$. [*Hint:* $f(x) = e^x f(x)/e^x$.]

12. Try to use L'Hospital's Rule to find the limit of $\dfrac{\tan x}{\sec x}$ as $x \to (\pi/2)-$. Then evaluate directly by changing to sines and cosines.

Section 6.4 Taylor's Theorem

A very useful technique in the analysis of real functions is the approximation of functions by polynomials. In this section we will prove a fundamental theorem in this area which goes back to Brook Taylor (1685–1731), although the remainder term was not provided until much later by Joseph-Louis Lagrange (1736–1813). Taylor's Theorem is a powerful result that has many applications. We will illustrate the versatility of Taylor's Theorem by briefly discussing some of its applications to numerical estimation, inequalities, extreme values of a function, and convex functions.

Taylor's Theorem can be regarded as an extension of the Mean Value Theorem to "higher order" derivatives. Whereas the Mean Value Theorem relates the values of a function and its first derivative, Taylor's Theorem provides a relation between the values of a function and its higher order derivatives.

Derivatives of order greater than one are obtained by a natural extension of the differentiation process. If the derivative $f'(x)$ of a function f exists at every point x in an interval I containing a point c, then we can consider the existence of the derivative of the function f' at the point c. In case f' has a derivative at the point c, we refer to the resulting number as the **second derivative** of f at c, and we denote this number by $f''(c)$ or by $f^{(2)}(c)$. In similar fashion we define the third derivative $f'''(c) = f^{(3)}(c), \cdots$, and the **nth derivative** $f^{(n)}(c)$, whenever these derivatives exist. It is noted that the existence of the nth derivative at c presumes the existence of the $(n-1)$st derivative in an interval containing c, but we do allow the possibility that c might be an endpoint of such an interval.

If a function f has an nth derivative at a point x_0, it is not difficult to construct an nth degree polynomial P_n such that $P_n(x_0) = f(x_0)$ and $P_n^{(k)}(x_0) = f^{(k)}(x_0)$ for $k = 1, 2, \cdots, n$. In fact, the polynomial

$$(1) \qquad P_n(x) := f(x_0) + f'(x_0)(x - x_0) + \frac{f''(x_0)}{2!}(x - x_0)^2$$
$$+ \cdots + \frac{f^{(n)}(x_0)}{n!}(x - x_0)^n$$

has the property that it and its derivatives up to order n agree with the function f and its derivatives up to order n, at the specified point x_0. This polynomial P_n is called the **nth Taylor polynomial for f at x_0**. It is natural to expect this polynomial to provide a reasonable approximation to f for points near x_0, but to gauge the quality of the approximation, it is necessary to have information concerning the remainder $R_n := f - P_n$. The following fundamental result provides such information.

6.4.1 Taylor's Theorem Let $n \in \mathbb{N}$, let $I := [a, b]$, and let $f : I \to \mathbb{R}$ be such that f and its derivatives $f', f'', \cdots, f^{(n)}$ are continuous on I and that $f^{(n+1)}$ exists on (a, b). If $x_0 \in I$, then for any x in I there exists a point c between x and x_0 such that

$$(2) \qquad f(x) = f(x_0) + f'(x_0)(x - x_0) + \frac{f''(x_0)}{2!}(x - x_0)^2$$
$$+ \cdots + \frac{f^{(n)}(x_0)}{n!}(x - x_0)^n + \frac{f^{(n+1)}(c)}{(n+1)!}(x - x_0)^{n+1}.$$

Proof. Let x_0 and x be given and let J denote the closed interval with endpoints x_0 and x. We define the function F on J by

$$F(t) := f(x) - f(t) - (x - t)f'(t) - \cdots - \frac{(x - t)^n}{n!}f^{(n)}(t)$$

for $t \in J$. Then an easy calculation shows that we have

$$F'(t) = -\frac{(x - t)^n}{n!}f^{(n+1)}(t).$$

If we define G on J by

$$G(t) := F(t) - \left(\frac{x - t}{x - x_0}\right)^{n+1} F(x_0)$$

for $t \in J$, then $G(x_0) = G(x) = 0$. An application of Rolle's Theorem 6.2.3 yields a point c between x and x_0 such that

$$0 = G'(c) = F'(c) + (n + 1)\frac{(x - c)^n}{(x - x_0)^{n+1}}F(x_0).$$

Hence, we obtain

$$F(x_0) = -\frac{1}{n+1}\frac{(x-x_0)^{n+1}}{(x-c)^n}F'(c)$$

$$= \frac{1}{n+1}\frac{(x-x_0)^{n+1}}{(x-c)^n}\frac{(x-c)^n}{n!}f^{(n+1)}(c) = \frac{f^{(n+1)}(c)}{(n+1)!}(x-x_0)^{n+1},$$

which implies the stated result. $\qquad$ Q.E.D.

We shall use the notation P_n for the nth Taylor polynomial (1) of f, and R_n for the remainder. Thus we may write the conclusion of Taylor's Theorem as $f(x) = P_n(x) + R_n(x)$ where R_n is given by

$$(3) \qquad\qquad R_n(x) := \frac{f^{(n+1)}(c)}{(n+1)!}(x-x_0)^{n+1}$$

for some point c between x and x_0. This formula for R_n is referred to as the **Lagrange form** (or the **derivative form**) of the remainder. Many other expressions for R_n are known; one is in terms of integration and will be given later. (See Theorem 7.3.18.)

Applications of Taylor's Theorem ────────────────────────────

The remainder term R_n in Taylor's Theorem can be used to estimate the error in approximating a function by its Taylor polynomial P_n. If the number n is prescribed, then the question of the accuracy of the approximation arises. On the other hand, if a certain accuracy is specified, then the question of finding a suitable value of n is germane. The following examples illustrate how one responds to these questions.

6.4.2 Examples (a) Use Taylor's Theorem with $n = 2$ to approximate $\sqrt[3]{1+x}$, $x > -1$.

We take the function $f(x) := (1+x)^{1/3}$, the point $x_0 = 0$, and $n = 2$. Since $f'(x) = \frac{1}{3}(1+x)^{-2/3}$ and $f''(x) = \frac{1}{3}\left(-\frac{2}{3}\right)(1+x)^{-5/3}$, we have $f'(0) = \frac{1}{3}$ and $f''(0) = -2/9$. Thus we obtain

$$f(x) = P_2(x) + R_2(x) = 1 + \tfrac{1}{3}x - \tfrac{1}{9}x^2 + R_2(x),$$

where $R_2(x) = \frac{1}{3!}f'''(c)x^3 = \frac{5}{81}(1+c)^{-8/3}x^3$ for some point c between 0 and x.

For example, if we let $x = 0.3$, we get the approximation $P_2(0.3) = 1.09$ for $\sqrt[3]{1.3}$. Moreover, since $c > 0$ in this case, then $(1+c)^{-8/3} < 1$ and so the error is at most

$$R_2(0.3) \le \frac{5}{81}\left(\frac{3}{10}\right)^3 = \frac{1}{600} < 0.17 \times 10^{-2}.$$

Hence, we have $|\sqrt[3]{1.3} - 1.09| < 0.5 \times 10^{-2}$, so that two decimal place accuracy is assured.

(b) Approximate the number e with error less than 10^{-5}.

We shall consider the function $g(x) := e^x$ and take $x_0 = 0$ and $x = 1$ in Taylor's Theorem. We need to determine n so that $|R_n(1)| < 10^{-5}$. To do so, we shall use the fact that $g'(x) = e^x$ and the initial bound of $e^x \le 3$ for $0 \le x \le 1$.

Since $g'(x) = e^x$, it follows that $g^{(k)}(x) = e^x$ for all $k \in \mathbb{N}$, and therefore $g^{(k)}(0) = 1$ for all $k \in \mathbb{N}$. Consequently the nth Taylor polynomial is given by

$$P_n(x) := 1 + x + \frac{x^2}{2!} + \cdots + \frac{x^n}{n!}$$

and the remainder for $x = 1$ is given by $R_n(1) = e^c/(n+1)!$ for some c satisfying $0 < c < 1$. Since $e^c < 3$, we seek a value of n such that $3/(n+1)! < 10^{-5}$. A calculation reveals that $9! = 362,880 > 3 \times 10^5$ so that the value $n = 8$ will provide the desired accuracy; moreover, since $8! = 40,320$, no smaller value of n will be certain to suffice. Thus, we obtain

$$e \approx P_8(1) = 1 + 1 + \frac{1}{2!} + \cdots + \frac{1}{8!} = 2.718\,28$$

with error less than 10^{-5}. □

Taylor's Theorem can also be used to derive inequalities.

6.4.3 Examples (a) $1 - \frac{1}{2}x^2 \le \cos x$ for all $x \in \mathbb{R}$.

Use $f(x) := \cos x$ and $x_0 = 0$ in Taylor's Theorem, to obtain

$$\cos x = 1 - \frac{1}{2}x^2 + R_2(x),$$

where for some c between 0 and x we have

$$R_2(x) = \frac{f'''(c)}{3!}x^3 = \frac{\sin c}{6}x^3.$$

If $0 \le x \le \pi$, then $0 \le c < \pi$; since c and x^3 are both positive, we have $R_2(x) \ge 0$. Also, if $-\pi \le x \le 0$, then $-\pi \le c \le 0$; since $\sin c$ and x^3 are both negative, we again have $R_2(x) \ge 0$. Therefore, we see that $1 - \frac{1}{2}x^2 \le \cos x$ for $|x| \le \pi$. If $|x| \ge \pi$, then we have $1 - \frac{1}{2}x^2 < -3 \le \cos x$ and the inequality is trivially valid. Hence, the inequality holds for all $x \in \mathbb{R}$.

(b) For any $k \in \mathbb{N}$, and for all $x > 0$, we have

$$x - \frac{1}{2}x^2 + \cdots - \frac{1}{2k}x^{2k} < \ln(1+x) < x - \frac{1}{2}x^2 + \cdots + \frac{1}{2k+1}x^{2k+1}.$$

Using the fact that the derivative of $\ln(1+x)$ is $1/(1+x)$ for $x > 0$, we see that the nth Taylor polynomial for $\ln(1+x)$ with $x_0 = 0$ is

$$P_n(x) = x - \frac{1}{2}x^2 + \cdots + (-1)^{n-1}\frac{1}{n}x^n$$

and the remainder is given by

$$R_n(x) = \frac{(-1)^n c^{n+1}}{n+1}x^{n+1}$$

for some c satisfying $0 < c < x$. Thus for any $x > 0$, if $n = 2k$ is even, then we have $R_{2k}(x) > 0$; and if $n = 2k+1$ is odd, then we have $R_{2k+1}(x) < 0$. The stated inequality then follows immediately. □

Relative Extrema _____

It was established in Theorem 6.2.1 that if a function $f : I \to \mathbb{R}$ is differentiable at a point c interior to the interval I, then a necessary condition for f to have a relative extremum at c is that $f'(c) = 0$. One way to determine whether f has a relative maximum or relative minimum [or neither] at c, is to use the First Derivative Test 6.2.8. Higher order derivatives, if they exist, can also be used in this determination, as we now show.

6.4.4 Theorem *Let I be an interval, let x_0 be an interior point of I, and let $n \geq 2$. Suppose that the derivatives $f', f'', \cdots, f^{(n)}$ exist and are continuous in a neighborhood of x_0 and that $f'(x_0) = \cdots = f^{(n-1)}(x_0) = 0$, but $f^{(n)}(x_0) \neq 0$.*

(i) *If n is even and $f^{(n)}(x_0) > 0$, then f has a relative minimum at x_0.*

(ii) *If n is even and $f^{(n)}(x_0) < 0$, then f has a relative maximum at x_0.*

(iii) *If n is odd, then f has neither a relative minimum nor relative maximum at x_0.*

Proof. Applying Taylor's Theorem at x_0, we find that for $x \in I$ we have

$$f(x) = P_{n-1}(x) + R_{n-1}(x) = f(x_0) + \frac{f^{(n)}(c)}{n!}(x - x_0)^n,$$

where c is some point between x_0 and x. Since $f^{(n)}$ is continuous, if $f^{(n)}(x_0) \neq 0$, then there exists an interval U containing x_0 such that $f^{(n)}(x)$ will have the same sign as $f^{(n)}(x_0)$ for $x \in U$. If $x \in U$, then the point c also belongs to U and consequently $f^{(n)}(c)$ and $f^{(n)}(x_0)$ will have the same sign.

(i) If n is even and $f^{(n)}(x_0) > 0$, then for $x \in U$ we have $f^{(n)}(c) > 0$ and $(x - x_0)^n \geq 0$ so that $R_{n-1}(x) \geq 0$. Hence, $f(x) \geq f(x_0)$ for $x \in U$, and therefore f has a relative minimum at x_0.

(ii) If n is even and $f^{(n)}(x_0) < 0$, then it follows that $R_{n-1}(x) \leq 0$ for $x \in U$, so that $f(x) \leq f(x_0)$ for $x \in U$. Therefore, f has a relative maximum at x_0.

(iii) If n is odd, then $(x - x_0)^n$ is positive if $x > x_0$ and negative if $x < x_0$. Consequently, if $x \in U$, then $R_{n-1}(x)$ will have opposite signs to the left and to the right of x_0. Therefore, f has neither a relative minimum nor a relative maximum at x_0. Q.E.D.

Convex Functions

The notion of convexity plays an important role in a number of areas, particularly in the modern theory of optimization. We shall briefly look at convex functions of one real variable and their relation to differentiation. The basic results, when appropriately modified, can be extended to higher dimensional spaces.

6.4.5 Definition Let $I \subseteq \mathbb{R}$ be an interval. A function $f : I \to \mathbb{R}$ is said to be **convex** on I if for any t satisfying $0 \leq t \leq 1$ and any points x_1, x_2 in I, we have

$$f\big((1 - t)x_1 + tx_2\big) \leq (1 - t)f(x_1) + tf(x_2).$$

Note that if $x_1 < x_2$, then as t ranges from 0 to 1, the point $(1 - t)x_1 + tx_2$ traverses the interval from x_1 to x_2. Thus if f is convex on I and if $x_1, x_2 \in I$, then the chord joining any two points $(x_1, f(x_1))$ and $(x_2, f(x_2))$ on the graph of f lies above the graph of f. (See Figure 6.4.1.)

A convex function need not be differentiable at every point, as the example $f(x) := |x|$, $x \in \mathbb{R}$, reveals. However, it can be shown that if I is an open interval and if $f : I \to \mathbb{R}$ is convex on I, then the left and right derivatives of f exist at every point of I. As a consequence, it follows that a convex function on an open interval is necessarily continuous. We will not verify the preceding assertions, nor will we develop many other interesting properties of convex functions. Rather, we will restrict ourselves to establishing the connection between a convex function f and its second derivative f'', assuming that f'' exists.

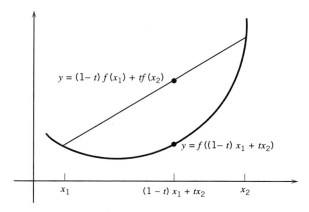

Figure 6.4.1   A convex function.

6.4.6 Theorem   *Let I be an open interval and let $f : I \to \mathbb{R}$ have a second derivative on I. Then f is a convex function on I if and only if $f''(x) \geq 0$ for all $x \in I$.*

Proof.   ($\Rightarrow$)   We will make use of the fact that the second derivative is given by the limit

(4) $$f''(a) = \lim_{h \to 0} \frac{f(a+h) - 2f(a) + f(a-h)}{h^2}$$

for each $a \in I$. (See Exercise 16.) Given $a \in I$, let h be such that $a + h$ and $a - h$ belong to I. Then $a = \frac{1}{2}((a+h) + (a-h))$, and since f is convex on I, we have

$$f(a) = f\left(\tfrac{1}{2}(a+h) + \tfrac{1}{2}(a-h)\right) \leq \tfrac{1}{2}f(a+h) + \tfrac{1}{2}f(a-h).$$

Therefore, we have $f(a+h) - 2f(a) + f(a-h) \geq 0$. Since $h^2 > 0$ for all $h \neq 0$, we see that the limit in (4) must be nonnegative. Hence, we obtain $f''(a) \geq 0$ for any $a \in I$.

   ($\Leftarrow$)   We will use Taylor's Theorem. Let x_1, x_2 be any two points of I, let $0 < t < 1$, and let $x_0 := (1-t)x_1 + tx_2$. Applying Taylor's Theorem to f at x_0 we obtain a point c_1 between x_0 and x_1 such that

$$f(x_1) = f(x_0) + f'(x_0)(x_1 - x_0) + \tfrac{1}{2}f''(c_1)(x_1 - x_0)^2,$$

and a point c_2 between x_0 and x_2 such that

$$f(x_2) = f(x_0) + f'(x_0)(x_2 - x_0) + \tfrac{1}{2}f''(c_2)(x_2 - x_0)^2.$$

If f'' is nonnegative on I, then the term

$$R := \tfrac{1}{2}(1-t)f''(c_1)(x_1 - x_0)^2 + \tfrac{1}{2}tf''(c_2)(x_2 - x_0)^2$$

is also nonnegative. Thus we obtain

$$\begin{aligned}
(1-t)f(x_1) + tf(x_2) &= f(x_0) + f'(x_0)\left((1-t)x_1 + tx_2 - x_0\right) \\
&\quad + \tfrac{1}{2}(1-t)f''(c_1)(x_1 - x_0)^2 + \tfrac{1}{2}tf''(c_2)(x_2 - x_0)^2 \\
&= f(x_0) + R \\
&\geq f(x_0) = f\left((1-t)x_1 + tx_2\right).
\end{aligned}$$

Hence, f is a convex function on I.   Q.E.D.

Newton's Method

It is often desirable to estimate a solution of an equation with a high degree of accuracy. The Bisection Method, used in the proof of the Location of Roots Theorem 5.3.5, provides one estimation procedure, but it has the disadvantage of converging to a solution rather slowly. A method that often results in much more rapid convergence is based on the geometric idea of successively approximating a curve by tangent lines. The method is named after its discoverer, Isaac Newton.

Let f be a differentiable function that has a zero at r and let x_1 be an initial estimate of r. The line tangent to the graph at $(x_1, f(x_1))$ has the equation $y = f(x_1) + f'(x_1)(x - x_1)$, and crosses the x-axis at the point

$$x_2 := x_1 - \frac{f(x_1)}{f'(x_1)}.$$

(See Figure 6.4.2.) If we replace x_1 by the second estimate x_2, then we obtain a point x_3, and so on. At the nth iteration we get the point x_{n+1} from the point x_n by the formula

$$x_{n+1} := x_n - \frac{f(x_n)}{f'(x_n)}.$$

Under suitable hypotheses, the sequence (x_n) will converge rapidly to a root of the equation $f(x) = 0$, as we now show. The key tool in establishing the rapid rate of convergence is Taylor's Theorem.

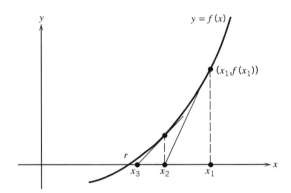

Figure 6.4.2 Newton's Method

6.4.7 Newton's Method Let $I := [a, b]$ and let $f : I \to \mathbb{R}$ be twice differentiable on I. Suppose that $f(a)f(b) < 0$ and that there are constants m, M such that $|f'(x)| \geq m > 0$ and $|f''(x)| \leq M$ for all $x \in I$ and let $K := M/2m$. Then there exists a subinterval I^* containing a zero r of f such that for any $x_1 \in I^*$ the sequence (x_n) defined by

(5) $\qquad\qquad x_{n+1} := x_n - \dfrac{f(x_n)}{f'(x_n)} \qquad$ for all $\quad n \in \mathbb{N}$,

belongs to I^* and (x_n) converges to r. Moreover

(6) $\qquad\qquad |x_{n+1} - r| \leq K\, |x_n - r|^2 \qquad$ for all $\quad n \in \mathbb{N}$.

Proof.   Since $f(a)f(b) < 0$, the numbers $f(a)$ and $f(b)$ have opposite signs; hence by Theorem 5.3.5 there exists $r \in I$ such that $f(r) = 0$. Since f' is never zero on I, it follows from Rolle's Theorem 6.2.3 that f does not vanish at any other point of I.

We now let $x' \in I$ be arbitrary; by Taylor's Theorem there exists a point c' between x' and r such that

$$0 = f(r) = f(x') + f'(x')(r - x') + \tfrac{1}{2}f''(c')(r - x')^2,$$

from which it follows that

$$-f(x') = f'(x')(r - x') + \tfrac{1}{2}f''(c')(r - x')^2.$$

If x'' is the number defined from x' by "the Newton procedure":

$$x'' := x' - \frac{f(x')}{f'(x')},$$

then an elementary calculation shows that

$$x'' = x' + (r - x') + \frac{1}{2}\frac{f''(c')}{f'(x')}(r - x')^2,$$

whence it follows that

$$x'' - r = \frac{1}{2}\frac{f''(c')}{f'(x')}\left(x' - r\right)^2.$$

Since $c' \in I$, the assumed bounds on f' and f'' hold and, setting $K := M/2m$, we obtain the inequality

(7) $$\left|x'' - r\right| \le K \left|x' - r\right|^2.$$

We now choose $\delta > 0$ so small that $\delta < 1/K$ and that the interval $I^* := [r - \delta, r + \delta]$ is contained in I. If $x_n \in I^*$, then $|x_n - r| \le \delta$ and it follows from (7) that $|x_{n+1} - r| \le K|x_n - r|^2 \le K\delta^2 < \delta$; hence $x_n \in I^*$ implies that $x_{n+1} \in I^*$. Therefore if $x_1 \in I^*$, we infer that $x_n \in I^*$ for all $n \in \mathbb{N}$. Also if $x_1 \in I^*$, then an elementary induction argument using (7) shows that $|x_{n+1} - r| < (K\delta)^n|x_1 - r|$ for $n \in \mathbb{N}$. But since $K\delta < 1$ this proves that $\lim(x_n) = r$.   Q.E.D.

6.4.8 Example   We will illustrate Newton's Method by using it to approximate $\sqrt{2}$.

If we let $f(x) := x^2 - 2$ for $x \in \mathbb{R}$, then we seek the positive root of the equation $f(x) = 0$. Since $f'(x) = 2x$, the iteration formula is

$$
\begin{aligned}
x_{n+1} &= x_n - \frac{f(x_n)}{f'(x_n)} \\
&= x_n - \frac{x_n^2 - 2}{2x_n} = \frac{1}{2}\left(x_n + \frac{2}{x_n}\right).
\end{aligned}
$$

If we take $x_1 := 1$ as our initial estimate, we obtain the successive values $x_2 = 3/2 = 1.5$, $x_3 = 17/12 = 1.416\,666 \cdots$, $x_4 = 577/408 = 1.414\,215 \cdots$, and $x_5 = 665\,857/470\,832 = 1.414\,213\,562\,374 \cdots$, which is correct to eleven places.   □

Remarks   **(a)** If we let $e_n := x_n - r$ be the error in approximating r, then inequality (6) can be written in the form $|Ke_{n+1}| \le |Ke_n|^2$. Consequently, if $|Ke_n| < 10^{-m}$ then

$|Ke_{n+1}| < 10^{-2m}$ so that the number of significant digits in Ke_n has been doubled. Because of this doubling, the sequence generated by Newton's Method is said to converge "quadratically".

(b) In practice, when Newton's Method is programmed for a computer, one often makes an initial guess x_1 and lets the computer run. If x_1 is poorly chosen, or if the root is too near the endpoint of I, the procedure may not converge to a zero of f. Two possible difficulties are illustrated in Figures 6.4.3 and 6.4.4. One familiar strategy is to use the Bisection Method to arrive at a fairly close estimate of the root and then to switch to Newton's Method for the coup de grâce.

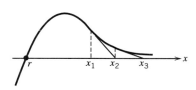

Figure 6.4.3 $x_n \to \infty$.

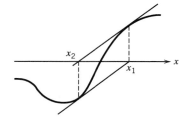

Figure 6.4.4 x_n oscillates
between x_1 and x_2.

Exercises for Section 6.4

1. Let $f(x) := \cos ax$ for $x \in \mathbb{R}$ where $a \neq 0$. Find $f^{(n)}(x)$ for $n \in \mathbb{N}$, $x \in \mathbb{R}$.

2. Let $g(x) := |x^3|$ for $x \in \mathbb{R}$. Find $g'(x)$ and $g''(x)$ for $x \in \mathbb{R}$, and $g'''(x)$ for $x \neq 0$. Show that $g'''(0)$ does not exist.

3. Use Induction to prove Leibniz's rule for the nth derivative of a product:
$$(fg)^{(n)}(x) = \sum_{k=0}^{n} \binom{n}{k} f^{(n-k)}(x) g^{(k)}(x).$$

4. Show that if $x > 0$, then $1 + \frac{1}{2}x - \frac{1}{8}x^2 \leq \sqrt{1+x} \leq 1 + \frac{1}{2}x$.

5. Use the preceding exercise to approximate $\sqrt{1.2}$ and $\sqrt{2}$. What is the best accuracy you can be sure of, using this inequality?

6. Use Taylor's Theorem with $n = 2$ to obtain more accurate approximations for $\sqrt{1.2}$ and $\sqrt{2}$.

7. If $x > 0$ show that $|(1 + x)^{1/3} - (1 + \frac{1}{3}x - \frac{1}{9}x^2)| \leq (5/81)x^3$. Use this inequality to approximate $\sqrt[3]{1.2}$ and $\sqrt[3]{2}$.

8. If $f(x) := e^x$, show that the remainder term in Taylor's Theorem converges to zero as $n \to \infty$, for each fixed x_0 and x. [*Hint:* See Theorem 3.2.11.]

9. If $g(x) := \sin x$, show that the remainder term in Taylor's Theorem converges to zero as $n \to \infty$ for each fixed x_0 and x.

10. Let $h(x) := e^{-1/x^2}$ for $x \neq 0$ and $h(0) := 0$. Show that $h^{(n)}(0) = 0$ for all $n \in \mathbb{N}$. Conclude that the remainder term in Taylor's Theorem for $x_0 = 0$ does *not* converge to zero as $n \to \infty$ for $x \neq 0$. [*Hint:* By L'Hospital's Rule, $\lim_{x \to 0} h(x)/x^k = 0$ for any $k \in \mathbb{N}$. Use Exercise 3 to calculate $h^{(n)}(x)$ for $x \neq 0$.]

11. If $x \in [0, 1]$ and $n \in \mathbb{N}$, show that
$$\left| \ln(1+x) - \left(x - \frac{x^2}{2} + \frac{x^3}{3} + \cdots + (-1)^{n-1} \frac{x^n}{n} \right) \right| < \frac{x^{n+1}}{n+1}.$$
Use this to approximate $\ln 1.5$ with an error less than 0.01. Less than 0.001.

12. We wish to approximate sin by a polynomial on $[-1, 1]$ so that the error is less than 0.001. Show that we have
$$\left| \sin x - \left(x - \frac{x^3}{6} + \frac{x^5}{120} \right) \right| < \frac{1}{5040} \qquad \text{for} \quad |x| \le 1.$$

13. Calculate e correct to 7 decimal places.

14. Determine whether or not $x = 0$ is a point of relative extremum of the following functions:
    (a) $f(x) := x^3 + 2$,   (b) $g(x) := \sin x - x$,
    (c) $h(x) := \sin x + \frac{1}{6}x^3$,   (d) $k(x) := \cos x - 1 + \frac{1}{2}x^2$.

15. Let f be continuous on $[a, b]$ and assume the second derivative f'' exists on (a, b). Suppose that the graph of f and the line segment joining the points $(a, f(a))$ and $(b, f(b))$ intersect at a point $(x_0, f(x_0))$ where $a < x_0 < b$. Show that there exists a point $c \in (a, b)$ such that $f''(c) = 0$.

16. Let $I \subseteq \mathbb{R}$ be an open interval, let $f : I \to \mathbb{R}$ be differentiable on I, and suppose $f''(a)$ exists at $a \in I$. Show that
$$f''(a) = \lim_{h \to 0} \frac{f(a+h) - 2f(a) + f(a-h)}{h^2}.$$
Give an example where this limit exists, but the function does not have a second derivative at a.

17. Suppose that $I \subseteq \mathbb{R}$ is an open interval and that $f''(x) \ge 0$ for all $x \in I$. If $c \in I$, show that the part of the graph of f on I is never below the tangent line to the graph at $(c, f(c))$.

18. Let $I \subseteq \mathbb{R}$ be an interval and let $c \in I$. Suppose that f and g are defined on I and that the derivatives $f^{(n)}, g^{(n)}$ exist and are continuous on I. If $f^{(k)}(c) = 0$ and $g^{(k)}(c) = 0$ for $k = 0, 1, \cdots, n-1$, but $g^{(n)}(c) \ne 0$, show that
$$\lim_{x \to c} \frac{f(x)}{g(x)} = \frac{f^{(n)}(c)}{g^{(n)}(c)}.$$

19. Show that the function $f(x) := x^3 - 2x - 5$ has a zero r in the interval $I := [2, 2.2]$. If $x_1 := 2$ and if we define the sequence (x_n) using the Newton procedure, show that $|x_{n+1} - r| \le (0.7)|x_n - r|^2$. Show that x_4 is accurate to within six decimal places.

20. Approximate the real zeros of $g(x) := x^4 - x - 3$.

21. Approximate the real zeros of $h(x) := x^3 - x - 1$. Apply Newton's Method starting with the initial choices (a) $x_1 := 2$, (b) $x_1 := 0$, (c) $x_1 := -2$. Explain what happens.

22. The equation $\ln x = x - 2$ has two solutions. Approximate them using Newton's Method. What happens if $x_1 := \frac{1}{2}$ is the initial point?

23. The function $f(x) = 8x^3 - 8x^2 + 1$ has two zeros in $[0, 1]$. Approximate them, using Newton's Method, with the starting points (a) $x_1 := \frac{1}{8}$, (b) $x_1 := \frac{1}{4}$. Explain what happens.

24. Approximate the solution of the equation $x = \cos x$, accurate to within six decimals.

CHAPTER 7

THE RIEMANN INTEGRAL

We have already mentioned the developments, during the 1630s, by Fermat and Descartes leading to analytic geometry and the theory of the derivative. However, the subject we know as calculus did not begin to take shape until the late 1660s when Isaac Newton created his theory of "fluxions" and invented the method of "inverse tangents" to find areas under curves. The reversal of the process for finding tangent lines to find areas was also discovered in the 1680s by Gottfried Leibniz, who was unaware of Newton's unpublished work and who arrived at the discovery by a very different route. Leibniz introduced the terminology "calculus differentialis" and "calculus integralis", since finding tangent lines involved differences and finding areas involved summations. Thus, they had discovered that integration, being a process of summation, was inverse to the operation of differentiation.

During a century and a half of development and refinement of techniques, calculus consisted of these paired operations and their applications, primarily to physical problems. In the 1850s, Bernhard Riemann adopted a new and different viewpoint. He separated the concept of integration from its companion, differentiation, and examined the motivating summation and limit process of finding areas by itself. He broadened the scope by considering all functions on an interval for which this process of "integration" could be defined: the class of "integrable" functions. The Fundamental Theorem of Calculus became a result that held only for a restricted set of integrable functions. The viewpoint of Riemann led others to invent other integration theories, the most significant being Lebesgue's theory of integration. But there have been some advances made in more recent times that extend even

---

Bernard Riemann

(Georg Friedrich) Bernard Riemann (1826–1866), the son of a poor Lutheran minister, was born near Hanover, Germany. To please his father, he enrolled (1846) at the University of Göttingen as a student of theology and philosophy, but soon switched to mathemtics. He interrupted his studies at Göttingen to study at Berlin under C. G. J. Jacobi, P. G. J. Dirichlet, and F. G. Eisenstein, but returned to Göttingen in 1849 to complete his thesis under Gauss. His thesis dealt with what are now called "Riemann surfaces". Gauss was so enthusiastic about Riemann's work that he arranged for him to become a  *privatdozent* at Göttingen in 1854. On admission as a *privatdozent*, Riemann was required to prove himself by delivering a probationary lecture before the entire faculty. As tradition dictated, he submitted three topics, the first two of which he was well prepared to discuss. To Riemann's surprise, Gauss chose that he should lecture on the third topic: "On the hypotheses that underlie the foundations of geometry". After its publication, this lecture had a profound effect on modern geometry.

Despite the fact that Riemann contracted tuberculosis and died at the age of 39, he made major contributions in many areas: the foundations of geometry, number theory, real and complex analysis, topology, and mathematical physics.

the Lebesgue theory to a considerable extent. We will give a brief introduction to these results in Chapter 10.

We begin by defining the concept of Riemann integrability of real-valued functions defined on a closed bounded interval of $\mathbb{R}$, using the Riemann sums familiar to the reader from calculus. This method has the advantage that it extends immediately to the case of functions whose values are complex numbers, or vectors in the space $\mathbb{R}^n$. In Section 7.2, we will establish the Riemann integrability of several important classes of functions: step functions, continuous functions, and monotone functions. However, we will also see that there are functions that are *not* Riemann integrable. The Fundamental Theorem of Calculus is the principal result in Section 7.3. We will present it in a form that is slightly more general than is customary and does not require the function to be a derivative at every point of the interval. A number of important consequences of the Fundamental Theorem are also given. In Section 7.3 we also give a statement of the definitive Lebesgue Criterion for Riemann integrability. This famous result is usually not given in books at this level, since its proof (given in Appendix C) is somewhat complicated. However, its statement is well within the reach of students, who will also comprehend the power of this result. The final section presents several methods of approximating integrals, a subject that has become increasingly important during this era of high-speed computers. While the proofs of these results are not particularly difficult, we defer them to Appendix D.

An interesting history of integration theory, including a chapter on the Riemann integral, is given in the book by Hawkins cited in the References.

Section 7.1   Riemann Integral

We will follow the procedure commonly used in calculus courses and define the Riemann integral as a kind of limit of the Riemann sums as the norm of the partitions tend to 0. Since we assume that the reader is familiar—at least informally—with the integral from a calculus course, we will not provide a motivation of the integral, or disuss its interpretation as the "area under the graph", or its many applications to physics, engineering, economics, etc. Instead, we will focus on the purely mathematical aspects of the integral.

However, we first recall some basic terms that will be frequently used.

Partitions and Tagged Partitions

If $I := [a, b]$ is a closed bounded interval in $\mathbb{R}$, then a **partition** of I is a finite, ordered set $\mathcal{P} := (x_0, x_1, \cdots, x_{n-1}, x_n)$ of points in I such that

$$a = x_0 < x_1 < \cdots < x_{n-1} < x_n = b.$$

(See Figure 7.1.1.) The points of $\mathcal{P}$ are used to divide $I = [a, b]$ into non-overlapping subintervals

$$I_1 := [x_0, x_1], \quad I_2 := [x_1, x_2], \cdots, \quad I_n := [x_{n-1}, x_n].$$

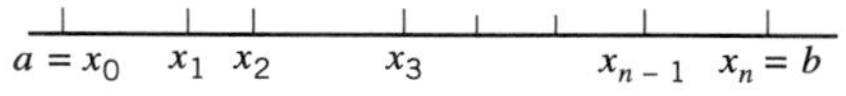

Figure 7.1.1   A partition of $[a, b]$.

Often we will denote the partition $\mathcal{P}$ by the notation $\mathcal{P} = \{[x_{i-1}, x_i]\}_{i=1}^n$. We define the **norm** (or **mesh**) of $\mathcal{P}$ to be the number

$$(1) \qquad \|\mathcal{P}\| := \max\{x_1 - x_0, x_2 - x_1, \cdots, x_n - x_{n-1}\}.$$

Thus the norm of a partition is merely the length of the largest subinterval into which the partition divides $[a, b]$. Clearly, many partitions have the same norm, so the partition is *not* a function of the norm.

If a point t_i has been selected from each subinterval $I_i = [x_{i-1}, x_i]$, for $i = 1, 2, \cdots, n$, then the points are called **tags** of the subintervals I_i. A set of ordered pairs

$$\dot{\mathcal{P}} := \{([x_{i-1}, x_i], t_i)\}_{i=1}^n$$

of subintervals and corresponding tags is called a **tagged partition** of I; see Figure 7.1.2. (The dot over the $\mathcal{P}$ indicates that a tag has been chosen for each subinterval.) The tags can be chosen in a wholly arbitrary fashion; for example, we can choose the tags to be the left endpoints, or the right endpoints, or the midpoints of the subintervals, etc. Note that an endpoint of a subinterval can be used as a tag for two consecutive subintervals. Since each tag can be chosen in infinitely many ways, each partition can be tagged in infinitely many ways. The norm of a tagged partition is defined as for an ordinary partition and does not depend on the choice of tags.

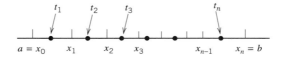

Figure 7.1.2   A tagged partition of $[a, b]$

If $\dot{\mathcal{P}}$ is the tagged partition given above, we define the **Riemann sum** of a function $f : [a, b] \to \mathbb{R}$ corresponding to $\dot{\mathcal{P}}$ to be the number

$$(2) \qquad S(f; \dot{\mathcal{P}}) := \sum_{i=1}^n f(t_i)(x_i - x_{i-1}).$$

We will also use this notation when $\dot{\mathcal{P}}$ denotes a *subset* of a partition, and not the entire partition.

The reader will perceive that if the function f is positive on $[a, b]$, then the Riemann sum (2) is the sum of the areas of n rectangles whose bases are the subintervals $I_i = [x_{i-1}, x_i]$ and whose heights are $f(t_i)$. (See Figure 7.1.3.)

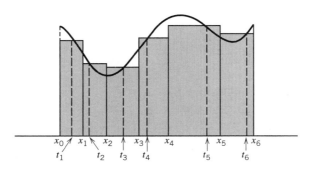

Figure 7.1.3   A Riemann sum.

Definition of the Riemann Integral ———————————————————————

We now define the Riemann integral of a function f on an interval $[a, b]$.

7.1.1 Definition   A function $f : [a, b] \to \mathbb{R}$ is said to be **Riemann integrable** on $[a, b]$ if there exists a number $L \in \mathbb{R}$ such that for every $\varepsilon > 0$ there exists $\delta_\varepsilon > 0$ such that if $\dot{\mathcal{P}}$ is any tagged partition of $[a, b]$ with $\|\dot{\mathcal{P}}\| < \delta_\varepsilon$, then

$$|S(f; \dot{\mathcal{P}}) - L| < \varepsilon.$$

The set of all Riemann-integrable functions on $[a, b]$ will be denoted by $\mathcal{R}[a, b]$.

Remark   It is sometimes said that the integral L is "the limit" of the Riemann sums $S(f; \dot{\mathcal{P}})$ as the norm $\|\dot{\mathcal{P}}\| \to 0$. However, since $S(f; \dot{\mathcal{P}})$ is not a function of $\|\dot{\mathcal{P}}\|$, this limit is not of the type that we have studied before.

First we will show that if $f \in \mathcal{R}[a, b]$, then the number L is uniquely determined. It will be called the **Riemann integral of** f over $[a, b]$. Instead of L, we will usually write

$$L = \int_a^b f \quad \text{or} \quad \int_a^b f(x)\, dx.$$

It should be understood that any letter other than x can be used in the latter expression, so long as it does not cause any ambiguity.

7.1.2 Theorem   *If $f \in \mathcal{R}[a, b]$, then the value of the integral is uniquely determined.*

Proof.   Assume that L' and L'' both satisfy the definition and let $\varepsilon > 0$. Then there exists $\delta'_{\varepsilon/2} > 0$ such that if $\dot{\mathcal{P}}_1$ is any tagged partion with $\|\dot{\mathcal{P}}_1\| < \delta'_{\varepsilon/2}$, then

$$|S(f; \dot{\mathcal{P}}_1) - L'| < \varepsilon/2.$$

Also there exists $\delta''_{\varepsilon/2} > 0$ such that if $\dot{\mathcal{P}}_2$ is any tagged partition with $\|\dot{\mathcal{P}}_2\| < \delta''_{\varepsilon/2}$, then

$$|S(f; \dot{\mathcal{P}}_2) - L''| < \varepsilon/2.$$

Now let $\delta_\varepsilon := \min\{\delta'_{\varepsilon/2}, \delta''_{\varepsilon/2}\} > 0$ and let $\dot{\mathcal{P}}$ be a tagged partition with $\|\dot{\mathcal{P}}\| < \delta_\varepsilon$. Since both $\|\dot{\mathcal{P}}\| < \delta'_{\varepsilon/2}$ and $\|\dot{\mathcal{P}}\| < \delta''_{\varepsilon/2}$, then

$$|S(f; \dot{\mathcal{P}}) - L'| < \varepsilon/2 \quad \text{and} \quad |S(f; \dot{\mathcal{P}}) - L''| < \varepsilon/2,$$

whence it follows from the Triangle Inequality that

$$\begin{aligned}
|L' - L''| &= |L' - S(f; \dot{\mathcal{P}}) + S(f; \dot{\mathcal{P}}) - L''| \\
&\leq |L' - S(f; \dot{\mathcal{P}})| + |S(f; \dot{\mathcal{P}}) - L''| \\
&< \varepsilon/2 + \varepsilon/2 = \varepsilon.
\end{aligned}$$

Since $\varepsilon > 0$ is arbitrary, it follows that $L' = L''$.   Q.E.D.

Some Examples ———————————————————————————————————

If we use only the definition, in order to show that a function f is Riemann integrable we must (i) know (or guess correctly) the value L of the integral, and (ii) construct a δ_ε that will suffice for an arbitrary $\varepsilon > 0$. The determination of L is sometimes done by calculating Riemann sums and guessing what L must be. The determination of δ_ε is likely to be difficult.

In actual practice, we usually show that $f \in \mathcal{R}[a, b]$ by making use of some of the theorems that will be given later.

7.1.3 Examples   (a)   Every constant function on $[a, b]$ is in $\mathcal{R}[a, b]$.

Let $f(x) := k$ for all $x \in [a, b]$. If $\dot{\mathcal{P}} := \{([x_{i-1}, x_i], t_i)\}_{i=1}^n$ is any tagged partition of $[a, b]$, then it is clear that

$$S(f; \dot{\mathcal{P}}) = \sum_{i=1}^n k(x_i - x_{i-1}) = k(b - a).$$

Hence, for any $\varepsilon > 0$, we can choose $\delta_\varepsilon := 1$ so that if $\|\dot{\mathcal{P}}\| < \delta_\varepsilon$, then

$$|S(f; \dot{\mathcal{P}}) - k(b - a)| = 0 < \varepsilon.$$

Since $\varepsilon > 0$ is arbitrary, we conclude that $f \in \mathcal{R}[a, b]$ and $\int_a^b f = k(b - a)$.

(b)   Let $g : [0, 3] \to \mathbb{R}$ be defined by $g(x) := 2$ for $0 \le x \le 1$, and $g(x) := 3$ for $1 < x \le 3$. A preliminary investigation, based on the graph of g (see Figure 7.1.4), suggests that we might expect that $\int_0^3 g = 8$.

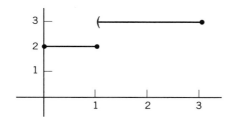

Figure 7.1.4   Graph of g.

Let $\dot{\mathcal{P}}$ be a tagged partition of $[0, 3]$ with norm $< \delta$; we will show how to determine δ in order to ensure that $|S(g; \dot{\mathcal{P}}) - 8| < \varepsilon$. Let $\dot{\mathcal{P}}_1$ be the subset of $\dot{\mathcal{P}}$ having its tags in $[0, 1]$ where $g(x) = 2$, and let $\dot{\mathcal{P}}_2$ be the subset of $\dot{\mathcal{P}}$ with its tags in $(1, 3]$ where $g(x) = 3$. It is obvious that we have

(3) $$S(g; \dot{\mathcal{P}}) = S(g; \dot{\mathcal{P}}_1) + S(g; \dot{\mathcal{P}}_2).$$

Since $\|\dot{\mathcal{P}}\| < \delta$, if $u \in [0, 1 - \delta]$ and $u \in [x_{i-1}, x_i]$, then $x_{i-1} \le 1 - \delta$ so that $x_i < x_{i-1} + \delta \le 1$, whence the tag $t_i \in [0, 1]$. Therefore, the interval $[0, 1 - \delta]$ is contained in the union of all subintervals in $\dot{\mathcal{P}}$ with tags $t_i \in [0, 1]$. Similarly, this union is contained in $[0, 1 + \delta]$. (Why?) Since $g(t_i) = 2$ for these tags, we have

$$2(1 - \delta) \le S(g; \dot{\mathcal{P}}_1) \le 2(1 + \delta).$$

A similar argument shows that the union of all subintervals with tags $t_i \in (1, 3]$ contains the interval $[1 + \delta, 3]$ of length $2 - \delta$, and is contained in $[1 - \delta, 3]$ of length $2 + \delta$. Therefore,

$$3(2 - \delta) \le S(g; \dot{\mathcal{P}}_2) \le 3(2 + \delta).$$

Adding these inequalities and using equation (3), we have

$$8 - 5\delta \le S(g; \dot{\mathcal{P}}) = S(g; \dot{\mathcal{P}}_1) + S(g; \dot{\mathcal{P}}_2) \le 8 + 5\delta,$$

whence it follows that

$$|S(g; \dot{P}) - 8| \le 5\delta.$$

To have this final term $< \varepsilon$, we are led to take $\delta_\varepsilon < \varepsilon/5$.

Making such a choice (for example, if we take $\delta_\varepsilon := \varepsilon/10$), we can retrace the argument and see that $|S(g; \dot{P}) - 8| < \varepsilon$ when $\|\dot{P}\| < \delta_\varepsilon$. Since $\varepsilon > 0$ is arbitrary, we have proved that $g \in \mathcal{R}[0, 3]$ and that $\int_0^3 g = 8$, as predicted.

(c) Let $h(x) := x$ for $x \in [0, 1]$; we will show that $h \in \mathcal{R}[0, 1]$.

We will employ a "trick" that enables us to guess the value of the integral by considering a particular choice of the tag points. Indeed, if $\{I_i\}_{i=1}^n$ is any partition of $[0, 1]$ and we choose the tag of the interval $I_i = [x_{i-1}, x_i]$ to be the midpoint $q_i := \frac{1}{2}(x_{i-1} + x_i)$, then the contribution of this term to the Riemann sum corresponding to the tagged partition $\dot{Q} := \{(I_i, q_i)\}_{i=1}^n$ is

$$h(q_i)(x_i - x_{i-1}) = \tfrac{1}{2}(x_i + x_{i-1})(x_i - x_{i-1}) = \tfrac{1}{2}(x_i^2 - x_{i-1}^2).$$

If we add these terms and note that the sum telescopes, we obtain

$$S(h; \dot{Q}) = \sum_{i=1}^n \tfrac{1}{2}(x_i^2 - x_{i-1}^2) = \tfrac{1}{2}(1^2 - 0^2) = \tfrac{1}{2}.$$

Now let $\dot{P} := \{(I_i, t_i)\}_{i=1}^n$ be an arbitrary tagged partition of $[0, 1]$ with $\|\dot{P}\| < \delta$ so that $x_i - x_{i-1} < \delta$ for $i = 1, \cdots, n$. Also let $\dot{Q}$ have the same partition points, but where we choose the tag q_i to be the midpoint of the interval I_i. Since both t_i and q_i belong to this interval, we have $|t_i - q_i| < \delta$. Using the Triangle Inequality, we deduce

$$|S(h; \dot{P}) - S(h; \dot{Q})| = \left| \sum_{i=1}^n t_i(x_i - x_{i-1}) - \sum_{i-1}^n q_i(x_i - x_{i-1}) \right|$$

$$\le \sum_{i=1}^n |t_i - q_i|(x_i - x_{i-1}) < \delta \sum_{i=1}^n (x_i - x_{i-1}) = \delta(x_n - x_0) = \delta.$$

Since $S(h; \dot{Q}) = \frac{1}{2}$, we infer that if $\dot{P}$ is any tagged partition with $\|\dot{P}\| < \delta$, then

$$|S(h; \dot{P}) - \tfrac{1}{2}| < \delta.$$

Therefore we are led to take $\delta_\varepsilon \le \varepsilon$. If we choose $\delta_\varepsilon := \varepsilon$, we can retrace the argument to conclude that $h \in \mathcal{R}[0, 1]$ and $\int_0^1 h = \int_0^1 x\,dx = \frac{1}{2}$.

(d) Let $F(x) := 1$ for $x = \frac{1}{5}, \frac{2}{5}, \frac{3}{5}, \frac{4}{5}$, and $F(x) := 0$ elsewhere on $[0, 1]$. We will show that $F \in \mathcal{R}[0, 1]$ and that $\int_0^1 F = 0$.

Here there are four points where F is not 0, each of which can belong to two subintervals in a given tagged partition $\dot{P}$. Only these terms will make a nonzero contribution to $S(F; \dot{P})$. Therefore we choose $\delta_\varepsilon := \varepsilon/8$.

If $\|\dot{P}\| < \delta_\varepsilon$, let $\dot{P}_0$ be the subset of $\dot{P}$ with tags different from $\frac{1}{5}, \frac{2}{5}, \frac{3}{5}, \frac{4}{5}$, and let $\dot{P}_1$ be the subset of $\dot{P}$ with tags at these points. Since $S(F; \dot{P}_0) = 0$, it is seen that $S(F; \dot{P}) = S(F; \dot{P}_0) + S(F; \dot{P}_1) = S(F; \dot{P}_1)$. Since there are at most 8 terms in $S(F; \dot{P}_1)$ and each term is $< 1 \cdot \delta_\varepsilon$, we conclude that $0 \le S(F; \dot{P}) = S(F; \dot{P}_1) < 8\delta_\varepsilon = \varepsilon$. Thus $F \in \mathcal{R}[0, 1]$ and $\int_0^1 F = 0$.

(e) Let $G(x) := 1/n$ for $x = 1/n$ ($n \in \mathbb{N}$), and $G(x) := 0$ elsewhere on $[0, 1]$.

Given $\varepsilon > 0$, let E_ε be the (finite) set of points where $G(x) \ge \varepsilon$, let n_ε be the number of points in E_ε, and let $\delta_\varepsilon := \varepsilon/(2n_\varepsilon)$. Let $\dot{P}$ be a tagged partition such that $\|\dot{P}\| < \delta_\varepsilon$. Let

$\dot{P}_0$ be the subset of $\mathcal{P}$ with tags outside of E_ε and let $\dot{P}_1$ be the subset of $\dot{P}$ with tags in E_ε. As in (d), we have

$$0 \le S(G; \dot{P}) = S(G; \dot{P}_1) < (2n_\varepsilon)\delta_\varepsilon = \varepsilon.$$

Since $\varepsilon > 0$ is arbitrary, we conclude that $G \in \mathcal{R}[0, 1]$ and $\int_0^1 G = 0$.   □

Some Properties of the Integral _____

The difficulties involved in determining the value of the integral and of δ_ε suggest that it would be very useful to have some general theorems. The first result in this direction enables us to form certain algebraic combinations of integrable functions.

7.1.4 Theorem   *Suppose that f and g are in $\mathcal{R}[a, b]$. Then:*

(a)   *If $k \in \mathbb{R}$, the function kf is in $\mathcal{R}[a, b]$ and*

$$\int_a^b kf = k \int_a^b f.$$

(b)   *The function $f + g$ is in $\mathcal{R}[a, b]$ and*

$$\int_a^b (f + g) = \int_a^b f + \int_a^b g.$$

(c)   *If $f(x) \le g(x)$ for all $x \in [a, b]$, then*

$$\int_a^b f \le \int_a^b g.$$

Proof.   If $\dot{P} = \{([x_{i-1}, x_i], t_i)\}_{i=1}^n$ is a tagged partition of $[a, b]$, then it is an easy exercise to show that

$$S(kf; \dot{P}) = kS(f; \dot{P}), \qquad S(f + g; \dot{P}) = S(f; \dot{P}) + S(g; \dot{P}),$$

$$S(f; \dot{P}) \le S(g; \dot{P}).$$

We leave it to the reader to show that the assertion (a) follows from the first equality. As an example, we will complete the proofs of (b) and (c).

Given $\varepsilon > 0$, we can use the argument in the proof of the Uniqueness Theorem 7.1.2 to construct a number $\delta_\varepsilon > 0$ such that if $\dot{P}$ is any tagged partition with $\|\dot{P}\| < \delta_\varepsilon$, then both

(4) $\left| S(f; \dot{P}) - \int_a^b f \right| < \varepsilon/2$ and $\left| S(g; \dot{P}) - \int_a^b g \right| < \varepsilon/2.$

To prove (b), we note that

$$\left| S(f + g; \dot{P}) - \left(\int_a^b f + \int_a^b g \right) \right| = \left| S(f; \dot{P}) + S(g; \dot{P}) - \int_a^b f - \int_a^b g \right|$$

$$\le \left| S(f; \dot{P}) - \int_a^b f \right| + \left| S(g; \dot{P}) - \int_a^b g \right|$$

$$< \varepsilon/2 + \varepsilon/2 = \varepsilon.$$

Since $\varepsilon > 0$ is arbitrary, we conclude that $f + g \in \mathcal{R}[a, b]$ and that its integral is the sum of the integrals of f and g.

To prove (c), we note that the Triangle Inequality applied to (4) implies

$$\int_a^b f - \varepsilon/2 < S(f;\dot{P}) \quad \text{and} \quad S(g;\dot{P}) < \int_a^b g + \varepsilon/2.$$

If we use the fact that $S(f;\dot{P}) \le S(g;\dot{P})$, we have

$$\int_a^b f \le \int_a^b g + \varepsilon.$$

But, since $\varepsilon > 0$ is arbitrary, we conclude that $\int_a^b f \le \int_a^b g$.   Q.E.D.

Boundedness Theorem ───

We now show that an unbounded function cannot be Riemann integrable.

7.1.5 Theorem *If $f \in \mathcal{R}[a,b]$, then f is bounded on $[a,b]$.*

Proof.   Assume that f is an unbounded function in $\mathcal{R}[a,b]$ with integral L. Then there exists $\delta > 0$ such that if $\dot{P}$ is any tagged partition of $[a,b]$ with $\|\dot{P}\| < \delta$, then we have $|S(f;\dot{P}) - L| < 1$, which implies that

(5) $|S(f;\dot{P})| < |L| + 1.$

Now let $\mathcal{Q} = \{[x_{i-1}, x_i]\}_{n=1}^n$ be a partition of $[a,b]$ with $\|\mathcal{Q}\| < \delta$. Since $|f|$ is not bounded on $[a,b]$, then there exists at least one subinterval in $\mathcal{Q}$, say $[x_{k-1}, x_k]$, on which $|f|$ is not bounded—for, if $|f|$ is bounded on each subinterval $[x_{i-1}, x_i]$ by M_i, then it is bounded on $[a,b]$ by $\max\{M_1, \cdots, M_n\}$.

We will now pick tags for $\mathcal{Q}$ that will provide a contradiction to (5). We tag $\mathcal{Q}$ by $t_i := x_i$ for $i \ne k$ and we pick $t_k \in [x_{k-1}, x_k]$ such that

$$\left| f(t_k)(x_k - x_{k-1}) \right| > |L| + 1 + \left| \sum_{i \ne k} f(t_i)(x_i - x_{i-1}) \right|.$$

From the Triangle Inequality (in the form $|A + B| \ge |A| - |B|$), we have

$$|S(f;\dot{\mathcal{Q}})| \ge |f(t_k)(x_k - x_{k-1})| - \left| \sum_{i \ne k} f(t_i)(x_i - x_{i-1}) \right| > |L| + 1,$$

which contradicts (5).   Q.E.D.

We will close this section with an example of a function that is discontinuous at every rational number and is not monotone, but is Riemann integrable nevertheless.

7.1.6 Example We consider Thomae's function $h : [0,1] \to \mathbb{R}$ defined, as in Example 5.1.6(h), by $h(x) := 0$ if $x \in [0,1]$ is irrational, $h(0) := 1$ and by $h(x) := 1/n$ if $x \in [0,1]$ is the rational number $x = m/n$ where $m, n \in \mathbb{N}$ have no common integer factors except 1. It was seen in 5.1.6(h) that h is continuous at every irrational number and discontinuous at every rational number in $[0,1]$. We will now show that $h \in \mathcal{R}[0,1]$.

Let $\varepsilon > 0$; then the set $E_\varepsilon := \{x \in [0,1] : h(x) \ge \varepsilon/2\}$ is a finite set. We let n_ε be the number of elements in E_ε and let $\delta_\varepsilon := \varepsilon/(4n_\varepsilon)$. If $\dot{P}$ is a tagged partition with $\|\dot{P}\| < \delta_\varepsilon$, let $\dot{P}_1$ be the subset of $\dot{P}$ having tags in E_ε and $\dot{P}_2$ be the subset of $\dot{P}$ having tags elsewhere in $[0,1]$. We observe that $\dot{P}_1$ has at most $2n_\varepsilon$ intervals whose total length is $< 2n_\varepsilon \delta_\varepsilon = \varepsilon/2$

and that $0 < h(t_i) \le 1$ for every tag in $\dot{P}_1$. Also the total lengths of the subintervals in $\dot{P}_2$ is ≤ 1 and $h(t_i) < \varepsilon/2$ for every tag in $\dot{P}_2$. Therefore we have

$$|S(h; \dot{P})| = S(h; \dot{P}_1) + S(h; \dot{P}_2) < 1 \cdot 2n_\varepsilon \delta_\varepsilon + (\varepsilon/2) \cdot 1 = \varepsilon.$$

Since $\varepsilon > 0$ is arbitrary, we infer that $h \in \mathcal{R}[0, 1]$ with integral 0.   □

Exercises for Section 7.1

1. If $I := [0, 4]$, calculate the norms of the following partitions:
   (a) $\mathcal{P}_1 := (0, 1, 2, 4)$,
   (b) $\mathcal{P}_2 := (0, 2, 3, 4)$,
   (c) $\mathcal{P}_3 := (0, 1, 1.5, 2, 3.4, 4)$,
   (d) $\mathcal{P}_4 := (0, .5, 2.5, 3.5, 4)$.

2. If $f(x) := x^2$ for $x \in [0, 4]$, calculate the following Riemann sums, where $\dot{P}_i$ has the same partition points as in Exercise 1, and the tags are selected as indicated.
   (a) $\mathcal{P}_1$ with the tags at the left endpoints of the subintervals.   *tag – where function is evaluated*
   (b) $\mathcal{P}_1$ with the tags at the right endpoints of the subintervals.
   (c) $\mathcal{P}_2$ with the tags at the left endpoints of the subintervals.
   (d) $\mathcal{P}_2$ with the tags at the right endpoints of the subintervals.

3. Show that $f : [a, b] \to \mathbb{R}$ is Riemann integrable on $[a, b]$ if and only if there exists $L \in \mathbb{R}$ such that for every $\varepsilon > 0$ there exists $\delta_\varepsilon > 0$ such that if $\dot{P}$ is any tagged partition with norm $\|\dot{P}\| \le \delta_\varepsilon$, then $|S(f; \dot{P}) - L| \le \varepsilon$.

4. Let $\dot{P}$ be a tagged parition of $[0, 3]$.
   (a) Show that the union U_1 of all subintervals in $\dot{P}$ with tags in $[0, 1]$ satisfies $[0, 1 - \|\dot{P}\|] \subseteq U_1 \subseteq [0, 1 + \|\dot{P}\|]$.
   (b) Show that the union U_2 of all subintervals in $\dot{P}$ with tags in $[1, 2]$ satisfies $[1 + \|\dot{P}\|, 2 - \|\dot{P}\|] \subseteq U_2 \subseteq [1 - \|\dot{P}\|, 2 + \|\dot{P}\|]$.

5. Let $\dot{P} := \{(I_i, t_i)\}_{i=1}^n$ be a tagged partition of $[a, b]$ and let $c_1 < c_2$.
   (a) If u belongs to a subinterval I_i whose tag satisfies $c_1 \le t_i \le c_2$, show that $c_1 - \|\dot{P}\| \le u \le c_2 + \|\dot{P}\|$.
   (b) If $v \in [a, b]$ and satisfies $c_1 + \|\dot{P}\| \le v \le c_2 - \|\dot{P}\|$, then the tag t_i of any subinterval I_i that contains v satisfies $t_i \in [c_1, c_2]$.

6. (a) Let $f(x) := 2$ if $0 \le x < 1$ and $f(x) := 1$ if $1 \le x \le 2$. Show that $f \in \mathcal{R}[0, 2]$ and evaluate its integral.
   (b) Let $h(x) := 2$ if $0 \le x < 1$, $h(1) := 3$ and $h(x) := 1$ if $1 < x \le 2$. Show that $h \in \mathcal{R}[0, 2]$ and evaluate its integral.

7. Use Mathematical Induction and Theorem 7.1.4 to show that if $f_1, \cdots, f_n$ are in $\mathcal{R}[a, b]$ and if $k_1, \cdots, k_n \in \mathbb{R}$, then the linear combination $f = \sum_{i=1}^n k_i f_i$ belongs to $\mathcal{R}[a, b]$ and $\int_a^b f = \sum_{i=1}^n k_i \int_a^b f_i$.

8. If $f \in \mathcal{R}[a, b]$ and $|f(x)| \le M$ for all $x \in [a, b]$, show that $|\int_a^b f| \le M(b - a)$.

9. If $f \in \mathcal{R}[a, b]$ and if $(\dot{P}_n)$ is any sequence of tagged partitions of $[a, b]$ such that $\|\dot{P}_n\| \to 0$, prove that $\int_a^b f = \lim_n S(f; \dot{P}_n)$.

10. Let $g(x) := 0$ if $x \in [0, 1]$ is rational and $g(x) := 1/x$ if $x \in [0, 1]$ is irrational. Explain why $g \notin \mathcal{R}[0, 1]$. However, show that there exists a sequence $(\dot{P}_n)$ of tagged partitions of $[a, b]$ such that $\|\dot{P}_n\| \to 0$ and $\lim_n S(g; \dot{P}_n)$ exists.

11. Suppose that f is bounded on $[a, b]$ and that there exists two sequences of tagged partitions of $[a, b]$ such that $\|\dot{P}_n\| \to 0$ and $\|\dot{Q}_n\| \to 0$, but such that $\lim_n S(f; \dot{P}_n) \neq \lim_n S(f; \dot{Q}_n)$. Show that f is not in $\mathcal{R}[a, b]$.

12. Consider the Dirichlet function, introduced in Example 5.1.5(g), defined by $f(x) := 1$ for $x \in [0, 1]$ rational and $f(x) := 0$ for $x \in [0, 1]$ irrational. Use the preceding exercise to show that f is *not* Riemann integrable on $[0, 1]$.

13. Suppose that $f : [a, b] \to \mathbb{R}$ and that $f(x) = 0$ except for a finite number of points $c_1, \cdots, c_n$ in $[a, b]$. Prove that $f \in \mathcal{R}[a, b]$ and that $\int_a^b f = 0$.

14. If $g \in \mathcal{R}[a, b]$ and if $f(x) = g(x)$ except for a finite number of points in $[a, b]$, prove that $f \in \mathcal{R}[a, b]$ and that $\int_a^b f = \int_a^b g$.

15. Suppose that $c \leq d$ are points in $[a, b]$. If $\varphi : [a, b] \to \mathbb{R}$ satisfies $\varphi(x) = \alpha > 0$ for $x \in [c, d]$ and $\varphi(x) = 0$ elsewhere in $[a, b]$, prove that $\varphi \in \mathcal{R}[a, b]$ and that $\int_a^b \varphi = \alpha(d - c)$. [*Hint:* Given $\varepsilon > 0$ let $\delta_\varepsilon := \varepsilon/4\alpha$ and show that if $\|\dot{\mathcal{P}}\| < \delta_\varepsilon$ then we have $\alpha(d - c - 2\delta_\varepsilon) \leq S(\varphi; \dot{\mathcal{P}}) \leq \alpha(d - c + 2\delta_\varepsilon)$.]

16. Let $0 \leq a < b$, let $Q(x) := x^2$ for $x \in [a, b]$ and let $\mathcal{P} := \{[x_{i-1}, x_i]\}_{i=1}^n$ be a partition of $[a, b]$. For each i, let q_i be the positive *square root* of

$$\tfrac{1}{3}(x_i^2 + x_i x_{i-1} + x_{i-1}^2).$$

   (a) Show that q_i satisfies $0 \leq x_{i-1} \leq q_i \leq x_i$.
   (b) Show that $Q(q_i)(x_i - x_{i-1}) = \tfrac{1}{3}(x_i^3 - x_{i-1}^3)$.
   (c) If $\dot{Q}$ is the tagged partition with the same subintervals as $\mathcal{P}$ and the tags q_i, show that $S(Q; \dot{Q}) = \tfrac{1}{3}(b^3 - a^3)$.
   (d) Use the argument in Example 7.1.3(c) to show that $Q \in \mathcal{R}[a, b]$ and

$$\int_a^b Q = \int_a^b x^2 \, dx = \tfrac{1}{3}(b^3 - a^3).$$

17. Let $0 \leq a < b$ and $m \in \mathbb{N}$, let $M(x) := x^m$ for $x \in [a, b]$ and let $\mathcal{P} := \{[x_{i-1}, x_i]\}_{i=1}^n$ be a partition of $[a, b]$. For each i, let q_i be the positive mth *root* of

$$\frac{1}{m+1} \left(x_i^m + x_i^{m-1} x_{i-1} + \cdots + x_i x_{i-1}^{m-1} + x_{i-1}^m \right).$$

   (a) Show that q_i satisfies $0 \leq x_{i-1} \leq q_i \leq x_i$.
   (b) Show that $M(q_i)(x_i - x_{i-1}) = \frac{1}{m+1} \left(x_i^{m+1} - x_{i-1}^{m+1} \right)$.
   (c) If $\dot{Q}$ is the tagged partition with the same subintervals as $\mathcal{P}$ and the tags q_i, show that $S(M; \dot{Q}) = \frac{1}{m+1}(b^{m+1} - a^{m+1})$.
   (d) Use the argument in Example 7.1.3(c) to show that $M \in \mathcal{R}[a, b]$ and

$$\int_a^b M = \int_a^b x^m \, dx = \frac{1}{m+1}(b^{m+1} - a^{m+1}).$$

18. If $f \in \mathcal{R}[a, b]$ and $c \in \mathbb{R}$, we define g on $[a + c, b + c]$ by $g(y) := f(y - c)$. Prove that $g \in \mathcal{R}[a + c, b + c]$ and that $\int_{a+c}^{b+c} g = \int_a^b f$. The function g is called the c-**translate** of f.

Section 7.2   Riemann Integrable Functions

We begin with a proof of the important Cauchy Criterion. We will then prove the Squeeze Theorem, which will be used to establish the Riemann integrability of several classes of functions (step functions, continuous functions, and monotone functions). Finally we will establish the Additivity Theorem.

   We have already noted that direct use of the definition requires that we know the value of the integral. The Cauchy Criterion removes this need, but at the cost of considering two Riemann sums, instead of just one.

7.2.1 Cauchy Criterion *A function $f : [a, b] \to \mathbb{R}$ belongs to $\mathcal{R}[a, b]$ if and only if for every $\varepsilon > 0$ there exists $\eta_\varepsilon > 0$ such that if $\dot{P}$ and $\dot{Q}$ are any tagged partitions of $[a, b]$ with $\|\dot{P}\| < \eta_\varepsilon$ and $\|\dot{Q}\| < \eta_\varepsilon$, then*

$$|S(f; \dot{P}) - S(f; \dot{Q})| < \varepsilon.$$

Proof. ($\Rightarrow$) If $f \in \mathcal{R}[a, b]$ with integral L, let $\eta_\varepsilon := \delta_\varepsilon/2 > 0$ be such that if $\dot{P}, \dot{Q}$ are tagged partitions such that $\|\dot{P}\| < \eta_\varepsilon$ and $\|\dot{Q}\| < \eta_\varepsilon$, then

$$|S(f; \dot{P}) - L| < \varepsilon/2 \qquad \text{and} \qquad |S(f; \dot{Q}) - L| < \varepsilon/2.$$

Therefore we have

$$\begin{aligned}
|S(f; \dot{P}) - S(f; \dot{Q})| &\le |S(f; \dot{P}) - L + L - S(f; \dot{Q})| \\
&\le |S(f; \dot{P}) - L| + |L - S(f; \dot{Q})| \\
&< \varepsilon/2 + \varepsilon/2 = \varepsilon.
\end{aligned}$$

($\Leftarrow$) For each $n \in \mathbb{N}$, let $\delta_n > 0$ be such that if $\dot{P}$ and $\dot{Q}$ are tagged partitions with norms $< \delta_n$, then

$$|S(f; \dot{P}) - S(f; \dot{Q})| < 1/n.$$

Evidently we may assume that $\delta_n \ge \delta_{n+1}$ for $n \in \mathbb{N}$; otherwise, we replace δ_n by $\delta_n' := \min\{\delta_1, \cdots, \delta_n\}$.

For each $n \in \mathbb{N}$, let $\dot{P}_n$ be a tagged partition with $\|\dot{P}_n\| < \delta_n$. Clearly, if $m > n$ then both $\dot{P}_m$ and $\dot{P}_n$ have norms $< \delta_n$, so that

(1) $$|S(f; \dot{P}_n) - S(f; \dot{P}_m)| < 1/n \qquad \text{for} \quad m > n.$$

Consequently, the sequence $(S(f; \dot{P}_m))_{m=1}^\infty$ is a Cauchy sequence in $\mathbb{R}$. Therefore (by Theorem 3.5.5) this sequence converges in $\mathbb{R}$ and we let $A := \lim_m S(f; \dot{P}_m)$.

Passing to the limit in (1) as $m \to \infty$, we have

$$|S(f; \dot{P}_n) - A| \le 1/n \qquad \text{for all} \quad n \in \mathbb{N}.$$

To see that A is the Riemann integral of f, given $\varepsilon > 0$, let $K \in \mathbb{N}$ satisfy $K > 2/\varepsilon$. If $\dot{Q}$ is any tagged partition with $\|\dot{Q}\| < \delta_K$, then

$$\begin{aligned}
|S(f; \dot{Q}) - A| &\le |S(f; \dot{Q}) - S(f; \dot{P}_K)| + |S(f; \dot{P}_K) - A| \\
&\le 1/K + 1/K < \varepsilon.
\end{aligned}$$

Since $\varepsilon > 0$ is arbitrary, then $f \in \mathcal{R}[a, b]$ with integral A.   Q.E.D.

We will now give two examples of the use of the Cauchy Criterion.

7.2.2 Examples (a) Let $g : [0, 3] \to \mathbb{R}$ be the function considered in Example 7.1.3(b). In that example we saw that if $\dot{P}$ is a tagged partition of $[0, 3]$ with norm $\|\dot{P}\| < \delta$, then

$$8 - 5\delta \le S(g; \dot{P}) \le 8 + 5\delta.$$

Hence if $\dot{Q}$ is another tagged partition with $\|\dot{Q}\| < \delta$, then

$$8 - 5\delta \le S(g; \dot{Q}) \le 8 + 5\delta.$$

If we subtract these two inequalities, we obtain

$$|S(g; \dot{P}) - S(g; \dot{Q})| \le 10\delta.$$

In order to make this final term $< \varepsilon$, we are led to employ the Cauchy Criterion with $\eta_\varepsilon := \varepsilon/20$. (We leave the details to the reader.)

(b) The Cauchy Criterion can be used to show that a function $f : [a, b] \to \mathbb{R}$ is *not* Riemann integrable. To do this we need to show that: *There exists* $\varepsilon_0 > 0$ *such that for any* $\eta > 0$ *there exists tagged partitions* $\dot{P}$ *and* $\dot{Q}$ *with* $\|\dot{P}\| < \eta$ *and* $\|\dot{Q}\| < \eta$ *such that* $|S(f; \dot{P}) - S(f; \dot{Q})| \geq \varepsilon_0$.

We will apply these remarks to the Dirichlet function, considered in 5.1.6(g), defined by $f(x) := 1$ if $x \in [0, 1]$ is rational and $f(x) := 0$ if $x \in [0, 1]$ is irrational.

Here we take $\varepsilon_0 := \frac{1}{2}$. If $\dot{P}$ is any partition all of whose tags are rational numbers then $S(f; \dot{P}) = 1$, while if $\dot{Q}$ is any tagged partition all of whose tags are irrational numbers then $S(f; \dot{Q}) = 0$. Since we are able to take such tagged partitions with arbitrarily small norms, we conclude that the Dirichlet function is *not* Riemann integrable.   □

The Squeeze Theorem

The next result will be used to establish the Riemann integrability of some important classes of functions.

7.2.3 Squeeze Theorem   *Let* $f : [a, b] \to \mathbb{R}$. *Then* $f \in \mathcal{R}[a, b]$ *if and only if for every* $\varepsilon > 0$ *there exist functions* α_ε *and* ω_ε *in* $\mathcal{R}[a, b]$ *with*

$$(2) \qquad\qquad \alpha_\varepsilon(x) \leq f(x) \leq \omega_\varepsilon(x) \qquad \textit{for all} \quad x \in [a, b],$$

and such that

$$(3) \qquad\qquad \int_a^b (\omega_\varepsilon - \alpha_\varepsilon) < \varepsilon.$$

Proof.   ($\Rightarrow$) Take $\alpha_\varepsilon = \omega_\varepsilon = f$ for all $\varepsilon > 0$.

($\Leftarrow$) Let $\varepsilon > 0$. Since α_ε and ω_ε belong to $\mathcal{R}[a, b]$, there exists $\delta_\varepsilon > 0$ such that if $\dot{P}$ is any tagged partition with $\|\dot{P}\| < \delta_\varepsilon$ then

$$\left| S(\alpha_\varepsilon; \dot{P}) - \int_a^b \alpha_\varepsilon \right| < \varepsilon \qquad \text{and} \qquad \left| S(\omega_\varepsilon; \dot{P}) - \int_a^b \omega_\varepsilon \right| < \varepsilon.$$

It follows from these inequalities that

$$\int_a^b \alpha_\varepsilon - \varepsilon < S(\alpha_\varepsilon; \dot{P}) \qquad \text{and} \qquad S(\omega_\varepsilon; \dot{P}) < \int_a^b \omega_\varepsilon + \varepsilon.$$

In view of inequality (2), we have $S(\alpha_\varepsilon; \dot{P}) \leq S(f; \dot{P}) \leq S(\omega_\varepsilon; \dot{P})$, whence

$$\int_a^b \alpha_\varepsilon - \varepsilon < S(f; \dot{P}) < \int_a^b \omega_\varepsilon + \varepsilon.$$

If $\dot{Q}$ is another tagged partition with $\|\dot{Q}\| < \delta_\varepsilon$, then we also have

$$\int_a^b \alpha_\varepsilon - \varepsilon < S(f; \dot{Q}) < \int_a^b \omega_\varepsilon + \varepsilon.$$

If we subtract these two inequalities and use (3), we conclude that

$$|S(f; \dot{P}) - S(f; \dot{Q})| < \int_a^b \omega_\varepsilon - \int_a^b \alpha_\varepsilon + 2\varepsilon$$

$$= \int_a^b (\omega_\varepsilon - \alpha_\varepsilon) + 2\varepsilon < 3\varepsilon.$$

Since $\varepsilon > 0$ is arbitrary, the Cauchy Criterion implies that $f \in \mathcal{R}[a, b]$.   Q.E.D.

Classes of Riemann Integrable Functions

The Squeeze Theorem is often used in connection with the class of step functions. It will be recalled from Definition 5.4.9 that a function $\varphi : [a, b] \to \mathbb{R}$ is a **step function** if it has only a finite number of distinct values, each value being assumed on one or more subintervals of $[a, b]$. For illustrations of step functions, see Figures 5.4.3 or 7.1.4.

7.2.4 Lemma *If J is a subinterval of $[a, b]$ having endpoints $c < d$ and if $\varphi_J(x) := 1$ for $x \in J$ and $\varphi_J(x) := 0$ elsewhere in $[a, b]$, then $\varphi_J \in \mathcal{R}[a, b]$ and $\int_a^b \varphi_J = d - c$.*

Proof. If $J = [c, d]$ with $c \leq d$, this is Exercise 7.1.15 and we can choose $\delta_\varepsilon := \varepsilon/4$. A similar proof can be given for the three other subintervals having these endpoints. Alternatively, we observe that we can write

$$\varphi_{[c,d)} = \varphi_{[c,d]} - \varphi_{[d,d]}, \quad \varphi_{(c,d]} = \varphi_{[c,d]} - \varphi_{[c,c]} \quad \text{and} \quad \varphi_{(c,d)} = \varphi_{[c,d)} - \varphi_{[c,c]}.$$

Since $\int_a^b \varphi_{[c,c]} = 0$, all four of these functions have integral equal to $d - c$. Q.E.D.

It is an important fact that any step function is Riemann integrable.

7.2.5 Theorem *If $\varphi : [a, b] \to \mathbb{R}$ is a step function, then $\varphi \in \mathcal{R}[a, b]$.*

Proof. Step functions of the type appearing in 7.2.4 are called "elementary step functions". In Exercise 5 it is shown that an arbitrary step function φ can be expressed as a linear combination of such elementary step functions:

(4) $$\varphi = \sum_{j=1}^m k_j \varphi_{J_j},$$

where J_j has endpoints $c_j < d_j$. The lemma and Theorem 7.1.4(a,b) imply that $\varphi \in \mathcal{R}[a, b]$ and that

(5) $$\int_a^b \varphi = \sum_{j=1}^m k_j (d_j - c_j).$$ Q.E.D.

We will now use the Squeeze Theorem to show that an arbitrary continuous function is Riemann integrable.

7.2.6 Theorem *If $f : [a, b] \to \mathbb{R}$ is continuous on $[a, b]$, then $f \in \mathcal{R}[a, b]$.*

Proof. It follows from Theorem 5.4.3 that f is uniformly continuous on $[a, b]$. Therefore, given $\varepsilon > 0$ there exists $\delta_\varepsilon > 0$ such that if $u, v \in [a, b]$ and $|u - v| < \delta_\varepsilon$, then we have $|f(u) - f(v)| < \varepsilon/(b - a)$.

Let $\mathcal{P} = \{I_i\}_{i=1}^n$ be a partition such that $\|\mathcal{P}\| < \delta_\varepsilon$, let $u_i \in I_i$ be a point where f attains its minimum value on I_i, and let $v_i \in I_i$ be a point where f attains its maximum value on I_i.

Let α_ε be the step function defined by $\alpha_\varepsilon(x) := f(u_i)$ for $x \in [x_{i-1}, x_i)$ $(i = 1, \cdots, n - 1)$ and $\alpha_\varepsilon(x) := f(u_n)$ for $x \in [x_{n-1}, x_n]$. Let ω_ε be defined similarly using the points v_i instead of the u_i. Then one has

$$\alpha_\varepsilon(x) \leq f(x) \leq \omega_\varepsilon(x) \qquad \text{for all} \quad x \in [a, b].$$

Moreover, it is clear that

$$0 \le \int_a^b (\omega_\varepsilon - \alpha_\varepsilon) = \sum_{i=1}^n \big(f(v_i) - f(u_i)\big)(x_i - x_{i-1})$$

$$< \sum_{i=1}^n \Big(\frac{\varepsilon}{b-a}\Big)(x_i - x_{i-1}) = \varepsilon.$$

Therefore it follows from the Squeeze Theorem that $f \in \mathcal{R}[a, b]$.   Q.E.D.

Monotone functions are not necessarily continuous at every point, but they are also Riemann integrable.

7.2.7 Theorem   If $f : [a, b] \to \mathbb{R}$ is monotone on $[a, b]$, then $f \in \mathcal{R}[a, b]$.

Proof.   Suppose that f is increasing on the interval $[a, b]$, $a < b$. If $\varepsilon > 0$ is given, we let $q \in \mathbb{N}$ be such that

$$h := \frac{f(b) - f(a)}{q} < \frac{\varepsilon}{b-a}.$$

Let $y_k := f(a) + kh$ for $k = 0, 1, \cdots, q$ and consider sets $A_k := f^{-1}([y_{k-1}, y_k))$ for $k = 1, \cdots, q - 1$ and $A_q := f^{-1}([y_{q-1}, y_q])$. The sets $\{A_k\}$ are pairwise disjoint and have union $[a, b]$. The Characterization Theorem 2.5.1 implies that each A_k is either (i) empty, (ii) contains a single point, or (iii) is a nondegenerate interval (not necessarily closed) in $[a, b]$. We discard the sets for which (i) holds and relabel the remaining ones. If we adjoin the endpoints to the remaining intervals $\{A_k\}$, we obtain closed intervals $\{I_k\}$. It is an exercise to show that the relabeled intervals $\{A_k\}_{k=1}^q$ are pairwise disjoint, satisfy $[a, b] = \bigcup_{k=1}^q A_k$ and that $f(x) \in [y_{k-1}, y_k]$ for $x \in A_k$.
We now define step functions α_ε and ω_ε on $[a, b]$ by setting

$$\alpha_\varepsilon(x) := y_{k-1} \qquad \text{and} \qquad \omega_\varepsilon(x) := y_k \qquad \text{for} \qquad x \in A_k.$$

It is clear that $\alpha_\varepsilon(x) \le f(x) \le \omega_\varepsilon(x)$ for all $x \in [a, b]$ and that

$$\int_a^b (\omega_\varepsilon - \alpha_\varepsilon) = \sum_{k=1}^q (y_k - y_{k-1})(x_k - x_{k-1})$$

$$= \sum_{k=1}^q h \cdot (x_k - x_{k-1}) = h \cdot (b - a) < \varepsilon.$$

Since $\varepsilon > 0$ is arbitrary, the Squeeze Theorem implies that $f \in \mathcal{R}[a, b]$.   Q.E.D.

The Additivity Theorem

We now return to arbitrary Riemann integrable functions. Our next result shows that the integral is an "additive function" of the interval over which the function is integrated. This property is no surprise, but its proof is a bit delicate and may be omitted on a first reading.

7.2.8 Additivity Theorem   Let $f : [a, b] \to \mathbb{R}$ and let $c \in (a, b)$. Then $f \in \mathcal{R}[a, b]$ if and only if its restrictions to $[a, c]$ and $[c, b]$ are both Riemann integrable. In this case

$$(6) \qquad \int_a^b f = \int_a^c f + \int_c^b f.$$

Proof.   ($\Leftarrow$) Suppose that the restriction f_1 of f to $[a, c]$, and the restriction f_2 of f to $[c, b]$ are Riemann integrable to L_1 and L_2, respectively. Then, given $\varepsilon > 0$ there exists $\delta' > 0$ such that if $\dot{\mathcal{P}}_1$ is a tagged partition of $[a, c]$ with $\|\dot{\mathcal{P}}_1\| < \delta'$, then $|S(f_1; \dot{\mathcal{P}}_1) - L_1| < \varepsilon/3$. Also there exists $\delta'' > 0$ such that if $\dot{\mathcal{P}}_2$ is a tagged partition of $[c, b]$ with $\|\dot{\mathcal{P}}_2\| < \delta''$ then $|S(f_2; \dot{\mathcal{P}}_2) - L_2| < \varepsilon/3$. If M is a bound for $|f|$, we define $\delta_\varepsilon := \min\{\delta', \delta'', \varepsilon/6M\}$ and let $\dot{\mathcal{P}}$ be a tagged partition of $[a, b]$ with $\|\dot{\mathcal{Q}}\| < \delta$. We will prove that

$$(7) \qquad\qquad\qquad \left| S(f; \dot{\mathcal{Q}}) - (L_1 + L_2) \right| < \varepsilon.$$

(i)   If c is a partition point of $\dot{\mathcal{Q}}$, we split $\dot{\mathcal{Q}}$ into a partition $\dot{\mathcal{Q}}_1$ of $[a, c]$ and a partition $\dot{\mathcal{Q}}_2$ of $[c, b]$. Since $S(f; \dot{\mathcal{Q}}) = S(f; \dot{\mathcal{Q}}_1) + S(f; \dot{\mathcal{Q}}_2)$, and since $\dot{\mathcal{Q}}_1$ has norm $< \delta'$ and $\dot{\mathcal{Q}}_2$ has norm $< \delta''$, the inequality (7) is clear.

(ii)   If c is not a partition point in $\dot{\mathcal{Q}} = \{(I_k, t_k)\}_{k=1}^m$, there exists $k \leq m$ such that $c \in (x_{k-1}, x_k)$. We let $\dot{\mathcal{Q}}_1$ be the tagged partition of $[a, c]$ defined by

$$\dot{\mathcal{Q}}_1 := \{(I_1, t_1), \cdots, (I_{k-1}, t_{k-1}), ([x_{k-1}, c], c)\},$$

and $\dot{\mathcal{Q}}_2$ be the tagged partition of $[c, b]$ defined by

$$\dot{\mathcal{Q}}_2 := \{([c, x_k], c), (I_{k+1}, t_{k+1}), \cdots, (I_m, t_m)\}.$$

A straightforward calculation shows that

$$S(f; \dot{\mathcal{Q}}) - S(f; \dot{\mathcal{Q}}_1) - S(f; \dot{\mathcal{Q}}_2) = f(t_k)(x_k - x_{k-1}) - f(c)(x_k - x_{k-1})$$
$$= (f(t_k) - f(c)) \cdot (x_k - x_{k-1}),$$

whence it follows that

$$|S(f; \dot{\mathcal{Q}}) - S(f; \dot{\mathcal{Q}}_1) - S(f; \dot{\mathcal{Q}}_2)| \leq 2M(x_k - x_{k-1}) < \varepsilon/3.$$

But since $\|\dot{\mathcal{Q}}_1\| < \delta \leq \delta'$ and $\|\dot{\mathcal{Q}}_2\| < \delta \leq \delta''$, it follows that

$$|S(f; \dot{\mathcal{Q}}_1) - L_1| < \varepsilon/3 \qquad \text{and} \qquad |S(f; \dot{\mathcal{Q}}_2) - L_2| < \varepsilon/3,$$

from which we obtain (7). Since $\varepsilon > 0$ is arbitrary, we infer that $f \in \mathcal{R}[a, b]$ and that (6) holds.

($\Rightarrow$) We suppose that $f \in \mathcal{R}[a, b]$ and, given $\varepsilon > 0$, we let $\eta_\varepsilon > 0$ satisfy the Cauchy Criterion 7.2.1. Let f_1 be the restriction of f to $[a, c]$ and let $\dot{\mathcal{P}}_1$, $\dot{\mathcal{Q}}_1$ be tagged partitions of $[a, c]$ with $\|\dot{\mathcal{P}}_1\| < \eta_\varepsilon$ and $\|\dot{\mathcal{Q}}_1\| < \eta_\varepsilon$. By adding additional partition points and tags from $[c, b]$, we can extend $\dot{\mathcal{P}}_1$ and $\dot{\mathcal{Q}}_1$ to tagged partitions $\dot{\mathcal{P}}$ and $\dot{\mathcal{Q}}$ of $[a, b]$ that satisfy $\|\dot{\mathcal{P}}\| < \eta_\varepsilon$ and $\|\dot{\mathcal{Q}}\| < \eta_\varepsilon$. If we use the *same* additional points and tags in $[c, b]$ for both $\dot{\mathcal{P}}$ and $\dot{\mathcal{Q}}$, then

$$S(f_1; \dot{\mathcal{P}}_1) - S(f_1; \dot{\mathcal{Q}}_1) = S(f; \dot{\mathcal{P}}) - S(f; \dot{\mathcal{Q}}).$$

Since both $\dot{\mathcal{P}}$ and $\dot{\mathcal{Q}}$ have norm $< \eta_\varepsilon$, then $|S(f_1; \dot{\mathcal{P}}_1) - S(f_1; \dot{\mathcal{Q}}_1)| < \varepsilon$. Therefore the Cauchy Condition shows that the restriction f_1 of f to $[a, c]$ is in $\mathcal{R}[a, c]$. In the same way, we see that the restriction f_2 of f to $[c, b]$ is in $\mathcal{R}[c, d]$.

The equality (6) now follows from the first part of the theorem.   Q.E.D.

7.2.9 Corollary   *If $f \in \mathcal{R}[a, b]$, and if $[c, d] \subseteq [a, b]$, then the restriction of f to $[c, d]$ is in $\mathcal{R}[c, d]$.*

Proof.   Since $f \in \mathcal{R}[a, b]$ and $c \in [a, b]$, it follows from the theorem that its restriction to $[c, b]$ is in $\mathcal{R}[c, b]$. But if $d \in [c, b]$, then another application of the theorem shows that the restriction of f to $[c, d]$ is in $\mathcal{R}[c, d]$.   Q.E.D.

7.2.10 Corollary   If $f \in \mathcal{R}[a, b]$ and if $a = c_0 < c_1 < \cdots < c_m = b$, then the restrictions of f to each of the subintervals $[c_{i-1}, c_i]$ are Riemann integrable and

$$\int_a^b f = \sum_{i=1}^m \int_{c_{i-1}}^{c_i} f.$$

Until now, we have considered the Riemann integral over an interval $[a, b]$ where $a < b$. It is convenient to have the integral defined more generally.

7.2.11 Definition   If $f \in \mathcal{R}[a, b]$ and if $\alpha, \beta \in [a, b]$ with $\alpha < \beta$, we define

$$\int_\beta^\alpha f := -\int_\alpha^\beta f \qquad \text{and} \qquad \int_\alpha^\alpha f := 0.$$

7.2.12 Theorem   If $f \in \mathcal{R}[a, b]$ and if α, β, γ are any numbers in $[a, b]$, then

(8)
$$\int_\alpha^\beta f = \int_\alpha^\gamma f + \int_\gamma^\beta f,$$

in the sense that the existence of any two of these integrals implies the existence of the third integral and the equality (8).

Proof.   If any two of the numbers α, β, γ are equal, then (8) holds. Thus we may suppose that all three of these numbers are distinct.

For the sake of symmetry, we introduce the expression

$$L(\alpha, \beta, \gamma) := \int_\alpha^\beta f + \int_\beta^\gamma f + \int_\gamma^\alpha f.$$

It is clear that (8) holds if and only if $L(\alpha, \beta, \gamma) = 0$. Therefore, to establish the assertion, we need to show that $L = 0$ for all six permutations of the arguments α, β and γ.

We note that the Additivity Theorem 7.2.8 implies that $L(\alpha, \beta, \gamma) = 0$ when $\alpha < \gamma < \beta$. But it is easily seen that both $L(\beta, \gamma, \alpha)$ and $L(\gamma, \alpha, \beta)$ equal $L(\alpha, \beta, \gamma)$. Moreover, the numbers

$$L(\beta, \alpha, \gamma), \qquad L(\alpha, \gamma, \beta), \qquad \text{and} \qquad L(\gamma, \beta, \alpha)$$

are all equal to $-L(\alpha, \beta, \gamma)$. Therefore, L vanishes for all possible configurations of these three points.   Q.E.D.

Exercises for Section 7.2

1. Let $f : [a, b] \to \mathbb{R}$. Show that $f \notin \mathcal{R}[a, b]$ if and only if there exists $\varepsilon_0 > 0$ such that for every $n \in \mathbb{N}$ there exist tagged partitions $\dot{\mathcal{P}}_n$ and $\dot{\mathcal{Q}}_n$ with $\|\dot{\mathcal{P}}_n\| < 1/n$ and $\|\dot{\mathcal{Q}}_n\| < 1/n$ such that $|S(f; \dot{\mathcal{P}}_n) - S(f; \dot{\mathcal{Q}}_n)| \geq \varepsilon_0$.

2. Consider the function h defined by $h(x) := x + 1$ for $x \in [0, 1]$ rational, and $h(x) := 0$ for $x \in [0, 1]$ irrational. Show that h is not Riemann integrable.

3. Let $H(x) := k$ for $x = 1/k$ $(k \in \mathbb{N})$ and $H(x) := 0$ elsewhere on $[0, 1]$. Use Exercise 1, or the argument in 7.2.2(b), to show that H is not Riemann integrable.

4. If $\alpha(x) := -x$ and $\omega(x) := x$ and if $\alpha(x) \le f(x) \le \omega(x)$ for all $x \in [0, 1]$, does it follow from the Squeeze Theorem 7.2.3 that $f \in \mathcal{R}[0, 1]$?

5. If J is any subinterval of $[a, b]$ and if $\varphi_J(x) := 1$ for $x \in J$ and $\varphi_J(x) := 0$ elsewhere on $[a, b]$, we say that φ_J is an *elementary step function* on $[a, b]$. Show that every step function is a linear combination of elementary step functions.

6. If $\psi : [a, b] \to \mathbb{R}$ takes on only a finite number of distinct values, is ψ a step function?

7. If $S(f; \dot{\mathcal{P}})$ is any Riemann sum of $f : [a, b] \to \mathbb{R}$, show that there exists a step function $\varphi : [a, b] \to \mathbb{R}$ such that $\int_a^b \varphi = S(f; \dot{\mathcal{P}})$.

8. Suppose that f is continuous on $[a, b]$, that $f(x) \ge 0$ for all $x \in [a, b]$ and that $\int_a^b f = 0$. Prove that $f(x) = 0$ for all $x \in [a, b]$.

9. Show that the continuity hypothesis in the preceding exercise cannot be dropped.

10. If f and g are continuous on $[a, b]$ and if $\int_a^b f = \int_a^b g$, prove that there exists $c \in [a, b]$ such that $f(c) = g(c)$.

11. If f is bounded by M on $[a, b]$ and if the restriction of f to every interval $[c, b]$ where $c \in (a, b)$ is Riemann integrable, show that $f \in \mathcal{R}[a, b]$ and that $\int_c^b f \to \int_a^b f$ as $c \to a+$. [*Hint:* Let $\alpha_c(x) := -M$ and $\omega_c(x) := M$ for $x \in [a, c)$ and $\alpha_c(x) := \omega_c(x) := f(x)$ for $x \in [c, b]$. Apply the Squeeze Theorem 7.2.3 for c sufficiently near a.]

12. Show that $g(x) := \sin(1/x)$ for $x \in (0, 1]$ and $g(0) := 0$ belongs to $\mathcal{R}[0, 1]$.

13. Give an example of a function $f : [a, b] \to \mathbb{R}$ that is in $\mathcal{R}[c, b]$ for every $c \in (a, b)$ but which is not in $\mathcal{R}[a, b]$.

14. Suppose that $f : [a, b] \to \mathbb{R}$, that $a = c_0 < c_1 < \cdots < c_m = b$ and that the restrictions of f to $[c_{i-1}, c_i]$ belong to $\mathcal{R}[c_{i-1}, c_i]$ for $i = 1, \cdots, m$. Prove that $f \in \mathcal{R}[a, b]$ and that the formula in Corollary 7.2.10 holds.

15. If f is bounded and there is a finite set E such that f is continuous at every point of $[a, b]\backslash E$, show that $f \in \mathcal{R}[a, b]$.

16. If f is continuous on $[a, b]$, $a < b$, show that there exists $c \in [a, b]$ such that we have $\int_a^b f = f(c)(b - a)$. This result is sometimes called the *Mean Value Theorem for Integrals*.

17. If f and g are continuous on $[a, b]$ and $g(x) > 0$ for all $x \in [a, b]$, show that there exists $c \in [a, b]$ such that $\int_a^b fg = f(c) \int_a^b g$. Show that this conclusion fails if we do not have $g(x) > 0$. (Note that this result is an extension of the preceding exercise.)

18. Let f be continuous on $[a, b]$, let $f(x) \ge 0$ for $x \in [a, b]$, and let $M_n := \left(\int_a^b f^n\right)^{1/n}$. Show that $\lim(M_n) = \sup\{f(x) : x \in [a, b]\}$.

19. Suppose that $a > 0$ and that $f \in \mathcal{R}[-a, a]$.
(a) If f is *even* (that is, if $f(-x) = f(x)$ for all $x \in [0, a]$), show that $\int_{-a}^a f = 2\int_0^a f$.
(b) If f is *odd* (that is, if $f(-x) = -f(x)$ for all $x \in [0, a]$), show that $\int_{-a}^a f = 0$.

20. Suppose that $f : [a, b] \to \mathbb{R}$ and that $n \in \mathbb{N}$. Let $\mathcal{P}_n$ be the partition of $[a, b]$ into n subintervals having equal lengths, so that $x_i := a + i(b - a)/n$ for $i = 0, 1, \cdots, n$. Let $L_n(f) := S(f; \dot{\mathcal{P}}_{n,l})$ and $R_n(f) := S(f; \dot{\mathcal{P}}_{n,r})$, where $\dot{\mathcal{P}}_{n,l}$ has its tags at the left endpoints, and $\dot{\mathcal{P}}_{n,r}$ has its tags at the right endpoints of the subintervals $[x_{i-1}, x_i]$.
(a) If f is increasing on $[a, b]$, show that $L_n(f) \le R_n(f)$ and that
$$0 \le R_n(f) - L_n(f) = \left(f(b) - f(a)\right) \cdot \frac{(b - a)}{n}.$$
(b) Show that $f(a)(b - a) \le L_n(f) \le \int_a^b f \le R_n(f) \le f(b)(b - a)$.
(c) If f is decreasing on $[a, b]$, obtain an inequality similar to that in (a).

(d) If $f \in \mathcal{R}[a, b]$ is not monotone, show that $\int_a^b f$ is not necessarily between $L_n(f)$ and $R_n(f)$.

21. If f is continuous on $[-a, a]$, show that $\int_{-a}^a f(x^2)\, dx = 2 \int_0^a f(x^2)\, dx$.

22. If f is continuous on $[-1, 1]$, show that $\int_0^{\pi/2} f(\cos x)\, dx = \int_0^{\pi/2} f(\sin x)\, dx = \frac{1}{2} \int_0^\pi f(\sin x)\, dx$.
    [*Hint:* Examine certain Riemann sums.]

Section 7.3   The Fundamental Theorem

We will now explore the connection between the notions of the derivative and the integral. In fact, there are *two* theorems relating to this problem: one has to do with integrating a derivative, and the other with differentiating an integral. These theorems, taken together, are called the Fundamental Theorem of Calculus. Roughly stated, they imply that the operations of differentiation and integration are inverse to each other. However, there are some subleties that should not be overlooked.

The Fundamental Theorem (First Form)

The First Form of the Fundamental Theorem provides a theoretical basis for the method of calculating an integral that the reader learned in calculus. It asserts that if a function f is the derivative of a function F, *and if f belongs to* $\mathcal{R}[a, b]$, then the integral $\int_a^b f$ can be calculated by means of the evaluation $F \big|_a^b := F(b) - F(a)$. A function F such that $F'(x) = f(x)$ for all $x \in [a, b]$ is called an **antiderivative** or a **primitive of** f on $[a, b]$. Thus, when f has an antiderivative, it is a very simple matter to calculate its integral.

In practice, it is convenient to allow some exceptional points c where $F'(c)$ does not exist in $\mathbb{R}$, or where it does not equal $f(c)$. It turns out that we can permit a *finite* number of such exceptional points.

7.3.1 Fundamental Theorem of Calculus (First Form)   *Suppose there is a **finite** set E in $[a, b]$ and functions $f, F : [a, b] \to \mathbb{R}$ such that:*

(a) *F is continuous on $[a, b]$,*

(b) *$F'(x) = f(x)$ for all $x \in [a, b] \backslash E$,*

(c) *f belongs to $\mathcal{R}[a, b]$.*

   Then we have

(1)
$$\int_a^b f = F(b) - F(a).$$

Proof.   We will prove the theorem in the case where $E := \{a, b\}$. The general case can be obtained by breaking the interval into the union of a finite number of intervals (see Exercise 1).

Let $\varepsilon > 0$ be given. Since $f \in \mathcal{R}[a, b]$ by assumption (c), there exists $\delta_\varepsilon > 0$ such that if $\dot{\mathcal{P}}$ is *any* tagged partition with $\|\dot{\mathcal{P}}\| < \delta_\varepsilon$, then

(2)
$$\left| S(f; \dot{\mathcal{P}}) - \int_a^b f \right| < \varepsilon.$$

If the subintervals in $\dot{\mathcal{P}}$ are $[x_{i-1}, x_i]$, then the Mean Value Theorem 6.2.4 applied to F on $[x_{i-1}, x_i]$ implies that there exists $u_i \in (x_{i-1}, x_i)$ such that

$$F(x_i) - F(x_{i-1}) = F'(u_i) \cdot (x_i - x_{i-1}) \qquad \text{for} \quad i = 1, \cdots, n.$$

If we add these terms, note the telescoping of the sum, and use the fact that $F'(u_i) = f(u_i)$, we obtain

$$F(b) - F(a) = \sum_{i=1}^{n} \left(F(x_i) - F(x_{i-1}) \right) = \sum_{i=1}^{n} f(u_i)(x_i - x_{i-1}).$$

Now let $\dot{\mathcal{P}}_u := \{([x_{i-1}, x_i], u_i)\}_{i=1}^{n}$, so the sum on the right equals $S(f; \dot{\mathcal{P}}_u)$. If we substitute $F(b) - F(a) = S(f; \dot{\mathcal{P}}_u)$ into (2), we conclude that

$$\left| F(b) - F(a) - \int_a^b f \right| < \varepsilon.$$

But, since $\varepsilon > 0$ is arbitrary, we infer that equation (1) holds.   Q.E.D.

Remark   If the function F is differentiable at every point of $[a, b]$, then (by Theorem 6.1.2) hypothesis (a) is automatically satisfied. If f is not defined for some point $c \in E$, we take $f(c) := 0$. Even if F is differentiable at every point of $[a, b]$, condition (c) is *not automatically satisfied,* since there exist functions F such that F' is not Riemann integrable. (See Example 7.3.2(e).)

7.3.2 Examples   (a)   If $F(x) := \frac{1}{2}x^2$ for all $x \in [a, b]$, then $F'(x) = x$ for all $x \in [a, b]$. Further, $f = F'$ is continuous so it is in $\mathcal{R}[a, b]$. Therefore the Fundamental Theorem (with $E = \emptyset$) implies that

$$\int_a^b x\, dx = F(b) - F(a) = \frac{1}{2}(b^2 - a^2).$$

(b)   If $G(x) := \text{Arctan } x$ for $x \in [a, b]$, then $G'(x) = 1/(x^2 + 1)$ for all $x \in [a, b]$; also G' is continuous, so it is in $\mathcal{R}[a, b]$. Therefore the Fundamental Theorem (with $E = \emptyset$) implies that

$$\int_a^b \frac{1}{x^2 + 1}\, dx = \text{Arctan } b - \text{Arctan } a.$$

(c)   If $A(x) := |x|$ for $x \in [-10, 10]$, then $A'(x) = -1$ if $x \in [-10, 0)$ and $A'(x) = +1$ for $x \in (0, 10]$. Recalling the definition of the signum function (in 4.1.10(b)), we have $A'(x) = \text{sgn}(x)$ for all $x \in [-10, 10]\backslash\{0\}$. Since the signum function is a step function, it belongs to $\mathcal{R}[-10, 10]$. Therefore the Fundamental Theorem (with $E = \{0\}$) implies that

$$\int_{-10}^{10} \text{sgn}(x)\, dx = A(10) - A(-10) = 10 - 10 = 0.$$

(d)   If $H(x) := 2\sqrt{x}$ for $x \in [0, b]$, then H is continuous on $[0, b]$ and $H'(x) = 1/\sqrt{x}$ for $x \in (0, b]$. Since $h := H'$ is not bounded on $(0, b]$, it does not belong to $\mathcal{R}[0, b]$ no matter how we define $h(0)$. Therefore, the Fundamental Theorem 7.3.1 does not apply. (However, we will see in Example 10.1.10(a) that h is *generalized* Riemann integrable on $[0, b]$.)

(e)   Let $K(x) := x^2 \cos(1/x^2)$ for $x \in (0, 1]$ and let $K(0) := 0$. It follows from the Product Rule 6.1.3(c) and the Chain Rule 6.1.6 that

$$K'(x) = 2x \cos(1/x^2) + (2/x) \sin(1/x^2) \qquad \text{for} \quad x \in (0, 1].$$

Further, as in Example 6.1.7(e), it can be shown that $K'(0) = 0$. Thus K is continuous and differentiable at *every point* of $[0, 1]$. Since it can be seen that the function K' is not bounded on $[0, 1]$, it does not belong to $\mathcal{R}[0, 1]$ and the Fundamental Theorem 7.3.1 does not apply to K'. (However, we will see from Theorem 10.1.9 that K' is *generalized Riemann integrable* on $[0, 1]$.)   □

The Fundamental Theorem (Second Form) ———————————————————————

We now turn to the Fundamental Theorem (Second Form) in which we wish to differentiate an integral involving a variable upper limit.

7.3.3 Definition   If $f \in \mathcal{R}[a, b]$, then the function defined by

(3)
$$F(z) := \int_a^z f \quad \text{for} \quad z \in [a, b],$$

is called the **indefinite integral** of f with **basepoint** a. (Sometimes a point other than a is used as a basepoint; see Exercise 6.)

We will first show that if $f \in \mathcal{R}[a, b]$, then its indefinite integral F satisfies a Lipschitz condition; hence F is continuous on $[a, b]$.

7.3.4 Theorem   *The indefinite integral F defined by (3) is continuous on $[a, b]$. In fact, if $|f(x)| \leq M$ for all $x \in [a, b]$, then $|F(z) - F(w)| \leq M|z - w|$ for all $z, w \in [a, b]$.*

Proof.   The Additivity Theorem 7.2.8 implies that if $z, w \in [a, b]$ and $w \leq z$, then

$$F(z) = \int_a^z f = \int_a^w f + \int_w^z f = F(w) + \int_w^z f,$$

whence we have

$$F(z) - F(w) = \int_w^z f.$$

Now if $-M \leq f(x) \leq M$ for all $x \in [a, b]$, then Theorem 7.1.4(c) implies that

$$-M(z - w) \leq \int_w^z f \leq M(z - w),$$

whence it follows that

$$|F(z) - F(w)| \leq \left| \int_w^z f \right| \leq M|z - w|,$$

as asserted.   Q.E.D.

We will now show that the indefinite integral F is differentiable at any point where f is continuous.

7.3.5 Fundamental Theorem of Calculus (Second Form)   Let $f \in \mathcal{R}[a, b]$ and let f be *continuous at a point $c \in [a, b]$. Then the indefinite integral, defined by (3), is differentiable at c and $F'(c) = f(c)$.*

Proof.   We will suppose that $c \in [a, b)$ and consider the right-hand derivative of F at c. Since f is continuous at c, given $\varepsilon > 0$ there exists $\eta_\varepsilon > 0$ such that if $c \leq x < c + \eta_\varepsilon$, then

(4)
$$f(c) - \varepsilon < f(x) < f(c) + \varepsilon.$$

Let h satisfy $0 < h < \eta_\varepsilon$. The Additivity Theorem 7.2.8 implies that f is integrable on the intervals $[a, c]$, $[a, c + h]$ and $[c, c + h]$ and that

$$F(c + h) - F(c) = \int_c^{c+h} f.$$

Now on the interval $[c, c + h]$ the function f satisfies inequality (4), so that (by Theorem 7.1.4(c)) we have

$$(f(c) - \varepsilon) \cdot h \leq F(c + h) - F(c) = \int_c^{c+h} f \leq (f(c) + \varepsilon) \cdot h.$$

If we divide by $h > 0$ and subtract $f(c)$, we obtain

$$\left| \frac{F(c + h) - F(c)}{h} - f(c) \right| \leq \varepsilon.$$

But, since $\varepsilon > 0$ is arbitrary, we conclude that the right-hand limit is given by

$$\lim_{h \to 0+} \frac{F(c + h) - F(c)}{h} = f(c).$$

It is proved in the same way that the left-hand limit of this difference quotient also equals $f(c)$ when $c \in (a, b]$, whence the assertion follows.   Q.E.D.

If f is continuous on all of $[a, b]$, we obtain the following result.

7.3.6 Theorem   *If f is continuous on $[a, b]$, then the indefinite integral F, defined by (3), is differentiable on $[a, b]$ and $F'(x) = f(x)$ for all $x \in [a, b]$.*

Theorem 7.3.6 can be summarized: *If f is continuous on $[a, b]$, then its indefinite integral is an antiderivative of f.* We will now see that, in general, the indefinite integral need not be an antiderivative (either because the derivative of the indefinite integral does not exist or does not equal $f(x)$).

7.3.7 Examples   (a)   If $f(x) := \operatorname{sgn} x$ on $[-1, 1]$, then $f \in \mathcal{R}[-1, 1]$ and has the indefinite integral $F(x) := |x| - 1$ with the basepoint -1. However, since $F'(0)$ does not exist, F is not an antiderivative of f on $[-1, 1]$.

(b)   If h denotes Thomae's function, considered in 7.1.6, then its indefinite integral $H(x) := \int_0^x h$ is identically 0 on $[0, 1]$. Here, the derivative of this indefinite integral exists at every point and $H'(x) = 0$. But $H'(x) \neq h(x)$ whenever $x \in \mathbb{Q} \cap [0, 1]$, so that H is not an antiderivative of h on $[0, 1]$.   □

Substitution Theorem ―――――――――――――――――――――――――――――――

The next theorem provides the justification for the "change of variable" method that is often used to evaluate integrals. This theorem is employed (usually implicitly) in the evaluation by means of procedures that involve the manipulation of "differentials", common in elementary courses.

7.3.8 Substitution Theorem   Let $J := [\alpha, \beta]$ and let $\varphi : J \to \mathbb{R}$ have a continuous derivative on J. If $f : I \to \mathbb{R}$ is continuous on an interval I containing $\varphi(J)$, then

(5)
$$\int_\alpha^\beta f(\varphi(t)) \cdot \varphi'(t)\, dt = \int_{\varphi(\alpha)}^{\varphi(\beta)} f(x)\, dx.$$

The proof of this theorem is based on the Chain Rule 6.1.6, and will be outlined in Exercise 15. The hypotheses that f and φ' are continuous are restrictive, but are used to ensure the existence of the Riemann integral on the left side of (5).

7.3.9 Examples   (a)   Consider the integral $\int_1^4 \dfrac{\sin \sqrt{t}}{\sqrt{t}}\, dt$.

Here we substitute $\varphi(t) := \sqrt{t}$ for $t \in [1, 4]$ so that $\varphi'(t) = 1/(2\sqrt{t})$ is continuous on $[1, 4]$. If we let $f(x) := 2 \sin x$, then the integrand has the form $(f \circ \varphi) \cdot \varphi'$ and the Substitution Theorem 7.3.8 implies that the integral equals $\int_1^2 2 \sin x\, dx = -2 \cos x|_1^2 = 2(\cos 1 - \cos 2)$.

(b)   Consider the integral $\int_0^4 \dfrac{\sin \sqrt{t}}{\sqrt{t}}\, dt$.

Since $\varphi(t) := \sqrt{t}$ does not have a continuous derivative on $[0, 4]$, the Substitution Theorem 7.3.8 is not applicable, at least with this substitution. (In fact, it is not obvious that this integral exists; however, we can apply Exercise 7.2.11 to obtain this conclusion. We could then apply the Fundamental Theorem 7.3.1 to $F(t) := -2 \cos \sqrt{t}$ with $E := \{0\}$ to evaluate this integral.)   □

We will give a more powerful Substitution Theorem for the *generalized* Riemann integral in Section 10.1.

Lebesgue's Integrability Criterion

We will now present a statement of the definitive theorem due to Henri Lebesgue (1875–1941) giving a necessary and sufficient condition for a function to be Riemann integrable, and will give some applications of this theorem. In order to state this result, we need to introduce the important notion of a null set.

Warning   Some people use the term "null set" as a synonym for the terms "empty set" or "void set" referring to $\varnothing$ ($=$ the set that has no elements). However, we will always use the term "null set" in conformity with our next definition, as is customary in the theory of integration.

7.3.10 Definition   (a)   A set $Z \subset \mathbb{R}$ is said to be a **null set** if for every $\varepsilon > 0$ there exists a countable collection $\{(a_k, b_k)\}_{k=1}^\infty$ of open intervals such that

(6)
$$Z \subseteq \bigcup_{k=1}^\infty (a_k, b_k) \qquad \text{and} \qquad \sum_{k=1}^\infty (b_k - a_k) \leq \varepsilon.$$

(b)   If $Q(x)$ is a statement about the point $x \in I$, we say that $Q(x)$ holds **almost everywhere** on I (or for **almost every** $x \in I$), if there exists a null set $Z \subset I$ such that $Q(x)$ holds for all $x \in I \backslash Z$. In this case we may write

$$Q(x) \quad \text{for a.e.} \quad x \in I.$$

It is trivial that any subset of a null set is also a null set, and it is easy to see that the union of two null sets is a null set. We will now give an example that may be very surprising.

7.3.11 Example   The $\mathbb{Q}_1$ of rational numbers in $[0, 1]$ is a null set.

We enumerate $\mathbb{Q}_1 = \{r_1, r_2, \cdots\}$. Given $\varepsilon > 0$, note that the open interval $J_1 := (r_1 - \varepsilon/4, r_1 + \varepsilon/4)$ contains r_1 and has length $\varepsilon/2$; also the open interval $J_2 := (r_2 - \varepsilon/8, r_2 + \varepsilon/8)$ contains r_2 and has length $\varepsilon/4$. In general, the open interval

$$J_k := \left(r_k - \frac{\varepsilon}{2^{k+1}}, r_k + \frac{\varepsilon}{2^{k+1}} \right)$$

contains the point r_k and has length $\varepsilon/2^k$. Therefore, the union $\bigcup_{k=1}^{\infty} J_k$ of these open intervals contains every point of $\mathbb{Q}_1$; moreover, the sum of the lengths is $\sum_{k=1}^{\infty} (\varepsilon/2^k) = \varepsilon$. Since $\varepsilon > 0$ is arbitrary, $\mathbb{Q}_1$ is a null set. $\qquad\square$

The argument just given can be modified to show that: *Every countable set is a null set.* However, it can be shown that there exist uncountable null sets in $\mathbb{R}$; for example, the Cantor set that will be introduced in Definition 11.1.10.

We now state Lebesgue's Integrability Criterion. It asserts that a bounded function on an interval is Riemann integrable if and only if its points of discontinuity form a null set.

7.3.12 Lebesgue's Integrability Criterion   *A bounded function $f : [a, b] \to \mathbb{R}$ is Riemann integrable if and only if it is continuous almost everywhere on $[a, b]$.*

A proof of this result will be given in Appendix C. However, we will apply Legesgue's Theorem here to some specific functions, and show that some of our previous results follow immediately from it. We shall also use this theorem to obtain the important Composition and Product Theorems.

7.3.13 Examples   **(a)**   The step function g in Example 7.1.3(b) is continuous at every point except the point $x = 1$. Therefore it follows from the Lebesgue Integrability Criterion that g is Riemann integrable.

In fact, since every step function has at most a finite set of points of discontinuity, then: *Every step function on $[a, b]$ is Riemann integrable.*

(b)   Since it was seen in Theorem 5.5.4 that the set of points of discontinuity of a monotone function is countable, we conclude that: *Every monotone function on $[a, b]$ is Riemann integrable.*

(c)   The function G in Example 7.1.3(e) is discontinuous precisely at the points $D := \{1, 1/2, \cdots, 1/n, \cdots\}$. Since this is a countable set, it is a null set and Lebesgue's Criterion implies that G is Riemann integrable.

(d)   The Dirichlet function was shown in Example 7.2.2(b) not to be Riemann integrable.

Note that it is discontinuous at *every* point of $[0, 1]$. Since it can be shown that the interval $[0, 1]$ is not a null set, Lebesgue's Criterion yields the same conclusion.

(e)   Let $h : [0, 1] \to \mathbb{R}$ be Thomae's function, defined in Examples 5.1.6(h) and 7.1.6.

In Example 5.1.6(h), we saw that h is continuous at every irrational number and is discontinuous at every rational number in $[0, 1]$. By Example 7.3.11, it is discontinuous on a null set, so Lebesgue's Criterion implies that Thomae's function is Riemann integrable on $[0, 1]$, as we saw in Example 7.1.6. $\qquad\square$

We now obtain a result that will enable us to take other combinations of Riemann integrable functions.

7.3.14 Composition Theorem   *Let $f \in \mathcal{R}[a, b]$ with $f([a, b]) \subseteq [c, d]$ and let $\varphi : [c, d]$ $\to \mathbb{R}$ be continuous. Then the composition $\varphi \circ f$ belongs to $\mathcal{R}[a, b]$.*

Proof.   If f is continuous at a point $u \in [a, b]$, then $\varphi \circ f$ is also continuous at u. Since the set D of points discontinuity of f is a null set, it follows that the set $D_1 \subseteq D$ of points of discontinuity of $\varphi \circ f$ is also a null set. Therefore the composition $\varphi \circ f$ also belongs to $\mathcal{R}[a, b]$.   Q.E.D.

It will be seen in Exercise 22 that the hypothesis that φ is continuous cannot be dropped. The next result is a corollary of the Composition Theorem.

7.3.15 Corollary   *Suppose that $f \in \mathcal{R}[a, b]$. Then its absolute value $|f|$ is in $\mathcal{R}[a, b]$, and*

$$\left| \int_a^b f \right| \leq \int_a^b |f| \leq M(b - a),$$

where $|f(x)| \leq M$ for all $x \in [a, b]$.

Proof.   We have seen in Theorem 7.1.5 that if f is integrable, then there exists M such that $|f(x)| \leq M$ for all $x \in [a, b]$. Let $\varphi(t) := |t|$ for $t \in [-M, M]$; then the Composition Theorem implies that $|f| = \varphi \circ f \in \mathcal{R}[a, b]$. The first inequality follows from the fact that $-|f| \leq f \leq |f|$ and 7.1.4(c), and the second from the fact that $|f(x)| \leq M$.   Q.E.D.

7.3.16 The Product Theorem   *If f and g belong to $\mathcal{R}[a, b]$, then the product fg belongs to $\mathcal{R}[a, b]$.*

Proof.   If $\varphi(t) := t^2$ for $t \in [-M, M]$, it follows from the Composition Theorem that $f^2 = \varphi \circ f$ belongs to $\mathcal{R}[a, b]$. Similarly, $(f + g)^2$ and g^2 belong to $\mathcal{R}[a, b]$. But since we can write the product as

$$fg = \tfrac{1}{2}\left[(f + g)^2 - f^2 - g^2\right],$$

it follows that $fg \in \mathcal{R}[a, b]$.   Q.E.D.

Integration by Parts

We will conclude this section with a rather general form of Integration by Parts for the Riemann integral, and Taylor's Theorem with the Remainder.

7.3.17 Integration by Parts   *Let F, G be differentiable on $[a, b]$ and let $f := F'$ and $g := G'$ belong to $\mathcal{R}[a, b]$. Then*

(7)
$$\int_a^b fG = FG \Big|_a^b - \int_a^b Fg.$$

Proof.   By Theorem 6.1.3(c), the derivative $(FG)'$ exists on $[a, b]$ and

$$(FG)' = F'G + FG' = fG + Fg.$$

Since F, G are continuous and f, g belong to $\mathcal{R}[a, b]$, the Product Theorem 7.3.16 implies that fG and Fg are integrable. Therefore the Fundamental Theorem 7.3.1 implies that

$$FG \Big|_a^b = \int_a^b (FG)' = \int_a^b fG + \int_a^b Fg,$$

from which (7) follows.   Q.E.D.

A special, but useful, case of this theorem is when f and g are continuous on $[a, b]$ and F, G are their indefinite integrals $F(x) := \int_a^x f$ and $G(x) := \int_a^x g$.

We close this section with a version of Taylor's Theorem for the Riemann Integral.

7.3.18 Taylor's Theorem with the Remainder   *Suppose that* $f', \cdots, f^{(n)}, f^{(n+1)}$ *exist on* $[a, b]$ *and that* $f^{(n+1)} \in \mathcal{R}[a, b]$. *Then we have*

(8)
$$f(b) = f(a) + \frac{f'(a)}{1!}(b - a) + \cdots + \frac{f^{(n)}(a)}{n!}(b - a)^n + R_n,$$

where the remainder is given by

(9)
$$R_n = \frac{1}{n!} \int_a^b f^{(n+1)}(t) \cdot (b - t)^n \, dt.$$

Proof.   Apply Integration by Parts to equation (9), with $F(t) := f^{(n)}(t)$ and $G(t) := (b - t)^n / n!$, so that $g(t) = -(b - t)^{n-1}/(n - 1)!$, to get

$$R_n = \frac{1}{n!} f^{(n)}(t) \cdot (b - t)^n \Big|_{t=a}^{t=b} + \frac{1}{(n-1)!} \int_a^b f^{(n)}(t) \cdot (b - a)^{n-1} \, dt$$

$$= -\frac{f^{(n)}(a)}{n!} \cdot (b - a)^n + \frac{1}{(n-1)!} \int_a^b f^{(n)}(t) \cdot (b - t)^{n-1} \, dt.$$

If we continue to integrate by parts in this way, we obtain (8).   Q.E.D.

Exercises for Section 7.3

1.   Extend the proof of the Fundamental Theorem 7.3.1 to the case of an arbitrary finite set E.

2.   If $n \in \mathbb{N}$ and $H_n(x) := x^{n+1}/(n + 1)$ for $x \in [a, b]$, show that the Fundamental Theorem 7.3.1 implies that $\int_a^b x^n \, dx = (b^{n+1} - a^{n+1})/(n + 1)$. What is the set E here?

3.   If $g(x) := x$ for $|x| \geq 1$ and $g(x) := -x$ for $|x| < 1$ and if $G(x) := \frac{1}{2}|x^2 - 1|$, show that $\int_{-2}^3 g(x) \, dx = G(3) - G(-2) = 5/2$.

4.   Let $B(x) := -\frac{1}{2}x^2$ for $x < 0$ and $B(x) := \frac{1}{2}x^2$ for $x \geq 0$. Show that $\int_a^b |x| \, dx = B(b) - B(a)$.

5.   Let $f : [a, b] \to \mathbb{R}$ and let $C \in \mathbb{R}$.
     (a)   If $\Phi : [a, b] \to \mathbb{R}$ is an antiderivative of f on $[a, b]$, show that $\Phi_C(x) := \Phi(x) + C$ is also an antiderivative of f on $[a, b]$.
     (b)   If Φ_1 and Φ_2 are antiderivatives of f on $[a, b]$, show that $\Phi_1 - \Phi_2$ is a constant function on $[a, b]$.

6.   If $f \in \mathcal{R}[a, b]$ and if $c \in [a, b]$, the function defined by $F_c(z) := \int_c^z f$ for $z \in [a, b]$ is called the **indefinite integral** of f with **basepoint** c. Find a relation between F_a and F_c.

7.   We have seen in Example 7.1.6 that Thomae's function is in $\mathcal{R}[0, 1]$ with integral equal to 0. Can the Fundamental Theorem 7.3.1 be used to obtain this conclusion? Explain your answer.

8.   Let $F(x)$ be defined for $x \geq 0$ by $F(x) := (n - 1)x - (n - 1)n/2$ for $x \in [n - 1, n), n \in \mathbb{N}$. Show that F is continuous and evaluate $F'(x)$ at points where this derivative exists. Use this result to evaluate $\int_a^b [\![x]\!] \, dx$ for $0 \leq a < b$, where $[\![x]\!]$ denotes the greatest integer in x, as defined in Exercise 5.1.4.

9.   Let $f \in \mathcal{R}[a, b]$ and define $F(x) := \int_a^x f$ for $x \in [a, b]$.
     (a)   Evaluate $G(x) := \int_c^x f$ in terms of F, where $c \in [a, b]$.
     (b)   Evaluate $H(x) := \int_x^b f$ in terms of F.
     (c)   Evaluate $S(x) := \int_x^{\sin x} f$ in terms of F.

10. Let $f : [a, b] \to \mathbb{R}$ be continuous on $[a, b]$ and let $v : [c, d] \to \mathbb{R}$ be differentiable on $[c, d]$ with $v([c, d]) \subseteq [a, b]$. If we define $G(x) := \int_a^{v(x)} f$, show that $G'(x) = f(v(x)) \cdot v'(x)$ for all $x \in [c, d]$.

11. Find $F'(x)$ when F is defined on $[0, 1]$ by:
    (a) $F(x) := \int_0^{x^2} (1 + t^3)^{-1} \, dt$.
    (b) $F(x) := \int_{x^2}^x \sqrt{1 + t^2} \, dt$.

12. Let $f : [0, 3] \to \mathbb{R}$ be defined by $f(x) := x$ for $0 \leq x < 1$, $f(x) := 1$ for $1 \leq x < 2$ and $f(x) := x$ for $2 \leq x \leq 3$. Obtain formulas for $F(x) := \int_0^x f$ and sketch the graphs of f and F. Where is F differentiable? Evaluate $F'(x)$ at all such points.

13. If $f : \mathbb{R} \to \mathbb{R}$ is continuous and $c > 0$, define $g : \mathbb{R} \to \mathbb{R}$ by $g(x) := \int_{x-c}^{x+c} f(t) \, dt$. Show that g is differentiable on $\mathbb{R}$ and find $g'(x)$.

14. If $f : [0, 1] \to \mathbb{R}$ is continuous and $\int_0^x f = \int_x^1 f$ for all $x \in [0, 1]$, show that $f(x) = 0$ for all $x \in [0, 1]$.

15. Use the following argument to prove the Substitution Theorem 7.3.8. Define $F(u) := \int_{\varphi(\alpha)}^u f(x) dx$ for $u \in I$, and $H(t) := F(\varphi(t))$ for $t \in J$. Show that $H'(t) = f(\varphi(t))\varphi'(t)$ for $t \in J$ and that

$$\int_{\varphi(\alpha)}^{\varphi(\beta)} f(x) \, dx = F(\varphi(\beta)) = H(\beta) = \int_\alpha^\beta f(\varphi(t))\varphi'(t) \, dt.$$

16. Use the Substitution Theorem 7.3.8 to evaluate the following integrals.
    (a) $\displaystyle\int_0^1 t\sqrt{1 + t^2} \, dt,$
    (b) $\displaystyle\int_0^2 t^2(1 + t^3)^{-1/2} \, dt = 4/3,$
    (c) $\displaystyle\int_1^4 \frac{\sqrt{1 + \sqrt{t}}}{\sqrt{t}} \, dt,$
    (d) $\displaystyle\int_1^4 \frac{\cos\sqrt{t}}{\sqrt{t}} \, dt = 2(\sin 2 - \sin 1).$

17. Sometimes the Substitution Theorem 7.3.8 cannot be applied but the following result, called the "Second Substitution Theorem" is useful. In addition to the hypotheses of 7.3.8, assume that $\varphi'(t) \neq 0$ for all $t \in J$, so the function $\psi : \varphi(J) \to \mathbb{R}$ inverse to φ exists and has derivative $\psi'(\varphi(t)) = 1/\varphi'(t)$. Then

$$\int_\alpha^\beta f(\varphi(t)) \, dt = \int_{\varphi(\alpha)}^{\varphi(\beta)} f(x)\psi'(x) \, dx.$$

To prove this, let $G(t) := \int_\alpha^t f(\varphi(s)) \, ds$ for $t \in J$, so that $G'(t) = f(\varphi(t))$. Note that $K(x) := G(\psi(x))$ is differentiable on the interval $\varphi(J)$ and that $K'(x) = G'(\psi(x))\psi'(x) = f(\varphi \circ \psi(x))\psi'(x) = f(x)\psi'(x)$. Calculate $G(\beta) = K(\varphi(\beta))$ in two ways to obtain the formula.

18. Apply the Second Substition Theorem to evaluate the following integrals.
    (a) $\displaystyle\int_1^9 \frac{dt}{2 + \sqrt{t}},$
    (b) $\displaystyle\int_1^3 \frac{dt}{t\sqrt{t + 1}} = \ln(3 + 2\sqrt{2}) - \ln 3,$
    (c) $\displaystyle\int_1^4 \frac{\sqrt{t} \, dt}{1 + \sqrt{t}},$
    (d) $\displaystyle\int_1^4 \frac{dt}{\sqrt{t}(t + 4)} = \text{Arctan}(1) - \text{Arctan}(1/2).$

19. Explain why Theorem 7.3.8 and/or Exercise 7.3.17 cannot be applied to evaluate the following integrals, using the indicated substitution.
    (a) $\displaystyle\int_0^4 \frac{\sqrt{t} \, dt}{1 + \sqrt{t}}$ $\varphi(t) = \sqrt{t},$
    (b) $\displaystyle\int_0^4 \frac{\cos\sqrt{t} \, dt}{\sqrt{t}}$ $\varphi(t) = \sqrt{t},$
    (c) $\displaystyle\int_{-1}^1 \sqrt{1 + 2|t|} \, dt$ $\varphi(t) = |t|,$
    (d) $\displaystyle\int_0^1 \frac{dt}{\sqrt{1 - t^2}}$ $\varphi(t) = \text{Arcsin}\, t.$

20. (a) If Z_1 and Z_2 are null sets, show that $Z_1 \cup Z_2$ is a null set.
    (b) More generally, if Z_n is a null set for each $n \in \mathbb{N}$, show that $\bigcup_{n=1}^\infty Z_n$ is a null set. [*Hint:* Given $\varepsilon > 0$ and $n \in \mathbb{N}$, let $\{J_k^n : k \in \mathbb{N}\}$ be a countable collection of open intervals whose union contains Z_n and the sum of whose lengths is $\leq \varepsilon/2^n$. Now consider the countable collection $\{J_k^n : n, k \in \mathbb{N}\}$.]

21. Let $f, g \in \mathcal{R}[a, b]$.

(a) If $t \in \mathbb{R}$, show that $\int_a^b (tf \pm g)^2 \geq 0$.

(b) Use (a) to show that $2|\int_a^b fg| \leq t \int_a^b f^2 + (1/t) \int_a^b g^2$ for $t > 0$.

(c) If $\int_a^b f^2 = 0$, show that $\int_a^b fg = 0$.

(d) Now prove that $\left| \int_a^b fg \right|^2 \leq (\int_a^b |fg|)^2 \leq (\int_a^b f^2) \cdot (\int_a^b g^2)$. This inequality is called the **Cauchy-Bunyakovsky-Schwarz Inequality** (or simply the **Schwarz Inequality**).

22. Let $h : [0, 1] \to \mathbb{R}$ be Thomae's function and let sgn be the signum function. Show that the composite function sgn $\circ\, h$ is not Riemann integrable on $[0, 1]$.

Section 7.4 Approximate Integration

The Fundamental Theorem of Calculus 7.3.1 yields an effective method of evaluating the integral $\int_a^b f$ *provided* we can find an antiderivative F such that $F'(x) = f(x)$ when $x \in [a, b]$. However, when we cannot find such an F, we may not be able to use the Fundamental Theorem. Nevertheless, *when f is continuous,* there are a number of techniques for approximating the Riemann integral $\int_a^b f$ by using sums that resemble the Riemann sums.

One very elementary procedure to obtain quick estimates of $\int_a^b f$, based on Theorem 7.1.4(c), is to note that if $g(x) \leq f(x) \leq h(x)$ for all $x \in [a, b]$, then

$$\int_a^b g \leq \int_a^b f \leq \int_a^b h.$$

If the integrals of g and h can be calculated, then we have bounds for $\int_a^b f$. Often these bounds are accurate enough for our needs.

For example, suppose we wish to estimate the value of $\int_0^1 e^{-x^2}\, dx$. It is easy to show that $e^{-x} \leq e^{-x^2} \leq 1$ for $x \in [0, 1]$, so that

$$\int_0^1 e^{-x}\, dx \leq \int_0^1 e^{-x^2}\, dx \leq \int_0^1 1\, dx.$$

Consequently, we have $1 - 1/e \leq \int_0^1 e^{-x^2}\, dx \leq 1$. If we use the mean of the bracketing values, we obtain the estimate $1 - 1/2e \approx 0.816$ for the integral with an error less than $1/2e < 0.184$. This estimate is crude, but it is obtained rapidly and may be quite satisfactory for our needs. If a better approximation is desired, we can attempt to find closer approximating functions g and h.

Taylor's Theorem 6.4.1 can be used to approximate e^{-x^2} by a polynomial. In using Taylor's Theorem, we must get bounds on the remainder term for our calculations to have significance. For example, if we apply Taylor's Theorem to e^{-y} for $0 \leq y \leq 1$, we get

$$e^{-y} = 1 - y + \tfrac{1}{2}y^2 - \tfrac{1}{6}y^3 + R_3,$$

where $R_3 = y^4 e^{-c}/24$ where c is some number with $0 \leq c \leq 1$. Since we have no better information as to the location of c, we must be content with the estimate $0 \leq R_3 \leq y^4/24$. Hence we have

$$e^{-x^2} = 1 - x^2 + \tfrac{1}{2}x^4 - \tfrac{1}{6}x^6 + R_3,$$

where $0 \leq R_3 \leq x^8/24$, for $x \in [0, 1]$. Therefore, we obtain

$$\int_0^1 e^{-x^2} \, dx = \int_0^1 (1 - x^2 + \tfrac{1}{2}x^4 - \tfrac{1}{6}x^6) \, dx + \int_0^1 R_3 \, dx$$

$$= 1 - \tfrac{1}{3} + \tfrac{1}{10} - \tfrac{1}{42} + \int_0^1 R_3 \, dx.$$

Since we have $0 \leq \int_0^1 R_3 \, dx \leq \tfrac{1}{9 \cdot 24} = \tfrac{1}{216} < 0.005$, it follows that

$$\int_0^1 e^{-x^2} \, dx \approx \tfrac{26}{35} \ (\approx 0.7429),$$

with an error less than 0.005.

Equal Partitions

If $f : [a, b] \to \mathbb{R}$ is continuous, we know that its Riemann integral exists. To find an approximate value for this integral with the minimum amount of calculation, it is convenient to consider partitions $\mathcal{P}_n$ of $[a, b]$ into n *equal* subintervals having length $h_n := (b - a)/n$. Hence $\mathcal{P}_n$ is the partition:

$$a < a + h_n < a + 2h_n < \cdots < a + nh_n = b.$$

If we pick our tag points to be the *left endpoints* and the *right endpoints* of the subintervals, we obtain the n**th left approximation** given by

$$L_n(f) := h_n \sum_{k=0}^{n-1} f(a + kh_n),$$

and the n**th right approximation** given by

$$R_n(f) := h_n \sum_{k=1}^{n} f(a + kh_n).$$

It should be noted that it is almost as easy to evaluate *both* of these approximations as only one of them, since they differ only by the terms $f(a)$ and $f(b)$.

Unless we have reason to believe that one of $L_n(f)$ or $R_n(f)$ is closer to the actual value of the integral than the other one, we generally take their mean:

$$\tfrac{1}{2} \left(L_n(f) + R_n(f) \right),$$

which is readily seen to equal

(1) $$T_n(f) := h_n (\tfrac{1}{2} f(a) + \sum_{k=1}^{n-1} f(a + kh_n) + \tfrac{1}{2} f(b)),$$

as a reasonable approximation to $\int_a^b f$.

However, we note that if f is *increasing* on $[a, b]$, then it is clear from a sketch of the graph of f that

(2) $$L_n(f) \leq \int_a^b f \leq R_n(f).$$

In this case, we readily see that

$$\left| \int_a^b f - T_n(f) \right| \leq \tfrac{1}{2} \left(R_n(f) - L_n(f) \right)$$

$$= \tfrac{1}{2} h_n \left(f(b) - f(a) \right) = (f(b) - f(a)) \cdot \frac{(b - a)}{2n}.$$

An error estimate such as this is useful, since it gives an upper bound for the error of the approximation in terms of quantities that are known at the outset. In particular, it can be used to determine how large we should choose n in order to have an approximation that will be correct to within a specified error $\varepsilon > 0$.

The above discussion was valid for the case that f is increasing on $[a, b]$. If f is decreasing, then the inequalities in (2) should be reversed. We can summarize both cases in the following statement.

7.4.1 Theorem   *If $f : [a, b] \to \mathbb{R}$ is monotone and if $T_n(f)$ is given by (1), then*

$$(3) \qquad \left| \int_a^b f - T_n(f) \right| \le |f(b) - f(a)| \cdot \frac{(b - a)}{2n}.$$

7.4.2 Example   If $f(x) := e^{-x^2}$ on $[0, 1]$, then f is decreasing. It follows from (3) that if $n = 8$, then $|\int_0^1 e^{-x^2}\, dx - T_8(f)| \le (1 - e^{-1})/16 < 0.04$, and if $n = 16$, then $|\int_0^1 e^{-x^2}\, dx - T_{16}(f)| \le (1 - e^{-1})/32 < 0.02$. Actually, the approximation is considerable better, as we will see in Example 7.4.5.   □

The Trapezoidal Rule _____

The method of numerical integration called the "Trapezoidal Rule" is based on approximating the continuous function $f : [a, b] \to \mathbb{R}$ by a piecewise linear continuous function. Let $n \in \mathbb{N}$ and, as before, let $h_n := (b - a)/n$ and consider the partition $\mathcal{P}_n$. We approximate f by the piecewise linear function g_n that passes through the points $(a + kh_n,\, f(a + kh_n))$, where $k = 0, 1, \cdots, n$. It seems reasonable that the integral $\int_a^b f$ will be "approximately equal to" the integral $\int_a^b g_n$ when n is sufficiently large (provided that f is reasonably smooth).

Since the area of a trapezoid with horizontal base h and vertical sides l_1 and l_2 is known to be $\frac{1}{2} h(l_1 + l_2)$, we have

$$\int_{a+kh_n}^{a+(k+1)h_n} g_n = \tfrac{1}{2} h_n \cdot \left[f(a + kh_n) + f(a + (k + 1)h_n) \right],$$

for $k = 0, 1, \cdots, n - 1$. Summing these terms and noting that each partition point in $\mathcal{P}_n$ except a and b belongs to two adjacent subintervals, we obtain

$$\int_a^b g_n = h_n \left(\tfrac{1}{2} f(a) + f(a + h_n) + \cdots + f(a + (k - 1)h_n) + \tfrac{1}{2} f(b) \right).$$

But the term on the right is precisely $T_n(f)$, found in (1) as the mean of $L_n(f)$ and $R_n(f)$. We call $T_n(f)$ the *n*th **Trapezoidal Approximation** of f.

In Theorem 7.4.1 we obtained an error estimate in the case where f is monotone; we now state one without this restriction on f, but in terms of the second derivative f'' of f.

7.4.3 Theorem   *Let f, f' and f'' be continuous on $[a, b]$ and let $T_n(f)$ be the *n*th Trapezoidal Approximation (1). Then there exists $c \in [a, b]$ such that*

$$(4) \qquad T_n(f) - \int_a^b f = \frac{(b - a)h_n^2}{12} \cdot f''(c).$$

A proof of this result will be given in Appendix D; it depends on a number of results we have obtained in Chapters 5 and 6.

The equality (4) is interesting in that it can give both an upper bound and a lower bound for the difference $T_n(f) - \int_a^b f$. For example, if $f''(x) \geq A > 0$ for all $x \in [a, b]$, then (4) implies that this difference always exceeds $\frac{1}{12} A(b - a)h_n^2$. If we only have $f''(x) \geq 0$ for $x \in [a, b]$, which is the case when f is **convex** (= **concave upward**), then the Trapezoidal Approximation is always *too large*. The reader should draw a figure to visualize this.

However, it is usually the upper bound that is of greater interest.

7.4.4 Corollary   Let f, f' and f'' be continuous, and let $|f''(x)| \leq B_2$ for all $x \in [a, b]$. Then

$$(5) \qquad \left| T_n(f) - \int_a^b f \right| \leq \frac{(b - a)h_n^2}{12} \cdot B_2 = \frac{(b - a)^3}{12n^2} \cdot B_2.$$

When an upper bound B_2 can be found, (5) can be used to determine how large n must be chosen in order to be certain of a desired accuracy.

7.4.5 Example   If $f(x) := e^{-x^2}$ on $[0, 1]$, then a calculation shows that $f''(x) = 2e^{-x^2}(2x^2 - 1)$, so that we can take $B_2 = 2$. Thus, if $n = 8$, then

$$\left| T_8(f) - \int_0^1 f \right| \leq \frac{2}{12 \cdot 64} = \frac{1}{384} < 0.003.$$

On the other hand, if $n = 16$, then we have

$$\left| T_{16}(f) - \int_0^1 f \right| \leq \frac{2}{12 \cdot 256} = \frac{1}{1536} < 0.00066.$$

Thus, the accuracy in this case is considerably better than predicted in Example 7.4.2   □

The Midpoint Rule

One obvious method of approximating the integral of f is to take the Riemann sums evaluated at the *midpoints* of the subintervals. Thus, if $\mathcal{P}_n$ is the equally spaced partition given before, the **Midpoint Approximation** of f is given by

$$M_n(f) := h_n \left(f(a + \tfrac{1}{2}h_n) + f(a + \tfrac{3}{2}h_n) + \cdots + f(a(n - \tfrac{1}{2})h_n) \right)$$
$$(6) \qquad = h_n \sum_{k=1}^n f\left(a + (k - \tfrac{1}{2})h_n\right).$$

Another method might be to use piecewise linear functions that are *tangent* to the graph of f at the midpoints of these subintervals. At first glance, it seems as if we would need to know the slope of the tangent line to the graph of f at each of the midpoints $a + (k - \tfrac{1}{2}h_n)$ ($k = 1, 2, \cdots, n$). However, it is an exercise in geometry to show that the *area* of the trapezoid whose top is this tangent line at the midpoint $a + (k - \tfrac{1}{2})h_n$ is equal to the *area* of the rectangle whose height is $f(a + (k - \tfrac{1}{2})h_n)$. (See Figure 7.4.1.) Thus, this area is given by (6), and the "Tangent Trapezoid Rule" turns out to be the same as the "Midpoint Rule". We now state a theorem showing that the Midpoint Rule gives better accuracy than the Trapezoidal Rule by a factor of 2.

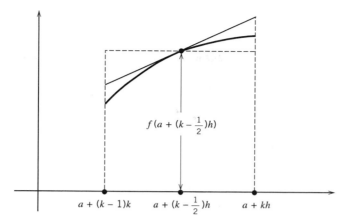

$$f\left(a + (k - \tfrac{1}{2})h\right)$$

$$a + (k - 1)k \qquad a + (k - \tfrac{1}{2})h \qquad a + kh$$

Figure 7.4.1   The tangent trapezoid.

7.4.6 Theorem   *Let f, f', and f'' be continuous on $[a, b]$ and let $M_n(f)$ be the nth Midpoint Approximation (6). Then there exists $\gamma \in [a, b]$ such that*

(7)
$$\int_a^b f - M_n(f) = \frac{(b - a)h_n^2}{24} \cdot f''(\gamma).$$

The proof of this result is in Appendix D.

As in the case with Theorem 7.4.3, formula (7) can be used to give both an upper bound and a lower bound for the difference $\int_a^b f - M_n(f)$, although it is an upper bound that is usually of greater interest. In contrast with the Trapezoidal Rule, if the function is **convex,** then the Midpoint Approximation is always too small.

The next result is parallel to Corollary 7.4.4.

7.4.7 Corollary   *Let f, f', and f'' be continuous, and let $|f''(x)| \leq B_2$ for all $x \in [a, b]$. Then*

(8)
$$\left| M_n(f) - \int_a^b f \right| \leq \frac{(b - a)h_n^2}{24} \cdot B_2 = \frac{(b - a)^3}{24n^2} \cdot B_2.$$

Simpson's Rule _____

The final approximation procedure that we will consider usually gives a better approximation than either the Trapezoidal or the Midpoint Rule and requires essentially no extra calculation. However, the convexity (or the concavity) of f does not give any information about the error for this method.

Whereas the Trapezoidal and Midpoint Rules were based on the approximation of f by piecewise linear functions, Simpson's Rule approximates the graph of f by parabolic arcs. To help motivate the formula, the reader may show that if three points

$$(-h, y_0), \quad (0, y_1), \quad \text{and} \quad (h, y_2)$$

are given, then the quadratic function $q(x) := Ax^2 + Bx + C$ that passes through these points has the property that

$$\int_{-h}^h q = \tfrac{1}{3}h(y_0 + 4y_1 + y_2).$$

Now let f be a continuous function on $[a, b]$ and let $n \in \mathbb{N}$ be *even*, and let $h_n :=$ $(b - a)/n$. On each "double subinterval"

$$[a, a + 2h_n], \quad [a + 2h_n, a + 4h_n], \quad \cdots, \quad [b - 2h_n, b],$$

we approximate f by $n/2$ quadratic functions that agree with f at the points

$$y_0 := f(a), \quad y_1 := f(a + h_n), \quad y_2 := f(a + 2h_n), \quad \cdots, \quad y_n := f(b).$$

These considerations lead to the *n*th **Simpson Approximation,** defined by

$$S_n(f) := \tfrac{1}{3} h_n \big(f(a) + 4f(a + h_n) + 2f(a + 2h_n) + 4f(a + 3h_n)$$

(9)
$$\qquad + 2f(a + 4h_n) + \cdots + 2f(b - 2h_n) + 4f(b - h_n) + f(b) \big).$$

Note that the coefficients of the values of f at the $n + 1$ partition points follow the pattern $1, 4, 2, 4, 2, \cdots, 4, 2, 4, 1$.

We now state a theorem that gives an estimate about the accuracy of the Simpson approximation; it involves the *fourth derivative* of f.

7.4.8 Theorem   Let $f, f', f'', f^{(3)}$ and $f^{(4)}$ be continuous on $[a, b]$ and let $n \in \mathbb{N}$ be *even*. If $S_n(f)$ is the nth Simpson Approximation (9), then there exists $c \in [a, b]$ such that

(10)
$$S_n(f) - \int_a^b f = \frac{(b - a)h_n^4}{180} \cdot f^{(4)}(c).$$

A proof of this result is given in Appendix D.

The next result is parallel to Corollaries 7.4.4 and 7.4.7.

7.4.9 Corollary   Let $f, f', f'', f^{(3)}$ and $f^{(4)}$ be continuous on $[a, b]$ and let $|f^{(4)}(x)| \le B_4$ for all $x \in [a, b]$. Then

(11)
$$\left| S_n(f) - \int_a^b f \right| \le \frac{(b - a)h_n^4}{180} \cdot B_4 = \frac{(b - a)^5}{180n^4} \cdot B_4.$$

Successful use of the estimate (11) depends on being able to find an upper bound for the fourth derivative.

7.4.10 Example   If $f(x) := 4e^{-x^2}$ on $[0, 1]$ then a calculation shows that

$$f^{(4)}(x) = 4e^{-x^2}(4x^4 - 12x^2 + 3),$$

whence it follows that $|f^{(4)}(x)| \le 20$ for $x \in [0, 1]$, so we can take $B_4 = 20$. It follows from (11) that if $n = 8$ then

$$\left| S_8(f) - \int_0^1 f \right| \le \frac{1}{180 \cdot 8^4} \cdot 20 = \frac{1}{36,864} < 0.000\,03$$

and that if $n = 16$ then

$$\left| S_{16}(f) - \int_0^1 f \right| \le \frac{1}{589,824} < 0.000\,001\,7. \qquad \square$$

Remark   The nth Midpoint Approximation $M_n(f)$ can be used to "step up" to the $(2n)$th Trapezoidal and Simpson Approximations by using the formulas

$$T_{2n}(f) = \tfrac{1}{2}M_n(f) + \tfrac{1}{2}T_n(f) \qquad \text{and} \qquad S_{2n}(f) = \tfrac{2}{3}M_n(f) + \tfrac{1}{3}T_n(f),$$

that are given in the Exercises. Thus once the initial Trapezoidal Approximation $T_1 = T_1(f)$ has been calculated, only the Midpoint Approximations $M_n = M_n(f)$ need be found. That is, we employ the following sequence of calculations:

$$T_1 = \tfrac{1}{2}(b-a)\big(f(a) + f(b)\big);$$

$$M_1 = (b-a)f\left(\tfrac{1}{2}(a+b)\right), \qquad T_2 = \tfrac{1}{2}M_1 + \tfrac{1}{2}T_1, \qquad S_2 = \tfrac{2}{3}M_1 + \tfrac{1}{3}T_1;$$

$$M_2, \qquad T_4 = \tfrac{1}{2}M_2 + \tfrac{1}{2}T_2, \qquad S_4 = \tfrac{2}{3}M_2 + \tfrac{1}{3}T_2;$$

$$M_4, \qquad T_8 = \tfrac{1}{2}M_4 + \tfrac{1}{2}T_4, \qquad S_8 = \tfrac{2}{3}M_4 + \tfrac{1}{3}T_4;$$

$$\cdots \qquad \cdots \qquad \cdots \qquad \cdots \qquad \cdots$$

Exercises for Section 7.4

1. Use the Trapezoidal Approximation with $n = 4$ to evaluate $\ln 2 = \int_1^2 (1/x)\,dx$. Show that $0.6866 \le \ln 2 \le 0.6958$ and that

    $$0.0013 < \frac{1}{768} \le T_4 - \ln 2 \le \frac{1}{96} < 0.0105.$$

2. Use the Simpson Approximation with $n = 4$ to evaluate $\ln 2 = \int_1^2 (1/x)\,dx$. Show that $0.6927 \le \ln 2 \le 0.6933$ and that

    $$0.000\,016 < \frac{1}{2^5} \cdot \frac{1}{1920} \le S_4 - \ln 2 \le \frac{1}{1920} < 0.000\,521.$$

3. Let $f(x) := (1 + x^2)^{-1}$ for $x \in [0, 1]$. Show that $f''(x) = 2(3x^2 - 1)(1 + x^2)^{-3}$ and that $|f''(x)| \le 2$ for $x \in [0, 1]$. Use the Trapezoidal Approximation with $n = 4$ to evaluate $\pi/4 = \int_0^1 f(x)\,dx$. Show that $|T_4(f) - (\pi/4)| \le 1/96 < 0.0105$.

4. If the Trapezoidal Approximation $T_n(f)$ is used to approximate $\pi/4$ as in Exercise 3, show that we must take $n \ge 409$ in order to be sure that the error is less than 10^{-6}.

5. Let f be as in Exercise 3. Show that $f^{(4)}(x) = 24(5x^4 - 10x^2 + 1)(1 + x^2)^{-5}$ and that $|f^{(4)}(x)| \le 96$ for $x \in [0, 1]$. Use Simpson's Approximation with $n = 4$ to evaluate $\pi/4$. Show that $|S_4(f) - (\pi/4)| \le 1/480 < 0.0021$.

6. If the Simpson Approximation $S_n(f)$ is used to approximate $\pi/4$ as in Exercise 5, show that we must take $n \ge 28$ in order to be sure that the error is less than 10^{-6}.

7. If p is a polynomial of degree at most 3, show that the Simpson Approximations are exact.

8. Show that if $f''(x) \ge 0$ on $[a, b]$ (that is, if f is convex on $[a, b]$), then for any natural numbers m, n we have $M_n(f) \le \int_a^b f(x)\,dx \le T_m(f)$. If $f''(x) \le 0$ on $[a, b]$, this inequality is reversed.

9. Show that $T_{2n}(f) = \tfrac{1}{2}[M_n(f) + T_n(f)]$.

10. Show that $S_{2n}(f) = \tfrac{2}{3}M_n(f) + \tfrac{1}{3}T_n(f)$.

11. Show that one has the estimate $\left|S_n(f) - \int_a^b f(x)\,dx\right| \le \left[(b-a)^2/18n^2\right]B_2$, where $B_2 \ge |f''(x)|$ for all $x \in [a, b]$.

12. Note that $\int_0^1 (1 - x^2)^{1/2}\, dx = \pi/4$. Explain why the error estimates given by formulas (4), (7), and (10) cannot be used. Show that if $h(x) = (1 - x^2)^{1/2}$ for x in $[0, 1]$, then $T_n(h) \le \pi/4 \le M_n(h)$. Calculate $M_8(h)$ and $T_8(h)$.

13. If h is as in Exercise 12, explain why $K := \int_0^{1/\sqrt{2}} h(x)\, dx = \pi/8 + 1/4$. Show that $\left|h''(x)\right| \le 2^{3/2}$ and that $\left|h^{(4)}(x)\right| \le 9 \cdot 2^{7/2}$ for $x \in [0, 1/\sqrt{2}]$. Show that $|K - T_n(h)| \le 1/12n^2$ and that $\left|K - S_n(h)\right| \le 1/10n^4$. Use these results to calculate π.

In Exercises 14–20, approximate the indicated integrals, giving estimates for the error. Use a calculator to obtain a high degree of precision.

14. $\displaystyle \int_0^2 (1 + x^4)^{1/2}\, dx.$

15. $\displaystyle \int_0^2 (4 + x^3)^{1/2}\, dx.$

16. $\displaystyle \int_0^1 \frac{dx}{1 + x^3}.$

17. $\displaystyle \int_0^\pi \frac{\sin x}{x}\, dx.$

18. $\displaystyle \int_0^{\pi/2} \frac{dx}{1 + \sin x}.$

19. $\displaystyle \int_0^{\pi/2} \sqrt{\sin x}\, dx.$

20. $\displaystyle \int_0^1 \cos\left(x^2\right)\, dx.$

CHAPTER 8

SEQUENCES OF FUNCTIONS

In previous chapters we have often made use of sequences of real numbers. In this chapter we shall consider sequences whose terms are *functions* rather than real numbers. Sequences of functions arise naturally in real analysis and are especially useful in obtaining approximations to a given function and defining new functions from known ones.

In Section 8.1 we will introduce two different notions of convergence for a sequence of functions: pointwise convergence and uniform convergence. The latter type of convergence is very important, and will be the main focus of our attention. The reason for this focus is the fact that, as is shown in Section 8.2, uniform convergence "preserves" certain properties in the sense that if each term of a uniformly convergent sequence of functions possesses these properties, then the limit function also possesses the properties.

In Section 8.3 we will apply the concept of uniform convergence to define and derive the basic properties of the exponential and logarithmic functions. Section 8.4 is devoted to a similar treatment of the trigonometric functions.

Section 8.1 Pointwise and Uniform Convergence

Let $A \subseteq \mathbb{R}$ be given and suppose that for each $n \in \mathbb{N}$ there is a function $f_n : A \to \mathbb{R}$; we shall say that (f_n) is a **sequence of functions** on A to $\mathbb{R}$. Clearly, for each $x \in A$, such a sequence gives rise to a sequence of real numbers, namely the sequence

$$(1) \qquad\qquad (f_n(x)),$$

obtained by evaluating each of the functions at the point x. For certain values of $x \in A$ the sequence (1) may converge, and for other values of $x \in A$ this sequence may diverge. For each $x \in A$ for which the sequence (1) converges, there is a uniquely determined real number $\lim(f_n(x))$. In general, the value of this limit, when it exists, will depend on the choice of the point $x \in A$. Thus, there arises in this way a function whose domain consists of all numbers $x \in A$ for which the sequence (1) converges.

8.1.1 Definition Let (f_n) be a sequence of functions on $A \subseteq \mathbb{R}$ to $\mathbb{R}$, let $A_0 \subseteq A$, and let $f : A_0 \to \mathbb{R}$. We say that the **sequence** (f_n) **converges on** A_0 **to** f if, for each $x \in A_0$, the sequence $(f_n(x))$ converges to $f(x)$ in $\mathbb{R}$. In this case we call f the **limit on** A_0 **of the sequence** (f_n). When such a function f exists, we say that the sequence (f_n) **is convergent on** A_0, or that (f_n) **converges pointwise on** A_0.

It follows from Theorem 3.1.4 that, except for a possible modification of the domain A_0, the limit function is uniquely determined. Ordinarily we choose A_0 to be the largest set possible; that is, we take A_0 to be the set of all $x \in A$ for which the sequence (1) is convergent in $\mathbb{R}$.

In order to symbolize that the sequence (f_n) converges on A_0 to f, we sometimes write

$$f = \lim(f_n) \quad \text{on} \quad A_0, \quad \text{or} \quad f_n \to f \quad \text{on} \quad A_0.$$

Sometimes, when f_n and f are given by formulas, we write

$$f(x) = \lim f_n(x) \quad \text{for} \quad x \in A_0, \quad \text{or} \quad f_n(x) \to f(x) \quad \text{for} \quad x \in A_0.$$

8.1.2 Examples   (a) $\lim(x/n) = 0$ for $x \in \mathbb{R}$.

For $n \in \mathbb{N}$, let $f_n(x) := x/n$ and let $f(x) := 0$ for $x \in \mathbb{R}$. By Example 3.1.6(a), we have $\lim(1/n) = 0$. Hence it follows from Theorem 3.2.3 that

$$\lim(f_n(x)) = \lim(x/n) = x \lim(1/n) = x \cdot 0 = 0$$

for all $x \in \mathbb{R}$. (See Figure 8.1.1.)

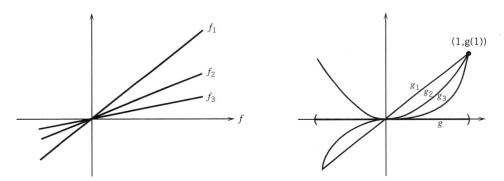

Figure 8.1.1 $f_n(x) = x/n$.   **Figure 8.1.2** $g_n(x) = x^n$.

(b) $\lim(x^n)$.

Let $g_n(x) := x^n$ for $x \in \mathbb{R}$, $n \in \mathbb{N}$. (See Figure 8.1.2.) Clearly, if $x = 1$, then the sequence $(g_n(1)) = (1)$ converges to 1. It follows from Example 3.1.11(b) that $\lim(x^n) = 0$ for $0 \le x < 1$ and it is readily seen that this is also true for $-1 < x < 0$. If $x = -1$, then $g_n(-1) = (-1)^n$, and it was seen in Example 3.2.8(b) that the sequence is divergent. Similarly, if $|x| > 1$, then the sequence (x^n) is not bounded, and so it is not convergent in $\mathbb{R}$. We conclude that if

$$g(x) := \begin{cases} 0 & \text{for} \quad -1 < x < 1, \\ 1 & \text{for} \quad x = 1, \end{cases}$$

then the sequence (g_n) converges to g on the set $(-1, 1]$.

(c) $\lim\left((x^2 + nx)/n\right) = x$ for $x \in \mathbb{R}$.

Let $h_n(x) := (x^2 + nx)/n$ for $x \in \mathbb{R}$, $n \in \mathbb{N}$, and let $h(x) := x$ for $x \in \mathbb{R}$. (See Figure 8.1.3.) Since we have $h_n(x) = (x^2/n) + x$, it follows from Example 3.1.6(a) and Theorem 3.2.3 that $h_n(x) \to x = h(x)$ for all $x \in \mathbb{R}$.

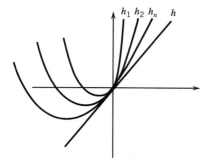

Figure 8.1.3 $h_n(x) = (x^2 + nx)/n$,

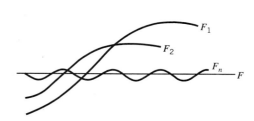

Figure 8.1.4 $F_n(x) = \sin(nx + n)/n$.

(d) $\lim\big((1/n)\sin(nx + n)\big) = 0$ for $x \in \mathbb{R}$.

Let $F_n(x) := (1/n)\sin(nx + n)$ for $x \in \mathbb{R}$, $n \in \mathbb{N}$, and let $F(x) := 0$ for $x \in \mathbb{R}$. (See Figure 8.1.4.) Since $|\sin y| \leq 1$ for all $y \in \mathbb{R}$ we have

(2)
$$\big|F_n(x) - F(x)\big| = \left|\frac{1}{n}\sin(nx + n)\right| \leq \frac{1}{n}$$

for all $x \in \mathbb{R}$. Therefore it follows that $\lim(F_n(x)) = 0 = F(x)$ for all $x \in \mathbb{R}$. The reader should note that, given any $\varepsilon > 0$, if n is sufficiently large, then $|F_n(x) - F(x)| < \varepsilon$ for all values of x simultaneously!   □

Partly to reinforce Definition 8.1.1 and partly to prepare the way for the important notion of uniform convergence, we reformulate Definition 8.1.1 as follows.

8.1.3 Lemma   *A sequence* (f_n) *of functions on* $A \subseteq \mathbb{R}$ *to* $\mathbb{R}$ *converges to a function* $f : A_0 \to \mathbb{R}$ *on* A_0 *if and only if for each* $\varepsilon > 0$ *and each* $x \in A_0$ *there is a natural number* $K(\varepsilon, x)$ *such that if* $n \geq K(\varepsilon, x)$, *then*

(3)
$$\big|f_n(x) - f(x)\big| < \varepsilon.$$

We leave it to the reader to show that this is equivalent to Definition 8.1.1. We wish to emphasize that the value of $K(\varepsilon, x)$ will depend, in general, on *both* $\varepsilon > 0$ and $x \in A_0$. The reader should confirm the fact that in Examples 8.1.2(a–c), the value of $K(\varepsilon, x)$ required to obtain an inequality such as (3) does depend on both $\varepsilon > 0$ and $x \in A_0$. The intuitive reason for this is that the convergence of the sequence is "significantly faster" at some points than it is at others. However, in Example 8.1.2(d), as we have seen in inequality (2), if we choose n sufficiently large, we can make $|F_n(x) - F(x)| < \varepsilon$ for *all* values of $x \in \mathbb{R}$. It is precisely this rather subtle difference that distinguishes between the notion of the "pointwise convergence" of a sequence of functions (as defined in Definition 8.1.1) and the notion of "uniform convergence".

Uniform Convergence _____

8.1.4 Definition   A sequence (f_n) of functions on $A \subseteq \mathbb{R}$ to $\mathbb{R}$ **converges uniformly on** $A_0 \subseteq A$ to a function $f : A_0 \to \mathbb{R}$ if for each $\varepsilon > 0$ there is a natural number $K(\varepsilon)$ (depending on ε but **not** on $x \in A_0$) such that if $n \geq K(\varepsilon)$, then

(4)
$$\big|f_n(x) - f(x)\big| < \varepsilon \qquad \text{for all} \quad x \in A_0.$$

In this case we say that the sequence (f_n) is **uniformly convergent on** A_0. Sometimes we write

$$f_n \rightrightarrows f \quad \text{on} \quad A_0, \qquad \text{or} \qquad f_n(x) \rightrightarrows f(x) \qquad \text{for} \quad x \in A_0.$$

It is an immediate consequence of the definitions that if the sequence (f_n) is uniformly convergent on A_0 to f, then this sequence also converges pointwise on A_0 to f in the sense of Definition 8.1.1. That the converse is not always true is seen by a careful examination of Examples 8.1.2(a–c); other examples will be given below.

It is sometimes useful to have the following necessary and sufficient condition for a sequence (f_n) to *fail* to converge uniformly on A_0 to f.

8.1.5 Lemma   *A sequence* (f_n) *of functions on* $A \subseteq \mathbb{R}$ *to* $\mathbb{R}$ *does not converge uniformly on* $A_0 \subseteq A$ *to a function* $f : A_0 \to \mathbb{R}$ *if and only if for some* $\varepsilon_0 > 0$ *there is a subsequence* (f_{n_k}) *of* (f_n) *and a sequence* (x_k) *in* A_0 *such that*

$$(5) \qquad \left| f_{n_k}(x_k) - f(x_k) \right| \geq \varepsilon_0 \qquad \text{for all} \quad k \in \mathbb{N}.$$

The proof of this result requires only that the reader negate Definition 8.1.4; we leave this to the reader as an important exercise. We now show how this result can be used.

8.1.6 Examples   **(a)**   Consider Example 8.1.2(a). If we let $n_k := k$ and $x_k := k$, then $f_{n_k}(x_k) = 1$ so that $|f_{n_k}(x_k) - f(x_k)| = |1 - 0| = 1$. Therefore the sequence (f_n) does not converge uniformly on $\mathbb{R}$ to f.

(b)   Consider Example 8.1.2(b). If $n_k := k$ and $x_k := \left(\frac{1}{2}\right)^{1/k}$, then

$$\left| g_{n_k}(x_k) - g(x_k) \right| = \left| \tfrac{1}{2} - 0 \right| = \tfrac{1}{2}.$$

Therefore the sequence (g_n) does not converge uniformly on $(-1, 1]$ to g.

(c)   Consider Example 8.1.2(c). If $n_k := k$ and $x_k := -k$, then $h_{n_k}(x_k) = 0$ and $h(x_k) = -k$ so that $|h_{n_k}(x_k) - h(x_k)| = k$. Therefore the sequence (h_n) does not converge uniformly on $\mathbb{R}$ to h. $\square$

The Uniform Norm

In discussing uniform convergence, it is often convenient to use the notion of the uniform norm on a set of bounded functions.

8.1.7 Definition   If $A \subseteq \mathbb{R}$ and $\varphi : A \to \mathbb{R}$ is a function, we say that φ is **bounded on** A if the set $\varphi(A)$ is a bounded subset of $\mathbb{R}$. If φ is bounded we define the **uniform norm of** φ **on** A by

$$(6) \qquad \|\varphi\|_A := \sup\{|\varphi(x)| : x \in A\}.$$

Note that it follows that if $\varepsilon > 0$, then

$$(7) \qquad \|\varphi\|_A \leq \varepsilon \qquad \Longleftrightarrow \qquad |\varphi(x)| \leq \varepsilon \qquad \text{for all} \quad x \in A.$$

8.1.8 Lemma   *A sequence* (f_n) *of bounded functions on* $A \subseteq \mathbb{R}$ *converges uniformly on* A *to* f *if and only if* $\|f_n - f\|_A \to 0$.

Proof. ($\Rightarrow$) If (f_n) converges uniformly on A to f, then by Definition 8.1.4, given any $\varepsilon > 0$ there exists $K(\varepsilon)$ such that if $n \geq K(\varepsilon)$ and $x \in A$ then

$$\left| f_n(x) - f(x) \right| \leq \varepsilon.$$

From the definition of supremum, it follows that $\| f_n - f \|_A \leq \varepsilon$ whenever $n \geq K(\varepsilon)$. Since $\varepsilon > 0$ is arbitrary this implies that $\| f_n - f \|_A \to 0$.

($\Leftarrow$) If $\| f_n - f \|_A \to 0$, then given $\varepsilon > 0$ there is a natural number $H(\varepsilon)$ such that if $n \geq H(\varepsilon)$ then $\| f_n - f \|_A \leq \varepsilon$. It follows from (7) that $|f_n(x) - f(x)| \leq \varepsilon$ for all $n \geq H(\varepsilon)$ and $x \in A$. Therefore (f_n) converges uniformly on A to f.　　　　　Q.E.D.

We now illustrate the use of Lemma 8.1.8 as a tool in examining a sequence of bounded functions for uniform convergence.

8.1.9 Examples　**(a)**　We cannot apply Lemma 8.1.8 to the sequence in Example 8.1.2(a) since the function $f_n(x) - f(x) = x/n$ is not bounded on $\mathbb{R}$.

For the sake of illustration, let $A := [0, 1]$. Although the sequence (x/n) did not converge uniformly on $\mathbb{R}$ to the zero function, we shall show that the convergence is uniform on A. To see this, we observe that

$$\| f_n - f \|_A = \sup \{ |x/n - 0| : 0 \leq x \leq 1 \} = \frac{1}{n}$$

so that $\| f_n - f \|_A \to 0$. Therefore (f_n) is uniformly convergent on A to f.

(b)　Let $g_n(x) := x^n$ for $x \in A := [0, 1]$ and $n \in \mathbb{N}$, and let $g(x) := 0$ for $0 \leq x < 1$ and $g(1) := 1$. The functions $g_n(x) - g(x)$ are bounded on A and

$$\| g_n - g \|_A = \sup \left\{ \begin{matrix} x^n & \text{for} & 0 \leq x < 1 \\ 0 & \text{for} & x = 1 \end{matrix} \right\} = 1$$

for any $n \in \mathbb{N}$. Since $\| g_n - g \|_A$ does *not* converge to 0, we infer that the sequence (g_n) does *not* converge uniformly on A to g.

(c)　We cannot apply Lemma 8.1.8 to the sequence in Example 8.1.2(c) since the function $h_n(x) - h(x) = x^2/n$ is not bounded on $\mathbb{R}$.

Instead, let $A := [0, 8]$ and consider

$$\| h_n - h \|_A = \sup \{ x^2/n : 0 \leq x \leq 8 \} = 64/n.$$

Therefore, the sequence (h_n) converges uniformly on A to h.

(d)　If we refer to Example 8.1.2(d), we see from (2) that $\| F_n - F \|_{\mathbb{R}} \leq 1/n$. Hence (F_n) converges uniformly on $\mathbb{R}$ to F.

(e)　Let $G(x) := x^n(1 - x)$ for $x \in A := [0, 1]$. Then the sequence $(G_n(x))$ converges to $G(x) := 0$ for each $x \in A$. To calculate the uniform norm of $G_n - G = G_n$ on A, we find the derivative and solve

$$G_n'(x) = x^{n-1}\big(n - (n+1)x\big) = 0$$

to obtain the point $x_n := n/(n+1)$. This is an interior point of $[0, 1]$, and it is easily verified by using the First Derivative Test 6.2.8 that G_n attains a maximum on $[0, 1]$ at x_n. Therefore, we obtain

$$\| G_n \|_A = G_n(x_n) = (1 + 1/n)^{-n} \cdot \frac{1}{n+1},$$

which converges to $(1/e) \cdot 0 = 0$. Thus we see that convergence is uniform on A.　　　□

By making use of the uniform norm, we can obtain a necessary and sufficient condition for uniform convergence that is often useful.

8.1.10 Cauchy Criterion for Uniform Convergence   *Let* (f_n) *be a sequence of bounded functions on* $A \subseteq \mathbb{R}$. *Then this sequence converges uniformly on* A *to a bounded function* f *if and only if for each* $\varepsilon > 0$ *there is a number* $H(\varepsilon)$ *in* $\mathbb{N}$ *such that for all* $m, n \geq H(\varepsilon)$, *then* $\| f_m - f_n \|_A \leq \varepsilon$.

Proof.   ($\Rightarrow$) If $f_n \rightrightarrows f$ on A, then given $\varepsilon > 0$ there exists a natural number $K\left(\frac{1}{2}\varepsilon\right)$ such that if $n \geq K\left(\frac{1}{2}\varepsilon\right)$ then $\| f_n - f \|_A \leq \frac{1}{2}\varepsilon$. Hence, if both $m, n \geq K\left(\frac{1}{2}\varepsilon\right)$, then we conclude that

$$\left| f_m(x) - f_n(x) \right| \leq \left| f_m(x) - f(x) \right| + \left| f_n(x) - f(x) \right| \leq \tfrac{1}{2}\varepsilon + \tfrac{1}{2}\varepsilon = \varepsilon$$

for all $x \in A$. Therefore $\| f_m - f_n \|_A \leq \varepsilon$, for $m, n \geq K\left(\frac{1}{2}\varepsilon\right) =: H(\varepsilon)$.

($\Leftarrow$) Conversely, suppose that for $\varepsilon > 0$ there is $H(\varepsilon)$ such that if $m, n \geq H(\varepsilon)$, then $\| f_m - f_n \|_A \leq \varepsilon$. Therefore, for each $x \in A$ we have

$$(8) \qquad \left| f_m(x) - f_n(x) \right| \leq \| f_m - f_n \|_A \leq \varepsilon \qquad \text{for} \quad m, n \geq H(\varepsilon).$$

It follows that $(f_n(x))$ is a Cauchy sequence in $\mathbb{R}$; therefore, by Theorem 3.5.5, it is a convergent sequence. We define $f : A \to \mathbb{R}$ by

$$f(x) := \lim(f_n(x)) \qquad \text{for} \quad x \in A.$$

If we let $n \to \infty$ in (8), it follows from Theorem 3.2.6 that for each $x \in A$ we have

$$\left| f_m(x) - f(x) \right| \leq \varepsilon \qquad \text{for} \quad m \geq H(\varepsilon).$$

Therefore the sequence (f_n) converges uniformly on A to f. Q.E.D.

Exercises for Section 8.1

1.   Show that $\lim(x/(x + n)) = 0$ for all $x \in \mathbb{R}$, $x \geq 0$.

2.   Show that $\lim(nx/(1 + n^2 x^2)) = 0$ for all $x \in \mathbb{R}$.

3.   Evaluate $\lim(nx/(1 + nx))$ for $x \in \mathbb{R}$, $x \geq 0$.

4.   Evaluate $\lim(x^n/(1 + x^n))$ for $x \in \mathbb{R}$, $x \geq 0$.

5.   Evaluate $\lim((\sin nx)/(1 + nx))$ for $x \in \mathbb{R}$, $x \geq 0$.

6.   Show that $\lim(\text{Arctan } nx) = (\pi/2)\text{sgn } x$ for $x \in \mathbb{R}$.

7.   Evaluate $\lim(e^{-nx})$ for $x \in \mathbb{R}$, $x \geq 0$.

8.   Show that $\lim(xe^{-nx}) = 0$ for $x \in \mathbb{R}$, $x \geq 0$.

9.   Show that $\lim(x^2 e^{-nx}) = 0$ and that $\lim(n^2 x^2 e^{-nx}) = 0$ for $x \in \mathbb{R}$, $x \geq 0$.

10. Show that $\lim\left((\cos \pi x)^{2n}\right)$ exists for all $x \in \mathbb{R}$. What is its limit?

11. Show that if $a > 0$, then the convergence of the sequence in Exercise 1 is uniform on the interval $[0, a]$, but is not uniform on the interval $[0, \infty)$.

12. Show that if $a > 0$, then the convergence of the sequence in Exercise 2 is uniform on the interval $[a, \infty)$, but is not uniform on the interval $[0, \infty)$.

13. Show that if $a > 0$, then the convergence of the sequence in Exercise 3 is uniform on the interval $[a, \infty)$, but is not uniform on the interval $[0, \infty)$.

14. Show that if $0 < b < 1$, then the convergence of the sequence in Exercise 4 is uniform on the interval $[0, b]$, but is not uniform on the interval $[0, 1]$.

15. Show that if $a > 0$, then the convergence of the sequence in Exercise 5 is uniform on the interval $[a, \infty)$, but is not uniform on the interval $[0, \infty)$.

16. Show that if $a > 0$, then the convergence of the sequence in Exercise 6 is uniform on the interval $[a, \infty)$, but is not uniform on the interval $(0, \infty)$.

17. Show that if $a > 0$, then the convergence of the sequence in Exercise 7 is uniform on the interval $[a, \infty)$, but is not uniform on the interval $[0, \infty)$.

18. Show that the convergence of the sequence in Exercise 8 is uniform on $[0, \infty)$.

19. Show that the sequence $(x^2 e^{-nx})$ converges uniformly on $[0, \infty)$.

20. Show that if $a > 0$, then the sequence $(n^2 x^2 e^{-nx})$ converges uniformly on the interval $[a, \infty)$, but that it does not converge uniformly on the interval $[0, \infty)$.

21. Show that if (f_n), (g_n) converge uniformly on the set A to f, g, respectively, then $(f_n + g_n)$ converges uniformly on A to $f + g$.

22. Show that if $f_n(x) := x + 1/n$ and $f(x) := x$ for $x \in \mathbb{R}$, then (f_n) converges uniformly on $\mathbb{R}$ to f, but the sequence (f_n^2) does not converge uniformly on $\mathbb{R}$. (Thus the product of uniformly convergent sequences of functions may not converge uniformly.)

23. Let (f_n), (g_n) be sequences of bounded functions on A that converge uniformly on A to f, g, respectively. Show that $(f_n g_n)$ converges uniformly on A to fg.

24. Let (f_n) be a sequence of functions that converges uniformly to f on A and that satisfies $|f_n(x)| \le M$ for all $n \in \mathbb{N}$ and all $x \in A$. If g is continuous on the interval $[-M, M]$, show that the sequence $(g \circ f_n)$ converges uniformly to $g \circ f$ on A.

Section 8.2 Interchange of Limits

It is often useful to know whether the limit of a sequence of functions is a continuous function, a differentiable function, or a Riemann integrable function. Unfortunately, it is not always the case that the limit of a sequence of functions possesses these useful properties.

8.2.1 Example **(a)** Let $g_n(x) := x^n$ for $x \in [0, 1]$ and $n \in \mathbb{N}$. Then, as we have noted in Example 8.1.2(b), the sequence (g_n) converges pointwise to the function

$$g(x) := \begin{cases} 0 & \text{for} \quad 0 \le x < 1, \\ 1 & \text{for} \quad x = 1. \end{cases}$$

Although all of the functions g_n are continuous at $x = 1$, the limit function g is not continuous at $x = 1$. Recall that it was shown in Example 8.1.6(b) that this sequence does not converge uniformly to g on $[0, 1]$.

(b) Each of the functions $g_n(x) = x^n$ in part (a) has a continuous derivative on $[0, 1]$. However, the limit function g does not have a derivative at $x = 1$, since it is not continuous at that point.

(c) Let $f_n : [0, 1] \to \mathbb{R}$ be defined for $n \ge 2$ by

$$f_n(x) := \begin{cases} n^2 x & \text{for} \quad 0 \le x \le 1/n, \\ -n^2 (x - 2/n) & \text{for} \quad 1/n \le x \le 2/n, \\ 0 & \text{for} \quad 2/n \le x \le 1. \end{cases}$$

(See Figure 8.2.1.) It is clear that each of the functions f_n is continuous on $[0, 1]$; hence it is Riemann integrable. Either by means of a direct calculation, or by referring to the significance of the integral as an area, we obtain

$$\int_0^1 f_n(x) \, dx = 1 \qquad \text{for} \quad n \geq 2.$$

The reader may show that $f_n(x) \to 0$ for all $x \in [0, 1]$; hence the limit function f vanishes identically and is continuous (and hence integrable), and $\int_0^1 f(x) \, dx = 0$. Therefore we have the uncomfortable situation that:

$$\int_0^1 f(x) \, dx = 0 \neq 1 = \lim \int_0^1 f_n(x) \, dx.$$

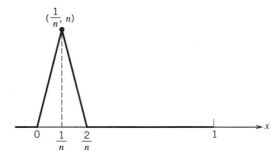

Figure 8.2.1   Example 8.2.1(c).

(d)   Those who consider the functions f_n in part (c) to be "artificial" may prefer to consider the sequence (h_n) defined by $h_n(x) := 2nxe^{-nx^2}$ for $x \in [0, 1]$, $n \in \mathbb{N}$. Since $h_n = H_n'$, where $H_n(x) := -e^{-nx^2}$, the Fundamental Theorem 7.3.1 gives

$$\int_0^1 h_n(x) \, dx = H_n(1) - H_n(0) = 1 - e^{-n}.$$

It is an exercise to show that $h(x) := \lim(h_n(x)) = 0$ for all $x \in [0, 1]$; hence

$$\int_0^1 h(x) \, dx \neq \lim \int_0^1 h_n(x) \, dx. \qquad \square$$

Although the extent of the discontinuity of the limit function in Example 8.2.1(a) is not very great, it is evident that more complicated examples can be constructed that will produce more extensive discontinuity. In any case, we must abandon the hope that the limit of a convergent sequence of continuous [respectively, differentiable, integrable] functions will be continuous [respectively, differentiable, integrable].

It will now be seen that the additional hypothesis of uniform convergence is sufficient to guarantee that the limit of a sequence of continuous functions is continuous. Similar results will also be established for sequences of differentiable and integrable functions.

Interchange of Limit and Continuity _____

8.2.2 Theorem   *Let (f_n) be a sequence of continuous functions on a set $A \subseteq \mathbb{R}$ and suppose that (f_n) converges uniformly on A to a function $f : A \to \mathbb{R}$. Then f is continuous on A.*

14. Show that if $0 < b < 1$, then the convergence of the sequence in Exercise 4 is uniform on the interval $[0, b]$, but is not uniform on the interval $[0, 1]$.

15. Show that if $a > 0$, then the convergence of the sequence in Exercise 5 is uniform on the interval $[a, \infty)$, but is not uniform on the interval $[0, \infty)$.

16. Show that if $a > 0$, then the convergence of the sequence in Exercise 6 is uniform on the interval $[a, \infty)$, but is not uniform on the interval $(0, \infty)$.

17. Show that if $a > 0$, then the convergence of the sequence in Exercise 7 is uniform on the interval $[a, \infty)$, but is not uniform on the interval $[0, \infty)$.

18. Show that the convergence of the sequence in Exercise 8 is uniform on $[0, \infty)$.

19. Show that the sequence $(x^2 e^{-nx})$ converges uniformly on $[0, \infty)$.

20. Show that if $a > 0$, then the sequence $(n^2 x^2 e^{-nx})$ converges uniformly on the interval $[a, \infty)$, but that it does not converge uniformly on the interval $[0, \infty)$.

21. Show that if (f_n), (g_n) converge uniformly on the set A to f, g, respectively, then $(f_n + g_n)$ converges uniformly on A to $f + g$.

22. Show that if $f_n(x) := x + 1/n$ and $f(x) := x$ for $x \in \mathbb{R}$, then (f_n) converges uniformly on $\mathbb{R}$ to f, but the sequence (f_n^2) does not converge uniformly on $\mathbb{R}$. (Thus the product of uniformly convergent sequences of functions may not converge uniformly.)

23. Let (f_n), (g_n) be sequences of bounded functions on A that converge uniformly on A to f, g, respectively. Show that $(f_n g_n)$ converges uniformly on A to fg.

24. Let (f_n) be a sequence of functions that converges uniformly to f on A and that satisfies $|f_n(x)| \leq M$ for all $n \in \mathbb{N}$ and all $x \in A$. If g is continuous on the interval $[-M, M]$, show that the sequence $(g \circ f_n)$ converges uniformly to $g \circ f$ on A.

Section 8.2   Interchange of Limits

It is often useful to know whether the limit of a sequence of functions is a continuous function, a differentiable function, or a Riemann integrable function. Unfortunately, it is not always the case that the limit of a sequence of functions possesses these useful properties.

8.2.1 Example   **(a)**   Let $g_n(x) := x^n$ for $x \in [0, 1]$ and $n \in \mathbb{N}$. Then, as we have noted in Example 8.1.2(b), the sequence (g_n) converges pointwise to the function

$$g(x) := \begin{cases} 0 & \text{for} \quad 0 \leq x < 1, \\ 1 & \text{for} \quad x = 1. \end{cases}$$

Although all of the functions g_n are continuous at $x = 1$, the limit function g is not continuous at $x = 1$. Recall that it was shown in Example 8.1.6(b) that this sequence does not converge uniformly to g on $[0, 1]$.

(b)   Each of the functions $g_n(x) = x^n$ in part (a) has a continuous derivative on $[0, 1]$. However, the limit function g does not have a derivative at $x = 1$, since it is not continuous at that point.

(c)   Let $f_n : [0, 1] \to \mathbb{R}$ be defined for $n \geq 2$ by

$$f_n(x) := \begin{cases} n^2 x & \text{for} \quad 0 \leq x \leq 1/n, \\ -n^2 (x - 2/n) & \text{for} \quad 1/n \leq x \leq 2/n, \\ 0 & \text{for} \quad 2/n \leq x \leq 1. \end{cases}$$

(See Figure 8.2.1.) It is clear that each of the functions f_n is continuous on $[0, 1]$; hence it is Riemann integrable. Either by means of a direct calculation, or by referring to the significance of the integral as an area, we obtain

$$\int_0^1 f_n(x)\,dx = 1 \qquad \text{for} \quad n \geq 2.$$

The reader may show that $f_n(x) \to 0$ for all $x \in [0, 1]$; hence the limit function f vanishes identically and is continuous (and hence integrable), and $\int_0^1 f(x)\,dx = 0$. Therefore we have the uncomfortable situation that:

$$\int_0^1 f(x)\,dx = 0 \neq 1 = \lim \int_0^1 f_n(x)\,dx.$$

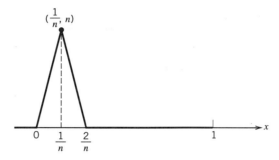

Figure 8.2.1   Example 8.2.1(c).

(d)   Those who consider the functions f_n in part (c) to be "artificial" may prefer to consider the sequence (h_n) defined by $h_n(x) := 2nxe^{-nx^2}$ for $x \in [0, 1]$, $n \in \mathbb{N}$. Since $h_n = H_n'$, where $H_n(x) := -e^{-nx^2}$, the Fundamental Theorem 7.3.1 gives

$$\int_0^1 h_n(x)\,dx = H_n(1) - H_n(0) = 1 - e^{-n}.$$

It is an exercise to show that $h(x) := \lim(h_n(x)) = 0$ for all $x \in [0, 1]$; hence

$$\int_0^1 h(x)\,dx \neq \lim \int_0^1 h_n(x)\,dx. \qquad \square$$

Although the extent of the discontinuity of the limit function in Example 8.2.1(a) is not very great, it is evident that more complicated examples can be constructed that will produce more extensive discontinuity. In any case, we must abandon the hope that the limit of a convergent sequence of continuous [respectively, differentiable, integrable] functions will be continuous [respectively, differentiable, integrable].

It will now be seen that the additional hypothesis of uniform convergence is sufficient to guarantee that the limit of a sequence of continuous functions is continuous. Similar results will also be established for sequences of differentiable and integrable functions.

Interchange of Limit and Continuity ─────────────────────────────

8.2.2 Theorem   Let (f_n) be a sequence of continuous functions on a set $A \subseteq \mathbb{R}$ and suppose that (f_n) converges uniformly on A to a function $f : A \to \mathbb{R}$. Then f is continuous on A.

Proof. By hypothesis, given $\varepsilon > 0$ there exists a natural number $H := H\left(\frac{1}{3}\varepsilon\right)$ such that if $n \geq H$ then $|f_n(x) - f(x)| < \frac{1}{3}\varepsilon$ for all $x \in A$. Let $c \in A$ be arbitrary; we will show that f is continuous at c. By the Triangle Inequality we have

$$|f(x) - f(c)| \leq |f(x) - f_H(x)| + |f_H(x) - f_H(c)| + |f_H(c) - f(c)|$$
$$\leq \tfrac{1}{3}\varepsilon + |f_H(x) - f_H(c)| + \tfrac{1}{3}\varepsilon.$$

Since f_H is continuous at c, there exists a number $\delta := \delta\left(\frac{1}{3}\varepsilon, c, f_H\right) > 0$ such that if $|x - c| < \delta$ and $x \in A$, then $|f_H(x) - f_H(c)| < \frac{1}{3}\varepsilon$. Therefore, if $|x - c| < \delta$ and $x \in A$, then we have $|f(x) - f(c)| < \varepsilon$. Since $\varepsilon > 0$ is arbitrary, this establishes the continuity of f at the arbitrary point $c \in A$. (See Figure 8.2.2.) Q.E.D.

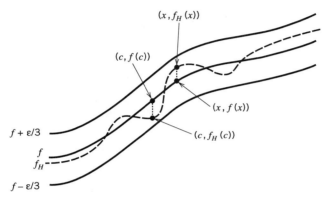

Figure 8.2.2 $|f(x) - f(c)| < \varepsilon$.

Remark Although the uniform convergence of the sequence of continuous functions is sufficient to guarantee the continuity of the limit function, it is *not* necessary. (See Exercise 2.)

Interchange of Limit and Derivative _____

We mentioned in Section 6.1 that Weierstrass showed that the function defined by the series

$$f(x) := \sum_{k=0}^{\infty} 2^{-k} \cos(3^k x)$$

is continuous at every point but does not have a derivative at any point in $\mathbb{R}$. By considering the partial sums of this series, we obtain a sequence of functions (f_n) that possess a derivative at every point and are uniformly convergent to f. Thus, even though the sequence of differentiable functions (f_n) is uniformly convergent, it does not follow that the limit function is differentiable.

We now show that if the *sequence of derivatives* (f_n') is uniformly convergent, then all is well. If one adds the hypothesis that the derivatives are continuous, then it is possible to give a short proof, based on the integral. (See Exercise 11.) However, if the derivatives are not assumed to be continuous, a somewhat more delicate argument is required.

8.2.3 Theorem *Let $J \subseteq \mathbb{R}$ be a bounded interval and let (f_n) be a sequence of functions on J to $\mathbb{R}$. Suppose that there exists $x_0 \in J$ such that $(f_n(x_0))$ converges, and that the sequence (f_n') of derivatives exists on J and converges uniformly on J to a function g.*

Then the sequence (f_n) converges uniformly on J to a function f that has a derivative at every point of J and $f' = g$.

Proof.   Let $a < b$ be the endpoints of J and let $x \in J$ be arbitrary. If $m, n \in \mathbb{N}$, we apply the Mean Value Theorem 6.2.4 to the difference $f_m - f_n$ on the interval with endpoints x_0, x. We conclude that there exists a point y (depending on m, n) such that

$$f_m(x) - f_n(x) = f_m(x_0) - f_n(x_0) + (x - x_0)\{f_m'(y) - f_n'(y)\}.$$

Hence we have

(1) $$\|f_m - f_n\|_J \leq |f_m(x_0) - f_n(x_0)| + (b - a)\|f_m' - f_n'\|_J.$$

From Theorem 8.1.10, it follows from (1) and the hypotheses that $(f_n(x_0))$ is convergent and that (f_n') is uniformly convergent on J, that (f_n) is uniformly convergent on J. We denote the limit of the sequence (f_n) by f. Since the f_n are all continuous and the convergence is uniform, it follows from Theorem 8.2.2 that f is continuous on J.

To establish the existence of the derivative of f at a point $c \in J$, we apply the Mean Value Theorem 6.2.4 to $f_m - f_n$ on an interval with end points c, x. We conclude that there exists a point z (depending on m, n) such that

$$\{f_m(x) - f_n(x)\} - \{f_m(c) - f_n(c)\} = (x - c)\{f_m'(z) - f_n'(z)\}.$$

Hence, if $x \neq c$, we have

$$\left| \frac{f_m(x) - f_m(c)}{x - c} - \frac{f_n(x) - f_n(c)}{x - c} \right| \leq \|f_m' - f_n'\|_J.$$

Since (f_n') converges uniformly on J, if $\varepsilon > 0$ is given there exists $H(\varepsilon)$ such that if $m, n \geq H(\varepsilon)$ and $x \neq c$, then

(2) $$\left| \frac{f_m(x) - f_m(c)}{x - c} - \frac{f_n(x) - f_n(c)}{x - c} \right| \leq \varepsilon.$$

If we take the limit in (2) with respect to m and use Theorem 3.2.6, we have

$$\left| \frac{f(x) - f(c)}{x - c} - \frac{f_n(x) - f_n(c)}{x - c} \right| \leq \varepsilon.$$

provided that $x \neq c, n \geq H(\varepsilon)$. Since $g(c) = \lim(f_n'(c))$, there exists $N(\varepsilon)$ such that if $n \geq N(\varepsilon)$, then $|f_n'(c) - g(c)| < \varepsilon$. Now let $K := \sup\{H(\varepsilon), N(\varepsilon)\}$. Since $f_K'(c)$ exists, there exists $\delta_K(\varepsilon) > 0$ such that if $0 < |x - c| < \delta_K(\varepsilon)$, then

$$\left| \frac{f_K(x) - f_K(c)}{x - c} - f_K'(c) \right| < \varepsilon.$$

Combining these inequalities, we conclude that if $0 < |x - c| < \delta_K(\varepsilon)$, then

$$\left| \frac{f(x) - f(c)}{x - c} - g(c) \right| < 3\varepsilon.$$

Since $\varepsilon > 0$ is arbitrary, this shows that $f'(c)$ exists and equals $g(c)$. Since $c \in J$ is arbitrary, we conclude that $f' = g$ on J.   Q.E.D.

Interchange of Limit and Integral

We have seen in Example 8.2.1(c) that if (f_n) is a sequence $\mathcal{R}[a, b]$ that converges on $[a, b]$ to a function f in $\mathcal{R}[a, b]$, then it need not happen that

(3) $$\int_a^b f = \lim_{n \to \infty} \int_a^b f_n.$$

We will now show that *uniform convergence* of the sequence is sufficient to guarantee that this equality holds.

8.2.4 Theorem   *Let (f_n) be a sequence of functions in $\mathcal{R}[a, b]$ and suppose that (f_n) converges **uniformly** on $[a, b]$ to f. Then $f \in \mathcal{R}[a, b]$ and (3) holds.*

Proof.   It follows from the Cauchy Criterion 8.1.10 that, given $\varepsilon > 0$ there exists $H(\varepsilon)$ such that if $m > n \geq H(\varepsilon)$ then

$$-\varepsilon \leq f_m(x) - f_n(x) \leq \varepsilon \qquad \text{for} \quad x \in [a.b].$$

Theorem 7.1.4 implies that

$$-\varepsilon(b - a) \leq \int_a^b f_m - \int_a^b f_n \leq \varepsilon(b - a).$$

Since $\varepsilon > 0$ is arbitrary, the sequence $(\int_a^b f_m)$ is a Cauchy sequence in $\mathbb{R}$ and therefore converges to some number, say $A \in \mathbb{R}$.

We now show $f \in \mathcal{R}[a, b]$ with integral A. If $\varepsilon > 0$ is given, let $K(\varepsilon)$ be such that if $m > K(\varepsilon)$, then $|f_m(x) - f(x)| < \varepsilon$ for all $x \in [a, b]$. If $\dot{\mathcal{P}} := \{([x_{i-1}, x_i], t_i)\}_{i=1}^n$ is any tagged partition of $[a, b]$ and if $m > K(\varepsilon)$, then

$$\left| S(f_m; \dot{\mathcal{P}}) - S(f; \dot{\mathcal{P}}) \right| = \left| \sum_{i=1}^n \{f_m(t_i) - f(t_i)\}(x_i - x_{i-1}) \right|$$

$$\leq \sum_{i=1}^n |f_m(t_i) - f(t_i)|(x_i - x_{i-1})$$

$$\leq \sum_{i=1}^n \varepsilon(x_i - x_{i-1}) = \varepsilon(b - a).$$

We now choose $r \geq K(\varepsilon)$ such that $|\int_a^b f_r - A| < \varepsilon$ and we let $\delta_{r,\varepsilon} > 0$ be such that $|\int_a^b f_r - S(f_r; \dot{\mathcal{P}})| < \varepsilon$ whenever $\|\dot{\mathcal{P}}\| < \delta_{r,\varepsilon}$. Then we have

$$\left| S(f; \dot{\mathcal{P}}) - A \right| \leq \left| S(f; \dot{\mathcal{P}}) - S(f_r; \dot{\mathcal{P}}) \right| + \left| S(f_r; \dot{\mathcal{P}}) - \int_a^b f_r \right| + \left| \int_a^b f_r - A \right|$$

$$\leq \varepsilon(b - a) + \varepsilon + \varepsilon = \varepsilon(b - a + 2).$$

But since $\varepsilon > 0$ is arbitrary, it follows that $f \in \mathcal{R}[a, b]$ and $\int_a^b f = A$.   Q.E.D.

The hypothesis of uniform convergence is a very stringent one and restricts the utility of this result. In Section 10.4 we will obtain some far-reaching generalizations of Theorem 8.2.4. For the present, we will state a result that does not require the uniformity of the convergence, but does require that the limit function be Riemann integrable. The proof is omitted.

8.2.5 Bounded Convergence Theorem   *Let (f_n) be a sequence in $\mathcal{R}[a, b]$ that converges on $[a, b]$ to a function $f \in \mathcal{R}[a, b]$. Suppose also that there exists $B > 0$ such that $|f_n(x)| \leq B$ for all $x \in [a, b], n \in \mathbb{N}$. Then equation (3) holds.*

Dini's Theorem ———

We will end this section with a famous theorem due to Ulisse Dini (1845–1918) which gives a partial converse to Theorem 8.2.2 when the sequence is monotone. We will present a proof using nonconstant gauges (see Section 5.5).

8.2.6 Dini's Theorem   *Suppose that* (f_n) *is a monotone sequence of continuous functions on* $I := [a, b]$ *that converges on* I *to a continuous function* f. *Then the convergence of the sequence is uniform.*

Proof.   We suppose that the sequence (f_n) is decreasing and let $g_m := f_m - f$. Then (g_m) is a decreasing sequence of continuous functions converging on I to the 0-function. We will show that the convergence is uniform on I.

Given $\varepsilon > 0$, $t \in I$, there exists $m_{\varepsilon,t} \in \mathbb{N}$ such that $0 \leq g_{m_{\varepsilon,t}}(t) < \varepsilon/2$. Since $g_{m_{\varepsilon,t}}$ is continuous at t, there exists $\delta_\varepsilon(t) > 0$ such that $0 \leq g_{m_{\varepsilon,t}}(x) < \varepsilon$ for all $x \in I$ satisfying $|x - t| \leq \delta_\varepsilon(t)$. Thus, δ_ε is a gauge on I, and if $\dot{\mathcal{P}} = \{(I_i, t_i)\}_{i=1}^n$ is a δ_ε-fine partition, we set $M_\varepsilon := \max\{m_{\varepsilon,t_1}, \cdots, m_{\varepsilon,t_n}\}$. If $m \geq M_\varepsilon$ and $x \in I$, then (by Lemma 5.5.3) there exists an index i with $|x - t_i| \leq \delta_\varepsilon(t_i)$ and hence

$$0 \leq g_m(x) \leq g_{m,t_i}(x) < \varepsilon.$$

Therefore, the sequence (g_m) converges uniformly to the 0-function.   Q.E.D.

It will be seen in the exercises that we cannot drop any one of the three hypotheses: (i) the functions f_n are continuous, (ii) the limit function f is continuous, (iii) I is a closed bounded interval.

Exercises for Section 8.2 ————————————————————————————————————

1. Show that the sequence $((x^n/(1 + x^n)))$ does not converge uniformly on $[0, 2]$ by showing that the limit function is not continuous on $[0, 2]$.

2. Prove that the sequence in Example 8.2.1(c) is an example of a sequence of continuous functions that converges nonuniformly to a continuous limit.

3. Construct a sequence of functions on $[0, 1]$ each of which is discontinuous at every point of $[0, 1]$ and which converges uniformly to a function that is continuous at every point.

4. Suppose (f_n) is a sequence of continuous functions on an interval I that converges uniformly on I to a function f. If $(x_n) \subseteq I$ converges to $x_0 \in I$, show that $\lim(f_n(x_n)) = f(x_0)$.

5. Let $f : \mathbb{R} \to \mathbb{R}$ be uniformly continuous on $\mathbb{R}$ and let $f_n(x) := f(x + 1/n)$ for $x \in \mathbb{R}$. Show that (f_n) converges uniformly on $\mathbb{R}$ to f.

6. Let $f_n(x) := 1/(1 + x)^n$ for $x \in [0, 1]$. Find the pointwise limit f of the sequence (f_n) on $[0, 1]$. Does (f_n) converge uniformly to f on $[0, 1]$?

7. Suppose the sequence (f_n) converges uniformly to f on the set A, and suppose that each f_n is bounded on A. (That is, for each n there is a constant M_n such that $|f_n(x)| \leq M_n$ for all $x \in A$.) Show that the function f is bounded on A.

8. Let $f_n(x) := nx/(1 + nx^2)$ for $x \in A := [0, \infty)$, Show that each f_n is bounded on A, but the pointwise limit f of the sequence is not bounded on A. Does (f_n) converge uniformly to f on A?

9. Let $f_n(x) := x^n/n$ for $x \in [0, 1]$. Show that the sequence (f_n) of differentiable functions converges uniformly to a differentiable function f on $[0, 1]$, and that the sequence (f_n') converges on $[0, 1]$ to a function g, but that $g(1) \neq f'(1)$.

10. Let $g_n(x) := e^{-nx}/n$ for $x \geq 0$, $n \in \mathbb{N}$. Examine the relation between $\lim(g_n)$ and $\lim(g_n')$.

11. Let $I := [a, b]$ and let (f_n) be a sequence of functions on $I \to \mathbb{R}$ that converges on I to f. Suppose that each derivative f_n' is continuous on I and that the sequence (f_n') is uniformly convergent to g on I. Prove that $f(x) - f(a) = \int_a^x g(t)\,dt$ and that $f'(x) = g(x)$ for all $x \in I$.

12. Show that $\lim \int_1^2 e^{-nx^2}\,dx = 0$.

13. If $a > 0$, show that $\lim \int_a^{\pi} (\sin nx)/(nx)\,dx = 0$. What happens if $a = 0$?

14. Let $f_n(x) := nx/(1 + nx)$ for $x \in [0, 1]$. Show that (f_n) converges nonuniformly to an integrable function f and that $\int_0^1 f(x)\,dx = \lim \int_0^1 f_n(x)\,dx$.

15. Let $g_n(x) := nx(1 - x)^n$ for $x \in [0, 1]$, $n \in \mathbb{N}$. Discuss the convergence of (g_n) and $(\int_0^1 g_n\,dx)$.

16. Let $\{r_1, r_2, \cdots, r_n \cdots\}$ be an enumeration of the rational numbers in $I := [0, 1]$, and let $f_n : I \to \mathbb{R}$ be defined to be 1 if $x = r_1, \cdots, r_n$ and equal to 0 otherwise. Show that f_n is Riemann integrable for each $n \in \mathbb{N}$, that $f_1(x) \leq f_2(x) \leq \cdots \leq f_n(x) \leq \cdots$ and that $f(x) := \lim(f_n(x))$ is the Dirichlet function, which is not Riemann integrable on $[0, 1]$.

17. Let $f_n(x) := 1$ for $x \in (0, 1/n)$ and $f_n(x) := 0$ elsewhere in $[0, 1]$. Show that (f_n) is a decreasing sequence of discontinuous functions that converges to a continuous limit function, but the convergence is not uniform on $[0, 1]$.

18. Let $f_n(x) := x^n$ for $x \in [0, 1]$, $n \in \mathbb{N}$. Show that (f_n) is a decreasing sequence of continuous functions that converges to a function that is not continuous, but the convergence is not uniform on $[0, 1]$.

19. Let $f_n(x) := x/n$ for $x \in [0, \infty)$, $n \in \mathbb{N}$. Show that (f_n) is a decreasing sequence of continuous functions that converges to a continuous limit function, but the convergence is not uniform on $[0, \infty)$.

20. Give an example of a decreasing sequence (f_n) of continuous functions on $[0, 1)$ that converges to a continuous limit function, but the convergence is not uniform on $[0, 1)$.

Section 8.3   The Exponential and Logarithmic Functions

We will now introduce the exponential and logarithmic functions and will derive some of their most important properties. In earlier sections of this book we assumed some familiarity with these functions for the purpose of discussing examples. However, it is necessary at some point to place these important functions on a firm foundation in order to establish their existence and determine their basic properties. We will do that here. There are several alternative approaches one can take to accomplish this goal. We will proceed by first proving the existence of a function that has *itself* as derivative. From this basic result, we obtain the main properties of the exponential function. The logarithm function is then introduced as the inverse of the exponential function, and this inverse relation is used to derive the properties of the logarithm function.

The Exponential Function

We begin by establishing the key existence result for the exponential function.

8.3.1 Theorem   *There exists a function $E : \mathbb{R} \to \mathbb{R}$ such that:*

(i) $E'(x) = E(x)$ for all $x \in \mathbb{R}$.

(ii) $E(0) = 1$.

Proof.   We inductively define a sequence (E_n) of continuous functions as follows:

(1) $$E_1(x) := 1 + x,$$

(2) $$E_{n+1}(x) := 1 + \int_0^x E_n(t)\, dt,$$

for all $n \in \mathbb{N}$, $x \in \mathbb{R}$. Clearly E_1 is continuous on $\mathbb{R}$ and hence is integrable over any bounded interval. If E_n has been defined and is continuous on $\mathbb{R}$, then it is integrable over any bounded interval, so that E_{n+1} is well-defined by the above formula. Moreover, it follows from the Fundamental Theorem (Second Form) 7.3.5 that E_{n+1} is differentiable at any point $x \in \mathbb{R}$ and that

(3) $$E'_{n+1}(x) = E_n(x) \qquad \text{for} \quad n \in \mathbb{N}.$$

An Induction argument (which we leave to the reader) shows that

(4) $$E_n(x) = 1 + \frac{x}{1!} + \frac{x^2}{2!} + \cdots + \frac{x^n}{n!} \qquad \text{for} \quad x \in \mathbb{R}.$$

Let $A > 0$ be given; then if $|x| \leq A$ and $m > n > 2A$, we have

(5)
$$
\big| E_m(x) - E_n(x) \big| = \left| \frac{x^{n+1}}{(n+1)!} + \cdots + \frac{x^m}{m!} \right|
$$
$$
\leq \frac{A^{n+1}}{(n+1)!} \left[1 + \frac{A}{n} + \cdots + \left(\frac{A}{n} \right)^{m-n-1} \right]
$$
$$
< \frac{A^{n+1}}{(n+1)!}\, 2.
$$

Since $\lim(A^n/n!) = 0$, it follows that the sequence (E_n) converges uniformly on the interval $[-A, A]$ where $A > 0$ is arbitrary. In particular this means that $(E_n(x))$ converges for each $x \in \mathbb{R}$. We define $E : \mathbb{R} \to \mathbb{R}$ by

$$E(x) := \lim E_n(x) \qquad \text{for} \quad x \in \mathbb{R}.$$

Since each $x \in \mathbb{R}$ is contained inside some interval $[-A, A]$, it follows from Theorem 8.2.2 that E is continuous at x. Moreover, it is clear from (1) and (2) that $E_n(0) = 1$ for all $n \in \mathbb{N}$. Therefore $E(0) = 1$, which proves (ii).

On any interval $[-A, A]$ we have the uniform convergence of the sequence (E_n). In view of (3), we also have the uniform convergence of the sequence (E'_n) of derivatives. It therefore follows from Theorem 8.2.3 that the limit function E is differentiable on $[-A, A]$ and that

$$E'(x) = \lim(E'_n(x)) = \lim(E_{n-1}(x)) = E(x)$$

for all $x \in [-A, A]$. Since $A > 0$ is arbitrary, statement (i) is established.   Q.E.D.

8.3.2 Corollary   *The function E has a derivative of every order and $E^{(n)}(x) = E(x)$ for all $n \in \mathbb{N}$, $x \in \mathbb{R}$.*

Proof.   If $n = 1$, the statement is merely property (i). It follows for arbitrary $n \in \mathbb{N}$ by Induction.   Q.E.D.

8.3.3 Corollary   *If $x > 0$, then $1 + x < E(x)$.*

Proof.   It is clear from (4) that if $x > 0$, then the sequence $(E_n(x))$ is strictly increasing. Hence $E_1(x) < E(x)$ for all $x > 0$.   Q.E.D.

It is next shown that the function E, whose existence was established in Theorem 8.3.1, is unique.

8.3.4 Theorem   *The function $E : \mathbb{R} \to \mathbb{R}$ that satisfies (i) and (ii) of Theorem 8.3.1 is unique.*

Proof.   Let E_1 and E_2 be two functions on $\mathbb{R}$ to $\mathbb{R}$ that satisfy properties (i) and (ii) of Theorem 8.3.1 and let $F := E_1 - E_2$. Then

$$F'(x) = E_1'(x) - E_2'(x) = E_1(x) - E_2(x) = F(x)$$

for all $x \in \mathbb{R}$ and

$$F(0) = E_1(0) - E_2(0) = 1 - 1 = 0.$$

It is clear (by Induction) that F has derivatives of all orders and indeed that $F^{(n)}(x) = F(x)$ for $n \in \mathbb{N}, x \in \mathbb{R}$.

Let $x \in \mathbb{R}$ be arbitrary, and let I_x be the closed interval with endpoints $0, x$. Since F is continuous on I_x, there exists $K > 0$ such that $|F(t)| \leq K$ for all $t \in I_x$. If we apply Taylor's Theorem 6.4.1 to F on the interval I_x and use the fact that $F^{(k)}(0) = F(0) = 0$ for all $k \in \mathbb{N}$, it follows that for each $n \in \mathbb{N}$ there is a point $c_n \in I_x$ such that

$$F(x) = F(0) + \frac{F'(0)}{1!} x + \cdots + \frac{F^{(n-1)}}{(n-1)!} x^{n-1} + \frac{F^{(n)}(c_n)}{n!} x^n$$

$$= \frac{F(c_n)}{n!} x^n.$$

Therefore we have

$$|F(x)| \leq \frac{K |x|^n}{n!} \qquad \text{for all} \quad n \in \mathbb{N}.$$

But since $\lim(|x|^n / n!) = 0$, we conclude that $F(x) = 0$. Since $x \in \mathbb{R}$ is arbitrary, we infer that $E_1(x) - E_2(x) = F(x) = 0$ for all $x \in \mathbb{R}$.   Q.E.D.

The standard terminology and notation for the function E (which we now know exists and is unique) is given in the following definition.

8.3.5 Definition   The unique function $E : \mathbb{R} \to \mathbb{R}$ such that $E'(x) = E(x)$ for all $x \in \mathbb{R}$ and $E(0) = 1$, is called the **exponential function**. The number $e := E(1)$ is called **Euler's number**. We will frequently write

$$\exp(x) := E(x) \qquad \text{or} \qquad e^x := E(x) \qquad \text{for} \quad x \in \mathbb{R}.$$

The number e can be obtained as a limit, and thereby approximated, in several different ways. [See Exercises 1 and 10, and Example 3.3.6.]

The use of the notation e^x for $E(x)$ is justified by property (v) in the next theorem, where it is noted that if r is a rational number, then $E(r)$ and e^r coincide. (Rational exponents were discussed in Section 5.6.) Thus, the function E can be viewed as extending the idea of exponentiation from rational numbers to arbitrary real numbers. For a definition of a^x for $a > 0$ and arbitrary $x \in \mathbb{R}$, see Definition 8.3.10.

8.3.6 Theorem   *The exponential function satisfies the following properties:*

(iii) $E(x) \neq 0$ *for all* $x \in \mathbb{R}$;

(iv) $E(x + y) = E(x)E(y)$ *for all* $x, y \in \mathbb{R}$;

(v) $E(r) = e^r$ *for all* $r \in \mathbb{Q}$.

Proof.   (iii) Let $\alpha \in \mathbb{R}$ be such that $E(\alpha) = 0$, and let J_α be the closed interval with endpoints 0, α. Let $K \geq |E(t)|$ for all $t \in J_\alpha$. Taylor's Theorem 6.4.1, implies that for each $n \in \mathbb{N}$ there exists a point $c_n \in J_\alpha$ such that

$$1 = E(0) = E(\alpha) + \frac{E'(\alpha)}{1!}(-\alpha) + \cdots + \frac{E^{(n-1)}(\alpha)}{(n-1)!}(-\alpha)^{n-1}$$

$$+ \frac{E^{(n)}(\alpha)}{(n)!}(-\alpha)^n = \frac{E(c_n)}{n!}(-\alpha)^n .$$

Thus we have $0 < 1 \leq (K/n!)|\alpha|^n$ for $n \in \mathbb{N}$. But since $\lim(|\alpha|^n/n!) = 0$, this is a contradiction.

(iv) Let y be fixed; by (iii) we have $E(y) \neq 0$. Let $G : \mathbb{R} \to \mathbb{R}$ be defined by

$$G(x) := \frac{E(x + y)}{E(y)} \qquad \text{for} \quad x \in \mathbb{R}.$$

Evidently we have $G'(x) = E'(x + y)/E(y) = E(x + y)/E(y) = G(x)$ for all $x \in \mathbb{R}$, and $G(0) = E(0 + y)/E(y) = 1$. It follows from the uniqueness of E, proved in Theorem 8.3.4, that $G(x) = E(x)$ for all $x \in \mathbb{R}$. Hence $E(x + y) = E(x)E(y)$ for all $x \in \mathbb{R}$. Since $y \in \mathbb{R}$ is arbitrary, we obtain (iv).

(v) It follows from (iv) and Induction that if $n \in \mathbb{N}$, $x \in \mathbb{R}$, then

$$E(nx) = E(x)^n.$$

If we let $x = 1/n$, this relation implies that

$$e = E(1) = E\left(n \cdot \frac{1}{n}\right) = \left(E\left(\frac{1}{n}\right)\right)^n ,$$

whence it follows that $E(1/n) = e^{1/n}$. Also we have $E(-m) = 1/E(m) = 1/e^m = e^{-m}$ for $m \in \mathbb{N}$. Therefore, if $m \in \mathbb{Z}$, $n \in \mathbb{N}$, we have

$$E(m/n) = \left(E(1/n)\right)^m = \left(e^{1/n}\right)^m = e^{m/n}.$$

This establishes (v).   Q.E.D.

8.3.7 Theorem   *The exponential function E is strictly increasing on $\mathbb{R}$ and has range equal to $\{y \in \mathbb{R} : y > 0\}$. Further, we have*

(vi) $\lim\limits_{x \to -\infty} E(x) = 0$ *and* $\lim\limits_{x \to \infty} E(x) = \infty$.

Proof.   We know that $E(0) = 1 > 0$ and $E(x) \neq 0$ for all $x \in \mathbb{R}$. Since E is continuous on $\mathbb{R}$, it follows from Bolzano's Intermediate Value Theorem 5.3.7 that $E(x) > 0$ for all $x \in \mathbb{R}$. Therefore $E'(x) = E(x) > 0$ for $x \in \mathbb{R}$, so that E is strictly increasing on $\mathbb{R}$.

It follows from Corollary 8.3.3 that $2 < e$ and that $\lim_{x \to \infty} E(x) = \infty$. Also, if $z > 0$, then since $0 < E(-z) = 1/E(z)$ it follows that $\lim_{x \to -\infty} E(x) = 0$. Therefore, by the Intermediate Value Theorem 5.3.7, every $y \in \mathbb{R}$ with $y > 0$ belongs to the range of E.   Q.E.D.

The Logarithm Function

We have seen that the exponential function E is a strictly increasing differentiable function with domain $\mathbb{R}$ and range $\{y \in \mathbb{R} : y > 0\}$. (See Figure 8.3.1.) It follows that $\mathbb{R}$ has an inverse function.

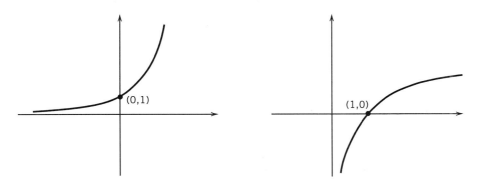

Figure 8.3.1   Graph of E.   **Figure 8.3.2**   Graph of L.

8.3.8 Definition   The function inverse to $E : \mathbb{R} \to \mathbb{R}$ is called the **logarithm** (or the **natural logarithm**). (See Figure 8.3.2.) It will be denoted by L, or by ln.

Since E and L are inverse functions, we have

$$(L \circ E)(x) = x \qquad \text{for all} \quad x \in \mathbb{R}$$

and

$$(E \circ L)(y) = y \qquad \text{for all} \quad y \in \mathbb{R}, y > 0.$$

These formulas may also be written in the form

$$\ln e^x = x, \qquad e^{\ln y} = y.$$

8.3.9 Theorem   *The logarithm is a strictly increasing function L with domain $\{x \in \mathbb{R} : x > 0\}$ and range $\mathbb{R}$. The derivative of L is given by*

(vii) $L'(x) = 1/x$ *for $x > 0$.*
The logarithm satisfies the functional equation

(viii) $L(xy) = L(x) + L(y)$ *for $x > 0, y > 0$.*
Moreover, we have

(ix) $L(1) = 0$ *and* $L(e) = 1$,

(x) $L(x^r) = rL(x)$ *for* $x > 0, r \in \mathbb{Q}$.

(xi) $\lim_{x \to 0+} L(x) = -\infty$ *and* $\lim_{x \to \infty} L(x) = \infty$.

Proof.   That L is strictly increasing with domain $\{x \in \mathbb{R} : x > 0\}$ and range $\mathbb{R}$ follows from the fact that E is strictly increasing with domain $\mathbb{R}$ and range $\{y \in \mathbb{R} : y > 0\}$.

(vii) Since $E'(x) = E(x) > 0$, it follows from Theorem 6.1.9 that L is differentiable on $(0, \infty)$ and that

$$L'(x) = \frac{1}{(E' \circ L)(x)} = \frac{1}{(E \circ L)(x)} = \frac{1}{x} \quad \text{for} \quad x \in (0, \infty).$$

(viii) If $x > 0$, $y > 0$, let $u := L(x)$ and $v := L(y)$. Then we have $x = E(u)$ and $y = E(v)$. It follows from property (iv) of Theorem 8.3.6 that

$$xy = E(u)E(v) = E(u + v),$$

so that $L(xy) = (L \circ E)(u + v) = u + v = L(x) + L(y)$. This establishes (viii).

The properties in (ix) follow from the relations $E(0) = 1$ and $E(1) = e$.

(x) This result follows from (viii) and Mathematical Induction for $n \in \mathbb{N}$, and is extended to $r \in \mathbb{Q}$ by arguments similar to those in the proof of 8.3.6(v).

To establish property (xi), we first note that since $2 < e$, then $\lim(e^n) = \infty$ and $\lim(e^{-n}) = 0$. Since $L(e^n) = n$ and $L(e^{-n}) = -n$ it follows from the fact that L is strictly increasing that

$$\lim_{x \to \infty} L(x) = \lim L(e^n) = \infty \quad \text{and} \quad \lim_{x \to 0+} L(x) = \lim L(e^{-n}) = -\infty. \qquad \text{Q.E.D.}$$

Power Functions _____

In Definition 5.6.6, we discussed the power function $x \mapsto x^r$, $x > 0$, where r is a rational number. By using the exponential and logarithm functions, we can extend the notion of power functions from rational to arbitrary real powers.

8.3.10 Definition   If $\alpha \in \mathbb{R}$ and $x > 0$, the number x^α is defined to be

$$x^\alpha := e^{\alpha \ln x} = E(\alpha L(x)).$$

The function $x \mapsto x^\alpha$ for $x > 0$ is called the **power function** with exponent α.

Note   If $x > 0$ and $\alpha = m/n$ where $m \in \mathbb{Z}$, $n \in \mathbb{N}$, then we defined $x^\alpha := (x^m)^{1/n}$ in Section 5.6. Hence we have $\ln x^\alpha = \alpha \ln x$, whence $x^\alpha = e^{\ln x^\alpha} = e^{\alpha \ln x}$. Hence Definition 8.3.10 is consistent with the definition given in Section 5.6.

We now state some properties of the power functions. Their proofs are immediate consequences of the properties of the exponential and logarithm functions and will be left to the reader.

8.3.11 Theorem   *If $\alpha \in \mathbb{R}$ and x, y belong to $(0, \infty)$, then:*

(a) $1^\alpha = 1$,   (b) $x^\alpha > 0$,

(c) $(xy)^\alpha = x^\alpha y^\alpha$,   (d) $(x/y)^\alpha = x^\alpha/y^\alpha$.

8.3.12 Theorem   *If $\alpha, \beta \in \mathbb{R}$ and $x \in (0, \infty)$, then:*

(a) $x^{\alpha+\beta} = x^\alpha x^\beta$,   (b) $(x^\alpha)^\beta = x^{\alpha\beta} = (x^\beta)^\alpha$,

(c) $x^{-\alpha} = 1/x^\alpha$,   (d)   *if $\alpha < \beta$, then $x^\alpha < x^\beta$ for $x > 1$.*

The next result concerns the differentiability of the power functions.

8.3.13 Theorem   *Let $\alpha \in \mathbb{R}$. Then the function $x \mapsto x^\alpha$ on $(0, \infty)$ to $\mathbb{R}$ is continuous and differentiable, and*

$$Dx^\alpha = \alpha x^{\alpha - 1} \qquad \text{for} \quad x \in (0, \infty).$$

Proof.   By the Chain Rule we have

$$Dx^\alpha = De^{\alpha \ln x} = e^{\alpha \ln x} \cdot D(\alpha \ln x)$$
$$= x^\alpha \cdot \frac{\alpha}{x} = \alpha x^{\alpha - 1} \qquad \text{for} \quad x \in (0, \infty). \qquad\qquad \text{Q.E.D.}$$

It will be seen in an exercise that if $\alpha > 0$, the power function $x \mapsto x^\alpha$ is strictly increasing on $(0, \infty)$ to $\mathbb{R}$, and that if $\alpha < 0$, the function $x \mapsto x^\alpha$ is strictly decreasing. (What happens if $\alpha = 0$?)

The graphs of the functions $x \mapsto x^\alpha$ on $(0, \infty)$ to $\mathbb{R}$ are similar to those in Figure 5.6.8.

The Function $\log_a$

If $a > 0$, $a \neq 1$, it is sometimes useful to define the function $\log_a$.

8.3.14 Definition   Let $a > 0$, $a \neq 1$. We define

$$\log_a(x) := \frac{\ln x}{\ln a} \qquad \text{for} \quad x \in (0, \infty).$$

For $x \in (0, \infty)$, the number $\log_a(x)$ is called the **logarithm of x to the base** a. The case $a = e$ yields the logarithm (or natural logarithm) function of Definition 8.3.8. The case $a = 10$ gives the base 10 logarithm (or common logarithm) function $\log_{10}$ often used in computations. Properties of the functions $\log_a$ will be given in the exercises.

Exercises for Section 8.3

1. Show that if $x > 0$ and if $n > 2x$, then

    $$\left| e^x - \left(1 + \frac{x}{1!} + \cdots + \frac{x^n}{n!} \right) \right| < \frac{2x^{n+1}}{(n+1)!}$$

    Use this formula to show that $2\frac{2}{3} < e < 2\frac{3}{4}$, hence e is not an integer.

2. Calculate e correct to 5 decimal places.

3. Show that if $0 \leq x \leq a$ and $n \in \mathbb{N}$, then

    $$1 + \frac{x}{1!} + \cdots + \frac{x^n}{n!} \leq e^x \leq 1 + \frac{x}{1!} + \cdots + \frac{x^{n-1}}{(n-1)!} + \frac{e^a x^n}{n!}.$$

4. Show that if $n \geq 2$, then

    $$0 < en! - \left(1 + 1 + \frac{1}{2!} + \cdots + \frac{1}{n!} \right) n! < \frac{e}{n+1} < 1.$$

    Use this inequality to prove that e is not a rational number.

5. If $x \geq 0$ and $n \in \mathbb{N}$, show that

    $$\frac{1}{x+1} = 1 - x + x^2 - x^3 + \cdots + (-x)^{n-1} + \frac{(-x)^n}{1+x}.$$

    Use this to show that

    $$\ln(x+1) = x - \frac{x^2}{2} + \frac{x^3}{3} - \cdots + (-1)^{n-1} \frac{x^n}{n} + \int_0^x \frac{(-t)^n}{1+t}\, dt.$$

and that

$$\left| \ln(x+1) - \left(x - \frac{x^2}{2} + \frac{x^3}{3} - \cdots + (-1)^{n-1}\frac{x^n}{n} \right) \right| \le \frac{x^{n+1}}{n+1}.$$

6. Use the formula in the preceding exercise to calculate $\ln 1.1$ and $\ln 1.4$ accurate to four decimal places. How large must one choose n in this inequality to calculate $\ln 2$ accurate to four decimal places?

7. Show that $\ln(e/2) = 1 - \ln 2$. Use this result to calculate $\ln 2$ accurate to four decimal places.

8. Let $f : \mathbb{R} \to \mathbb{R}$ be such that $f'(x) = f(x)$ for all $x \in \mathbb{R}$. Show that there exists $K \in \mathbb{R}$ such that $f(x) = Ke^x$ for all $x \in \mathbb{R}$.

9. Let $a_k > 0$ for $k = 1, \cdots, n$ and let $A := (a_1 + \cdots + a_n)/n$ be the arithmetic mean of these numbers. For each k, put $x_k := a_k/A - 1$ in the inequality $1 + x \le e^x$ (valid for $x \ge 0$). Multiply the resulting terms to prove the Arithmetic–Geometric Mean Inequality

(6)
$$\left(a_1 \cdots a_n \right)^{1/n} \le \frac{1}{n}(a_1 + \cdots + a_n).$$

Moreover, show that equality holds in (6) if and only if $a_1 = a_2 = \cdots = a_n$.

10. Evaluate $L'(1)$ by using the sequence $(1 + 1/n)$ and the fact that $e = \lim\left((1 + 1/n)^n \right)$.

11. Establish the assertions in Theorem 8.3.11.

12. Establish the assertions in Theorem 8.3.12.

13. (a) Show that if $\alpha > 0$, then the function $x \mapsto x^\alpha$ is strictly increasing on $(0, \infty)$ to $\mathbb{R}$ and that $\lim_{x \to 0+} x^\alpha = 0$ and $\lim_{x \to \infty} x^\alpha = \infty$.
    (b) Show that if $\alpha < 0$, then the function $x \mapsto x^\alpha$ is strictly decreasing on $(0, \infty)$ to $\mathbb{R}$ and that $\lim_{x \to 0+} x^\alpha = \infty$ and $\lim_{x \to \infty} x^\alpha = 0$.

14. Prove that if $a > 0$, $a \ne 1$, then $a^{\log_a x} = x$ for all $x \in (0, \infty)$ and $\log_a(a^y) = y$ for all $y \in \mathbb{R}$. Therefore the function $x \mapsto \log_a x$ on $(0, \infty)$ to $\mathbb{R}$ is inverse to the function $y \mapsto a^y$ on $\mathbb{R}$.

15. If $a > 0$, $a \ne 1$, show that the function $x \mapsto \log_a x$ is differentiable on $(0, \infty)$ and that $D \log_a x = 1/(x \ln a)$ for $x \in (0, \infty)$.

16. If $a > 0$, $a \ne 1$, and x and y belong to $(0, \infty)$, prove that $\log_a(xy) = \log_a x + \log_a y$.

17. If $a > 0$, $a \ne 1$, and $b > 0$, $b \ne 1$, show that

$$\log_a x = \left(\frac{\ln b}{\ln a} \right) \log_b x \qquad \text{for} \quad x \in (0, \infty).$$

In particular, show that $\log_{10} x = (\ln e/\ln 10) \ln x = (\log_{10} e) \ln x$ for $x \in (0, \infty)$.

Section 8.4   The Trigonometric Functions

Along with the exponential and logarithmic functions, there is another very important collection of transcendental functions known as the "trigonometric functions". These are the sine, cosine, tangent, cotangent, secant, and cosecant functions. In elementary courses, they are usually introduced on a geometric basis in terms of either triangles or the unit circle. In this section, we introduce the trigonometric functions in an analytical manner and then establish some of their basic properties. In particular, the various properties of the trigonometric functions that were used in examples in earlier parts of this book will be derived rigorously in this section.

It suffices to deal with the sine and cosine since the other four trigonometric functions are defined in terms of these two. Our approach to the sine and cosine is similar in spirit to our approach to the exponential function in that we first establish the existence of functions that satisfy certain differentiation properties.

8.4.1 Theorem   *There exist functions $C : \mathbb{R} \to \mathbb{R}$ and $S : \mathbb{R} \to \mathbb{R}$ such that*

(i) $C''(x) = -C(x)$ and $S''(x) = -S(x)$ for all $x \in \mathbb{R}$

(ii) $C(0) = 1, C'(0) = 0$, and $S(0) = 0, S'(0) = 1$.

Proof.   We define the sequences (C_n) and (S_n) of continuous functions inductively as follows:

$$(1) \qquad\qquad C_1(x) := 1, \qquad S_1(x) := x,$$

$$(2) \qquad\qquad S_n(x) := \int_0^x C_n(t)\, dt,$$

$$(3) \qquad\qquad C_{n+1}(x) := 1 - \int_0^x S_n(t)\, dt,$$

for all $n \in \mathbb{N}, x \in \mathbb{R}$.

One sees by Induction that the functions C_n and S_n are continuous on $\mathbb{R}$ and hence they are integrable over any bounded interval; thus these functions are well-defined by the above formulas. Moreover, it follows from the Fundamental Theorem 7.3.5 that S_n and C_{n+1} are differentiable at every point and that

$$(4) \qquad S_n'(x) = C_n(x) \qquad \text{and} \qquad C_{n+1}'(x) = -S_n(x) \qquad \text{for} \quad n \in \mathbb{N}, x \in \mathbb{R}.$$

Induction arguments (which we leave to the reader) show that

$$C_{n+1}(x) = 1 - \frac{x^2}{2!} + \frac{x^4}{4!} - \cdots + (-1)^n \frac{x^{2n}}{(2n)!},$$

$$S_{n+1}(x) = x - \frac{x^3}{3!} + \frac{x^5}{5!} - \cdots + (-1)^n \frac{x^{2n+1}}{(2n+1)!}.$$

Let $A > 0$ be given. Then if $|x| \le A$ and $m > n > 2A$, we have that (since $A/2n < 1/4$):

$$(5) \qquad \left| C_m(x) - C_n(x) \right| = \left| \frac{x^{2n}}{(2n)!} - \frac{x^{2n+2}}{(2n+2)!} + \cdots \pm \frac{x^{2m-2}}{(2m-2)!} \right|$$

$$\le \frac{A^{2n}}{(2n)!} \left[1 + \left(\frac{A}{2n} \right)^2 + \cdots + \left(\frac{A}{2n} \right)^{2m-2n-2} \right]$$

$$< \frac{A^{2n}}{(2n)!} \left(\frac{16}{15} \right).$$

Since $\lim(A^{2n}/(2n)!) = 0$, the sequence (C_n) converges uniformly on the interval $[-A, A]$, where $A > 0$ is arbitrary. In particular, this means that $(C_n(x))$ converges for each $x \in \mathbb{R}$. We define $C : \mathbb{R} \to \mathbb{R}$ by

$$C(x) := \lim C_n(x) \qquad \text{for} \quad x \in \mathbb{R}.$$

It follows from Theorem 8.2.2 that C is continuous on $\mathbb{R}$ and, since $C_n(0) = 1$ for all $n \in \mathbb{N}$, that $C(0) = 1$.

If $|x| \le A$ and $m \ge n > 2A$, it follows from (2) that

$$S_m(x) - S_n(x) = \int_0^x \left\{ C_m(t) - C_n(t) \right\} dt.$$

If we use (5) and Corollary 7.3.15, we conclude that

$$|S_m(x) - S_n(x)| \le \frac{A^{2n}}{(2n)!}\left(\frac{16}{15}A\right),$$

whence the sequence (S_n) converges uniformly on $[-A, A]$. We define $S : \mathbb{R} \to \mathbb{R}$ by

$$S(x) := \lim S_n(x) \qquad \text{for} \quad x \in \mathbb{R}.$$

It follows from Theorem 8.2.2 that S is continuous on $\mathbb{R}$ and, since $S_n(0) = 0$ for all $n \in \mathbb{N}$, that $S(0) = 0$.

Since $C_n'(x) = -S_{n-1}(x)$ for $n > 1$, it follows from the above that the sequence (C_n') converges uniformly on $[-A, A]$. Hence by Theorem 8.2.3, the limit function C is differentiable on $[-A, A]$ and

$$C'(x) = \lim C_n'(x) = \lim(-S_{n-1}(x)) = -S(x) \qquad \text{for} \quad x \in [-A, A].$$

Since $A > 0$ is arbitrary, we have

(6) $$C'(x) = -S(x) \qquad \text{for} \quad x \in \mathbb{R}.$$

A similar argument, based on the fact that $S_n'(x) = C_n(x)$, shows that S is differentiable on $\mathbb{R}$ and that

(7) $$S'(x) = C(x) \qquad \text{for all} \quad x \in \mathbb{R}.$$

It follows from (6) and (7) that

$$C''(x) = -(S(x))' = -C(x) \qquad \text{and} \qquad S''(x) = (C(x))' = -S(x)$$

for all $x \in \mathbb{R}$. Moreover, we have

$$C'(0) = -S(0) = 0, \qquad S'(0) = C(0) = 1.$$

Thus statements (i) and (ii) are proved.   Q.E.D.

8.4.2 Corollary   *If C, S are the functions in Theorem 8.4.1, then*

(iii) $C'(x) = -S(x)$ *and* $S'(x) = C(x)$ *for* $x \in \mathbb{R}$.

Moreover, these functions have derivatives of all orders.

Proof.   The formulas (iii) were established in (6) and (7). The existence of the higher order derivatives follows by Induction.   Q.E.D.

8.4.3 Corollary   *The functions C and S satisfy the Pythagorean Identity:*

(iv) $(C(x))^2 + (S(x))^2 = 1$ *for* $x \in \mathbb{R}$.

Proof.   Let $f(x) := (C(x))^2 + (S(x))^2$ for $x \in \mathbb{R}$, so that

$$f'(x) = 2C(x)(-S(x)) + 2S(x)(C(x)) = 0 \qquad \text{for} \quad x \in \mathbb{R}.$$

Thus it follows that $f(x)$ is a constant for all $x \in \mathbb{R}$. But since $f(0) = 1 + 0 = 1$, we conclude that $f(x) = 1$ for all $x \in \mathbb{R}$.   Q.E.D.

We next establish the uniqueness of the functions C and S.

8.4.4 Theorem   *The functions C and S satisfying properties (i) and (ii) of Theorem 8.4.1 are unique.*

Proof. Let C_1 and C_2 be two functions on $\mathbb{R}$ to $\mathbb{R}$ that satisfy $C_j''(x) = -C_j(x)$ for all $x \in \mathbb{R}$ and $C_j(0) = 1, C_j'(0) = 0$ for $j = 1, 2$. If we let $D := C_1 - C_2$, then $D''(x) = -D(x)$ for $x \in \mathbb{R}$ and $D(0) = 0$ and $D^{(k)}(0) = 0$ for all $k \in \mathbb{N}$.

Now let $x \in \mathbb{R}$ be arbitrary, and let I_x be the interval with endpoints $0, x$. Since $D = C_1 - C_2$ and $T := S_1 - S_2 = C_2' - C_1'$ are continuous on I_x, there exists $K > 0$ such that $|D(t)| \leq K$ and $|T(t)| \leq K$ for all $t \in I_x$. If we apply Taylor's Theorem 6.4.1 to D on I_x and use the fact that $D(0) = 0$, $D^{(k)}(0) = 0$ for $k \in \mathbb{N}$, it follows that for each $n \in \mathbb{N}$ there is a point $c_n \in I_x$ such that

$$D(x) = D(0) + \frac{D'(0)}{1!}x + \cdots + \frac{D^{(n-1)}(0)}{(n-1)!}x^{n-1} + \frac{D^{(n)}(c_n)}{n!}x^n$$
$$= \frac{D^{(n)}(c_n)}{n!}x^n.$$

Now either $D^{(n)}(c_n) = \pm D(c_n)$ or $D^{(n)}(c_n) = \pm T(c_n)$. In either case we have

$$|D(x)| \leq \frac{K |x|^n}{n!}.$$

But since $\lim(|x|^n/n!) = 0$, we conclude that $D(x) = 0$. Since $x \in \mathbb{R}$ is arbitrary, we infer that $C_1(x) - C_2(x) = 0$ for all $x \in \mathbb{R}$.

A similar argument shows that if S_1 and S_2 are two functions on $\mathbb{R} \to \mathbb{R}$ such that $S_j''(x) = -S_j(x)$ for all $x \in \mathbb{R}$ and $S_j(0) = 0$, $S_j'(0) = 1$ for $j = 1, 2$, then we have $S_1(x) = S_2(x)$ for all $x \in \mathbb{R}$.   Q.E.D.

Now that existence and uniqueness of the functions C and S have been established, we shall give these functions their familiar names.

8.4.5 Definition   The unique functions $C : \mathbb{R} \to \mathbb{R}$ and $S : \mathbb{R} \to \mathbb{R}$ such that $C''(x) = -C(x)$ and $S''(x) = -S(x)$ for all $x \in \mathbb{R}$ and $C(0) = 1, C'(0) = 0$, and $S(0) = 0$, $S'(0) = 1$, are called the **cosine function** and the **sine function**, respectively. We ordinarily write

$$\cos x := C(x) \qquad \text{and} \qquad \sin x := S(x) \qquad \text{for} \quad x \in \mathbb{R}.$$

The differentiation properties in (i) of Theorem 8.4.1 do not by themselves lead to uniquely determined functions. We have the following relationship.

8.4.6 Theorem   *If $f : \mathbb{R} \to \mathbb{R}$ is such that*

$$f''(x) = -f(x) \qquad \text{for} \quad x \in \mathbb{R},$$

then there exist real numbers α, β such that

$$f(x) = \alpha C(x) + \beta S(x) \qquad \text{for} \quad x \in \mathbb{R}.$$

Proof.   Let $g(x) := f(0)C(x) + f'(0)S(x)$ for $x \in \mathbb{R}$. It is readily seen that $g''(x) = -g(x)$ and that $g(0) = f(0)$, and since

$$g'(x) = -f(0)S(x) + f'(0)C(x),$$

that $g'(0) = f'(0)$. Therefore the function $h := f - g$ is such that $h''(x) = -h(x)$ for all $x \in \mathbb{R}$ and $h(0) = 0$, $h'(0) = 0$. Thus it follows from the proof of the preceding theorem that $h(x) = 0$ for all $x \in \mathbb{R}$. Therefore $f(x) = g(x)$ for all $x \in \mathbb{R}$.   Q.E.D.

We shall now derive a few of the basic properties of the cosine and sine functions.

8.4.7 Theorem   *The function C is even and S is odd in the sense that*

(v) $C(-x) = C(x)$ *and* $S(-x) = -S(x)$ *for* $x \in \mathbb{R}$.

If $x, y \in \mathbb{R}$, *then we have the "addition formulas"*

(vi) $C(x + y) = C(x)C(y) - S(x)S(y)$, $S(x + y) = S(x)C(y) + C(x)S(y)$.

Proof.   (v) If $\varphi(x) := C(-x)$ for $x \in \mathbb{R}$, then a calculation shows that $\varphi''(x) = -\varphi(x)$ for $x \in \mathbb{R}$. Moreover, $\varphi(0) = 1$ and $\varphi'(0) = 0$ so that $\varphi = C$. Hence, $C(-x) = C(x)$ for all $x \in \mathbb{R}$. In a similar way one shows that $S(-x) = -S(x)$ for all $x \in \mathbb{R}$.
     (vi) Let $y \in \mathbb{R}$ be given and let $f(x) := C(x + y)$ for $x \in \mathbb{R}$. A calculation shows that $f''(x) = -f(x)$ for $x \in \mathbb{R}$. Hence, by Theorem 8.4.6, there exists real numbers α, β such that

$$f(x) = C(x + y) = \alpha C(x) + \beta S(x) \qquad \text{and}$$
$$f'(x) = -S(x + y) = -\alpha S(x) + \beta C(x)$$

for $x \in \mathbb{R}$. If we let $x = 0$, we obtain $C(y) = \alpha$ and $-S(y) = \beta$, whence the first formula in (vi) follows. The second formula is proved similarly.   Q.E.D.

The following inequalities were used earlier (for example, in 4.2.8).

8.4.8 Theorem   *If* $x \in \mathbb{R}$, $x \geq 0$, *then we have*

(vii) $-x \leq S(x) \leq x$;   **(viii)** $1 - \frac{1}{2}x^2 \leq C(x) \leq 1$;

(ix) $x - \frac{1}{6}x^3 \leq S(x) \leq x$;   **(x)** $1 - \frac{1}{2}x^2 \leq C(x) \leq 1 - \frac{1}{2}x^2 + \frac{1}{24}x^4$.

Proof.   Corollary 8.4.3 implies that $-1 \leq C(t) \leq 1$ for $t \in \mathbb{R}$, so that if $x \geq 0$, then

$$-x \leq \int_0^x C(t)\, dt \leq x,$$

whence we have (vii). If we integrate (vii), we obtain

$$-\tfrac{1}{2}x^2 \leq \int_0^x S(t)\, dt \leq \tfrac{1}{2}x^2,$$

whence we have

$$-\tfrac{1}{2}x^2 \leq -C(x) + 1 \leq \tfrac{1}{2}x^2.$$

Thus we have $1 - \frac{1}{2}x^2 \leq C(x)$, which implies (viii).
     Inequality (ix) follows by integrating (viii), and (x) follows by integrating (ix).   Q.E.D.

The number π is obtained via the following lemma.

8.4.9 Lemma   *There exists a root* γ *of the cosine function in the interval* $(\sqrt{2}, \sqrt{3})$. *Moreover* $C(x) > 0$ *for* $x \in [0, \gamma)$. *The number* 2γ *is the smallest positive root of* S.

Proof.   Inequality (x) of Theorem 8.4.8 implies that C has a root between the positive root $\sqrt{2}$ of $x^2 - 2 = 0$ and the smallest positive root of $x^4 - 12x^2 + 24 = 0$, which is $\sqrt{6 - 2\sqrt{3}} < \sqrt{3}$. We let γ be the smallest such root of C.

It follows from the second formula in (vi) with $x = y$ that $S(2x) = 2S(x)C(x)$. This relation implies that $S(2\gamma) = 0$, so that 2γ is a positive root of S. The same relation implies that if $2\delta > 0$ is the smallest positive root of S, then $C(\delta) = 0$. Since γ is the smallest positive root of C, we have $\delta = \gamma$.   Q.E.D.

8.4.10 Definition   Let $\pi := 2\gamma$ denote the smallest positive root of S.

Note   The inequality $\sqrt{2} < \gamma < \sqrt{6 - 2\sqrt{3}}$ implies that $2.828 < \pi < 3.185$.

8.4.11 Theorem   *The functions C and S have period 2π in the sense that*

(xi) $C(x + 2\pi) = C(x)$ *and* $S(x + 2\pi) = S(x)$ *for* $x \in \mathbb{R}$.

Moreover we have

(xii) $S(x) = C\left(\frac{1}{2}\pi - x\right) = -C\left(x + \frac{1}{2}\pi\right), \quad C(x) = S\left(\frac{1}{2}\pi - x\right) = S\left(x + \frac{1}{2}\pi\right)$ *for all* $x \in \mathbb{R}$.

Proof.   (xi) Since $S(2x) = 2S(x)C(x)$ and $S(\pi) = 0$, then $S(2\pi) = 0$. Further, if $x = y$ in (vi), we obtain $C(2x) = (C(x))^2 - (S(x))^2$. Therefore $C(2\pi) = 1$. Hence (vi) with $y = 2\pi$ gives

$$C(x + 2\pi) = C(x)C(2\pi) - S(x)S(2\pi) = C(x),$$

and

$$S(x + 2\pi) = S(x)C(2\pi) + C(x)S(2\pi) = S(x).$$

(xii) We note that $C(\frac{1}{2}\pi) = 0$, and it is an exercise to show that $S(\frac{1}{2}\pi) = 1$. If we employ these together with formulas (vi), the desired relations are obtained.   Q.E.D.

Exercises for Section 8.4

1. Calculate $\cos(.2)$, $\sin(.2)$ and $\cos 1$, $\sin 1$ correct to four decimal places.

2. Show that $|\sin x| \le 1$ and $|\cos x| \le 1$ for all $x \in \mathbb{R}$.

3. Show that property (vii) of Theorem 8.4.8 does not hold if $x < 0$, but that we have $|\sin x| \le |x|$ for all $x \in \mathbb{R}$. Also show that $|\sin x - x| \le |x|^3/6$ for all $x \in \mathbb{R}$.

4. Show that if $x > 0$ then

$$1 - \frac{x^2}{2} + \frac{x^4}{24} - \frac{x^6}{720} \le \cos x \le 1 - \frac{x^2}{2} + \frac{x^4}{24}.$$

   Use this inequality to establish a lower bound for π.

5. Calculate π by approximating the smallest positive zero of $\sin$. (Either bisect intervals or use Newton's Method of Section 6.4.)

6. Define the sequence (c_n) and (s_n) inductively by $c_1(x) := 1$, $s_1(x) := x$, and

$$s_n(x) := \int_0^x c_n(t)\, dt, \qquad c_{n+1}(x) := 1 + \int_0^x s_n(t)\, dt$$

   for all $n \in \mathbb{N}$, $x \in \mathbb{R}$. Reason as in the proof of Theorem 8.4.1 to conclude that there exist functions $c : \mathbb{R} \to \mathbb{R}$ and $s : \mathbb{R} \to \mathbb{R}$ such that (j) $c''(x) = c(x)$ and $s''(x) = s(x)$ for all $x \in \mathbb{R}$, and (jj) $c(0) = 1$, $c'(0) = 0$ and $s(0) = 0$, $s'(0) = 1$. Moreover, $c'(x) = s(x)$ and $s'(x) = c(x)$ for all $x \in \mathbb{R}$.

7. Show that the functions c, s in the preceding exercise have derivatives of all orders, and that they satisfy the identity $(c(x))^2 - (s(x))^2 = 1$ for all $x \in \mathbb{R}$. Moreover, they are the unique functions satisfying (j) and (jj). (The functions c, s are called the **hyperbolic cosine** and **hyperbolic sine functions**, respectively.)

8. If $f : \mathbb{R} \to \mathbb{R}$ is such that $f''(x) = f(x)$ for all $x \in \mathbb{R}$, show that there exist real numbers α, β such that $f(x) = \alpha c(x) + \beta s(x)$ for all $x \in \mathbb{R}$. Apply this to the functions $f_1(x) := e^x$ and $f_2(x) := e^{-x}$ for $x \in \mathbb{R}$. Show that $c(x) = \frac{1}{2}(e^x + e^{-x})$ and $s(x) = \frac{1}{2}(e^x - e^{-x})$ for $x \in \mathbb{R}$.

9. Show that the functions c, s in the preceding exercises are even and odd, respectively, and that
$$c(x + y) = c(x)c(y) + s(x)s(y), \qquad s(x + y) = s(x)c(y) + c(x)s(y),$$
for all $x, y \in \mathbb{R}$.

10. Show that $c(x) \geq 1$ for all $x \in \mathbb{R}$, that both c and s are strictly increasing on $(0, \infty)$, and that
$$\lim_{x \to \infty} c(x) = \lim_{x \to \infty} s(x) = \infty.$$

CHAPTER 9

INFINITE SERIES

In Section 3.7 we gave a brief introduction to the theory of infinite series. The reader will do well to look over that section at this time, since we will not repeat the definitions and results given there.

Instead, in Section 9.1 we will introduce the important notion of the "absolute convergence" of a series. In Section 9.2 we will present some "tests" for absolute convergence that will probably be familiar to the reader from calculus. The third section gives a discussion of series that are not absolutely convergent. In the final section we study series of functions and will establish the basic properties of power series which are very important in applications.

Section 9.1   Absolute Convergence

We have already met (in Section 3.7) a number of infinite series that are convergent and others that are divergent. For example, in Example 3.7.6(b) we saw that the **harmonic series:**

$$\sum_{n=1}^{\infty} \frac{1}{n}$$

is divergent since its sequence of partial sums $s_n := \frac{1}{1} + \frac{1}{2} + \cdots + \frac{1}{n}$ $(n \in \mathbb{N})$ is unbounded. On the other hand, we saw in Example 3.7.6(f) that the **alternating harmonic series:**

$$\sum_{n=1}^{\infty} \frac{(-1)^{n+1}}{n}$$

is convergent because of the subtraction that takes place. Since

$$\left| \frac{(-1)^{n+1}}{n} \right| = \frac{1}{n},$$

these two series illustrate the fact that a series $\sum x_n$ may be convergent, but the series $\sum |x_n|$ obtained by taking the absolute values of the terms may be divergent. This observation leads us to an important definition.

9.1.1 Definition   Let $X := (x_n)$ be a sequence in $\mathbb{R}$. We say that the series $\sum x_n$ is **absolutely convergent** if the series $\sum |x_n|$ is convergent in $\mathbb{R}$. A series is said to be **conditionally** (or **nonabsolutely**) **convergent** if it is convergent, but it is not absolutely convergent.

It is trivial that a series of *positive terms* is absolutely convergent if and only if it is convergent. We have noted above that the alternating harmonic series is conditionally convergent.

9.1.2 Theorem   *If a series in $\mathbb{R}$ is absolutely convergent, then it is convergent.*

Proof.   Since $\sum |x_n|$ is convergent, the Cauchy Criterion 3.7.4 implies that, given $\varepsilon > 0$ there exists $M(\varepsilon) \in \mathbb{N}$ such that if $m > n \geq M(\varepsilon)$, then

$$|x_{n+1}| + |x_{n+1}| + \cdots + |x_m| < \varepsilon.$$

However, by the Triangle Inequality, the left side of this expression dominates

$$|s_m - s_n| = |x_{n+1} + x_{n+2} + \cdots + x_m|.$$

Since $\varepsilon > 0$ is arbitrary, Cauchy's Criterion implies that $\sum x_n$ converges.   Q.E.D.

Grouping of Series _____

Given a series $\sum x_n$, we can construct many other series $\sum y_k$ by leaving the order of the terms x_n fixed, but inserting parentheses that group together finite numbers of terms. For example, the series indicated by

$$1 - \frac{1}{2} + \left(\frac{1}{3} - \frac{1}{4} \right) + \left(\frac{1}{5} - \frac{1}{6} + \frac{1}{7} \right) - \frac{1}{8} + \left(\frac{1}{9} - \cdots + \frac{1}{13} \right) - \cdots$$

is obtained by **grouping** the terms in the alternating harmonic series. It is an interesting fact that such grouping does not affect the convergence or the value of a convergent series.

9.1.3 Theorem   *If a series $\sum x_n$ is convergent, then any series obtained from it by grouping the terms is also convergent and to the same value.*

Proof.   Suppose that we have

$$y_1 := x_1 + \cdots + x_{k_1}, \qquad y_2 := x_{k_1+1} + \cdots + x_{k_2}, \qquad \cdots.$$

If s_n denotes the nth partial sum of $\sum x_n$ and t_k denotes the kth partial sum of $\sum y_k$, then we have

$$t_1 = y_1 = s_{k_1}, \qquad t_2 = y_1 + y_2 = s_{k_2}, \qquad \cdots$$

Thus, the sequence (t_k) of partial sums of the grouped series $\sum y_k$ is a subsequence of the sequence (s_n) of partial sums of $\sum x_n$. Since this latter series was assumed to be convergent, so is the grouped series $\sum y_k$.   Q.E.D.

It is clear that the converse to this theorem is not true. Indeed, the grouping

$$(1 - 1) + (1 - 1) + (1 - 1) + \cdots$$

produces a convergent series from $\sum_{n=0}^{\infty} (-1)^n$, which was seen to be divergent in Example 3.7.2(b) since the terms do not approach 0.

Rearrangements of Series

Loosely speaking, a "rearrangement" of a series is another series that is obtained from the given one by using all of the terms exactly once, but scrambling the order in which the terms are taken. For example, the harmonic series has rearrangements

$$\frac{1}{2} + \frac{1}{1} + \frac{1}{4} + \frac{1}{3} + \cdots + \frac{1}{2n} + \frac{1}{2n-1} + \cdots,$$

$$\frac{1}{1} + \frac{1}{2} + \frac{1}{4} + \frac{1}{3} + \frac{1}{5} + \frac{1}{7} + \cdots.$$

The first rearrangement is obtained from the harmonic series by interchanging the first and second terms, the third and fourth terms, and so forth. The second rearrangement is obtained from the harmonic series by taking one "odd term", two "even terms", three "odd terms", and so forth. It is obvious that there are infinitely many other possible rearrangements of the harmonic series.

9.1.4 Definition A series $\sum y_k$ in $\mathbb{R}$ is a **rearrangement** of a series $\sum x_n$ if there is a bijection f of $\mathbb{N}$ onto $\mathbb{N}$ such that $y_k = x_{f(k)}$ for all $k \in \mathbb{N}$.

While grouping series does not affect the convergence of a series, making rearrangements may do so. If fact, there is a remarkable observation, due to Riemann, that if $\sum s_n$ is a conditionally convergent series in $\mathbb{R}$, and if $c \in \mathbb{R}$ is arbitrary, then there is a rearrangement of $\sum x_n$ that converges to c.

To prove this assertion, we first note that a conditionally convergent series must contain infinitely many positive terms and infinitely many negative terms (see Exercise 1), and that both the series of positive terms and the series of negative terms diverge (see Exercise 2). To construct a series converging to c, we take positive terms until the partial sum is greater than c, then we take negative terms until the partial sum is less than c, then we take positive terms until the partial sum is greater than c, then we take negative terms, etc.

In our manipulations with series, we generally want to be sure that rearrangements will not affect the convergence or the value of the series. That is why the following result is important.

9.1.5 Rearrangement Theorem *Let $\sum x_n$ be an absolutely convergent series in $\mathbb{R}$. Then any rearrangement $\sum y_k$ of $\sum x_n$ converges to the same value.*

Proof. Suppose that $\sum x_n$ converges to $x \in \mathbb{R}$. Thus, if $\varepsilon > 0$, let N be such that if $n, q > N$ and $s_n := x_1 + \cdots + x_n$, then

$$|x - s_n| < \varepsilon \quad \text{and} \quad \sum_{k=N+1}^{q} |x_k| < \varepsilon.$$

Let $M \in \mathbb{N}$ be such that all of the terms $x_1, \cdots, x_N$ are contained as summands in $t_M := y_1 + \cdots + y_M$. It follows that if $m \geq M$, then $t_m - s_n$ is the sum of a finite number of terms x_k with index $k > N$. Hence, for some $q > N$, we have

$$|t_m - s_n| \leq \sum_{k=N+1}^{q} |x_k| < \varepsilon.$$

Therefore, if $m \geq M$, then we have

$$|t_m - x| \leq |t_m - s_n| + |s_n - x| < \varepsilon + \varepsilon = 2\varepsilon.$$

Since $\varepsilon > 0$ is arbitrary, we conclude that $\sum y_k$ converges to x.   Q.E.D.

Exercises for Section 9.1

1. Show that if a convergent series contains only a finite number of negative terms, then it is absolutely convergent.

2. Show that if a series is conditionally convergent, then the series obtained from its positive terms is divergent, and the series obtained from its negative terms is divergent.

3. If $\sum a_n$ is conditionally convergent, give an argument to show that there exists a rearrangement whose partial sums diverge to ∞.

4. Where is the fact that the series $\sum x_n$ is absolutely convergent used in the proof of 9.1.5?

5. If $\sum a_n$ is absolutely convergent, is it true that every rearrangement of $\sum a_n$ is also absolutely convergent?

6. Find an explicit expression for the nth partial sum of $\sum_{n=2}^{\infty} \ln(1 - 1/n^2)$ to show that this series converges to $-\ln 2$. Is this convergence absolute?

7. (a) If $\sum a_n$ is absolutely convergent and (b_n) is a bounded sequence, show that $\sum a_n b_n$ is absolutely convergent.
    (b) Give an example to show that if the convergence of $\sum a_n$ is conditional and (b_n) is a bounded sequence, then $\sum a_n b_n$ may diverge.

8. Give an example of a convergent series $\sum a_n$ such that $\sum a_n^2$ is not convergent. (Compare this with Exercise 3.7.8)

9. If (a_n) is a decreasing sequence of strictly positive numbers and if $\sum a_n$ is convergent, show that $\lim(na_n) = 0$.

10. Give an example of a divergent series $\sum a_n$ with (a_n) decreasing and such that $\lim(na_n) = 0$.

11. If (a_n) is a sequence and if $\lim(n^2 a_n)$ exists in $\mathbb{R}$, show that $\sum a_n$ is absolutely convergent.

12. Let $a > 0$. Show that the series $\sum(1 + a^n)^{-1}$ is divergent if $0 < a \leq 1$ and is convergent if $a > 1$.

13. (a) Does the series $\sum_{n=1}^{\infty} \left(\dfrac{\sqrt{n+1} - \sqrt{n}}{\sqrt{n}} \right)$ converge?

     (b) Does the series $\sum_{n=1}^{\infty} \left(\dfrac{\sqrt{n+1} - \sqrt{n}}{n} \right)$ converge?

14. If (a_{n_k}) is a subsequence of (a_n), then the series $\sum a_{n_k}$ is called a **subseries** of $\sum a_n$. Show that $\sum a_n$ is absolutely convergent if and only if every subseries of it is convergent.

15. Let $a : \mathbb{N} \times \mathbb{N} \to \mathbb{R}$ and write $a_{ij} := a(i, j)$. If $A_i := \sum_{j=1}^{\infty} a_{ij}$ for each $i \in \mathbb{N}$ and if $A := \sum_{i=1}^{\infty} A_i$, we say that A is an **iterated sum** of the a_{ij} and write $A = \sum_{i=1}^{\infty} \sum_{j=1}^{\infty} a_{ij}$. We define the other iterated sum, denoted by $\sum_{j=1}^{\infty} \sum_{i=1}^{\infty} a_{ij}$, in a similar way.
     Suppose $a_{ij} \geq 0$ for $i, j \in \mathbb{N}$. If (c_k) is any enumeration of $\{a_{ij} : i, j \in \mathbb{N}\}$, show that the following statements are equivalent:
     (i)   The interated sum $\sum_{i=1}^{\infty} \sum_{j=1}^{\infty} a_{ij}$ converges to B.
     (ii) The series $\sum_{k=1}^{\infty} c_k$ converges to C.
     In this case, we have $B = C$.

16.   The preceding exercise may fail if the terms are not positive. For example, let $a_{ij} := +1$ if $i - j = 1, a_{ij} := -1$ if $i - j = -1$, and $a_{ij} := 0$ elsewhere. Show that the iterated sums

$$\sum_{i=1}^{\infty} \sum_{j=1}^{\infty} a_{ij} \qquad \text{and} \qquad \sum_{j=1}^{\infty} \sum_{i=1}^{\infty} a_{ij}$$

both exist but are not equal.

Section 9.2   Tests for Absolute Convergence

In Section 3.7 we gave some results concerning the convergence of infinite series; namely, the nth Term Test, the fact that a series of positive terms is convergent if and only if its sequence of partial sums is bounded, the Cauchy Criterion, and the Comparison and Limit Comparison Tests.

We will now give some additional results that may be familiar from calculus. These results are particularly useful in establishing absolute convergence.

9.2.1 Limit Comparison Test, II.   *Suppose that $X := (x_n)$ and $Y := (y_n)$ are nonzero real sequences and suppose that the following limit exists in $\mathbb{R}$:*

$$r := \lim \left| \frac{x_n}{y_n} \right|. \tag{1}$$

(a)   *If $r \neq 0$, then $\sum x_n$ is absolutely convergent if and only if $\sum y_n$ is absolutely convergent.*
(b)   *If $r = 0$ and if $\sum y_n$ is absolutely convergent, then $\sum x_n$ is absolutely convergent.*

Proof.   This result follows immediately from Theorem 3.7.8.   Q.E.D.

The Root and Ratio Tests ───

The following test is due to Cauchy.

9.2.2 Root Test   *Let $X := (x_n)$ be a sequence in $\mathbb{R}$.*
(a)   *If there exist $r \in \mathbb{R}$ with $r < 1$ and $K \in \mathbb{N}$ such that*

$$|x_n|^{1/n} \leq r \qquad \text{for} \quad n \geq K, \tag{2}$$

then the series $\sum x_n$ is absolutely convergent.
(b)   *If there exists $K \in \mathbb{N}$ such that*

$$|x_n|^{1/n} \geq 1 \qquad \text{for} \quad n \geq K, \tag{3}$$

then the series $\sum x_n$ is divergent.

Proof.   (a) If (2) holds, then we have $|x_n| \leq r^n$ for $n \geq K$. Since the geometric series $\sum r^n$ is convergent for $0 \leq r < 1$, the Comparison Test 3.7.7 implies that $\sum |x_n|$ is convergent.
    (b) If (3) holds, then $|x_n| \geq 1$ for $n \geq K$, so the terms do not approach 0 and the nth Term Test 3.7.3 applies.   Q.E.D.

In calculus courses, one often meets the following version of the Root Test.

9.2.3 Corollary   *Let $X := (x_n)$ be a sequence in $\mathbb{R}$ and suppose that the limit*

$$\text{(4)} \qquad\qquad r := \lim |x_n|^{1/n}$$

exists in $\mathbb{R}$. Then $\sum x_n$ is absolutely convergent when $r < 1$ and is divergent when $r > 1$.

Proof.   If the limit in (4) exists and $r < 1$, then there exist r_1 with $r < r_1 < 1$ and $K \in \mathbb{N}$ such that $|x_n|^{1/n} \le r_1$ for $n > K$. In this case we can apply 9.2.2(a).

If $r > 1$, then there exists $K \in \mathbb{N}$ such that $|x_n|^{1/n} > 1$ for $n \ge K$ and the nth Term Test applies.   Q.E.D.

Note   No conclusion is possible in Corollary 9.2.3 when $r = 1$, for either convergence or divergence is possible. See Example 9.2.7(b).

Our next test is due to D'Alembert.

9.2.4 Ratio Test   *Let $X := (x_n)$ be a sequence of nonzero real numbers.*

(a)   *If there exist $r \in \mathbb{R}$ with $0 < r < 1$ and $K \in \mathbb{N}$ such that*

$$\text{(5)} \qquad\qquad \left| \frac{x_{n+1}}{x_n} \right| \le r \qquad \text{for} \quad n \ge K,$$

then the series $\sum x_n$ is absolutely convergent.

(b)   *If there exists $K \in \mathbb{N}$ such that*

$$\text{(6)} \qquad\qquad \left| \frac{x_{n+1}}{x_n} \right| \ge 1 \qquad \text{for} \quad n \ge K,$$

then the series $\sum x_n$ is divergent.

Proof.   (a) If (5) holds, an Induction argument shows that $|x_{K+m}| \le |x_K| r^m$ for $m \in \mathbb{N}$. Thus, for $n \ge K$ the terms in $\sum |x_n|$ are dominated by a fixed multiple of the terms in the geometric series $\sum r^m$ with $0 < r < 1$. The Comparison Test 3.7.7 then implies that $\sum |x_n|$ is convergent.

(b) If (6) holds, an Induction argument shows that $|x_{K+m}| \ge |x_K|$ for $m \in \mathbb{N}$ and the nth Term Test applies.   Q.E.D.

Once again we have a familiar result from calculus.

9.2.5 Corollary   *Let $X := (x_n)$ be a nonzero sequence in $\mathbb{R}$ and suppose that the limit*

$$\text{(7)} \qquad\qquad r := \lim \left| \frac{x_{n+1}}{x_n} \right|$$

exists in $\mathbb{R}$. Then $\sum x_n$ is absolutely convergent when $r < 1$ and is divergent when $r > 1$.

Proof.   If $r < 1$ and if $r < r_1 < 1$, then there exists $K \in \mathbb{R}$ such that $|x_{n+1}/x_n| < r_1$ for $n \ge K$. Thus Theorem 9.2.4(a) applies to give the absolute convergence of $\sum x_n$.

If $r > 1$, then there exists $K \in \mathbb{N}$ such that $|x_{n+1}/x_n| > 1$ for $n \ge K$, whence it follows that $|x_k|$ does not converge to 0 and the nth Term Test applies.   Q.E.D.

Note   No conclusion is possible in Corollary 9.2.5 when $r = 1$, for either convergence or divergence is possible. See Example 9.2.7(c).

The Integral Test

The next test—a very powerful one—uses the notion of the improper integral, which is defined as follows: If f is in $\mathcal{R}[a, b]$ for every $b > a$ and if the limit $\lim_{b \to \infty} \int_a^b f(t)\, dt$ exists in $\mathbb{R}$, then the **improper integral** $\int_a^\infty f(t)\, dt$ is defined to be this limit.

9.2.6 Integral Test   Let f be a positive, decreasing function on $\{t : t \geq 1\}$. Then the series $\sum_{k=1}^{\infty} f(k)$ converges if and only if the improper integral

$$\int_1^\infty f(t)\, dt = \lim_{b \to \infty} \int_1^b f(t)\, dt$$

exists. In the case of convergence, the partial sum $s_n = \sum_{k=1}^n f(k)$ and the sum $s = \sum_{k=1}^\infty f(k)$ satisfy the estimate

(8)
$$\int_{n+1}^\infty f(t)\, dt \leq s - s_n \leq \int_n^\infty f(t)\, dt.$$

Proof.   Since f is positive and decreasing on the interval $[k - 1, k]$, we have

(9)
$$f(k) \leq \int_{k-1}^k f(t)\, dt \leq f(k - 1).$$

By adding this inequality for $k = 2, 3, \cdots, n$, we obtain

$$s_n - f(1) \leq \int_1^n f(t)\, dt \leq s_{n-1},$$

which shows that either both or neither of the limits

$$\lim_{n \to \infty} s_n \qquad \text{and} \qquad \lim_{n \to \infty} \int_1^n f(t)\, dt$$

exist. If they exist, then on adding (9) for $k = n + 1, \cdots, m$, we obtain

$$s_m - s_n \leq \int_n^m f(t)\, dt \leq s_{m-1} - s_{n-1},$$

whence it follows that

$$\int_{n+1}^{m+1} f(t)\, dt \leq s_m - s_n \leq \int_n^m f(t)\, dt.$$

If we take the limit in this last inequality as $m \to \infty$, we obtain (8).   Q.E.D.

We will now show how the results in Theorems 9.2.1–9.2.6 can be applied to the p-series, which were introduced in Example 3.7.6(d,e).

9.2.7 Examples   (a)   Consider the case $p = 2$; that is, the series $\sum 1/n^2$. We compare it with the convergent series $\sum 1/(n(n + 1))$ of Example 3.7.2(c). Since

$$\left| \frac{1}{n^2} \div \frac{1}{n(n + 1)} \right| = \frac{n + 1}{n} = 1 + \frac{1}{n} \to 1,$$

the Limit Comparison Test 9.2.1 implies that $\sum 1/n^2$ is convergent.

(b)   We demonstrate the failure of the Root Test for the p-series. Note that

$$\left| \frac{1}{n^p} \right|^{1/n} = \frac{1}{(n^p)^{1/n}} = \frac{1}{(n^{1/n})^p}.$$

Since (see Example 3.1.11(d)) we know that $n^{1/n} \to 1$, we have $r = 1$ in Corollary 9.2.3, and the theorem does not give any information.

(c)   We apply the Ratio Test to the p-series. Since

$$\left| \frac{1}{(n+1)^p} \div \frac{1}{n^p} \right| = \frac{n^p}{(n+1)^p} = \frac{1}{(1+1/n)^p} \to 1,$$

the Ratio Test, in the form of Corollary 9.2.5, does not give any information.

(d)   Finally, we apply the Integral Test to the p-series. Let $f(t) := 1/t^p$ for $t \geq 1$ and recall that

$$\int_1^n \frac{1}{t} \, dt = \ln n - \ln 1,$$

$$\int_1^n \frac{1}{t^p} \, dt = \frac{1}{1-p} \left(\frac{1}{n^{p-1}} - 1 \right) \qquad \text{for} \quad p \neq 1.$$

From these relations we see that the p-series converges if $p > 1$ and diverges if $p \leq 1$, as we have seen before in 3.7.6(d,e). $\square$

Raabe's Test

If the limits $\lim |x_n|^{1/n}$ and $\lim(|x_{n+1}/x_n|)$ that are used in Corollaries 9.2.3 and 9.2.5 equal 1, we have seen that these tests do not give any information about the convergence or divergence of the series. In this case it is often useful to employ a more delicate test. Here is one that is frequently useful.

9.2.8 Raabe's Test   *Let $X := (x_n)$ be a sequence of nonzero real numbers.*

(a)   *If there exist numbers $a > 1$ and $K \in \mathbb{N}$ such that*

$$(10) \qquad \left| \frac{x_{n+1}}{x_n} \right| \leq 1 - \frac{a}{n} \qquad \text{for} \quad n \geq K,$$

then $\sum x_n$ is absolutely convergent.

(b)   *If there exist real numbers $a \leq 1$ and $K \in \mathbb{N}$ such that*

$$(11) \qquad \left| \frac{x_{n+1}}{x_n} \right| \geq 1 - \frac{a}{n} \qquad \text{for} \quad n \geq K,$$

then $\sum x_n$ is not absolutely convergent.

Proof.   (a) If the inequality (10) holds, then we have (after replacing n by k and multiplying)

$$k|x_{k+1}| \leq (k-1)|x_k| - (a-1)|x_k| \qquad \text{for} \quad k \geq K.$$

On reorganizing the inequality, we have

$$(12) \qquad (k-1)|x_k| - k|x_{k+1}| \geq (a-1)|x_k| > 0 \qquad \text{for} \quad k \geq K,$$

from which we deduce that the sequence $(k|x_{k+1}|)$ is decreasing for $k \geq K$. If we add (12) for $k = K, \cdots, n$ and note that the left side telescopes, we get

$$(K-1)|x_K| - n|x_{n+1}| \geq (a-1)\big(|x_K| + \cdots + |x_n|\big).$$

This shows (why?) that the partial sums of $\sum |x_n|$ are bounded and establishes the absolute convergence of the series.

(b) If the relation (11) holds for $n \geq K$, then since $a \leq 1$, we have

$$n|x_{n+1}| \geq (n - a)|x_n| \geq (n - 1)|x_n| \qquad \text{for} \quad n \geq K.$$

Therefore the sequence $(n|x_{n+1}|)$ is increasing for $n \geq K$ and there exists a number $c > 0$ such that $|x_{n+1}| > c/n$ for $n \geq K$. But since the harmonic series $\sum 1/n$ diverges, the series $\sum |x_n|$ also diverges.

Q.E.D.

In the application of Raabe's Test, it is often convenient to use the following limiting form.

9.2.9 Corollary *Let $X := (x_n)$ be a nonzero sequence in $\mathbb{R}$ and let*

$$(13) \qquad a := \lim\left(n\left(1 - \left|\frac{x_{n+1}}{x_n}\right|\right)\right),$$

whenever this limit exists. Then $\sum x_n$ is absolutely convergent when $a > 1$ and is not absolutely convergent when $a < 1$.

Proof. Suppose the limit in (13) exists and that $a > 1$. If a_1 is any number with $a > a_1 > 1$, then there exists $K \in \mathbb{N}$ such that $a_1 < n(1 - |x_{n+1}/x_n|)$ for $n > K$. Therefore $|x_{n+1}/x_n| < 1 - a_1/n$ for $n \geq K$ and Raabe's Test 9.2.8(a) applies.

The case where $a < 1$ is similar and is left to the reader.

Q.E.D.

Note There is no conclusion when $a = 1$; either convergence or divergence is possible, as the reader can show.

9.2.10 Examples **(a)** We reconsider the p-series in the light of Raabe's Test. Applying L'Hospital's Rule when $p \geq 1$, we obtain (why?)

$$a = \lim\left(n\left[1 - \frac{n^p}{(n+1)^p}\right]\right) = \lim\left(n\left[\frac{(n+1)^p - n^p}{(n+1)^p}\right]\right)$$

$$= \lim\left(\frac{(1+1/n)^p - 1}{1/n}\right) \cdot \lim\left(\frac{1}{(1+1/n)^p}\right) = p \cdot 1 = p.$$

We conclude that if $p > 1$ then the p-series is convergent, and if $0 < p < 1$ then the series is divergent (since the terms are positive). However, if $p = 1$ (the harmonic series!), Corollary 9.2.9 yields no information.

(b) We now consider $\sum_{n=1}^{\infty} \frac{n}{n^2 + 1}$.

An easy calculation shows that $\lim(x_{n+1}/x_n) = 1$, so that Corollary 9.2.5 does not apply. Also, we have $\lim(n(1 - x_{n+1}/x_n)) = 1$, so that Corollary 9.2.9 does not apply either. However, it is an exercise to establish the inequality $x_{n+1}/x_n \geq (n-1)/n$, whence it follows from Raabe's Test 9.2.8(b) that the series is divergent. (Of course, the Integral Test, or the Limit Comparison Test with $(y_n) = (1/n)$, can be applied here.) $\square$

Although the limiting form 9.2.9 of Rabbe's Test is much easier to apply, Example 9.2.10(b) shows that the form 9.2.8 is stronger than 9.2.9.

Exercises for Section 9.2

1. Establish the convergence or the divergence of the series whose nth term is:

   (a) $\dfrac{1}{(n+1)(n+2)}$,

   (b) $\dfrac{n}{(n+1)(n+2)}$,

   (c) $2^{-1/n}$,

   (d) $n/2^n$.

2. Establish the convergence or divergence of the series whose nth term is:

   (a) $(n(n+1))^{-1/2}$,

   (b) $(n^2(n+1))^{-1/2}$,

   (c) $n!/n^n$,

   (d) $(-1)^n n/(n+1)$.

3. Discuss the convergence or the divergence of the series with nth term (for sufficiently large n) given by

   (a) $(\ln n)^{-p}$,

   (b) $(\ln n)^{-n}$,

   (c) $(\ln n)^{-\ln n}$,

   (d) $(\ln n)^{-\ln \ln n}$,

   (e) $(n \ln n)^{-1}$,

   (f) $\left(n(\ln n)(\ln \ln n)^2\right)^{-1}$.

4. Discuss the convergence or the divergence of the series with nth term

   (a) $2^n e^{-n}$,

   (b) $n^n e^{-n}$,

   (c) $e^{-\ln n}$,

   (d) $(\ln n) e^{-\sqrt{n}}$,

   (e) $n! e^{-n}$,

   (f) $n! e^{-n^2}$.

5. Show that the series $1/1^2 + 1/2^3 + 1/3^2 + 1/4^3 + \cdots$ is convergent, but that both the Ratio and the Root Tests fail to apply.

6. If a and b are positive numbers, then $\sum (an+b)^{-p}$ converges if $p > 1$ and diverges if $p \le 1$.

7. Discuss the series whose nth term is

   (a) $\dfrac{n!}{3 \cdot 5 \cdot 7 \cdots (2n+1)}$,

   (b) $\dfrac{(n!)^2}{(2n)!}$,

   (c) $\dfrac{2 \cdot 4 \cdots (2n)}{3 \cdot 5 \cdots (2n+1)}$,

   (d) $\dfrac{2 \cdot 4 \cdots (2n)}{5 \cdot 7 \cdots (2n+3)}$.

8. Let $0 < a < 1$ and consider the series

   $$a^2 + a + a^4 + a^3 + \cdots + a^{2n} + a^{2n-1} + \cdots.$$

   Show that the Root Test applies, but that the Ratio Test does not apply.

9. If $r \in (0,1)$ satisfies (2) in the Root Test 9.2.2, show that the partial sums s_n of $\sum x_n$ approximate its limit s according to the estimate $|s - s_n| \le r^{n+1}/(1-r)$ for $n \ge K$.

10. If $r \in (0,1)$ satisfies (5) in the Ratio Test 9.2.4, show that $|s - s_n| \le r |x_n|/(1-r)$ for $n \ge K$.

11. If $a > 1$ satisfies (10) in Raabe's Test 9.2.8, show that $|s - s_n| \le n |x_n|/(a-1)$ for $n \ge K$.

12. For each of the series in Exercise 1 that converge, estimate the remainder if only four terms are taken. If only ten terms are taken. If we wish to determine the sum of the series within $1/1000$, how many terms should be taken?

13. Answer the questions posed in Exercise 12 for the series given in Exercise 2.

14. Show that the series $1 + \frac{1}{2} - \frac{1}{3} + \frac{1}{4} + \frac{1}{5} - \frac{1}{6} + + - \cdots$ is divergent.

15. For $n \in \mathbb{N}$, let c_n be defined by $c_n := \frac{1}{1} + \frac{1}{2} + \cdots + 1/n - \ln n$. Show that (c_n) is a decreasing sequence of positive numbers. The limit C of this sequence is called **Euler's Constant** and is approximately equal to 0.577. Show that if we put

   $$b_n := \frac{1}{1} - \frac{1}{2} + \frac{1}{3} - \cdots - \frac{1}{2n},$$

   then the sequence (b_n) converges to $\ln 2$. [*Hint:* $b_n = c_{2n} - c_n + \ln 2$.]

16. Let $\{n_1, n_2, \cdots\}$ denote the collection of natural numbers that do not use the digit 6 in their decimal expansion. Show that $\sum 1/n_k$ converges to a number less than 80. If $\{m_1, m_2, \cdots\}$ is the collection of numbers that end in 6, then $\sum 1/m_k$ diverges. If $\{p_1, p_2, \cdots, \}$ is the collection of numbers that do not end in 6, then $\sum 1/p_k$ diverges.

17. If $p > 0, q > 0$, show that the series

$$\sum \frac{(p+1)(p+2)\cdots(p+n)}{(q+1)(q+2)\cdots(q+n)}$$

converges for $q > p+1$ and diverges for $q \leq p+1$.

18. Suppose that none of the numbers a, b, c is a negative integer or zero. Prove that the **hypergeometric series**

$$\frac{ab}{1!c} + \frac{a(a+1)b(b+1)}{2!c\,(c+1)} + \frac{a(a+1)(a+2)b(b+1)(b+2)}{3!c(c+1)(c+2)} + \cdots$$

is absolutely convergent for $c > a + b$ and divergent for $c < a + b$.

19. Let $a_n > 0$ and suppose that $\sum a_n$ converges. Construct a convergent series $\sum b_n$ with $b_n > 0$ such that $\lim(a_n/b_n) = 0$; hence $\sum b_n$ converges less rapidly than $\sum a_n$. [*Hint:* Let (A_n) be the partial sums of $\sum a_n$ and A its limit. Define $b_1 := \sqrt{A} - \sqrt{A - A_1}$ and $b_n := \sqrt{A - A_{n-1}} - \sqrt{A - A_n}$ for $n \geq 1$.]

20. Let (a_n) be a decreasing sequence of real numbers converging to 0 and suppose that $\sum a_n$ diverges. Construct a divergent series $\sum b_n$ with $b_n > 0$ such that $\lim(b_n/a_n) = 0$; hence $\sum b_n$ diverges less rapidly than $\sum a_n$. [*Hint:* Let $b_n := a_n/\sqrt{A_n}$ where A_n is the nth partial sum of $\sum a_n$.]

Section 9.3   Tests for Nonabsolute Convergence

The convergence tests that were discussed in the preceding section were primarily directed to establishing the absolute convergence of a series. Since there are many series, such as

(1)
$$\sum_{n=1}^{\infty} \frac{(-1)^{n+1}}{n}, \qquad \sum_{n=1}^{\infty} \frac{(-1)^{n+1}}{\sqrt{n}},$$

that are convergent but not absolutely convergent, it is desirable to have some tests for this phenomenon. In this short section we shall present first the test for alternating series and then tests for more general series due to Dirichlet and Abel.

Alternating Series

The most familiar test for nonabsolutely convergent series is the one due to Leibniz that is applicable to series that are "alternating" in the following sense.

9.3.1 Definition   A sequence $X := (x_n)$ of nonzero real numbers is said to be **alternating** if the terms $(-1)^{n+1}x_n, n \in \mathbb{N}$, are all positive (or all negative) real numbers. If the sequence $X = (x_n)$ is alternating, we say that the series $\sum x_n$ it generates is an **alternating series**.

In the case of an alternating series, it is useful to set $x_n = (-1)^{n+1}z_n$ [or $x_n = (-1)^n z_n$], where $z_n > 0$ for all $n \in \mathbb{N}$.

9.3.2 Alternating Series Test   *Let $Z := (z_n)$ be a decreasing sequence of strictly positive numbers with $\lim(z_n) = 0$. Then the alternating series $\sum (-1)^{n+1}z_n$ is convergent.*

Proof.   Since we have

$$s_{2n} = (z_1 - z_2) + (z_3 - z_4) + \cdots + (z_{2n-1} - z_{2n}),$$

and since $z_k - z_{k+1} \geq 0$, it follows that the subsequence (s_{2n}) of partial sums is increasing. Since

$$s_{2n} = z_1 - (z_2 - z_3) - \cdots - (z_{2n-2} - z_{2n-1}) - z_{2n},$$

it also follows that $s_{2n} \leq z_1$ for all $n \in \mathbb{N}$. It follows from the Monotone Convergence Theorem 3.3.2 that the subsequence (s_{2n}) converges to some number $s \in \mathbb{R}$.

We now show that the entire sequence (s_n) converges to s. Indeed, if $\varepsilon > 0$, let K be such that if $n \geq K$ then $|s_{2n} - s| \leq \frac{1}{2}\varepsilon$ and $|z_{2n+1}| \leq \frac{1}{2}\varepsilon$. It follows that if $n \geq K$ then

$$\begin{aligned} |s_{2n+1} - s| &= |s_{2n} + z_{2n+1} - s| \\ &\leq |s_{2n} - s| + |z_{2n+1}| \leq \tfrac{1}{2}\varepsilon + \tfrac{1}{2}\varepsilon = \varepsilon. \end{aligned}$$

Therefore every partial sum of an odd number of terms is also within ε of s if n is large enough. Since $\varepsilon > 0$ is arbitrary, the convergence of (s_n) and hence of $\sum (-1)^{n+1} z_n$ is established.   Q.E.D.

Note   It is an exercise to show that if s is the sum of the alternating series and if s_n is its nth partial sum, then

(2) $$|s - s_n| \leq z_{n+1}.$$

It is clear that this Alternating Series Test establishes the convergence of the two series already mentioned, in (1).

The Dirichlet and Abel Tests

We will now present two other tests of wide applicability. They are based on the following lemma, which is sometimes called the **partial summation formula,** since it corresponds to the familiar formula for integration by parts.

9.3.3 Abel's Lemma   *Let* $X := (x_n)$ *and* $Y := (y_n)$ *be sequences in* $\mathbb{R}$ *and let the partial sums of* $\sum y_n$ *be denoted by* (s_n) *with* $s_0 := 0$. *If* $m > n$, *then*

(3) $$\sum_{k=n+1}^{m} x_k y_k = (x_m s_m - x_{n+1} s_n) + \sum_{k=n+1}^{m-1} (x_k - x_{k+1}) s_k.$$

Proof.   Since $y_k = s_k - s_{k-1}$ for $k = 1, 2, \cdots$, the left side of (3) is seen to be equal to $\sum_{k=n+1}^{m} x_k(s_k - s_{k-1})$. If we collect the terms multiplying $s_n, s_{n+1}, \cdots, s_m$, we obtain the right side of (3).   Q.E.D.

We now apply Abel's Lemma to obtain tests for convergence of series of the form $\sum x_n y_n$.

9.3.4 Dirichlet's Test   *If* $X := (x_n)$ *is a decreasing sequence with* $\lim x_n = 0$, *and if the partial sums* (s_n) *of* $\sum y_n$ *are bounded, then the series* $\sum x_n y_n$ *is convergent.*

Proof.   Let $|s_n| \leq B$ for all $n \in \mathbb{N}$. If $m > n$, it follows from Abel's Lemma 9.3.3 and the fact that $x_k - x_{k+1} \geq 0$ that

$$\left| \sum_{k=n+1}^{m} x_k y_k \right| \leq (x_m + x_{n+1})B + \sum_{k=n+1}^{m-1} (x_k - x_{k+1})B$$
$$= [(x_m + x_{n+1}) + (x_{n+1} - x_m)]B$$
$$= 2x_{n+1}B.$$

Since $\lim(x_k) = 0$, the convergence of $\sum x_k y_k$ follows from the Cauchy Convergence Criterion 3.7.4.   Q.E.D.

9.3.5 Abel's Test   *If $X := (x_n)$ is a convergent monotone sequence and the series $\sum y_n$ is convergent, then the series $\sum x_n y_n$ is also convergent.*

Proof.   If (x_n) is decreasing with limit x, let $u_n := x_n - x$, $n \in \mathbb{N}$, so that (u_n) decreases to 0. Then $x_n = x + u_n$, whence $x_n y_n = x y_n + u_n y_n$. It follows from the Dirichlet Test 9.3.4 that $\sum u_n y_n$ is convergent and, since $\sum x y_n$ converges (because of the assumed convergence of the series $\sum y_n$), we conclude that $\sum x_n y_n$ is convergent.

If (x_n) is increasing with limit x, let $v_n := x - x_n$, $n \in \mathbb{N}$, so that (v_n) decreases to 0. Here $x_n = x - v_n$, whence $x_n y_n = x y_n - v_n y_n$, and the argument proceeds as before.   Q.E.D.

9.3.6 Examples   (a)   Since we have

$$2 \left(\sin \tfrac{1}{2}x \right) \left(\cos x + \cdots + \cos nx \right) = \sin \left(n + \tfrac{1}{2} \right) x - \sin \tfrac{1}{2}x,$$

it follows that if $x \neq 2k\pi$ $(k \in \mathbb{N})$, then

$$|\cos x + \cdots + \cos nx| = \frac{\left| \sin \left(n + \tfrac{1}{2} \right) x - \sin \tfrac{1}{2}x \right|}{\left| 2 \sin \tfrac{1}{2}x \right|} \leq \frac{1}{\left| \sin \tfrac{1}{2}x \right|}.$$

Hence Dirichlet's Test implies that if (a_n) is decreasing with $\lim (a_n) = 0$, then the series $\sum_{n=1}^{\infty} a_n \cos nx$ converges provided $x \neq 2k\pi$.

(b)   Since we have

$$2 \left(\sin \tfrac{1}{2}x \right) \left(\sin x + \cdots + \sin nx \right) = \cos \tfrac{1}{2}x - \cos \left(n + \tfrac{1}{2} \right) x,$$

it follows that if $x \neq 2k\pi$ $(k \in \mathbb{N})$, then

$$|\sin x + \cdots + \sin nx| \leq \frac{1}{\left| \sin \tfrac{1}{2}x \right|}.$$

As before, if (a_n) is decreasing and if $\lim(a_n) = 0$, then the series $\sum_{n=1}^{\infty} a_n \sin nx$ converges for $x \neq 2k\pi$ (and it also converges for these values).   □

Exercises for Section 9.3

1.   Test the following series for convergence and for absolute convergence.

(a) $\displaystyle\sum_{n=1}^{\infty} \frac{(-1)^{n+1}}{n^2 + 1}$,

(b) $\displaystyle\sum_{n=1}^{\infty} \frac{(-1)^{n+1}}{n + 1}$,

(c) $\displaystyle\sum_{n=1}^{\infty} \frac{(-1)^{n+1} n}{n + 2}$,

(d) $\displaystyle\sum_{n=1}^{\infty} (-1)^{n+1} \frac{\ln n}{n}$.

2. If s_n is the nth partial sum of the alternating series $\sum_{n=1}^{\infty} (-1)^{n+1} z_n$, and if s denotes the sum of this series, show that $|s - s_n| \leq z_{n+1}$.

3. Give an example to show that the Alternating Series Test 9.3.2 may fail if (z_n) is not a decreasing sequence.

4. Show that the Alternating Series Test is a consequence of Dirichlet's Test 9.3.4.

5. Consider the series
$$1 - \frac{1}{2} - \frac{1}{3} + \frac{1}{4} + \frac{1}{5} - \frac{1}{6} - \frac{1}{7} + + - - \cdots,$$
where the signs come in pairs. Does it converge?

6. Let $a_n \in \mathbb{R}$ for $n \in \mathbb{N}$ and let $p < q$. If the series $\sum a_n / n^p$ is convergent, show that the series $\sum a_n / n^q$ is also convergent.

7. If p and q are positive numbers, show that $\sum (-1)^n (\ln n)^p / n^q$ is a convergent series.

8. Discuss the series whose nth term is:

(a) $(-1)^n \dfrac{n^n}{(n+1)^{n+1}}$,

(b) $\dfrac{n^n}{(n+1)^{n+1}}$,

(c) $(-1)^n \dfrac{(n+1)^n}{n^n}$,

(d) $\dfrac{(n+1)^n}{n^{n+1}}$.

9. If the partial sums of $\sum a_n$ are bounded, show that the series $\sum_{n=1}^{\infty} a_n e^{-nt}$ converges for $t > 0$.

10. If the partial sums s_n of $\sum_{n=1}^{\infty} a_n$ are bounded, show that the series $\sum_{n=1}^{\infty} a_n / n$ converges to $\sum_{n=1}^{\infty} s_n / n(n+1)$.

11. Can Dirichlet's Test be applied to establish the convergence of
$$1 - \frac{1}{2} - \frac{1}{3} + \frac{1}{4} + \frac{1}{5} + \frac{1}{6} - \cdots$$
where the number of signs increases by one in each "block"? If not, use another method to establish the convergence of this series.

12. Show that the hypotheses that the sequence $X := (x_n)$ is decreasing in Dirichlet's Test 9.3.4 can be replaced by the hypothesis that $\sum_{n=1}^{\infty} |x_n - x_{n+1}|$ is convergent.

13. If (a_n) is a bounded decreasing sequence and (b_n) is a bounded increasing sequence and if $x_n := a_n + b_n$ for $n \in \mathbb{N}$, show that $\sum_{n=1}^{\infty} |x_n - x_{n+1}|$ is convergent.

14. Show that if the partial sums s_n of the series $\sum_{k=1}^{\infty} a_k$ satisfy $|s_n| \leq M n^r$ for some $r < 1$, then the series $\sum_{n=1}^{\infty} a_n / n$ converges.

15. Suppose that $\sum a_n$ is a convergent series of real numbers. Either prove that $\sum b_n$ converges or give a counter-example, when we define b_n by

(a) a_n / n,

(b) $\sqrt{a_n}/n$ $(a_n \geq 0)$,

(c) $a_n \sin n$,

(d) $\sqrt{a_n/n}$ $(a_n \geq 0)$,

(e) $n^{1/n} a_n$,

(f) $a_n / (1 + |a_n|)$.

Section 9.4   Series of Functions

Because of their frequent appearance and importance, we now present a discussion of infinite series of functions. Since the convergence of an infinite series is handled by examining the sequence of partial sums, questions concerning series of functions are answered by examining corresponding questions for sequences of functions. For this reason, a portion of the present section is merely a translation of facts already established for sequences of functions into series terminology. However, in the second part of the section, where we discuss power series, some new features arise because of the special character of the functions involved.

9.4.1 Definition   If (f_n) is a sequence of functions defined on a subset D of $\mathbb{R}$ with values in $\mathbb{R}$, the sequence of **partial sums** (s_n) of the infinite series $\sum f_n$ is defined for x in D by

$$
\begin{aligned}
s_1(x) &:= f_1(x), \\
s_2(x) &:= s_1(x) + f_2(x)
\end{aligned}
$$
....
$$
s_{n+1}(x) := s_n(x) + f_{n+1}(x)
$$
....

In case the sequence (s_n) of functions converges on D to a function f, we say that the infinite series of functions $\sum f_n$ **converges** to f on D. We will often write

$$
\sum f_n \qquad \text{or} \qquad \sum_{n=1}^{\infty} f_n
$$

to denote either the series or the limit function, when it exists.

If the series $\sum |f_n(x)|$ converges for each x in D, we say that $\sum f_n$ is **absolutely convergent** on D. If the sequence (s_n) of partial sums is uniformly convergent on D to f, we say that $\sum f_n$ is **uniformly convergent** on D, or that it **converges to f uniformly on D**.

One of the main reasons for the interest in uniformly convergent series of functions is the validity of the following results which give conditions justifying the change of order of the summation and other limiting operations.

9.4.2 Theorem   *If f_n is continuous on $D \subseteq \mathbb{R}$ to $\mathbb{R}$ for each $n \in \mathbb{N}$ and if $\sum f_n$ converges to f uniformly on D, then f is continuous on D.*

This is a direct translation of Theorem 8.2.2 for series. The next result is a translation of Theorem 8.2.4.

9.4.3 Theorem   *Suppose that the real-valued functions $f_n, n \in \mathbb{N}$, are Riemann integrable on the interval $J := [a, b]$. If the series $\sum f_n$ converges to f uniformly on J, then f is Riemann integrable and*

(1)
$$
\int_a^b f = \sum_{n=1}^{\infty} \int_a^b f_n.
$$

Next we turn to the corresponding theorem pertaining to differentiation. Here we assume the uniform convergence of the series obtained after term-by-term differentiation of the given series. This result is an immediate consequence of Theorem 8.2.3.

9.4.4 Theorem   *For each $n \in \mathbb{N}$, let f_n be a real-valued function on $J := [a, b]$ that has a derivative f_n' on J. Suppose that the series $\sum f_n$ converges for at least one point of J and that the series of derivatives $\sum f_n'$ converges uniformly on J.*

Then there exists a real-valued function f on J such that $\sum f_n$ converges uniformly on J to f. In addition, f has a derivative on J and $f' = \sum f_n'$.

Tests for Uniform Convergence ───────────────────────────────

Since we have stated some consequences of uniform convergence of series, we shall now present a few tests that can be used to establish uniform convergence.

9.4.5 Cauchy Criterion   *Let (f_n) be a sequence of functions on $D \subseteq \mathbb{R}$ to $\mathbb{R}$. The series $\sum f_n$ is uniformly convergent on D if and only if for every $\varepsilon > 0$ there exists an $M(\varepsilon)$ such that if $m > n \geq M(\varepsilon)$, then*

$$\left| f_{n+1}(x) + \cdots + f_m(x) \right| < \varepsilon \qquad \text{for all} \quad x \in D.$$

9.4.6 Weierstrass M-Test   *Let (M_n) be a sequence of positive real numbers such that $\left| f_n(x) \right| \leq M_n$ for $x \in D$, $n \in \mathbb{N}$. If the series $\sum M_n$ is convergent, then $\sum f_n$ is uniformly convergent on D.*

Proof.   If $m > n$, we have the relation

$$\left| f_{n+1}(x) + \cdots + f_m(x) \right| \leq M_{n+1} + \cdots + M_m \qquad \text{for} \quad x \in D.$$

Now apply 3.7.4, 9.4.5, and the convergence of $\sum M_n$.   Q.E.D.

In Appendix E we will use the Weierstrass M-Test to construct two interesting examples.

Power Series ───

We shall now turn to a discussion of power series. This is an important class of series of functions and enjoys properties that are *not* valid for general series of functions.

9.4.7 Definition   A series of real functions $\sum f_n$ is said to be a **power series around** $x = c$ if the function f_n has the form

$$f_n(x) = a_n(x - c)^n,$$

where a_n and c belong to $\mathbb{R}$ and where $n = 0, 1, 2, \cdots$.

For the sake of simplicity of our notation, we shall treat only the case where $c = 0$. This is no loss of generality, however, since the translation $x' = x - c$ reduces a power series around c to a power series around 0. Thus, whenever we refer to a power series, we shall mean a series of the form

(2) $$\sum_{n=0}^{\infty} a_n x^n = a_0 + a_1 x + \cdots + a_n x^n + \cdots.$$

Even though the functions appearing in (2) are defined over all of $\mathbb{R}$, it is not to be expected that the series (2) will converge for all x in $\mathbb{R}$. For example, by using the Ratio Test 9.2.4, we can show that the series

$$\sum_{n=0}^{\infty} n!\, x^n, \qquad \sum_{n=0}^{\infty} x^n, \qquad \sum_{n=0}^{\infty} x^n/n!,$$

converge for x in the sets

$$\{0\}, \qquad \{x \in \mathbb{R} : |x| < 1\}, \qquad \mathbb{R},$$

respectively. Thus, the set on which a power series converges may be small, medium, or large. However, an arbitrary subset of $\mathbb{R}$ cannot be the precise set on which a power series converges, as we shall show.

If (b_n) is a bounded sequence of nonnegative real numbers, then we define the **limit superior** of (b_n) to be the infimum of those numbers v such that $b_n \leq v$ for all sufficiently large $n \in \mathbb{N}$. This infimum is uniquely determined and is denoted by $\lim \sup(b_n)$. The only facts we need to know are (i) that if $v > \lim \sup(b_n)$, then $b_n \leq v$ for all sufficiently large $n \in \mathbb{N}$, and (ii) that if $w < \lim \sup(b_n)$, then $w \leq b_n$ for infinitely many $n \in \mathbb{N}$.

9.4.8 Definition   Let $\sum a_n x^n$ be a power series. If the sequence $(|a_n|^{1/n})$ is bounded, we set $\rho := \lim \sup(|a_n|^{1/n})$; if this sequence is not bounded we set $\rho = +\infty$. We define the **radius of convergence** of $\sum a_n x^n$ to be given by

$$R := \begin{cases} 0 & \text{if} \quad \rho = +\infty, \\ 1/\rho & \text{if} \quad 0 < \rho < +\infty, \\ +\infty & \text{if} \quad \rho = 0. \end{cases}$$

The **interval of convergence** is the open interval $(-R, R)$.

We shall now justify the term "radius of convergence".

9.4.9 Cauchy-Hadamard Theorem   *If R is the radius of convergence of the power series $\sum a_n x^n$, then the series is absolutely convergent if $|x| < R$ and is divergent if $|x| > R$.*

Proof.   We shall treat only the case where $0 < R < +\infty$, leaving the cases $R = 0$ and $R = +\infty$ as exercises. If $0 < |x| < R$, then there exists a positive number $c < 1$ such that $|x| < cR$. Therefore $\rho < c/|x|$ and so it follows that if n is sufficiently large, then $|a_n|^{1/n} \leq c/|x|$. This is equivalent to the statement that

(3) $$|a_n x^n| \leq c^n$$

for all sufficiently large n. Since $c < 1$, the absolute convergence of $\sum a_n x^n$ follows from the Comparison Test 3.7.7.

If $|x| > R = 1/\rho$, then there are infinitely many $n \in \mathbb{N}$ for which $|a_n|^{1/n} > 1/|x|$. Therefore, $|a_n x^n| > 1$ for infinitely many n, so that the sequence $(a_n x^n)$ does not converge to zero.   Q.E.D.

Remark   It will be noted that the Cauchy-Hadamard Theorem makes no statement as to whether the power series converges when $|x| = R$. Indeed, anything can happen, as the examples

$$\sum x^n, \qquad \sum \frac{1}{n} x^n, \qquad \sum \frac{1}{n^2} x^n,$$

show. Since $\lim(n^{1/n}) = 1$, each of these power series has radius of convergence equal to 1. The first power series converges at neither of the points $x = -1$ and $x = +1$; the second series converges at $x = -1$ but diverges at $x = +1$; and the third power series converges at both $x = -1$ and $x = +1$. (Find a power series with $R = 1$ that converges at $x = +1$ but diverges at $x = -1$.)

It is an exercise to show that the radius of convergence of the series $\sum a_n x^n$ is also given by

(4)
$$\lim \left| \frac{a_n}{a_{n+1}} \right|,$$

provided this limit exists. Frequently, it is more convenient to use (4) than Definition 9.4.8.

The argument used in the proof of the Cauchy-Hadamard Theorem yields the uniform convergence of the power series on any fixed closed and bounded interval in the interval of convergence $(-R, R)$.

9.4.10 Theorem   *Let R be the radius of convergence of $\sum a_n x^n$ and let K be a closed and bounded interval contained in the interval of convergence $(-R, R)$. Then the power series converges uniformly on K.*

Proof.   The hypothesis on $K \subseteq (-R, R)$ implies that there exists a positive constant $c < 1$ such that $|x| < cR$ for all $x \in K$. (Why?) By the argument in 9.4.9, we infer that for sufficiently large n, the estimate (3) holds for all $x \in K$. Since $c < 1$, the uniform convergence of $\sum a_n x^n$ on K is a direct consequence of the Weierstrass M-test with $M_n := c^n$.   Q.E.D.

9.4.11 Theorem   *The limit of a power series is continuous on the interval of convergence. A power series can be integrated term-by-term over any closed and bounded interval contained in the interval of convergence.*

Proof.   If $|x_0| < R$, then the preceding result asserts that $\sum a_n x^n$ converges uniformly on any closed and bounded neighborhood of x_0 contained in $(-R, R)$. The continuity at x_0 then follows from Theorem 9.4.2, and the term-by-term integration is justified by Theorem 9.4.3.   Q.E.D.

We now show that a power series can be differentiated term-by-term. Unlike the situation for general series, we do not need to assume that the differentiated series is uniformly convergent. Hence this result is stronger than Theorem 9.4.4.

9.4.12 Differentiation Theorem   *A power series can be differentiated term-by-term within the interval of convergence. In fact, if*

$$f(x) = \sum_{n=0}^{\infty} a_n x^n, \qquad then \qquad f'(x) = \sum_{n=1}^{\infty} n a_n x^{n-1} \qquad for \quad |x| < R.$$

Both series have the same radius of convergence.

Proof.   Since $\lim(n^{1/n}) = 1$, the sequence $(|na_n|^{1/n})$ is bounded if and only if the sequence $(|a_n|^{1/n})$ is bounded. Moreover, it is easily seen that

$$\limsup \left(|na_n|^{1/n} \right) = \limsup \left(|a_n|^{1/n} \right).$$

Therefore, the radius of convergence of the two series is the same, so the formally differentiated series is uniformly convergent on each closed and bounded interval contained in the interval of convergence. We can then apply Theorem 9.4.4 to conclude that the formally differentiated series converges to the derivative of the given series.   Q.E.D.

Remark It is to be observed that the theorem makes no assertion about the endpoints of the interval of convergence. If a series is convergent at an endpoint, then the differentiated series may or may not be convergent at this point. For example, the series $\sum_{n=1}^{\infty} x^n/n^2$ converges at both endpoints $x = -1$ and $x = +1$. However, the differentiated series given by $\sum_{n=1}^{\infty} x^{n-1}/n$ converges at $x = -1$ but diverges at $x = +1$.

By repeated application of the preceding result, we conclude that if $k \in \mathbb{N}$ then $\sum_{n=0}^{\infty} a_n x^n$ can be differentiated term-by-term k times to obtain

$$(5) \qquad \sum_{n=k}^{\infty} \frac{n!}{(n-k)!} a_n x^{n-k}.$$

Moreover, this series converges absolutely to $f^{(k)}(x)$ for $|x| < R$ and uniformly over any closed and bounded interval in the interval of convergence. If we substitute $x = 0$ in (5), we obtain the important formula

$$f^{(k)}(0) = k! a_k.$$

9.4.13 Uniqueness Theorem *If $\sum a_n x^n$ and $\sum b_n x^n$ converge on some interval $(-r, r), r > 0$, to the same function f, then*

$$a_n = b_n \qquad \text{for all} \quad n \in \mathbb{N}.$$

Proof. Our preceding remarks show that $n! a_n = f^{(n)}(0) = n! b_n$ for all $n \in \mathbb{N}$. Q.E.D.

Taylor Series ───────────────────────────────────────

If a function f has derivatives of all orders at a point c in $\mathbb{R}$, then we can calculate the Taylor coefficients by $a_0 := f(c), a_n := f^{(n)}(c)/n!$ for $n \in \mathbb{N}$ and in this way obtain a power series with these coefficients. However, it is not necessarily true that the resulting power series converges to the function f in an interval about c. (See Exercise 12 for an example.) The issue of convergence is resolved by the remainder term R_n in Taylor's Theorem 6.4.1. We will write

$$(6) \qquad f(x) = \sum_{n=0}^{\infty} \frac{f^{(n)}(c)}{n!}(x-c)^n$$

for $|x - c| < R$ if and only if the sequence $(R_n(x))$ of remainders converges to 0 for each x in some interval $\{x: |x - c| < R\}$. In this case we say that the power series (6) is the **Taylor expansion** of f at c. We observe that the Taylor polynomials for f discussed in Section 6.4 are just the partial sums of the Taylor expansion (6) of f. (Recall that $0! := 1$.)

9.4.14 Examples (a) If $f(x) := \sin x, x \in \mathbb{R}$, we have $f^{(2n)}(x) = (-1)^n \sin x$ and $f^{(2n+1)}(x) = (-1)^n \cos x$ for $n \in \mathbb{N}, x \in \mathbb{R}$. Evaluating at $c = 0$, we get the Taylor coefficients $a_{2n} = 0$ and $a_{2n+1} = (-1)^n/(2n+1)!$ for $n \in \mathbb{N}$. Since $|\sin x| \le 1$ and $|\cos x| \le 1$ for all x, then $|R_n(x)| \le |x|^n/n!$ for $n \in \mathbb{N}$ and $x \in \mathbb{R}$. Since $\lim(R_n(x)) = 0$ for each $x \in \mathbb{R}$, we obtain the Taylor expansion

$$\sin x = \sum_{n=0}^{\infty} \frac{(-1)^n}{(2n+1)!} x^{2n+1} \qquad \text{for all} \quad x \in \mathbb{R}.$$

An application of Theorem 9.4.12 gives us the Taylor expansion

$$\cos x = \sum_{n=0}^{\infty} \frac{(-1)^n}{(2n)!} x^{2n} \qquad \text{for all} \quad x \in \mathbb{R}.$$

(b) If $g(x) := e^x, x \in \mathbb{R}$, then $g^{(n)}(x) = e^x$ for all $n \in \mathbb{N}$, and hence the Taylor coefficients are given by $a_n = 1/n!$ for $n \in \mathbb{N}$. For a given $x \in \mathbb{R}$, we have $\left|R_n(x)\right| \le e^{|x|}|x|^n /n!$ and therefore $(R_n(x))$ tends to 0 as $n \to \infty$. Therefore, we obtain the Taylor expansion

(7)
$$e^x = \sum_{n=0}^{\infty} \frac{1}{n!} x^n \qquad \text{for all} \quad x \in \mathbb{R}.$$

We can obtain the Taylor expansion at an arbitrary $c \in \mathbb{R}$ by the device of replacing x by $x - c$ in (7) and noting that

$$e^x = e^c \cdot e^{x-c} = e^c \sum_{n=0}^{\infty} \frac{1}{n!} (x-c)^n = \sum_{n=0}^{\infty} \frac{e^c}{n!} (x-c)^n \qquad \text{for} \quad x \in \mathbb{R}. \qquad \square$$

Exercises for Section 9.4

1. Discuss the convergence and the uniform convergence of the series $\sum f_n$, where $f_n(x)$ is given by:

   (a) $(x^2 + n^2)^{-1}$,   (b) $(nx)^{-2}$ $(x \neq 0)$,
   (c) $\sin(x/n^2)$,   (d) $(x^n + 1)^{-1}$ $(x \neq 0)$,
   (e) $x^n/(x^n + 1)$ $(x \ge 0)$,   (f) $(-1)^n(n + x)^{-1}$ $(x \ge 0)$.

2. If $\sum a_n$ is an absolutely convergent series, then the series $\sum a_n \sin nx$ is absolutely and uniformly convergent.

3. Let (c_n) be a decreasing sequence of positive numbers. If $\sum c_n \sin nx$ is uniformly convergent, then $\lim(nc_n) = 0$.

4. Discuss the cases $R = 0$, $R = +\infty$ in the Cauchy-Hadamard Theorem 9.4.9.

5. Show that the radius of convergence R of the power series $\sum a_n x^n$ is given by $\lim \left(\left|a_n/a_{n+1}\right|\right)$ whenever this limit exists. Give an example of a power series where this limit does not exist.

6. Determine the radius of convergence of the series $\sum a_n x^n$, where a_n is given by:

   (a) $1/n^n$,   (b) $n^\alpha/n!$,
   (c) $n^n/n!$,   (d) $(\ln n)^{-1}$, $n \ge 2$,
   (e) $(n!)^2 / (2n)!$,   (f) $n^{-\sqrt{n}}$.

7. If $a_n := 1$ when n is the square of a natural number and $a_n := 0$ otherwise, find the radius of convergence of $\sum a_n x^n$. If $b_n := 1$ when $n = m!$ for $m \in \mathbb{N}$ and $b_n := 0$ otherwise, find the radius of convergence of the series $\sum b_n x^n$.

8. Prove in detail that $\limsup(\left|na_n\right|^{1/n}) = \limsup(\left|a_n\right|^{1/n})$.

9. If $0 < p \le \left|a_n\right| \le q$ for all $n \in \mathbb{N}$, find the radius of convergence of $\sum a_n x^n$.

10. Let $f(x) = \sum a_n x^n$ for $|x| < R$. If $f(x) = f(-x)$ for all $|x| < R$, show that $a_n = 0$ for all odd n.

11. Prove that if f is defined for $|x| < r$ and if there exists a constant B such that $\left|f^{(n)}(x)\right| \le B$ for all $|x| < r$ and $n \in \mathbb{N}$, then the Taylor series expansion

$$\sum_{n=0}^{\infty} \frac{f^{(n)}(0)}{n!} x^n$$

converges to $f(x)$ for $|x| < r$.

12. Prove by Induction that the function given by $f(x) := e^{-1/x^2}$ for $x \neq 0$, $f(0) := 0$, has derivatives of all orders at every point and that all of these derivatives vanish at $x = 0$. Hence this function is not given by its Taylor expansion about $x = 0$.

13. Give an example of a function which is equal to its Taylor series expansion about $x = 0$ for $x \geq 0$, but which is not equal to this expansion for $x < 0$.

14. Use the Lagrange form of the remainder to justify the general Binomial Expansion
$$(1+x)^m = \sum_{n=0}^{\infty} \binom{m}{n} x^n \qquad \text{for} \quad 0 \leq x < 1.$$

15. (Geometric series) Show directly that if $|x| < 1$, then $1/(1-x) = \sum_{n=0}^{\infty} x^n$.

16. Show by integrating the series for $1/(1+x)$ that if $|x| < 1$, then
$$\ln(1+x) = \sum_{n=1}^{\infty} \frac{(-1)^{n+1}}{n} x^n.$$

17. Show that if $|x| < 1$, then $\text{Arctan } x = \sum_{n=0}^{\infty} \frac{(-1)^n}{2n+1} x^{2n+1}$.

18. Show that if $|x| < 1$, then $\text{Arcsin } x = \sum_{n=0}^{\infty} \frac{1 \cdot 3 \cdots (2n-1)}{2 \cdot 4 \cdots 2n} \cdot \frac{x^{2n+1}}{2n+1}$.

19. Find a series expansion for $\int_0^x e^{-t^2} \, dt$ for $x \in \mathbb{R}$.

20. If $\alpha \in \mathbb{R}$ and $|k| < 1$, the integral $F(\alpha, k) := \int_0^{\alpha} \left(1 - k^2(\sin x)^2\right)^{-1/2} dx$ is called an **elliptic integral of the first kind**. Show that
$$F\left(\frac{\pi}{2}, k\right) = \frac{\pi}{2} \sum_{n=0}^{\infty} \left(\frac{1 \cdot 3 \cdots (2n-1)}{2 \cdot 4 \cdots 2n}\right)^2 k^{2n} \qquad \text{for} \quad |k| < 1.$$

CHAPTER 10

THE GENERALIZED RIEMANN INTEGRAL

In Chapter 7 we gave a rather complete discussion of the Riemann integral of a function on a closed bounded interval, defining the integral as the limit of Riemann sums of the function. This is the integral (and the approach) that the reader met in calculus courses; it is also the integral that is most frequently used in applications to engineering and other areas. We have seen that continuous and monotone functions on $[a, b]$ are Riemann integrable, so most of the functions arising in calculus are included in its scope.

However, by the end of the 19th century, some inadequacies in the Riemann theory of integration had become apparent. These failings came primarily from the fact that the collection of Riemann integrable functions became inconveniently small as mathematics developed. For example, the set of functions for which the Newton-Leibniz formula:

$$\int_a^b F' = F(b) - F(a)$$

holds, does not include *all* differentiable functions. Also, limits of sequences of Riemann integrable functions are not necessarily Riemann integrable. These inadequacies led others to invent other integration theories, the best known of which was due to Henri Lebesgue (1875–1941) and was developed at the very beginning of the 20th century. (For an account of the history of the development of the Lebesgue integral, the reader should consult the book of Hawkins given in the References.)

Indeed, the Lebesgue theory of integration has become pre-eminent in contemporary mathematical research, since it enables one to integrate a much larger collection of functions, and to take limits of integrals more freely. However, the Lebesgue integral also has several inadequacies and difficulties: (1) There exist functions F that are differentiable on $[a, b]$ but such that F' is not Lebesgue integrable. (2) Some "improper integrals", such as the important Dirichlet integral:

$$\int_0^\infty \frac{\sin x}{x}\, dx,$$

do not exist as Lebesgue integrals. (3) Most treatments of the Lebesgue integral have considerable prerequisites and are not easily within the reach of an undergraduate student of mathematics.

As important as the Lebesgue integral is, there are even more inclusive theories of integration. One of these was developed independently in the late 1950s by the Czech mathematician Jaroslav Kurzweil (b. 1926) and the English mathematician Ralph Henstock (b. 1923). Surprisingly, their approach is only slightly different from that used by Riemann, yet it yields an integral (which we will call the *generalized Riemann integral*) that includes both the Riemann and the Lebesgue integrals as special cases. Since the approach is so

Ralph Henstock and Jaroslav Kurzweil

Ralph Henstock (pictured on the left) was born on June 2, 1923, in Nottinghamshire, England, the son of a mineworker. At an early age he showed that he was a gifted scholar in mathematics and science. He entered St. John's College, Cambridge, in 1941, studying with J. D. Bernal, G. H. Hardy, and J. C. Burkhill and was classified Wrangler in Part II of the Tripos Exams in 1943. He earned his B.A. at Cambridge in 1944 and his Ph.D. at the University of London in 1948. His research is in the theory of summability, linear analysis, and integration theory. Most of his teaching has been in Northern
Ireland. He is presently an Emeritus Professor at the Coleraine Campus of the University of Ulster.

Jaroslav Kurzweil (pictured on the right) was born on May 7, 1926, in Prague. A student of V. Jarník, he has done a considerable amount of research in the theory of differential equations and the theory of integration, and also has had a serious interest in mathematical education. In 1964 he was awarded the Klement Gottwald State Prize, and in 1981 he was awarded the Bolzano medal of the Czechoslovak Academy of Sciences. Since 1989 he has been Director of the Mathematical Institute of the Czech Academy of Sciences in Prague and has had a profound influence on the mathematicians there.

similar to that of Riemann, it is technically much simpler than the usual Lebesgue integral—yet its scope is considerably greater; in particular, it includes functions that are derivatives, and also includes all "improper integrals".

In this chapter, we give an exposition of the generalized Riemann integral. In Section 10.1, it will be seen that the basic theory is almost exactly the same as for the ordinary Riemann integral. However, we have omitted the proofs of a few results when their proofs are unduly complicated. In the short Section 10.2, we indicate that improper integrals on $[a, b]$ are included in the generalized theory. We will introduce the class of Lebesgue integrable functions as those generalized integrable functions f whose absolute value $|f|$ is also generalized integrable; this is a very different approach to the Lebesgue integral than is usual, but it gives the same class of functions. In Section 10.3, we will integrate functions on *unbounded* closed intervals. In the final section, we discuss the limit theorems that hold for the generalized Riemann and Lebesgue integrals, and we will give some interesting applications of these theorems. We will also define what is meant by a "measurable function" and relate that notion to generalized integrability.

Readers wishing to study the proofs that are omitted here, should consult the first author's book, *A Modern Theory of Integration*, which we refer to as [MTI], or the books of DePree and Swartz, Gordon, and McLeod listed in the References.

Section 10.1   Definition and Main Properties

In Definition 5.5.2, we defined a **gauge** on $[a, b]$ to be a strictly positive function $\delta : [a, b] \to (0, \infty)$. Further, a tagged partition $\dot{\mathcal{P}} := \{(I_i, t_i)\}_{i=1}^{n}$ of $[a, b]$, where $I_i := [x_{i-1}, x_i]$, is said to be δ-**fine** in case

$$(1) \qquad t_i \in I_i \subseteq [t_i - \delta(t_i), t_i + \delta(t_i)] \qquad \text{for} \quad i = 1, \cdots, n.$$

This is shown in Figure 5.5.1. Note that (i) only a tagged partition can be δ-fine, and (ii) the δ-fineness of a tagged partition depends on the choice of the tags t_i and the values $\delta(t_i)$.

In Examples 5.5.4, we gave some specific examples of gauges, and in Theorem 5.5.5 we showed that if δ is any gauge on $[a, b]$, then there exist δ-fine tagged partitions of $[a, b]$.

We will define the generalized Riemann (or the "Henstock-Kurzweil") integral. It will be seen that the definition is *very similar* to that of the ordinary Riemann integral, and that many of the proofs are essentially the same. Indeed, the only difference between the definitions of these integrals is that the notion of smallness of a tagged partition is specified by a gauge, rather than its norm. It will be seen that this—apparently minor— difference results in a very much larger class of integrable functions. In order to avoid some complications, a few proofs will be omitted; they can be found in [MTI].

Before we begin our study, it is appropriate that we ask: *Why are gauges more useful than norms?* Briefly, the reason is that the norm of a partition is a rather coarse measure of the fineness of the partition, since it is merely the length of the largest subinterval in the partition. On the other hand, gauges can give one more delicate control of the subintervals in the partitions, by requiring the use of small subinterals when the function is varying rapidly but permitting the use of larger subintervals when the function is nearly constant. Moreover, gauges can be used to force specific points to be tags; this is often useful when unusual behavior takes place at such a point. Since gauges are more flexible than norms, their use permits a larger class of functions to become integrable.

10.1.1 Definition   A function $f : [a, b] \to \mathbb{R}$ is said to be ***generalized* Riemann integrable** on $[a, b]$ if there exists a number $L \in \mathbb{R}$ such that for every $\varepsilon > 0$ there exists a gauge δ_ε on $[a, b]$ such that if $\dot{\mathcal{P}}$ is any δ_ε-fine partition of $[a, b]$, then

$$|S(f; \dot{\mathcal{P}}) - L| < \varepsilon.$$

The collection of all generalized Riemann integrable functions will usually be denoted by $\mathcal{R}^*[a, b]$.

It will be shown that if $f \in \mathcal{R}^*[a, b]$, then the number L is uniquely determined; it will be called the **generalized Riemann integral of** f over $[a, b]$. It will also be shown that if $f \in \mathcal{R}[a, b]$, then $f \in \mathcal{R}^*[a, b]$ and the value of the two integrals is the same. Therefore, it will not cause any ambiguity if we also denote the generalized Riemann integral of $f \in \mathcal{R}^*[a, b]$ by the symbols

$$\int_a^b f \quad \text{or} \quad \int_a^b f(x)\, dx.$$

Our first result gives the uniqueness of the value of the generalized Riemann integral. Although its proof is almost identical to that of Theorem 7.1.2, we will write it out to show how gauges are used instead of norms of partitions.

10.1.2 Uniqueness Theorem   *If $f \in \mathcal{R}^*[a, b]$, then the value of the integral is uniquely determined.*

Proof.   Assume that L' and L'' both satisfy the definition and let $\varepsilon > 0$. Thus there exists a gauge $\delta'_{\varepsilon/2}$ such that if $\dot{\mathcal{P}}_1$ is any $\delta'_{\varepsilon/2}$-fine partition, then

$$|S(f; \dot{\mathcal{P}}_1) - L'| < \varepsilon/2.$$

Also there exists a gauge $\delta''_{\varepsilon/2}$ such that if $\dot{\mathcal{P}}_2$ is any $\delta''_{\varepsilon/2}$-fine partition, then

$$|S(f; \dot{\mathcal{P}}_2) - L''| < \varepsilon/2.$$

We define δ_ε by $\delta_\varepsilon(t) := \min\{\delta'_{\varepsilon/2}(t), \delta''_{\varepsilon/2}(t)\}$ for $t \in [a, b]$, so that δ_ε is a gauge on $[a, b]$. If $\dot{\mathcal{P}}$ is a δ_ε-fine partition, then the partition $\dot{\mathcal{P}}$ is both $\delta'_{\varepsilon/2}$-fine and $\delta''_{\varepsilon/2}$-fine, so that

$$|S(f; \dot{\mathcal{P}}) - L'| < \varepsilon/2 \qquad \text{and} \qquad |S(f; \dot{\mathcal{P}}) - L''| < \varepsilon/2,$$

whence it follows that

$$|L' - L''| \leq |L' - S(f; \dot{\mathcal{P}})| + |S(f; \dot{\mathcal{P}}) - L''|$$
$$< \varepsilon/2 + \varepsilon/2 = \varepsilon.$$

Since $\varepsilon > 0$ is arbitrary, it follows that $L' = L''$.   Q.E.D.

We now show that every Riemann integrable function f is also generalized Riemann integrable, and with the same value for the integral. This is done by using a gauge that is a *constant function*.

10.1.3 Consistency Theorem   *If $f \in \mathcal{R}[a, b]$ with integral L, then also $f \in \mathcal{R}^*[a, b]$ with integral L.*

Proof.   Given $\varepsilon > 0$, we need to construct an appropriate gauge on $[a, b]$. Since $f \in \mathcal{R}[a, b]$, there exists a number $\delta_\varepsilon > 0$ such that if $\dot{\mathcal{P}}$ is any tagged partition with $\|\dot{\mathcal{P}}\| < \delta_\varepsilon$, then $|S(f; \dot{\mathcal{P}}) - L| < \varepsilon$. We define the function $\delta^*_\varepsilon(t) := \frac{1}{4}\delta_\varepsilon$ for $t \in [a, b]$, so that δ^*_ε is a gauge on $[a, b]$.

If $\dot{\mathcal{P}} = \{(I_i, t_i)\}_{i=1}^n$, where $I_i := [x_{i-1}, x_i]$, is a δ^*_ε-fine partition, then since

$$I_i \subseteq [t_i - \delta^*_\varepsilon(t_i), t_i + \delta^*_\varepsilon(t_i)] = [t_i - \tfrac{1}{4}\delta_\varepsilon, t_i + \tfrac{1}{4}\delta_\varepsilon],$$

it is readily seen that $0 < x_i - x_{i-1} \leq \frac{1}{2}\delta_\varepsilon < \delta_\varepsilon$ for all $i = 1, \cdots, n$. Therefore this partition also satisfies $\|\dot{\mathcal{P}}\| < \delta_\varepsilon$ and consequently $|S(f; \dot{\mathcal{P}}) - L| < \varepsilon$.

Thus every δ^*_ε-fine partition $\dot{\mathcal{P}}$ also satisfies $|S(f; \dot{\mathcal{P}}) - L| < \varepsilon$. Since $\varepsilon > 0$ is arbitrary, it follows that f is generalized Riemann integrable to L.   Q.E.D.

From Theorems 7.2.5, 7.2.6 and 7.2.7, we conclude that: *Every step function, every continuous function and every monotone function belongs to $\mathcal{R}^*[a, b]$.* We will now show that Dirichlet's function, which was shown not to be Riemann integrable in 7.2.2(b) and 7.3.13(d), is *generalized* Riemann integrable.

10.1.4 Examples   **(a)**   The Dirichlet function f belongs to $\mathcal{R}^*[0, 1]$ and has integral 0.

We enumerate the rational numbers in $[0, 1]$ as $\{r_k\}_{k=1}^\infty$. Given $\varepsilon > 0$ we define $\delta_\varepsilon(r_k) := \varepsilon/2^{k+2}$ and $\delta_\varepsilon(x) := 1$ when x is irrational. Thus δ_ε is a gauge on $[0, 1]$ and if the partition $\dot{\mathcal{P}} := \{(I_i, t_i)\}_{i=1}^n$ is δ_ε-fine, then we have $x_i - x_{i-1} \leq 2\delta_\varepsilon(t_i)$. Since the only nonzero contributions to $S(f; \dot{\mathcal{P}})$ come from rational tags $t_i = r_k$, where

$$0 < f(r_k)(x_i - x_{i-1}) = 1 \cdot (x_i - x_{i-1}) \leq \frac{2\varepsilon}{2^{k+2}} = \frac{\varepsilon}{2^{k+1}},$$

and since each such tag can occur in at most two subintervals, we have

$$0 \leq S(f; \dot{\mathcal{P}}) < \sum_{k=1}^\infty \frac{2\varepsilon}{2^{k+1}} = \sum_{k=1}^\infty \frac{\varepsilon}{2^k} = \varepsilon.$$

Since $\varepsilon > 0$ is arbitrary, then $f \in \mathcal{R}^*[0, 1]$ and $\int_0^1 f = 0$.

(b)   Let $H : [0, 1] \to \mathbb{R}$ be defined by $H(1/k) := k$ for $k \in \mathbb{N}$ and $H(x) := 0$ elsewhere on $[0, 1]$.

Since H is not bounded on $[0, 1]$, it follows from the Boundedness Theorem 7.1.5 that it is not Riemann integrable on $[0, 1]$. We will now show that H is *generalized* Riemann integrable to 0.

In fact, given $\varepsilon > 0$, we define $\delta_\varepsilon(1/k) := \varepsilon/(k2^{k+2})$ and set $\delta_\varepsilon(x) := 1$ elsewhere on $[0, 1]$, so δ_ε is a gauge on $[0, 1]$. If $\dot{\mathcal{P}}$ is a δ_ε-fine partition of $[0, 1]$ then $x_i - x_{i-1} \le 2\delta_\varepsilon(t_i)$. Since the only nonzero contributions to $S(H; \dot{\mathcal{P}})$ come from tags $t_i = 1/k$, where

$$0 < H(1/k)(x_i - x_{i-1}) = k \cdot (x_i - x_{i-1}) \le k \cdot \frac{2\varepsilon}{k2^{k+2}} = \frac{\varepsilon}{2^{k+1}},$$

and since each such tag can occur in at most two subintervals, we have

$$0 \le S(H; \dot{\mathcal{P}}) < \sum_{k=1}^{\infty} \frac{\varepsilon}{2^k} = \varepsilon.$$

Since $\varepsilon > 0$ is arbitrary, then $H \in \mathcal{R}^*[0, 1]$ and $\int_0^1 H = 0$.   □

The next result is exactly similar to Theorem 7.1.4.

10.1.5 Theorem   *Suppose that f and g are in $\mathcal{R}^*[a, b]$. Then:*

(a)   *If $k \in \mathbb{R}$, the function kf is in $\mathcal{R}^*[a, b]$ and*

$$\int_a^b kf = k \int_a^b f.$$

(b)   *The function $f + g$ is in $\mathcal{R}^*[a, b]$ and*

$$\int_a^b (f + g) = \int_a^b f + \int_a^b g.$$

(c)   *If $f(x) \le g(x)$ for all $x \in [a, b]$, then*

$$\int_a^b f \le \int_a^b g.$$

Proof.   **(b)**   Given $\varepsilon > 0$, we can use the argument in the proof of the Uniqueness Theorem 10.1.2 to construct a gauge δ_ε on $[a, b]$ such that if $\dot{\mathcal{P}}$ is any δ_ε-fine partition of $[a, b]$, then

$$\left| S(f; \dot{\mathcal{P}}) - \int_a^b f \right| < \varepsilon/2 \quad \text{and} \quad \left| S(g; \dot{\mathcal{P}}) - \int_a^b g \right| < \varepsilon/2.$$

Since $S(f + g; \dot{\mathcal{P}}) = S(f; \dot{\mathcal{P}}) + S(g; \dot{\mathcal{P}})$, it follows as in the proof of Theorem 7.1.4(b) that

$$\left| S(f + g; \dot{\mathcal{P}}) - \left(\int_a^b f + \int_a^b g \right) \right| \le \left| S(f; \dot{\mathcal{P}}) - \int_a^b f \right| + \left| S(g; \dot{\mathcal{P}}) - \int_a^b g \right|$$
$$< \varepsilon/2 + \varepsilon/2 = \varepsilon.$$

Since $\varepsilon > 0$ is arbitrary, then $f + g \in \mathcal{R}^*[a, b]$ and its integral is the sum of the integrals of f and g.

The proofs of (a) and (c) are analogous and are left to the reader.   Q.E.D.

It might be expected that an argument similar to that given in Theorem 7.1.5 can be used to show that a function in $\mathcal{R}^*[a, b]$ is necessarily bounded. However, that is *not* the case; indeed, we have already seen an unbounded function in $\mathcal{R}^*[0, 1]$ in Example 10.1.4(b) and will encounter more later. However, it is a profitable exercise for the reader to determine exactly where the proof of Theorem 7.1.5 breaks down for a function in $\mathcal{R}^*[a, b]$.

The Cauchy Criterion _____

There is an analogous form for the Cauchy Criterion for functions in $\mathcal{R}^*[a, b]$. It is important because it eliminates the need to know the value of the integral. Its proof is essentially the same as that of 7.2.1.

10.1.6 Cauchy Criterion   *A function $f : [a, b] \to \mathbb{R}$ belongs to $\mathcal{R}^*[a, b]$ if and only if for every $\varepsilon > 0$ there exist a gauge η_ε on $[a, b]$ such that if $\dot{P}$ and $\dot{Q}$ are any partitions of $[a, b]$ that are η_ε-fine, then*

$$|S(f; \dot{P}) - S(f; \dot{Q})| < \varepsilon.$$

Proof.   ($\Rightarrow$) If $f \in \mathcal{R}^*[a, b]$ with integral L, let $\delta_{\varepsilon/2}$ be a gauge on $[a, b]$ such that if $\dot{P}$ and $\dot{Q}$ are $\delta_{\varepsilon/2}$-fine partitions of $[a, b]$, then

$$|S(f; \dot{P}) - L| < \varepsilon/2 \quad \text{and} \quad |S(f; \dot{Q}) - L| < \varepsilon/2.$$

We set $\eta_\varepsilon(t) := \delta_{\varepsilon/2}(t)$ for $t \in [a, b]$, so if $\dot{P}$ and $\dot{Q}$ are η_ε-fine, then

$$|S(f; \dot{P}) - S(f; \dot{Q})| \le |S(f; \dot{P}) - L| + |L - S(f; \dot{Q})|$$
$$< \varepsilon/2 + \varepsilon/2 = \varepsilon.$$

($\Leftarrow$) For each $n \in \mathbb{N}$, let δ_n be a gauge on $[a, b]$ such that if $\dot{P}$ and $\dot{Q}$ are partitions that are δ_n-fine, then

$$|S(f; \dot{P}) - S(f; \dot{Q})| < 1/n.$$

We may assume that $\delta_n(t) \ge \delta_{n+1}(t)$ for all $t \in [a, b]$ and $n \in \mathbb{N}$; otherwise, we replace δ_n by the gauge $\delta'_n(t) := \min\{\delta_1(t), \cdots, \delta_n(t)\}$ for all $t \in [a, b]$.

For each $n \in \mathbb{N}$, let $\dot{P}_n$ be a partition that is δ_n-fine. Clearly, if $m > n$ then both $\dot{P}_m$ and $\dot{P}_n$ are δ_n-fine, so that

(2) $$|S(f; \dot{P}_n) - S(f; \dot{P}_m)| < 1/n \quad \text{for} \quad m > n.$$

Consequently, the sequence $(S(f; \dot{P}_m))_{m=1}^\infty$ is a Cauchy sequence in $\mathbb{R}$, so it converges to some number A. Passing to the limit in (2) as $m \to \infty$, we have

$$|S(f; \dot{P}_n) - A| \le 1/n \quad \text{for all} \quad n \in \mathbb{N}.$$

To see that A is the generalized Riemann integral of f, given $\varepsilon > 0$, let $K \in \mathbb{N}$ satisfy $K > 2/\varepsilon$. If $\dot{Q}$ is a δ_K-fine partition, then

$$|S(f; \dot{Q}) - A| \le |S(f; \dot{Q}) - S(f; \dot{P}_K)| + |S(f; \dot{P}_K) - A|$$
$$\le 1/K + 1/K < \varepsilon.$$

Since $\varepsilon > 0$ is arbitrary, then $f \in \mathcal{R}^*[a, b]$ with integral A.   Q.E.D.

10.1.7 Squeeze Theorem   *Let* $f : [a, b] \to \mathbb{R}$. *Then* $f \in \mathcal{R}^*[a, b]$ *if and only if for every* $\varepsilon > 0$ *there exist functions* α_ε *and* ω_ε *in* $\mathcal{R}^*[a, b]$ *with*

$$\alpha_\varepsilon(x) \le f(x) \le \omega_\varepsilon(x) \qquad \text{for all} \quad x \in [a, b],$$

and such that

$$\int_a^b (\omega_\varepsilon - \alpha_\varepsilon) \le \varepsilon.$$

The proof of this result is exactly similar to the proof of Theorem 7.2.3, and will be left to the reader.

The Additivity Theorem

We now present a result quite analogous to Theorem 7.2.8. Its proof is a modification of the proof of that theorem, but since it is somewhat technical, the reader may choose to omit the proof on a first reading.

10.1.8 Additivity Theorem   *Let* $f : [a, b] \to \mathbb{R}$ *and let* $c \in (a, b)$. *Then* $f \in \mathcal{R}^*[a, b]$ *if and only if its restrictions to* $[a, c]$ *and* $[c, b]$ *are both generalized Riemann integrable. In this case*

$$(3) \qquad \int_a^b f = \int_a^c f + \int_c^b f.$$

Proof.   ($\Leftarrow$) Suppose that the restriction f_1 of f to $[a, c]$, and the restriction f_2 of f to $[c, b]$ are generalized Riemann integrable to L_1 and L_2, respectively. Then, given $\varepsilon > 0$ there exists a gauge δ' on $[a, c]$ such that if $\dot{\mathcal{P}}_1$ is a δ'-fine partition of $[a, c]$ then $|S(f_1; \dot{\mathcal{P}}_1) - L_1| < \varepsilon/2$. Also there exists a gauge δ'' on $[c, b]$ such that if $\dot{\mathcal{P}}_2$ is a δ''-fine partition of $[c, b]$ then $|S(f_2; \dot{\mathcal{P}}_2) - L_2| < \varepsilon/2$.

We now define a gauge δ_ε on $[a, b]$ by

$$\delta_\varepsilon(t) := \begin{cases} \min\{\delta'(t), \tfrac{1}{2}(c - t)\} & \text{for} \quad t \in [a, c), \\ \min\{\delta'(c), \delta''(c)\} & \text{for} \quad t = c, \\ \min\{\delta''(t), \tfrac{1}{2}(t - c)\} & \text{for} \quad t \in (c, b]. \end{cases}$$

(This gauge has the property that any δ_ε-fine partition must have c as a tag for any subinterval containing the point c.)

We will show that if $\dot{\mathcal{Q}}$ is any δ_ε-fine partition of $[a, b]$, then there exist a δ'-fine partition $\dot{\mathcal{Q}}_1$ of $[a, c]$ and a δ''-fine partition $\dot{\mathcal{Q}}_2$ of $[c, b]$ such that

$$(4) \qquad S(f; \dot{\mathcal{Q}}) = S(f_1; \dot{\mathcal{Q}}_1) + S(f_2; \dot{\mathcal{Q}}_2).$$

Case (i)   If c is a partition point of $\dot{\mathcal{Q}}$, then it belongs to two subintervals of $\dot{\mathcal{Q}}$ and is the tag for both of these subintervals. If $\dot{\mathcal{Q}}_1$ consists of the part of $\dot{\mathcal{Q}}$ having subintervals in $[a, c]$, then $\dot{\mathcal{Q}}_1$ is δ'-fine. Similarly, if $\dot{\mathcal{Q}}_2$ consists of the part of $\dot{\mathcal{Q}}$ having subintervals in $[c, b]$, then $\dot{\mathcal{Q}}_2$ is δ''-fine. The relation (4) is now clear.

Case (ii)   If c is not a partition point in $\dot{\mathcal{Q}} = \{(I_i, t_i)\}_{i=1}^m$, then it is the tag for some subinterval, say $[x_{k-1}, x_k]$. We replace the pair $([x_{k-1}, x_k], c)$ by the two pairs $([x_{k-1}, c], c)$ and $([c, x_k], c)$, and let $\dot{\mathcal{Q}}_1$ and $\dot{\mathcal{Q}}_2$ be the tagged partitions of $[a, c]$ and $[c, b]$ that result. Since $f(c)(x_k - x_{k-1}) = f(c)(c - x_{k-1}) + f(c)(x_k - c)$, it is seen that the relation (4) also holds.

In either case, equation (4) and the Triangle Inequality imply that

$$\left|S(f;\dot{\mathcal{Q}}) - (L_1 + L_2)\right| = \left|(S(f;\dot{\mathcal{Q}}_1) + S(f;\dot{\mathcal{Q}}_2)) - (L_1 + L_2)\right|$$
$$\leq \left|S(f;\dot{\mathcal{Q}}_1) - L_1\right| + \left|S(f;\dot{\mathcal{Q}}_2) - L_2\right|.$$

Since $\dot{\mathcal{Q}}_1$ is δ'-fine and $\dot{\mathcal{Q}}_2$ is δ''-fine, we conclude that

$$\left|S(f;\dot{\mathcal{Q}}) - (L_1 + L_2)\right| < \varepsilon.$$

Since $\varepsilon > 0$ is arbitrary, we infer that $f \in \mathcal{R}^*[a, b]$ and that (3) holds.

($\Rightarrow$) Suppose that $f \in \mathcal{R}^*[a, b]$ and, given $\varepsilon > 0$, let the gauge η_ε satisfy the Cauchy Criterion. Let f_1 be the restriction of f to $[a, c]$ and let $\dot{\mathcal{P}}_1, \dot{\mathcal{Q}}_1$ be η_ε-fine partitions of $[a, c]$. By adding additional partition points and tags from $[c, b]$, we can extend $\dot{\mathcal{P}}_1$ and $\dot{\mathcal{Q}}_1$ to η_ε-fine partitions $\dot{\mathcal{P}}$ and $\dot{\mathcal{Q}}$ of $[a, b]$. If we use the *same* additional points and tags in $[c, b]$ for both $\dot{\mathcal{P}}$ and $\dot{\mathcal{Q}}$, then

$$S(f;\dot{\mathcal{P}}) - S(f;\dot{\mathcal{Q}}) = S(f_1;\dot{\mathcal{P}}_1) - S(f_1;\dot{\mathcal{Q}}_1).$$

Since both $\dot{\mathcal{P}}$ and $\dot{\mathcal{Q}}$ are η_ε-fine, then $|S(f_1;\dot{\mathcal{P}}_1) - S(f_1;\dot{\mathcal{Q}}_1)| < \varepsilon$ also holds. Therefore the Cauchy Condition shows that the restriction f_1 of f to $[a, c]$ is in $\mathcal{R}^*[a, c]$. Similarly, the restriction f_2 of f to $[c, d]$ is in $\mathcal{R}^*[c, d]$.

The equality (3) now follows from the first part of the theorem. Q.E.D.

It is easy to see that results exactly similar to 7.2.9–7.2.12 hold for the generalized Riemann integral. We leave their statements to the reader, but will use these results freely.

The Fundamental Theorem (First Form)

We will now give versions of the Fundamental Theorems for the generalized Riemann integral. It will be seen that the First Form is *significantly stronger* than for the (ordinary) Riemann integral; indeed, we will show that the derivative of any function automatically belongs to $\mathcal{R}^*[a, b]$, so the integrability of the function becomes a conclusion, rather than a hypothesis.

10.1.9 The Fundamental Theorem of Calculus (First Form) *Suppose there exists a* **countable** *set E in $[a, b]$, and functions $f, F : [a, b] \to \mathbb{R}$ such that:*

(a) *F is continuous on $[a, b]$.*
(b) *$F'(x) = f(x)$ for all $x \in [a, b] \setminus E$.*
Then f belongs to $\mathcal{R}^*[a, b]$ and

$$(5) \qquad \int_a^b f = F(b) - F(a).$$

Proof. We will prove the theorem in the case where $E = \emptyset$, leaving the general case to be handled in the Exercises.

Thus, we assume that (b) holds for all $x \in [a, b]$. Since we wish to show that $f \in \mathcal{R}^*[a, b]$, given $\varepsilon > 0$, we need to construct a gauge δ_ε; this will be done by using the differentiability of F on $[a, b]$. If $t \in I$, since the derivative $f(t) = F'(t)$ exists, there exists $\delta_\varepsilon(t) > 0$ such that if $0 < |z - t| \leq \delta_\varepsilon(t)$, $z \in [a, b]$, then

$$\left|\frac{F(z) - F(t)}{z - t} - f(t)\right| < \tfrac{1}{2}\varepsilon.$$

If we multiply this inequality by $|z - t|$, we obtain

$$|F(z) - F(t) - f(t)(z - t)| \leq \tfrac{1}{2}\varepsilon|z - t|$$

whenever $z \in [t - \delta_\varepsilon(t), t + \delta_\varepsilon(t)] \cap [a, b]$. The function δ_ε is our desired gauge.

Now let $u, v \in [a, b]$ with $u < v$ satisfy $t \in [u, v] \subseteq [t - \delta_\varepsilon(t), t + \delta_\varepsilon(t)]$. If we subtract and add the term $F(t) - f(t) \cdot t$ and use the Triangle Inequality and the fact that $v - t \geq 0$ and $t - u \geq 0$, we get

$$|F(v) - F(u) - f(t)(v - u)|$$
$$\leq |F(v) - F(t) - f(t)(v - t)| + |F(t) - F(u) - f(t)(t - u)|$$
$$\leq \tfrac{1}{2}\varepsilon(v - t) + \tfrac{1}{2}\varepsilon(t - u) = \tfrac{1}{2}\varepsilon(v - u).$$

Therefore, if $t \in [u, v] \subseteq [t - \delta_\varepsilon(t), t + \delta_\varepsilon(t)]$, then we have

(6) $$|F(v) - F(u) - f(t)(v - u)| \leq \tfrac{1}{2}\varepsilon(v - u).$$

We will show that $f \in \mathcal{R}^*[a, b]$ with integral given by the telescoping sum

(7) $$F(b) - F(a) = \sum_{i=1}^{n}\{F(x_i) - F(x_{i-1})\}.$$

For, if the partition $\dot{\mathcal{P}} := \{([x_{i-1}, x_i], t_i)\}_{i=1}^{n}$ is δ_ε-fine, then

$$t_i \in [x_{i-1}, x_i] \subseteq [t_i - \delta_\varepsilon(t_i), t_i + \delta_\varepsilon(t_i)] \qquad \text{for} \quad i = 1, \cdots, n,$$

and so we can use (7), the Triangle Inequality, and (6) to obtain

$$|F(b) - F(a) - S(f; P)| = \left| \sum_{i=1}^{n}\{F(x_i) - F(x_{i-1}) - f(t_i)(x_i - x_{i-1})\} \right|$$
$$\leq \sum_{i=1}^{n}\left| F(x_i) - F(x_{i-1}) - f(t_i)(x_i - x_{i-1}) \right|$$
$$\leq \sum_{i=1}^{n} \tfrac{1}{2}\varepsilon(x_i - x_{i-1}) < \varepsilon(b - a).$$

Since $\varepsilon > 0$ is arbitrary, we conclude that $f \in \mathcal{R}^*[a, b]$ and (5) holds.   Q.E.D.

10.1.10 Examples   (a) If $H(x) := 2\sqrt{x}$ for $x \in [0, b]$, then H is continuous on $[0, b]$ and $H'(x) = 1/\sqrt{x}$ for $x \in (0, b]$. We define $h(x) := H'(x)$ for $x \in (0, b]$ and $h(0) := 0$. It follows from the Fundamental Theorem 10.1.9 with $E := \{0\}$ that h belongs to $\mathcal{R}^*[0, b]$ and that $\int_0^b h = H(b) - H(0) = H(b)$, which we write as

$$\int_0^b \frac{1}{\sqrt{x}}\, dx = 2\sqrt{b}.$$

(b) More generally, if $\alpha > 0$, let $H_\alpha(x) := x^\alpha/\alpha = e^{\alpha \ln x}/\alpha$ for $x \in (0, b]$ and let $H_\alpha(0) := 0$ so that H_α is continuous on $[0, b]$ and $H'_\alpha(x) = x^{\alpha-1}$ for all $x \in (0, b]$; see 8.3.10 and 8.3.13. We define $h_\alpha(x) := H'_\alpha(x)$ for $x \in (0, b]$ and $h_\alpha(0) := 0$.

Then Theorem 10.1.9 implies that $h_\alpha \in \mathcal{R}^*[0, b]$ and that $\int_0^b h_\alpha = H_\alpha(b) - H_\alpha(0) = H_\alpha(b)$, which we write as

$$\int_0^b x^{\alpha-1}\, dx = \frac{b^\alpha}{\alpha}.$$

(c) Let $L(x) := x \ln x - x$ for $x \in (0, b]$ and $L(0) := 0$. Then L is continuous on $[0, b]$ (use l'Hospital's Rule at $x = 0$), and it is seen that $L'(x) = \ln x$ for $x \in (0, b]$.

It follows from Theorem 10.1.9 with $E = \{0\}$ that the unbounded function $l(x) := \ln x$ for $x \in (0, b]$ and $l(0) := 0$ belongs to $\mathcal{R}^*[0, b]$ and that $\int_0^b l = L(b) - L(0)$, which we write as

$$\int_0^b \ln x \, dx = b \ln b - b.$$

(d)   Let $A(x) := \text{Arcsin}\, x$ for $x \in [-1, 1]$ so that A is continuous on $[-1, 1]$ and $A'(x) = 1/\sqrt{1 - x^2}$ for $x \in (-1, 1)$. We define $s(x) := A'(x)$ for $x \in (-1, 1)$ and let $s(-1) = s(1) := 0$.

Then Theorem 10.1.9 with $E = \{-1, 1\}$ implies that $s \in \mathcal{R}^*[-1, 1]$ and that $\int_{-1}^1 s = A(1) - A(-1) = \pi$, which we write as

$$\int_{-1}^1 \frac{dx}{\sqrt{1 - x^2}} = \text{Arcsin}\, 1 - \text{Arcsin}(-1) = \pi. \qquad \square$$

The Fundamental Theorem (Second Form)

We now turn to the Second Form of the Fundamental Theorem, in which we wish to differentiate the **indefinite integral** F of f, defined by:

(8) $$F(z) := \int_a^z f(x) \, dx \qquad \text{for} \quad z \in [a, b].$$

10.1.11 Fundamental Theorem of Calculus (Second Form)   *Let f belong to $\mathcal{R}^*[a, b]$ and let F be the indefinite integral of f. Then we have:*

(a)   *F is continuous on $[a, b]$.*

(b)   *There exists a null set Z such that if $x \in [a, b] \setminus Z$, then F is differentiable at x and $F'(x) = f(x)$.*

(c)   *If f is continuous at $c \in [a, b]$, then $F'(c) = f(c)$.*

Proof.   The proofs of (a) and (b) can be found in [MTI]. The proof of (c) is exactly as the proof of Theorem 7.3.5 except that we use Theorems 10.1.8 and 10.1.5(c).   Q.E.D.

We can restate conclusion (b) as: *The indefinite integral F of f is differentiable to f almost everywhere on $[a, b]$.*

Substitution Theorem

In view of the simplicity of the Fundamental Theorem 10.1.9, we can improve the theorem justifying the "substitution formula". The next result is a considerable strengthening of Theorem 7.3.8. The reader should write out the hypotheses in the case $E_f = E_\varphi = E = \emptyset$.

10.1.12 Substitution Theorem   (a)   *Let $I := [a, b]$ and $J := [\alpha, \beta]$, and let $F : I \to \mathbb{R}$ and $\varphi : J \to \mathbb{R}$ be continuous functions with $\varphi(J) \subseteq I$.*

(b)   *Suppose there exist sets $E_f \subset I$ and $E_\varphi \subset J$ such that $f(x) = F'(x)$ for $x \in I \setminus E_f$, that $\varphi'(t)$ exists for $t \in J \setminus E_\varphi$, and that $E := \varphi^{-1}(E_f) \cup E_\varphi$ is countable.*

(c)   *Set $f(x) := 0$ for $x \in E_f$ and $\varphi'(t) := 0$ for $t \in E_\varphi$.*
We conclude that $f \in \mathcal{R}^*(\varphi(J))$, that $(f \circ \varphi) \cdot \varphi' \in \mathcal{R}^*(J)$ and that

(9) $$\int_\alpha^\beta (f \circ \varphi) \cdot \varphi' = F \circ \varphi \Big|_\alpha^\beta = \int_{\varphi(\alpha)}^{\varphi(\beta)} f.$$

Proof. Since φ is continuous on J, Theorem 5.3.9 implies that $\varphi(J)$ is a closed interval in I. Also $\varphi^{-1}(E_f)$ is countable, whence $E_f \cap \varphi(J) = \varphi(\varphi^{-1}(E_f))$ is also countable. Since $f(x) = F'(x)$ for all $x \in \varphi(J) \setminus E_f$, the Fundamental Theorem 10.1.9 implies that $f \in \mathcal{R}^*(\varphi(J))$ and that

$$\int_{\varphi(\alpha)}^{\varphi(\beta)} f = F \Big|_{\varphi(\alpha)}^{\varphi(\beta)} = F(\varphi(\beta)) - F(\varphi(\alpha)).$$

If $t \in J \setminus E$, then $t \in J \setminus E_\varphi$ and $\varphi(t) \in I \setminus E_f$. Hence the Chain Rule 6.1.6 implies that

$$(F \circ \varphi)'(t) = f(\varphi(t)) \cdot \varphi'(t) \qquad \text{for} \quad t \in J \setminus E.$$

Since E is countable, the Fundamental Theorem implies that $(f \circ \varphi) \cdot \varphi' \in \mathcal{R}^*(J)$ and that

$$\int_\alpha^\beta (f \circ \varphi) \cdot \varphi' = F \circ \varphi \Big|_\alpha^\beta = F(\varphi(\beta)) - F(\varphi(\alpha)).$$

The conclusion follows by equating these two terms.   Q.E.D.

10.1.13 Examples   (a)   Consider the integral $\displaystyle\int_0^4 \frac{\cos \sqrt{t}}{\sqrt{t}}\, dt.$

Since the integrand is unbounded as $t \to 0+$, there is some doubt about the existence of the integral. Also, we have seen in Exercise 7.3.19(b) that Theorem 7.3.8 does not apply with $\varphi(t) := \sqrt{t}$. However, Theorem 10.1.12 applies.

Indeed, this substitution gives $\varphi'(t) = 1/(2\sqrt{t})$ for $t \in (0, 4]$ and we set $\varphi(0) := 0$. If we put $F(x) := 2 \sin x$, then $f(x) = F'(x) = 2 \cos x$ and the integrand has the form

$$f(\varphi(t)) \cdot \varphi'(t) = \left(2 \cos \sqrt{t}\right)\left(\frac{1}{2\sqrt{t}}\right) \qquad \text{for} \quad t \neq 0.$$

Thus, the Substitution Theorem 10.1.12 with $E_\varphi := \{0\}$, $E_f := \emptyset$, $E := \{0\}$ implies that

$$\int_{t=0}^{t=4} \frac{\cos \sqrt{t}}{\sqrt{t}}\, dt = \int_{x=0}^{x=2} 2 \cos x\, dx = 2 \sin 2.$$

(b)   Consider the integral $\displaystyle\int_0^1 \frac{dt}{\sqrt{t - t^2}} = \int_0^1 \frac{dt}{\sqrt{t}\sqrt{1 - t}}.$

Note that this integrand is unbounded as $t \to 0+$ and as $t \to 1-$. As in (a), we let $x = \varphi(t) := \sqrt{t}$ for $t \in [0, 1]$ so that $\varphi'(t) = 1/(2\sqrt{t})$ for $t \in (0, 1]$. Since $\sqrt{1 - t} = \sqrt{1 - x^2}$, the integrand takes the form

$$\frac{2}{\sqrt{1 - t}} \cdot \frac{1}{2\sqrt{t}} = \frac{2}{\sqrt{1 - x^2}} \cdot \varphi'(t),$$

which suggests $f(x) = 2/\sqrt{1 - x^2}$ for $x \neq 1$. Therefore, we are led to choose $F(x) := 2 \,\mathrm{Arcsin}\, x$ for $x \in [0, 1]$, since

$$\frac{2}{\sqrt{1 - x^2}} = F'(x) = (2 \,\mathrm{Arcsin}\, x)' \qquad \text{for} \quad x \in [0, 1).$$

Consequently, we have $E_\varphi = \{0\}$ and $E_f = \{1\}$, so that $E = \{0, 1\}$, and the Substitution Theorem yields

$$\int_{t=0}^{t=1} \frac{dt}{\sqrt{t}\sqrt{1 - t}} = \int_{x=0}^{x=1} \frac{2\, dx}{\sqrt{1 - x^2}} = 2 \,\mathrm{Arcsin}\, x \Big|_0^1 = 2 \,\mathrm{Arcsin}\, 1 = \pi. \qquad \square$$

Other formulations of the Substitution Theorem are given in [MTI].

The Multiplication Theorem

In Theorem 7.3.16 we saw that the product of two Riemann integrable functions is Riemann integrable. That result is *not* true for generalized Riemann integrable functions; see Exercises 18 and 20. However, we will state a theorem in this direction that is often useful. Its proof will be found in [MTI].

10.1.14 Multiplication Theorem *If $f \in \mathcal{R}^*[a, b]$ and if g is a monotone function on $[a, b]$, then the product $f \cdot g$ belongs to $\mathcal{R}^*[a, b]$.*

Integration by Parts

The following version of the formula for integration by parts is useful.

10.1.15 Integration by Parts Theorem *Let F and G be differentiable on $[a, b]$. Then $F'G$ belongs to $\mathcal{R}^*[a, b]$ if and only if FG' belongs to $\mathcal{R}^*[a, b]$. In this case we have*

$$(10) \qquad \int_a^b F'G = FG \Big|_a^b - \int_a^b FG'.$$

The proof uses Theorem 6.1.3(c); it will be left to the reader. In applications, we usually have $F'(x) = f(x)$ and $G'(x) = g(x)$ for all $x \in [a, b]$. It will be noted that we need to assume that one of the functions $fG = F'G$ and $Fg = FG'$ belongs to $\mathcal{R}^*[a, b]$.

The reader should contrast the next result with Theorem 7.3.18. Note that we do *not* need to assume the integrability of $f^{(n+1)}$.

10.1.16 Taylor's Theorem *Suppose that $f, f', f'', \cdots, f^{(n)}$ and $f^{(n+1)}$ exist on $[a, b]$. Then we have*

$$(11) \qquad f(b) = f(a) + \frac{f'(a)}{1!}(b - a) + \cdots + \frac{f^{(n)}(a)}{n!}(b - a)^n + R_n,$$

where the remainder is given by

$$(12) \qquad R_n = \frac{1}{n!} \int_a^b f^{(n+1)}(t) \cdot (b - t)^n \, dt.$$

Proof. Since $f^{(n+1)}$ is a derivative, it belongs to $\mathcal{R}^*[a, b]$. Moreover, since $t \mapsto (b - t)^n$ is monotone on $[a, b]$, the Multiplication Theorem 10.1.14 implies the integral in (12) exists. Integrating by parts repeatedly, we obtain (11).   Q.E.D.

Exercises for Section 10.1

1. Let δ be a gauge on $[a, b]$ and let $\dot{\mathcal{P}} = \{([x_{i-1}, x_i], t_i)\}_{i=1}^n$ be a δ-fine partition of $[a, b]$.
   (a) Show that $0 < x_i - x_{i-1} \leq 2\delta(t_i)$ for $i = 1, \cdots, n$.
   (b) If $\delta^* := \sup\{\delta(t) : t \in [a, b]\} < \infty$, show that $\|\dot{\mathcal{P}}\| \leq 2\delta^*$.
   (c) If $\delta_* := \inf\{\delta(t) : t \in [a, b]\}$ satisfies $\delta_* > 0$, and if $\dot{\mathcal{Q}}$ is a tagged partition of $[a, b]$ such that we have $\|\dot{\mathcal{Q}}\| \leq \delta_*$, show that $\dot{\mathcal{Q}}$ is δ-fine.
   (d) If $\varepsilon = 1$, show that the gauge δ_1 in Example 10.1.4(a) has the property that $\inf\{\delta_1(t) : t \in [0, 1]\} = 0$.

2. (a) If $\dot{\mathcal{P}}$ is a tagged partition of $[a, b]$, show that each tag can belong to *at most* two subintervals in $\dot{\mathcal{P}}$.
    (b) Are there tagged partitions in which every tag belongs to exactly two subintervals?

3. Let δ be a gauge on $[a, b]$ and let $\dot{\mathcal{P}}$ be a δ-fine partition of $[a, b]$.
    (a) Show that there exists a δ-fine partition $\dot{\mathcal{Q}}_1$ such that (i) no tag belongs to two subintervals in $\dot{\mathcal{Q}}_1$, and (ii) $S(f; \dot{\mathcal{Q}}_1) = S(f; \dot{\mathcal{P}})$ for any function f on $[a, b]$.
    (b) Does there exist a δ-fine partition $\dot{\mathcal{Q}}_2$ such that (j) every tag belongs to two subintervals in $\dot{\mathcal{Q}}_2$, and (jj) $S(f; \dot{\mathcal{Q}}_2) = S(f; \dot{\mathcal{P}})$ for any function f on $[a, b]$?
    (c) Show that there exists a δ-fine partition $\dot{\mathcal{Q}}_3$ such that (k) every tag is an endpoint of its subinterval, and (kk) $S(f; \dot{\mathcal{Q}}_3) = S(f; \dot{\mathcal{P}})$ for any function f on $[a, b]$.

4. If δ is defined on $[0, 2]$ by $\delta(t) := \frac{1}{2}|t - 1|$ for $x \neq 1$ and $\delta(1) := 0.01$, show that every δ-fine partition $\dot{\mathcal{P}}$ of $[0, 2]$ has $t = 1$ as a tag for at least one subinterval, and that the total length of the subintervals in $\dot{\mathcal{P}}$ having 1 as a tag is ≤ 0.02.

5. (a) Construct a gauge δ on $[0, 4]$ that will force the numbers $1, 2, 3$ to be tags of any δ-fine partition of this interval.
    (b) Given a gauge δ_1 on $[0, 4]$, construct a gauge δ_2 such that every δ_2-fine partition of $[0, 4]$ will (i) have the numbers $1, 2, 3$ in its collection of tags, and (ii) be δ_1-fine.

6. Show that $f \in \mathcal{R}^*[a, b]$ with integral L if and only if for every $\varepsilon > 0$ there exists a gauge γ_ε on $[a, b]$ such that if $\dot{\mathcal{P}} = \{([x_{i-1}, x_i], t_i)\}_{i=1}^n$ is any tagged partition such that $0 < x_i - x_{i-1} \leq \gamma_\varepsilon(t_i)$ for $i = 1, \cdots, n$, then $|S(f; \dot{\mathcal{P}}) - L| < \varepsilon$. (This provides an alternate—but equivalent—way of defining the generalized Riemann integral.)

7. Show that the following functions belong to $\mathcal{R}^*[0, 1]$ by finding a function F_k that is continuous on $[0, 1]$ and such that $F_k'(x) = f_k(x)$ for $x \in [0, 1] \setminus E_k$, for some finite set E_k.
    (a) $f_1(x) := (x + 1)/\sqrt{x}$ for $x \in (0, 1]$ and $f_1(0) := 0$.
    (b) $f_2(x) := x/\sqrt{1 - x}$ for $x \in [0, 1)$ and $f_2(1) := 0$.
    (c) $f_3(x) := \sqrt{x} \ln x$ for $x \in (0, 1]$ and $f_3(0) := 0$.
    (d) $f_4(x) := (\ln x)/\sqrt{x}$ for $x \in (0, 1]$ and $f_4(0) := 0$.
    (e) $f_5(x) := \sqrt{(1 + x)/(1 - x)}$ for $x \in [0, 1)$ and $f_5(1) := 0$.
    (f) $f_6(x) := 1/(\sqrt{x}\sqrt{2 - x})$ for $x \in (0, 1]$ and $f_6(0) := 0$.

8. Explain why the argument in Theorem 7.1.5 does not apply to show that a function in $\mathcal{R}^*[a, b]$ is bounded.

9. Let $f(x) := 1/x$ for $x \in (0, 1]$ and $f(0) := 0$; then f is continuous except at $x = 0$. Show that f does not belong to $\mathcal{R}^*[0, 1]$. [*Hint:* Compare f with $s_n(x) := 1$ on $(1/2, 1]$, $s_n(x) := 2$ on $(1/3, 1/2]$, $s_n(x) := 3$ on $(1/4, 1/3], \cdots, s_n(x) := n$ on $[0, 1/n]$.]

10. Let $k : [0, 1] \to \mathbb{R}$ be defined by $k(x) := 0$ if $x \in [0, 1]$ is 0 or is irrational, and $k(m/n) := n$ if $m, n \in \mathbb{N}$ have no common integer factors other than 1. Show that $k \in \mathcal{R}^*[0, 1]$ with integral equal to 0. Also show that k is not continuous at any point, and not bounded on any subinterval $[c, d]$ with $c < d$.

11. Let f be Dirichlet's function on $[0, 1]$ and $F(x) := 0$ for all $x \in [0, 1]$. Since $F'(x) = f(x)$ for all $x \in [0, 1] \setminus \mathbb{Q}$, show that the Fundamental Theorem 10.1.9 implies that $f \in \mathcal{R}^*[0, 1]$.

12. Let $M(x) := \ln |x|$ for $x \neq 0$ and $M(0) := 0$. Show that $M'(x) = 1/x$ for all $x \neq 0$. Explain why it does not follow that $\int_{-2}^2 (1/x)\, dx = \ln | -2| - \ln 2 = 0$.

13. Let $L_1(x) := x \ln |x| - x$ for $x \neq 0$ and $L_1(0) := 0$, and let $l_1(x) := \ln |x|$ if $x \neq 0$ and $l_1(0) := 0$. If $[a, b]$ is any interval, show that $l_1 \in \mathcal{R}^*[a, b]$ and that $\int_a^b \ln |x|\, dx = L_1(b) - L_1(a)$.

14. Let $E := \{c_1, c_2, \cdots\}$ and let F be continuous on $[a, b]$ and $F'(x) = f(x)$ for $x \in [a, b] \setminus E$ and $f(c_k) := 0$. We want to show that $f \in \mathcal{R}^*[a, b]$ and that equation (5) holds.
    (a) Given $\varepsilon > 0$ and $t \in [a, b] \setminus E$, let $\delta_\varepsilon(t)$ be defined as in the proof of 10.1.9. Choose $\delta_\varepsilon(c_k) > 0$ such that if $|z - c_k| < \delta_\varepsilon(c_k)$ and $z \in [a, b]$, then $|F(z) - F(c_k)| < \varepsilon/2^{k+2}$.

(b) Show that if the partition $\dot{\mathcal{P}}$ is δ_ε-fine and has a tag $t_i = c_k$, then we have $|F(x_i) - F(x_{i-1}) - f(c_k)(x_i - x_{i-1})| < \varepsilon/2^{k+1}$.

(c) Use the argument in 10.1.9 to get $|S(f; \dot{\mathcal{P}}) - (F(b) - F(a))| < \varepsilon(b - a + 1)$.

15. Show that the function $g_1(x) := x^{-1/2} \sin(1/x)$ for $x \in (0, 1]$ and $g_1(0) := 0$ belongs to $\mathcal{R}^*[0, 1]$. [*Hint:* Differentiate $C_1(x) := x^{3/2} \cos(1/x)$ for $x \in (0, 1]$ and $C_1(0) := 0$.]

16. Show that the function $g_2(x) := (1/x) \sin(1/x)$ for $x \in (0, 1]$ and $g_2(0) := 0$ belongs to $\mathcal{R}^*[0, 1]$. [*Hint:* Differentiate $C_2(x) := x \cos(1/x)$ for $x \in (0, 1]$ and $C_2(0) := 0$, and use the result for the cosine function that corresponds to Exercise 7.2.12.]

17. Use the Substitution Theorem 10.1.12 to evaluate the following integrals.

(a) $\displaystyle\int_{-3}^3 (2t + 1)\mathrm{sgn}(t^2 + t - 2)\, dt = 6,$ (b) $\displaystyle\int_0^4 \frac{\sqrt{t}\, dt}{1 + \sqrt{t}},$

(c) $\displaystyle\int_1^5 \frac{dt}{t\sqrt{t - 1}} = 2 \operatorname{Arctan} 2,$ (d) $\displaystyle\int_0^1 \sqrt{1 - t^2}\, dt.$

18. Give an example of a function $f \in \mathcal{R}^*[0, 1]$ whose square f^2 does not belong to $\mathcal{R}^*[0, 1]$.

19. Let $F(x) := x \cos(\pi/x)$ for $x \in (0, 1]$ and $F(0) := 0$. It will be seen that $f := F' \in \mathcal{R}^*[0, 1]$ but that its absolute value $|f| = |F'| \notin \mathcal{R}^*[0, 1]$. (Here $f(0) := 0$.)

(a) Show that F' and $|F'|$ are continuous on any interval $[c, 1], 0 < c < 1$ and $f \in \mathcal{R}^*[0, 1]$.

(b) If $a_k := 2/(2k + 1)$ and $b_k := 1/k$ for $k \in \mathbb{N}$, then the intervals $[a_k, b_k]$ are non-overlapping and $1/k \le \int_{a_k}^{b_k} |f|$.

(c) Since the series $\sum_{k=1}^\infty 1/k$ diverges, then $|f| \notin \mathcal{R}^*[0, 1]$.

20. Let f be as in Exercise 19 and let $m(x) := (-1)^k$ for $x \in [a_k, b_k]\, (k \in \mathbb{N})$, and $m(x) := 0$ elsewhere in $[0, 1]$. Show that $m \cdot f = |m \cdot f|$. Use Exercise 7.2.11 to show that the bounded functions m and $|m|$ belong to $\mathcal{R}[0, 1]$. Conclude that the product of a function in $\mathcal{R}^*[0, 1]$ and a bounded function in $\mathcal{R}[0, 1]$ may not belong to $\mathcal{R}^*[0, 1]$.

21. Let $\Phi(x) := x\, |\cos(\pi/x)|$ for $x \in (0, 1]$ and let $\Phi(0) := 0$. Then Φ is continuous on $[0, 1]$ and $\Phi'(x)$ exists for $x \notin E := \{0\} \cup \{a_k : k \in \mathbb{N}\}$, where $a_k := 2/(2k + 1)$. Let $\varphi(x) := \Phi'(x)$ for $x \notin E$ and $\varphi(x) := 0$ for $x \in E$. Show that φ is not bounded on $[0, 1]$. Using the Fundamental Theorem 10.1.9 with E *countable*, conclude that $\varphi \in \mathcal{R}^*[0, 1]$ and that $\int_a^b \varphi = \Phi(b) - \Phi(a)$ for $a, b \in [0, 1]$. As in Exercise 19, show that $|\varphi| \notin \mathcal{R}^*[0, 1]$.

22. Let $\Psi(x) := x^2 |\cos(\pi/x)|$ for $x \in (0, 1]$ and $\Psi(0) := 0$. Then Ψ is continuous on $[0, 1]$ and $\Psi'(x)$ exists for $x \notin E_1 := \{a_k\}$. Let $\psi(x) := \Psi'(x)$ for $x \notin E_1$ and $\psi(x) := 0$ for $x \in E_1$. Show that ψ is *bounded* on $[0, 1]$ and (using Exercise 7.2.11) that $\psi \in \mathcal{R}[0, 1]$. Show that $\int_a^b \psi = \Psi(b) - \Psi(a)$ for $a, b \in [0, 1]$. Also show that $|\psi| \in \mathcal{R}[0, 1]$.

23. If $f : [a, b] \to \mathbb{R}$ is continuous and if $p \in \mathcal{R}^*[a, b]$ does not change sign on $[a, b]$, and if $fp \in \mathcal{R}^*[a, b]$, then there exists $\xi \in [a, b]$ such that $\int_a^b fp = f(\xi) \int_a^b p$. (This is a generalization of Exercise 7.2.16; it is called the *First Mean Value Theorem* for integrals.)

24. Let $f \in \mathcal{R}^*[a, b]$, let g be monotone on $[a, b]$ and suppose that $f \ge 0$. Then there exists $\xi \in [a, b]$ such that $\int_a^b fg = g(a) \int_a^\xi f + g(b) \int_\xi^b f$. (This is a form of the *Second Mean Value Theorem* for integrals.)

Section 10.2 Improper and Lebesgue Integrals

We have seen in Theorem 7.1.5 that a function f in $\mathcal{R}[a, b]$ must be bounded on $[a, b]$ (although this need *not* be the case for a function in $\mathcal{R}^*[a, b]$). In order to integrate certain functions that have infinite limits at a point c in $[a, b]$, or which are highly oscillatory at such a point, one learns in calculus to take limits of integrals over subintervals, as the endpoints of these subintervals tend to the point c.

For example, the function $h(x) := 1/\sqrt{x}$ for $x \in (0, 1]$ and $h(0) := 0$ is unbounded on a neighborhood of the *left* endpoint of $[0, 1]$. However, it does belong to $\mathcal{R}[\gamma, 1]$ for every $\gamma \in (0, 1]$ and we define the "improper Riemann integral" of h on $[0, 1]$ to be the limit

$$\int_0^1 \frac{1}{\sqrt{x}} \, dx := \lim_{\gamma \to 0+} \int_\gamma^1 \frac{1}{\sqrt{x}} \, dx.$$

We would treat the oscillatory function $k(x) := \sin(1/x)$ for $x \in (0, 1]$ and $k(0) := 0$ in the same way.

One handles a function that becomes unbounded, or is highly oscillatory, at the *right* endpoint of the interval in a similar fashion. Furthermore, if a function g is unbounded, or is highly oscillatory, near some $c \in (a, b)$, then we define the "improper Riemann integral" to be

$$(1) \qquad \int_a^b g := \lim_{\alpha \to c-} \int_a^\alpha g + \lim_{\beta \to c+} \int_\beta^b g.$$

These limiting processes are not necessary when one deals with the *generalized* Riemann integral.

For example, we have seen in Example 10.1.10(a) that if $H(x) := 2\sqrt{x}$ for $x \in [0, 1]$ then $H'(x) = 1/\sqrt{x} =: h(x)$ for $x \in (0, 1]$ and the Fundamental Theorem 10.1.9 asserts that $h \in \mathcal{R}^*[0, 1]$ and that

$$\int_0^1 \frac{1}{\sqrt{x}} \, dx = H(1) - H(0) = 2.$$

This example is an instance of a remarkable theorem due to Heinrich Hake, which we now state in the case where the function becomes unbounded or is oscillatory near the *right* endpoint of the interval.

10.2.1 Hake's Theorem   If $f : [a, b] \to \mathbb{R}$, then $f \in \mathcal{R}^*[a, b]$ if and only if for every $\gamma \in (a, b)$ the restriction of f to $[a, \gamma]$ belongs to $\mathcal{R}^*[a, \gamma]$ and

$$(2) \qquad \lim_{\gamma \to b-} \int_a^\gamma f = A \in \mathbb{R}.$$

In this case $\displaystyle\int_a^b f = A$.

The idea of the proof of the ($\Leftarrow$) part of this result is to take an increasing sequence (γ_n) converging to b so that $f \in \mathcal{R}^*[a, \gamma_n]$ and $\lim_n \int_a^{\gamma_n} f = A$. In order to show that $f \in \mathcal{R}^*[a, b]$, we need to construct gauges on $[a, b]$. This is done by carefully "piecing together" gauges that work for the intervals $[\gamma_{i-1}, \gamma_i]$ to obtain a gauge on $[a, b]$. Since the details of this construction are somewhat delicate and not particularly informative, we will not go through them here but refer the reader to [MTI].

It is important to understand the significance of Hake's Theorem.

- It implies that the generalized Riemann integral *cannot be extended by taking limits* as in (2). Indeed, if a function f has the property that its restriction to every subinterval $[a, \gamma]$, where $\gamma \in (a, b)$, is generalized Riemann integrable and such that (2) holds, then f *already* belongs to $\mathcal{R}^*[a, b]$.

An alternative way of expressing this fact is that the generalized Riemann integral *does not need to be extended* by taking such limits.

* One can test a function for integrability on $[a, b]$ by examining its behavior on subintervals $[a, \gamma]$ with $\gamma < b$. Since it is usually difficult to establish that a function is in $\mathcal{R}^*[a, b]$ by using Definition 10.1.1, this fact gives us another tool for showing that a function is generalized Riemann integrable on $[a, b]$.

* It is often useful to evaluate the integral of a function by using (2).

We will use these observations to give an important example that provides insight into the set of generalized Riemann integrable functions.

10.2.2 Example   **(a)**   Let $\sum_{k=1}^{\infty} a_k$ be any series of real numbers converging to $A \in \mathbb{R}$. We will construct a function $\varphi \in \mathcal{R}^*[0, 1]$ such that

$$\int_0^1 \varphi = \sum_{k=1}^{\infty} a_k = A.$$

Indeed, we define $\varphi : [0, 1] \to \mathbb{R}$ to be the function that takes the values $2a_1$, $2^2 a_2$, $2^3 a_3, \cdots$ on the intervals $[0, \frac{1}{2})$, $[\frac{1}{2}, \frac{3}{4})$, $[\frac{3}{4}, \frac{7}{8})$, $\cdots$. (See Figure 10.2.1.) For convenience, let $c_k := 1 - 1/2^k$ for $k = 0, 1, \cdots$, then

$$\varphi(x) := \begin{cases} 2^k a_k & \text{for} \quad c_{k-1} \le x < c_k \ (k \in \mathbb{N}), \\ 0 & \text{for} \quad x = 1. \end{cases}$$

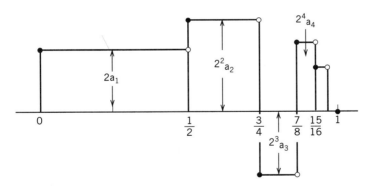

Figure 10.2.1   The graph of φ.

Clearly the restriction of φ to each interval $[0, \gamma]$ for $\gamma \in (0, 1)$, is a step function and therefore is integrable. In fact, if $\gamma \in [c_n, c_{n+1})$ then

$$\int_0^\gamma \varphi = (2a_1) \cdot \left(\frac{1}{2}\right) + (2^2 a_2) \cdot \left(\frac{1}{2^2}\right) + \cdots + (2^n a_n) \cdot \left(\frac{1}{2^n}\right) + r_\gamma$$
$$= a_1 + a_2 + \cdots + a_n + r_\gamma,$$

where $|r_\gamma| \le |a_{n+1}|$. But since the series is convergent, then $r_\gamma \to 0$ and so

$$\lim_{\gamma \to 1-} \int_0^\gamma \varphi = \lim_{n \to \infty} \sum_{k=1}^{n} a_k = A.$$

(b) If the series $\sum_{k=1}^{\infty} a_k$ is *absolutely convergent* in the sense of Definition 9.1.1, then it follows as in (a) that the function $|\varphi|$ also belongs to $\mathcal{R}^*[0, 1]$ and that

$$\int_0^1 |\varphi| = \sum_{k=1}^{\infty} |a_k|.$$

However, if the series $\sum_{k=1}^{\infty} |a_k|$ is not convergent, then the function $|\varphi|$ does not belong to $\mathcal{R}^*[0, 1]$.

Since there are many convergent series that are not absolutely convergent (for example, $\sum_{k=1}^{\infty} (-1)^k / k$), we have examples of *functions that belong to $\mathcal{R}^*[0, 1]$ but whose absolute values do not belong to $\mathcal{R}^*[0, 1]$.* We have already encountered such functions in Exercises 10.1.19 and 10.1.21.   □

The fact that there are generalized Riemann integrable functions whose absolute value is not generalized Riemann integrable is often summarized by saying that the generalized Riemann integral is not an "absolute integral". Thus, in passing to the generalized Riemann integral we lose an important property of the (ordinary) Riemann integral. But that is the price that one must pay in order to be able to integrate a much larger class of functions.

Lebesgue Integrable Functions

In view of the importance of the subset of functions in $\mathcal{R}^*[a, b]$ whose absolute values also belong to $\mathcal{R}^*[a, b]$, we will introduce the following definition.

10.2.3 Definition   A function $f \in \mathcal{R}^*[a, b]$ such that $|f| \in \mathcal{R}^*[a, b]$ is said to be **Lebesgue integrable** on $[a, b]$. The collection of all Lebesgue integrable functions on $[a, b]$ is denoted by $\mathcal{L}[a, b]$.

Note   The collection of all Lebesgue integrable functions is usually introduced in a *totally different* manner. One of the advantages of the generalized Riemann integral is that it includes the collection of Lebesgue integrable functions as a special—and easily identifiable—collection of functions.

It is clear that if $f \in \mathcal{R}^*[a, b]$ and if $f(x) \geq 0$ for all $x \in [a, b]$, then we have $|f| = f \in \mathcal{R}^*[a, b]$, so that $f \in \mathcal{L}[a, b]$. That is, a nonnegative function $f \in \mathcal{R}^*[a, b]$ belongs to $\mathcal{L}[a, b]$. The next result gives a more powerful test for a function in $\mathcal{R}^*[a, b]$ to belong to $\mathcal{L}[a, b]$.

10.2.4 Comparison Test   If $f, \omega \in \mathcal{R}^*[a, b]$ and $|f(x)| \leq \omega(x)$ for all $x \in [a, b]$, then $f \in \mathcal{L}[a, b]$ and

$$(3) \qquad \left| \int_a^b f \right| \leq \int_a^b |f| \leq \int_a^b \omega.$$

Partial Proof.   The fact that $|f| \in \mathcal{R}^*[a, b]$ is proved in [MTI]. Since $|f| \geq 0$, this implies that $f \in \mathcal{L}[a, b]$.

To establish (3), we note that $-|f| \leq f \leq |f|$ and 10.1.5(c) imply that

$$-\int_a^b |f| \leq \int_a^b f \leq \int_a^b |f|,$$

whence the first inequality in (3) follows. The second inequality follows from another application of 10.1.5(c).   Q.E.D.

The next result shows that constant multiples and sums of functions in $\mathcal{L}[a, b]$ also belong to $\mathcal{L}[a, b]$.

10.2.5 Theorem   *If* $f, g \in \mathcal{L}[a, b]$ *and if* $c \in \mathbb{R}$, *then* cf *and* $f + g$ *also belong to* $\mathcal{L}[a, b]$. *Moreover*

(4)
$$\int_a^b cf = c \int_a^b f \quad \text{and} \quad \int_a^b |f + g| \leq \int_a^b |f| + \int_a^b |g|.$$

Proof.   Since $|cf(x)| = |c||f(x)|$ for all $x \in [a, b]$, the hypothesis that $|f|$ belongs to $\mathcal{R}^*[a, b]$ implies that cf and $|cf|$ also belong to $\mathcal{R}^*[a, b]$, whence $cf \in \mathcal{L}[a, b]$.

The Triangle Inequality implies that $|f(x) + g(x)| \leq |f(x)| + |g(x)|$ for all $x \in [a, b]$. But since $\omega := |f| + |g|$ belongs to $\mathcal{R}^*[a, b]$, the Comparison Test 10.2.4 implies that $f + g$ belongs to $\mathcal{L}[a, b]$ and that

$$\int_a^b |f + g| \leq \int_a^b (|f| + |g|) = \int_a^b |f| + \int_a^b |g|. \qquad \text{Q.E.D.}$$

The next result asserts that one only needs to establish a one-sided inequality in order to show that a function $f \in \mathcal{R}^*[a, b]$ actually belongs to $\mathcal{L}[a, b]$.

10.2.6 Theorem   *If* $f \in \mathcal{R}^*[a, b]$, *the following assertions are equivalent:*

(a) $f \in \mathcal{L}[a, b]$.
(b)   *There exists* $\omega \in \mathcal{L}[a, b]$ *such that* $f(x) \leq \omega(x)$ *for all* $x \in [a, b]$.
(c)   *There exists* $\alpha \in \mathcal{L}[a, b]$ *such that* $\alpha(x) \leq f(x)$ *for all* $x \in [a, b]$.

Proof.   (a) $\Rightarrow$ (b) Let $\omega := f$.

(b) $\Rightarrow$ (a) Note that $f = \omega - (\omega - f)$. Since $\omega - f \geq 0$ and since $\omega - f$ belongs to $\mathcal{R}^*[a, b]$, it follows that $\omega - f \in \mathcal{L}[a, b]$. Now apply Theorem 10.2.5.

We leave the proof that (a) $\Longleftrightarrow$ (c) to the reader.   Q.E.D.

10.2.7 Theorem   *If* $f, g \in \mathcal{L}[a, b]$, *then the functions* $\max\{f, g\}$ *and* $\min\{f, g\}$ *also belong to* $\mathcal{L}[a, b]$.

Proof.   It follows from Exercise 2.2.16 that if $x \in [a, b]$, then

$$\max\{f(x), g(x)\} = \tfrac{1}{2}(f(x) + g(x) + |f(x) - g(x)|),$$
$$\min\{f(x), g(x)\} = \tfrac{1}{2}(f(x) + g(x) - |f(x) - g(x)|).$$

The assertions follow from these equations and Theorem 10.2.5.   Q.E.D.

In fact, the preceding result gives a useful conclusion about the maximum and the minimum of two functions in $\mathcal{R}^*[a, b]$.

10.2.8 Theorem   *Suppose that* f, g, α *and* ω *belong to* $\mathcal{R}^*[a, b]$. *If*

$$f \leq \omega, \, g \leq \omega \quad \text{or if} \quad \alpha \leq f, \, \alpha \leq g,$$

then $\max\{f, g\}$ *and* $\min\{f, g\}$ *also belong to* $\mathcal{R}^*[a, b]$.

Proof. Suppose that $f \leq \omega$ and $g \leq \omega$; then $\max\{f, g\} \leq \omega$. It follows from the first equality in the proof of Theorem 10.2.7 that

$$0 \leq |f - g| = 2\max\{f, g\} - f - g \leq 2\omega - f - g.$$

Since $2\omega - f - g \geq 0$, this function belongs to $\mathcal{L}[a, b]$. The Comparison Test 10.2.4 implies that $2\max\{f, g\} - f - g$ belongs to $\mathcal{L}[a, b]$, and so $\max\{f, g\}$ belongs to $\mathcal{R}^*[a, b]$.

The second part of the assertion is proved similarly.   Q.E.D.

The Seminorm in $\mathcal{L}[a, b]$

We will now define the "seminorm" of a function in $\mathcal{L}[a, b]$ and the "distance between" two such functions.

10.2.9 Definition   If $f \in \mathcal{L}[a, b]$, we define the **seminorm** of f to be

$$\|f\| := \int_a^b |f|.$$

If $f, g \in \mathcal{L}[a, b]$, we define the **distance between** f and g to be

$$\text{dist}(f, g) := \|f - g\| = \int_a^b |f - g|.$$

We now establish a few properties of the seminorm and distance functions.

10.2.10 Theorem   *The seminorm function satisfies:*

(i) $\|f\| \geq 0$ *for all* $f \in \mathcal{L}[a, b]$.
(ii)   *If* $f(x) = 0$ *for* $x \in [a, b]$, *then* $\|f\| = 0$.
(iii) *If* $f \in \mathcal{L}[a, b]$ *and* $c \in \mathbb{R}$, *then* $\|cf\| = |c| \cdot \|f\|$.
(iv)   *If* $f, g \in \mathcal{L}[a, b]$, *then* $\|f + g\| \leq \|f\| + \|g\|$.

Proof. Parts (i)–(iii) are easily seen. Part (iv) follows from the fact that $|f + g| \leq |f| + |g|$ and Theorem 10.1.5(c).   Q.E.D.

10.2.11 Theorem   *The distance function satisfies:*

(j) $\text{dist}(f, g) \geq 0$ *for all* $f, g \in \mathcal{L}[a, b]$.
(jj)   *If* $f(x) = g(x)$ *for* $x \in [a, b]$, *then* $\text{dist}(f, g) = 0$.
(jjj) $\text{dist}(f, g) = \text{dist}(g, f)$ *for all* $f, g \in \mathcal{L}[a, b]$.
(jv) $\text{dist}(f, h) \leq \text{dist}(f, g) + \text{dist}(g, h)$ *for all* $f, g, h \in \mathcal{L}[a, b]$.

These assertions follow from the corresponding ones in Theorem 10.2.10. Their proofs will be left as exercises.

Using the seminorm (or the distance function) we can define what we mean for a sequence of functions (f_n) in $\mathcal{L}[a, b]$ to converge to a function $f \in \mathcal{L}[a, b]$; namely, given any $\varepsilon > 0$ there exists $K(\varepsilon)$ such that if $n \geq K(\varepsilon)$ then

$$\|f_n - f\| = \text{dist}(f_n, f) < \varepsilon.$$

This notion of convergence can be used exactly as we have used the distance function in $\mathbb{R}$ for the convergence of sequences of real numbers.

We will conclude this section with a statement of the Completeness Theorem for $\mathcal{L}[a, b]$ (also called the Riesz-Fischer Theorem). It plays the same role in the space $\mathcal{L}[a, b]$ that the Completeness Property plays in $\mathbb{R}$.

10.2.12 Completeness Theorem   *A sequence* (f_n) *of functions in* $\mathcal{L}[a, b]$ *converges to a function* $f \in \mathcal{L}[a, b]$ *if and only if it has the property that for every* $\varepsilon > 0$ *there exists* $H(\varepsilon)$ *such that if* $m, n \geq H(\varepsilon)$ *then*

$$\| f_m - f_n \| = \operatorname{dist}(f_m, f_n) < \varepsilon.$$

The direction $(\Rightarrow)$ is very easy to prove and is left as an exercise. A proof of the direction $(\Leftarrow)$ is more involved, but can be based on the following idea: Find a subsequence $(g_k) := (f_{n_k})$ of (f_n) such that $\|g_{k+1} - g_k\| < 1/2^k$ and define $f(x) := g_1(x) + \sum_{k=1}^{\infty} \big(g_{k+1}(x) - g_k(x)\big)$, where this series is absolutely convergent, and $f(x) := 0$ elsewhere. It can then be shown that $f \in \mathcal{L}[a, b]$ and that $\| f_n - f \| \to 0$. (The details are given in [MTI].)

Exercises for Section 10.2

1. Show that Hake's Theorem 10.2.1 can be given the following sequential formulation: A function $f \in \mathcal{R}^*[a, b]$ if and only if there exists $A \in \mathbb{R}$ such that for any increasing sequence (c_n) in (a, b) with $c_n \to b$, then $f \in \mathcal{R}^*[a, c_n]$ and $\int_a^{c_n} f \to A$.

2. (a) Apply Hake's Theorem to conclude that $g(x) := 1/x^{2/3}$ for $x \in (0, 1]$ and $g(0) := 0$ belongs to $\mathcal{R}^*[0, 1]$.
    (b) Explain why Hake's Theorem does not apply to $f(x) := 1/x$ for $x \in (0, 1]$ and $f(0) := 0$ (which does not belong to $\mathcal{R}^*[0, 1]$).

3. Apply Hake's Theorem to $g(x) := (1 - x)^{-1/2}$ for $x \in [0, 1)$ and $g(1) := 0$.

4. Suppose that $f \in \mathcal{R}^*[a, c]$ for all $c \in (a, b)$ and that there exists $\gamma \in (a, b)$ and $\omega \in \mathcal{L}[\gamma, b]$ such that $|f(x)| \leq \omega(x)$ for $x \in [\gamma, b]$. Show that $f \in \mathcal{R}^*[a, b]$.

5. Show that the function $g_1(x) := x^{-1/2} \sin(1/x)$ for $x \in (0, 1]$ and $g_1(0) := 0$ belongs to $\mathcal{L}[0, 1]$. (This function was also considered in Exercise 10.1.15.)

6. Show that the following functions (properly defined when necessary) are in $\mathcal{L}[0, 1]$.
    (a) $\dfrac{x \ln x}{1 + x^2}$,
    (b) $\dfrac{\sin \pi x}{\ln x}$,
    (c) $(\ln x)(\ln(1 - x))$,
    (d) $\dfrac{\ln x}{\sqrt{1 - x^2}}$.

7. Determine whether the following integrals are convergent or divergent. (Define the integrands to be 0 where they are not already defined.)
    (a) $\displaystyle\int_0^1 \frac{\sin x\, dx}{x^{3/2}}$,
    (b) $\displaystyle\int_0^1 \frac{\cos x\, dx}{x^{3/2}}$,
    (c) $\displaystyle\int_0^1 \frac{\ln x\, dx}{x\sqrt{1 - x^2}}$,
    (d) $\displaystyle\int_0^1 \frac{\ln x\, dx}{1 - x}$,
    (e) $\displaystyle\int_0^1 (\ln x)(\sin(1/x))\, dx$,
    (f) $\displaystyle\int_0^1 \frac{dx}{\sqrt{x}(1 - x)}$.

8. If $f \in \mathcal{R}[a, b]$, show that $f \in \mathcal{L}[a, b]$.

9. If $f \in \mathcal{L}[a, b]$, show that f^2 is not necessarily in $\mathcal{L}[a, b]$.

10. If $f, g \in \mathcal{L}[a, b]$ and if g is bounded and monotone, show that $fg \in \mathcal{L}[a, b]$. More exactly, if $|g(x)| \leq B$, show that $\|fg\| \leq B\|f\|$.

11. (a) Give an example of a function $f \in \mathcal{R}^*[0, 1]$ such that $\max\{f, 0\}$ does not belong to $\mathcal{R}^*[0, 1]$.

    (b) Can you give an example of $f \in \mathcal{L}[0, 1]$ such that $\max\{f, 0\} \notin \mathcal{L}[0, 1]$?

12. Write out the details of the proof that $\min\{f, g\} \in \mathcal{R}^*[a, b]$ in Theorem 10.2.8 when $\alpha \leq f$ and $\alpha \leq g$.

13. Write out the details of the proofs of Theorem 10.2.11.

14. Give an $f \in \mathcal{L}[a, b]$ with f not identically 0, but such that $\|f\| = 0$.

15. If $f, g \in \mathcal{L}[a, b]$, show that $\big|\|f\| - \|g\|\big| \leq \|f \pm g\|$.

16. Establish the easy part of the Completeness Theorem 10.2.12.

17. If $f_n(x) := x^n$ for $n \in \mathbb{N}$, show that $f_n \in \mathcal{L}[0, 1]$ and that $\|f_n\| \to 0$. Thus $\|f_n - \theta\| \to 0$, where θ denotes the function identically equal to 0.

18. Let $g_n(x) := -1$ for $x \in [-1, -1/n)$, let $g_n(x) := nx$ for $x \in [-1/n, 1/n]$ and let $g_n(x) := 1$ for $x \in (1/n, 1]$. Show that $\|g_m - g_n\| \to 0$ as $m, n \to \infty$, so that the Completeness Theorem 10.2.12 implies that there exists $g \in \mathcal{L}[-1, 1]$ such that (g_n) converges to g in $\mathcal{L}[-1, 1]$. Find such a function g.

19. Let $h_n(x) := n$ for $x \in (0, 1/n)$ and $h_n(x) := 0$ elsewhere in $[0, 1]$. Does there exist $h \in \mathcal{L}[0, 1]$ such that $\|h_n - h\| \to 0$?

20. Let $k_n(x) := n$ for $x \in (0, 1/n^2)$ and $k_n(x) := 0$ elsewhere in $[0, 1]$. Does there exist $k \in \mathcal{L}[0, 1]$ such that $\|k_n - k\| \to 0$?

Section 10.3   Infinite Intervals

In the preceding two sections, we have discussed the integration of functions defined on *bounded* closed intervals $[a, b]$. However, in applications we often want to integrate functions defined on unbounded closed intervals, such as

$$[a, \infty), \qquad (-\infty, b], \qquad \text{or} \qquad (-\infty, \infty).$$

In calculus, the standard approach is to define an integral over $[a, \infty)$ as a limit:

$$\int_a^\infty f := \lim_{\gamma \to \infty} \int_a^\gamma f,$$

and to define integrals over the other infinite intervals similarly. In this section, we will treat the generalized Riemann integrable (and Lebesgue integrable) functions defined on infinite intervals.

In defining the generalized Riemann integral of a function f on $[a, \infty)$, we will adopt a somewhat different procedure from that in calculus. We note that if $\dot{\mathcal{Q}} := \{([x_0, x_1], t_1), \cdots, ([x_{n-1}, x_n], t_n), ([x_n, \infty], t_{n+1})\}$ is a tagged partition of $[a, \infty]$, then $x_0 = a$ and $x_{n+1} = \infty$ and the Riemann sum corresponding to $\dot{\mathcal{Q}}$ has the form:

(1) $$f(t_1)(x_1 - x_0) + \cdots + f(t_n)(x_n - x_{n-1}) + f(t_{n+1})(\infty - x_n).$$

Since the final term $f(t_{n+1})(\infty - x_n)$ in (1) is not meaningful, we wish to suppress this term. We can do this in two different ways: (i) define the Riemann sum to contain only the first n terms, or (ii) have a procedure that will enable us to deal with the symbols $\pm \infty$ in calculations in such a way that we eliminate the final term in (1).

We choose to adopt method (i): instead of dealing with partitions of $[a, \infty)$ into a finite number of non-overlapping intervals (one of which must necessarily have infinite length), we deal with certain **subpartitions** of $[a, \infty)$, which are finite collections of non-overlapping intervals of finite length whose union is properly contained in $[a, \infty)$.

We define a **gauge** on $[a, \infty]$ to be an ordered pair consisting of a strictly positive function δ defined on $[a, \infty)$ and a number $d^* > 0$. When we say that a tagged subpartition $\dot{\mathcal{P}} := \{([x_0, x_1], t_1), \cdots, ([x_{n-1}, x_n], t_n)\}$ is (δ, d^*)-**fine**, we mean that

(2)
$$[a, \infty) = \bigcup_{i=1}^{n} [x_{i-1}, x_i] \cup [x_n, \infty),$$

that

(3)
$$[x_{i-1}, x_i] \subseteq [t_i - \delta(t_i), t_i + \delta(t_i)] \qquad \text{for} \quad i = 1, \cdots, n,$$

and that

(4)
$$[x_n, \infty) \subseteq [1/d^*, \infty)$$

or, equivalently, that

(4′)
$$1/d^* \leq x_n.$$

Note Ordinarily we consider a gauge on $[a, \infty]$ to be a strictly positive function δ with domain $[a, \infty] := [a, \infty) \cup \{\infty\}$ where $\delta(\infty) := d^*$.

We will now define the generalized Riemann integral over $[a, \infty)$.

10.3.1 Definition **(a)** A function $f : [a, \infty) \to \mathbb{R}$ is said to be **generalized Riemann integrable** if there exists $A \in \mathbb{R}$ such that for every $\varepsilon > 0$ there exists a gauge δ_ε on $[a, \infty]$ such that if $\dot{\mathcal{P}}$ is any δ_ε-fine tagged subpartition of $[a, \infty)$, then $|S(f; \dot{\mathcal{P}}) - A| \leq \varepsilon$. In this case we write $f \in \mathcal{R}^*[a, \infty)$ and

$$\int_a^\infty f := A.$$

(b) A function $f : [a, \infty) \to \mathbb{R}$ is said to be **Lebesgue integrable** if both f and $|f|$ belong to $\mathcal{R}^*[a, \infty)$. In this case we write $f \in \mathcal{L}[a, \infty)$.

Of particular importance is the version of Hake's Theorem for functions in $\mathcal{R}^*[a, \infty)$. Other results for functions in $\mathcal{L}[a, \infty)$ will be given in the exercises.

10.3.2 Hake's Theorem *If $f : [a, \infty) \to \mathbb{R}$, then $f \in \mathcal{R}^*[a, \infty)$ if and only if for every $\gamma \in (a, \infty)$ the restriction of f to $[a, \gamma]$ belongs to $\mathcal{R}^*[a, \gamma]$ and*

(5)
$$\lim_{\gamma \to \infty} \int_a^\gamma f = A \in \mathbb{R}.$$

In this case $\int_a^\infty f = A.$

The idea of the proof of Hake's theorem is as before; the details are given in [MTI].

The generalized Riemann integral on the unbounded interval $[a, \infty)$ has the same properties as this integral on a bounded interval $[a, b]$ that were demonstrated in Section

10.1. They can be obtained by either modifying the proofs given there, or by using Hake's Theorem. We will give two examples.

10.3.3 Examples (a) If $f, g \in \mathcal{R}^*[a, \infty)$, then $f + g \in \mathcal{R}^*[a, \infty)$ and

$$\int_a^\infty (f + g) = \int_a^\infty f + \int_a^\infty g.$$

If $\varepsilon > 0$ is given let δ_f be a gauge on $[a, \infty]$ such that if $\dot{\mathcal{P}}$ is δ_f-fine, then $|S(f; \dot{\mathcal{P}}) - \int_a^\infty f| \le \varepsilon/2$, and there exists a gauge δ_g such that if $\dot{\mathcal{P}}$ is δ_g-fine, then $|S(g; \dot{\mathcal{P}}) - \int_a^\infty g| \le \varepsilon/2$. Now let $\delta_\varepsilon(t) := \min\{\delta_f(t), \delta_g(t)\}$ for $t \in [a, \infty]$ and argue as in the proof of 10.1.5(b).

(b) Let $f : [a, \infty) \to \mathbb{R}$ and let $c \in (a, \infty)$. Then $f \in \mathcal{R}^*[a, \infty)$ if and only if its restrictions to $[a, c]$ and $[c, \infty)$ are integrable. In this case,

$$(6) \qquad \int_a^\infty f = \int_a^c f + \int_c^\infty f.$$

We will prove ($\Leftarrow$) using Hake's Theorem. By hypothesis, the restriction of f to $[c, \infty)$ is integrable. Therefore, Hake's Theorem implies that for every $\gamma \in (c, \infty)$, the restriction of f to $[c, \gamma]$ is integrable and that

$$\int_c^\infty f = \lim_{\gamma \to \infty} \int_c^\gamma f.$$

If we apply the Additivity Theorem 10.1.8 to the interval $[a, \gamma] = [a, c] \cup [c, \gamma]$, we conclude that the restriction of f to $[a, \gamma]$ is integrable and that

$$\int_a^\gamma f = \int_a^c f + \int_c^\gamma f,$$

whence it follows that

$$\lim_{\gamma \to \infty} \int_a^\gamma f = \int_a^c f + \lim_{\gamma \to \infty} \int_c^\gamma f = \int_a^c f + \int_c^\infty f.$$

Another application of Hake's Theorem establishes (6). $\qquad \square$

10.3.4 Examples (a) Let $\alpha > 1$ and let $f_\alpha(x) := 1/x^\alpha$ for $x \in [1, \infty)$. We will show that $f_\alpha \in \mathcal{R}^*[1, \infty)$.

Indeed, if $\gamma \in (1, \infty)$ then the restriction of f_α to $[1, \gamma]$ is continuous and therefore belongs to $\mathcal{R}^*[1, \gamma]$. Moreover, we have

$$\int_1^\gamma \frac{1}{x^\alpha} dx = \frac{1}{1 - \alpha} \cdot x^{1-\alpha} \Big|_1^\gamma = \frac{1}{\alpha - 1} \cdot \left[1 - \frac{1}{\gamma^{\alpha-1}}\right].$$

But since the last term tends to $1/(\alpha - 1)$ as $\gamma \to \infty$, Hake's Theorem implies that $f_\alpha \in \mathcal{R}^*[1, \infty)$ and that

$$\int_1^\infty \frac{1}{x^\alpha} dx = \frac{1}{\alpha - 1} \qquad \text{when} \quad \alpha > 1.$$

(b) Let $\sum_{k=1}^\infty a_k$ be a series of real numbers that converges to $A \in \mathbb{R}$. We will construct a function $s \in \mathcal{R}^*[0, \infty)$ such that

$$\int_0^\infty s = \sum_{k=1}^\infty a_k = A.$$

Indeed, we define $s(x) := a_k$ for $x \in [k-1, k)$, $k \in \mathbb{N}$. It is clear that the restriction of s to every subinterval $[0, \gamma]$ is a step function, and therefore belongs to $\mathcal{R}^*[0, \gamma]$. Moreover, if $\gamma \in [n, n+1)$, then

$$\int_0^\gamma s = a_1 + \cdots + a_n + r_\gamma,$$

where $|r_\gamma| \le |a_{n+1}|$. But since the series is convergent, then $r_\gamma \to 0$ and so Hake's Theorem 10.3.2 implies that

$$\lim_{\gamma \to \infty} \int_0^\gamma s = \lim_{n \to \infty} \sum_{k=1}^n a_k = A.$$

(c) If the function s is defined as in (b), then $|s|$ has the value $|a_k|$ on the interval $[k-1, k)$, $k \in \mathbb{N}$. Thus s belongs to $\mathcal{L}[0, \infty)$ if and only if the series $\sum_{k=1}^\infty |a_k|$ is convergent; that is, if and only if $\sum_{k=1}^\infty a_k$ is *absolutely convergent*.

(d) Let $D(x) := (\sin x)/x$ for $x \in (0, \infty)$ and let $D(0) := 1$. We will consider the important **Dirichlet integral**:

$$\int_0^\infty D(x)\, dx = \int_0^\infty \frac{\sin x}{x}\, dx.$$

Since the restriction of D to every interval $[0, \gamma]$ is continuous, this restriction belongs to $\mathcal{R}^*[0, \gamma]$. To see that $\int_0^\gamma D(x)\, dx$ has a limit as $\gamma \to \infty$, we let $0 < \beta < \gamma$. An integration by parts shows that

$$\int_0^\gamma D(x)\, dx - \int_0^\beta D(x)\, dx = \int_\beta^\gamma \frac{\sin x}{x}\, dx$$

$$= -\frac{\cos x}{x}\Big|_\beta^\gamma - \int_\beta^\gamma \frac{\cos x}{x^2}\, dx.$$

But since $|\cos x| \le 1$, it is an exercise to show that the above terms approach 0 as $\beta < \gamma$ tend to ∞. Therefore the Cauchy Condition applies and Hake's Theorem implies that $D \in \mathcal{R}^*[0, \infty)$.

However, it will be seen in Exercise 13 that $|D|$ does *not* belong to $\mathcal{R}^*[0, \infty)$. Thus the function D does not belong to $\mathcal{L}[0, \infty)$. $\square$

We close this discussion of integrals over $[a, \infty)$ with a version of the Fundamental Theorem (First Form).

10.3.5 Fundamental Theorem *Suppose that E is a countable subset of $[a, \infty)$ and that $f, F : [a, \infty) \to \mathbb{R}$ are such that:*

(a) *F is continuous on $[a, \infty)$ and $\lim_{x \to \infty} F(x)$ exists.*

(b) *$F'(x) = f(x)$ for all $x \in (a, \infty)$, $x \notin E$.*

Then f belongs to $\mathcal{R}^*[a, \infty)$ and

(7)
$$\int_a^\infty f = \lim_{x \to \infty} F(x) - F(a).$$

Proof. If γ is any number in (a, ∞), we can apply the Fundamental Theorem 10.1.9 to the interval $[a, \gamma]$ to conclude that f belongs to $\mathcal{R}^*[a, \gamma]$ and

$$\int_a^\gamma f = F(\gamma) - F(a).$$

Letting $\gamma \to \infty$, we conclude from Hake's Theorem that $f \in \mathcal{R}^*[a, \infty)$ and that equation (7) holds.   Q.E.D.

Integrals over $(-\infty, b]$

We now discuss integration over closed intervals that are unbounded below.

Let $b \in \mathbb{R}$ and $g : (-\infty, b] \to \mathbb{R}$ be a function that is to be integrated over the infinite interval $(-\infty, b]$. By a **gauge** on $[-\infty, b]$ we mean an ordered pair consisting of a number $d_* > 0$ and a strictly positive function δ on $(-\infty, b)$. We say that a tagged subpartition $\dot{\mathcal{P}} := \{([x_0, x_1], t_1), ([x_1, x_2], t_2), \cdots, ([x_{n-1}, b], t_n)\}$ of $(-\infty, b]$ is (d_*, δ)-**fine** in case that

$$(-\infty, b] = (-\infty, x_0] \cup \bigcup_{i=1}^{n} [x_{i-1}, x_i],$$

that

$$[x_{i-1}, x_i] \subseteq [t_i - \delta(t_i), t_i + \delta(t_i)] \qquad \text{for} \quad i = 1, \cdots, n,$$

and that

$$(-\infty, x_0] \subseteq (-\infty, -1/d_*]$$

or, equivalently, that

$$x_0 \leq -1/d_*.$$

Note   Ordinarily we consider a gauge on $[-\infty, b]$ to be a strictly positive function δ with domain $[-\infty, b] := \{-\infty\} \cup (\infty, b]$ where $\delta(-\infty) := d_*$.

Here the Riemann sum of g for $\dot{\mathcal{P}}$ is $S(g; \dot{\mathcal{P}}) = \sum_{i=1}^{n} g(t_i)(x_i - x_{i-1})$.

Finally, we say that $g : (-\infty, b] \to \mathbb{R}$ is **generalized Riemann integrable** if there exists $B \in \mathbb{R}$ such that for every $\varepsilon > 0$ there exists a gauge δ_ε on $[-\infty, b]$ such that if $\dot{\mathcal{P}}$ is any δ_ε-fine subpartition of $(-\infty, b]$, then $|S(g; \dot{\mathcal{P}}) - B| \leq \varepsilon$. In this case we write $g \in \mathcal{R}^*(-\infty, b]$ and

$$\int_{-\infty}^{b} g = B.$$

Similarly, a function $g : (-\infty, b] \to \mathbb{R}$ is said to be **Lebesgue integrable** if both g and $|g|$ belong to $\mathcal{R}^*(-\infty, b]$. In this case we will write $g \in \mathcal{L}(-\infty, b]$.

The theorems valid for the integral over $[a, \infty]$ are obtained in this case as well. Their formulation will be left to the reader.

Integrals over $(-\infty, \infty)$

Let $h : (-\infty, \infty) \to \mathbb{R}$ be a function that we wish to integrate over the infinite interval $(-\infty, \infty)$. By a **gauge** on $(-\infty, \infty)$ we mean a triple consisting of a strictly positive function δ on $(-\infty, \infty)$ and two strictly positive numbers d_*, d^*. We say that a tagged subpartition $\dot{\mathcal{P}} := \{([x_0, x_1], t_1), ([x_1, x_2], t_2), \cdots, ([x_{n-1}, x_n], t_n)\}$ is (d_*, δ, d^*)-**fine** in case that

$$(-\infty, \infty) = (-\infty, x_0] \cup \bigcup_{i=1}^{n} [x_{i-1}, x_i] \cup [x_n, \infty),$$

that

$$[x_{i-1}, x_i] \subseteq [t_i - \delta(t_i), t_i + \delta(t_i)] \qquad \text{for} \quad i = 1, \cdots, n,$$

and that

$$(-\infty, x_0] \subseteq (-\infty, -1/d_*] \qquad \text{and} \qquad [x_n, \infty) \subseteq [1/d^*, \infty)$$

or, equivalently, that

$$x_0 \leq -1/d_* \qquad \text{and} \qquad 1/d^* \leq x_n.$$

Note Ordinarily we consider a gauge on $[-\infty, \infty]$ to be a strictly positive function δ with domain $[-\infty, \infty] := \{-\infty\} \cup (\infty, \infty) \cup \{\infty\}$ where $\delta(-\infty) := d_*$ and $\delta(\infty) := d^*$.

Here the Riemann sum of h for $\dot{\mathcal{P}}$ is $S(h; \dot{\mathcal{P}}) = \sum_{i=1}^{n} h(t_i)(x_i - x_{i-1})$.

Finally, we say that $h : (-\infty, \infty) \to \mathbb{R}$ is **generalized Riemann integrable** if there exists $C \in \mathbb{R}$ such that for every $\varepsilon > 0$ there exists a gauge δ_ε on $[-\infty, \infty]$ such that if $\dot{\mathcal{P}}$ is any δ_ε-fine subpartition of $(-\infty, \infty)$, then $|S(h; \dot{\mathcal{P}}) - C| \leq \varepsilon$. In this case we write $h \in \mathcal{R}^*(-\infty, \infty)$ and

$$\int_{-\infty}^{\infty} h = C.$$

Similarly, a function $h : (-\infty, \infty) \to \mathbb{R}$ is said to be **Lebesgue integrable** if both h and $|h|$ belong to $\mathcal{R}^*(-\infty, \infty)$. In this case we write $h \in \mathcal{L}(-\infty, \infty)$.

In view of its importance, we will state the version of Hake's Theorem that is valid for the integral over $(-\infty, \infty)$.

10.3.6 Hake's Theorem *If $h : (-\infty, \infty) \to \mathbb{R}$, then $h \in \mathcal{R}^*(-\infty, \infty)$ if and only if for every $\beta < \gamma$ in $(-\infty, \infty)$, the restriction of h to $[\beta, \gamma]$ is in $\mathcal{R}^*[\beta, \gamma]$ and*

$$\lim_{\substack{\beta \to -\infty \\ \gamma \to +\infty}} \int_{\beta}^{\gamma} h = C \in \mathbb{R}.$$

In this case $\int_{-\infty}^{\infty} h = C$.

As before, most of the theorems valid for the finite interval $[a, b]$ remain true. They are proved as before, or by using Hake's Theorem. We also state the first form of the Fundamental Theorem for this case.

10.3.7 Fundamental Theorem *Suppose that E is a countable subset of $(-\infty, \infty)$ and that $h, H : (-\infty, \infty) \to \mathbb{R}$ satisfy:*

(a) *H is continuous on $(-\infty, \infty)$ and the limits $\lim_{x \to \pm\infty} H(x)$ exist.*

(b) *$H'(x) = h(x)$ for all $x \in (-\infty, \infty)$, $x \notin E$.*

Then h belongs to $\mathcal{R}^*(-\infty, \infty)$ and

(8)
$$\int_{-\infty}^{\infty} h = \lim_{x \to \infty} H(x) - \lim_{y \to -\infty} H(y).$$

10.3.8 Examples   **(a)**   Let $h(x) := 1/(x^2 + 1)$ for $x \in (-\infty, \infty)$. If we let $H(x) :=$ Arctan x, then $H'(x) = h(x)$ for all $x \in (-\infty, \infty)$. Further, we have $\lim_{x \to \infty} H(x) = \frac{1}{2}\pi$ and $\lim_{x \to -\infty} H(x) = -\frac{1}{2}\pi$. Therefore it follows that

$$\int_{-\infty}^{\infty} \frac{1}{x^2 + 1}\, dx = \tfrac{1}{2}\pi - \left(-\tfrac{1}{2}\pi\right) = \pi.$$

(b)   Let $k(x) := |x|e^{-x^2}$ for $x \in (-\infty, \infty)$. If we let $K(x) := \frac{1}{2}(1 - e^{-x^2})$ for $x \geq 0$ and $K(x) := -\frac{1}{2}(1 - e^{-x^2})$ for $x < 0$, then it is seen that K is continuous on $(-\infty, \infty)$ and that $K'(x) = k(x)$ for $x \neq 0$. Further, $\lim_{x \to \infty} K(x) = \frac{1}{2}$ and $\lim_{x \to -\infty} K(x) = -\frac{1}{2}$. Therefore it follows that

$$\int_{-\infty}^{\infty} |x|e^{-x^2}\, dx = \tfrac{1}{2} - \left(-\tfrac{1}{2}\right) = 1. \qquad \Box$$

Exercises for Section 10.3

1. Let δ be a gauge on $[a, \infty]$. From Theorem 5.5.5, every bounded subinterval $[a, b]$ has a δ-fine partition. Now show that $[a, \infty]$ has a δ-fine partition.

2. Let $f \in \mathcal{R}^*[a, \gamma]$ for all $\gamma \geq a$. Show that $f \in \mathcal{R}^*[a, \infty)$ if and only if for every $\varepsilon > 0$ there exists $K(\varepsilon) \geq a$ such that if $q > p \geq K(\varepsilon)$, then $|\int_p^q f| < \varepsilon$.

3. Let f and $|f|$ belong to $\mathcal{R}^*[a, \gamma]$ for all $\gamma \geq a$. Show that $f \in \mathcal{L}[a, \infty)$ if and only if for every $\varepsilon > 0$ there exists $K(\varepsilon) \geq a$ such that if $q > p > K(\varepsilon)$ then $\int_p^q |f| < \varepsilon$.

4. Let f and $|f|$ belong to $\mathcal{R}^*[a, \gamma]$ for every $\gamma \geq a$. Show that $f \in \mathcal{L}[a, \infty)$ if and only if the set $V := \{\int_a^x |f| : x \geq a\}$ is bounded in $\mathbb{R}$.

5. If $f, g \in \mathcal{L}[a, \infty)$, show that $f + g \in \mathcal{L}[a, \infty)$. Moreover, if $\|h\| := \int_a^\infty |h|$ for any $h \in \mathcal{L}[a, \infty)$, show that $\|f + g\| \leq \|f\| + \|g\|$.

6. If $f(x) := 1/x$ for $x \in [1, \infty)$, show that $f \notin \mathcal{R}^*[1, \infty)$.

7. If f is continuous on $[1, \infty)$ and if $|f(x)| \leq K/x^2$ for $x \in [1, \infty)$, show that $f \in \mathcal{L}[1, \infty)$.

8. Let $f(x) := \cos x$ for $x \in [0, \infty)$. Show that $f \notin \mathcal{R}^*[0, \infty)$.

9. If $s > 0$, let $g(x) := e^{-sx}$ for $x \in [0, \infty)$.
    (a)   Use Hake's Theorem to show that $g \in \mathcal{L}[0, \infty)$ and $\int_0^\infty e^{-sx}\, dx = 1/s$.
    (b)   Use the Fundamental Theorem 10.3.5.

10. (a)   Use Integration by Parts and Hake's Theorem to show that $\int_0^\infty xe^{-sx}\, dx = 1/s^2$ for $s > 0$.
    (b)   Use the Fundamental Theorem 10.3.5.

11. Show that if $n \in \mathbb{N}$, $s > 0$, then $\int_0^\infty x^n e^{-sx}\, dx = n!/s^{n+1}$.

12. (a)   Show that the integral $\int_1^\infty x^{-1} \ln x\, dx$ does not converge.
    (b)   Show that if $\alpha > 1$, then $\int_1^\infty x^{-\alpha} \ln x\, dx = 1/(\alpha - 1)^2$.

13. (a)   Show that $\int_{n\pi}^{(n+1)\pi} |x^{-1} \sin x|\, dx > 1/4(n + 1)$.
    (b)   Show that $|D| \notin \mathcal{R}^*[0, \infty)$, where D is as in Example 10.3.4(d).

14. Show that the integral $\int_0^\infty (1/\sqrt{x}) \sin x\, dx$ converges. [*Hint:* Integrate by Parts.]

15. Establish the convergence of **Fresnel's integral** $\int_0^\infty \sin(x^2)\, dx$. [*Hint:* Use the Substitution Theorem 10.1.12.]

16. Establish the convergence or the divergence of the following integrals:

(a) $\int_0^\infty \dfrac{\ln x\, dx}{x^2 + 1}$,

(b) $\int_0^\infty \dfrac{\ln x\, dx}{\sqrt{x^2 + 1}}$,

(c) $\int_0^\infty \dfrac{dx}{x(x + 1)}$,

(d) $\int_0^\infty \dfrac{x\, dx}{(x + 1)^3}$,

(e) $\int_0^\infty \dfrac{dx}{\sqrt[3]{1 + x^3}}$,

(f) $\int_0^\infty \dfrac{\operatorname{Arctan} x\, dx}{x^{3/2} + 1}$.

17. Let $f, \varphi : [a, \infty) \to \mathbb{R}$. **Abel's Test** asserts that if $f \in \mathcal{R}^*[a, \infty)$ and φ is bounded and monotone on $[a, \infty)$, then $f\varphi \in \mathcal{R}^*[a, \infty)$.

(a) Show that Abel's Test does not apply to establish the convergence of $\int_0^\infty (1/x) \sin x\, dx$ by taking $\varphi(x) := 1/x$. However, it does apply if we take $\varphi(x) := 1/\sqrt{x}$ and use Exercise 14.

(b) Use Abel's Test and Exercise 15 to show the convergence of $\int_0^\infty (x/(x + 1)) \sin(x^2)\, dx$.

(c) Use Abel's Test and Exercise 14 to show the convergence of $\int_0^\infty x^{-3/2}(x + 1) \sin x\, dx$.

(d) Use Abel's Test to obtain the convergence of Exercise 16(f).

18. With the notation as in Exercise 17, the **Chartier-Dirichlet Test** asserts that if $f \in \mathcal{R}^*[a, \gamma]$ for all $\gamma \geq a$, if $F(x) := \int_a^x f$ is bounded on $[a, \infty)$, and if φ is monotone and $\lim\limits_{x \to \infty} \varphi(x) = 0$, then $f\varphi \in \mathcal{R}^*[a, \infty]$.

(a) Show that the integral $\int_0^\infty (1/x) \sin x\, dx$ converges.

(b) Show that $\int_2^\infty (1/\ln x) \sin x\, dx$ converges.

(c) Show that $\int_0^\infty (1/\sqrt{x}) \cos x\, dx$ converges.

(d) Show that the Chartier-Dirichlet Test does not apply to establish the convergence of $\int_0^\infty (x/(x + 1)) \sin(x^2)\, dx$ by taking $f(x) := \sin(x^2)$.

19. Show that the integral $\int_0^\infty \sqrt{x} \cdot \sin(x^2)\, dx$ is convergent, even though the integrand is not bounded as $x \to \infty$. [*Hint:* Make a substitution.]

20. Establish the convergence of the following integrals.

(a) $\int_{-\infty}^\infty e^{-|x|}\, dx$,

(b) $\int_{-\infty}^\infty (x - 2)e^{-|x|}\, dx$,

(c) $\int_{-\infty}^\infty e^{-x^2}\, dx$,

(d) $\int_{-\infty}^\infty \dfrac{2x\, dx}{e^x - e^{-x}}$.

Section 10.4 Convergence Theorems

We will conclude our discussion of the generalized Riemann integral with an indication of the convergence theorems that are available for it. It will be seen that the results are much stronger than those presented in Section 8.2 for the (ordinary) Riemann integral. Finally, we will introduce a "measurable" function on $[a, b]$ as the almost everywhere limit of a sequence of step functions. We will show that every integrable function is measurable, and that a measurable function on $[a, b]$ is generalized Riemann integrable if and only if it satisfies a two-sided boundedness condition.

We proved in Example 8.2.1(c) that if (f_k) is a sequence in $\mathcal{R}[a, b]$ that converges on $[a, b]$ to a function $f \in \mathcal{R}[a, b]$, then it need not happen that

(1)
$$\int_a^b f = \lim_{k \to \infty} \int_a^b f_k.$$

However, in Theorem 8.2.4 we saw that *uniform convergence* of the sequence is sufficient to guarantee that this equality holds. In fact, we will now show that this is even true for a sequence of *generalized* Riemann integrable functions.

10.4.1 Uniform Convergence Theorem   *Let (f_k) be a sequence in $\mathcal{R}^*[a, b]$ and suppose that (f_k) converges **uniformly** on $[a, b]$ to f. Then $f \in \mathcal{R}^*[a, b]$ and (1) holds.*

Proof.   Given $\varepsilon > 0$, there exists $K(\varepsilon)$ such that if $k \geq K(\varepsilon)$ and $x \in [a, b]$, then we have $|f_k(x) - f(x)| < \varepsilon$. Consequently, if $h, k \geq K(\varepsilon)$, then

$$-2\varepsilon < f_k(x) - f_h(x) < 2\varepsilon \qquad \text{for} \quad x \in [a, b].$$

Theorem 10.1.5 implies that

$$-2\varepsilon(b - a) < \int_a^b f_k - \int_a^b f_h < 2\varepsilon(b - a).$$

Since $\varepsilon > 0$ is arbitrary, the sequence $(\int_a^b f_k)$ is a Cauchy sequence in $\mathbb{R}$ and therefore converges to some number, say $A \in \mathbb{R}$. We will now show that $f \in \mathcal{R}^*[a, b]$ with integral A. For, if $\varepsilon > 0$ is given, let $K(\varepsilon)$ be as above. If $\dot{\mathcal{P}} := \{([x_{i-1}, x_i], t_i)\}_{i=1}^n$ is any tagged partition of $[a, b]$ and if $k \geq K(\varepsilon)$, then

$$\left| S(f_k; \dot{\mathcal{P}}) - S(f; \dot{\mathcal{P}}) \right| = \left| \sum_{i=1}^n \{f_k(t_i) - f(t_i)\}(x_i - x_{i-1}) \right|$$

$$\leq \sum_{i=1}^n |f_k(t_i) - f(t_i)|(x_i - x_{i-1})$$

$$< \sum_{i=1}^n \varepsilon(x_i - x_{i-1}) = \varepsilon(b - a).$$

Now fix $r \geq K(\varepsilon)$ such that $|\int_a^b f_r - A| < \varepsilon$ and let $\delta_{r,\varepsilon}$ be a gauge on $[a, b]$ such that $|\int_a^b f_r - S(f_r; \dot{\mathcal{P}})| < \varepsilon$ whenever $\dot{\mathcal{P}}$ is $\delta_{r,\varepsilon}$-fine. Then we have

$$\left| S(f; \dot{\mathcal{P}}) - A \right| \leq \left| S(f; \dot{\mathcal{P}}) - S(f_r; \dot{\mathcal{P}}) \right| + \left| S(f_r; \dot{\mathcal{P}}) - \int_a^b f_r \right| + \left| \int_a^b f_r - A \right|$$

$$< \varepsilon(b - a) + \varepsilon + \varepsilon = \varepsilon(b - a + 2).$$

But since $\varepsilon > 0$ is arbitrary, it follows that $f \in \mathcal{R}^*[a, b]$ and $\int_a^b f = A$.   Q.E.D.

It will be seen in Example 10.4.6(a) that the conclusion of 10.4.1 is false for an *infinite* interval.

Equi-integrability ———————————————————————————————

The hypothesis of uniform convergence in Theorem 10.4.1 is a very stringent one and restricts the utility of this result. Consequently, we now show that another type of uniformity condition can be used to obtain the desired limit. This notion is due to Jaroslav Kurzweil, as is Theorem 10.4.3.

10.4.2 Definition   A sequence (f_k) in $\mathcal{R}^*(I)$ is said to be **equi-integrable** if for every $\varepsilon > 0$ there exists a gauge δ_ε on I such that if $\dot{\mathcal{P}}$ is any δ_ε-fine partition of I and $k \in \mathbb{N}$, then $|S(f_k; \dot{\mathcal{P}}) - \int_I f_k| < \varepsilon$.

10.4.3 Equi-integrability Theorem   *If $(f_k) \in \mathcal{R}^*(I)$ is equi-integrable on I and if $f(x) = \lim f_k(x)$ for all $x \in I$, then $f \in \mathcal{R}^*(I)$ and*

(2)
$$\int_I f = \lim_{k \to \infty} \int_I f_k.$$

Proof.   We will treat the case $I = [a, b]$; the general case can be found in [MTI].

Given $\varepsilon > 0$, by the equi-integrability hypothesis, there exists a gauge δ_ε on I such that if $\dot{\mathcal{P}} := \{([x_{i-1}, x_i], t_i)\}_{i=1}^n$ is a δ_ε-fine partition of I, then we have $|S(f_k; \dot{\mathcal{P}}) - \int_I f_k| < \varepsilon$ for all $k \in \mathbb{N}$. Since $\dot{\mathcal{P}}$ has only a finite number of tags and since $f_k(t) \to f(t)$ for $t \in [a, b]$, there exists a K_ε such that if $h, k \geq K_\varepsilon$, then

(3)
$$\left| S(f_k; \dot{\mathcal{P}}) - S(f_h; \dot{\mathcal{P}}) \right| \leq \sum_{i=1}^n \left| f_k(t_i) - f_h(t_i) \right| (x_i - x_{i-1}) \leq \varepsilon(b - a).$$

If we let $h \to \infty$ in (3), we have

(4)
$$\left| S(f_k; \dot{\mathcal{P}}) - S(f; \dot{\mathcal{P}}) \right| \leq \varepsilon(b - a) \qquad \text{for} \quad k \geq K_\varepsilon.$$

Moreover, if $h, k \geq K_\varepsilon$, then the equi-integrability hypothesis and (3) give

$$\left| \int_I f_k - \int_I f_h \right| \leq \left| \int_I f_k - S(f_k; \dot{\mathcal{P}}) \right| + \left| S(f_k; \dot{\mathcal{P}}) - S(f_h; \dot{\mathcal{P}}) \right|$$
$$+ \left| S(f_h; \dot{\mathcal{P}}) - \int_I f_h \right| \leq \varepsilon + \varepsilon(b - a) + \varepsilon = \varepsilon(2 + b - a).$$

Since $\varepsilon > 0$ is arbitrary, then $(\int_I f_k)$ is a Cauchy sequence and converges to some $A \in \mathbb{R}$. If we let $h \to \infty$ in this last inequality, we obtain

(5)
$$\left| \int_I f_k - A \right| \leq \varepsilon(2 + b - a) \qquad \text{for} \quad k \geq K_\varepsilon.$$

We now show that $f \in \mathcal{R}^*(I)$ with integral A. Indeed, given $\varepsilon > 0$, if $\dot{\mathcal{P}}$ is a δ_ε-fine partition of I and $k \geq K_\varepsilon$, then

$$\left| S(f; \dot{\mathcal{P}}) - A \right| \leq \left| S(f; \dot{\mathcal{P}}) - S(f_k; \dot{\mathcal{P}}) \right| + \left| S(f_k; \dot{\mathcal{P}}) - \int_I f_k \right| + \left| \int_I f_k - A \right|$$
$$\leq \varepsilon(b - a) + \varepsilon + \varepsilon(2 + b - a) = \varepsilon(3 + 2b - 2a),$$

where we used (4) for the first term, the equi-integrability for the second, and (5) for the third. Since $\varepsilon > 0$ is arbitrary, $f \in \mathcal{R}^*(I)$ with integral A.   Q.E.D.

The Monotone and Dominated Convergence Theorems

Although the Equi-integrability Theorem is interesting, it is difficult to apply because it is not easy to construct the gauges δ_ε. We now state two very important theorems summarizing the most important convergence theorems for the integral that are often useful. McLeod [pp. 96–101] has shown that both of these theorems can be proved by using the Equi-integrability Theorem. However, those proofs require a delicate construction of the gauge functions. Direct proofs of these results are given in [MTI], but these proofs also use results not given here; therefore we will omit the proofs of these results.

We say that a sequence of functions on an interval $I \subseteq \mathbb{R}$ is **monotone increasing** if it satisfies $f_1(x) \leq f_2(x) \leq \cdots \leq f_k(x) \leq f_{k+1}(x) \leq \cdots$ for all $k \in \mathbb{N}$, $x \in I$. It is said to be **monotone decreasing** if it satisfies the opposite string of inequalities, and to be **monotone** if it is either monotone increasing or decreasing.

10.4.4 Monotone Convergence Theorem   *Let* (f_k) *be a monotone sequence of functions in* $\mathcal{R}^*(I)$ *such that* $f(x) = \lim f_k(x)$ *almost everywhere on* I. *Then* $f \in \mathcal{R}^*(I)$ *if and only if the sequence of integrals* $(\int_I f_k)$ *is bounded in* $\mathbb{R}$, *in which case*

$$\tag{6} \int_I f = \lim_{k \to \infty} \int_I f_k.$$

The next result is the most important theorem concerning the convergence of integrable functions. It is an extension of the celebrated "Lebesgue Dominated Convergence Theorem" from which it can also be proved.

10.4.5 Dominated Convergence Theorem   *Let* (f_n) *be a sequence in* $\mathcal{R}^*(I)$ *and let* $f(x) = \lim f_k(x)$ *almost everywhere on* I. *If there exist functions* α, ω *in* $\mathcal{R}^*(I)$ *such that*

$$\tag{7} \alpha(x) \le f_k(x) \le \omega(x) \qquad \text{for almost every} \quad x \in I,$$

then $f \in \mathcal{R}^*(I)$ *and*

$$\tag{8} \int_I f = \lim_{k \to \infty} \int_I f_k.$$

Moreover, if α *and* ω *belong to* $\mathcal{L}(I)$, *then* f_k *and* f *belong to* $\mathcal{L}(I)$ *and*

$$\tag{9} \|f_k - f\| = \int_I |f_k - f| \to 0.$$

Note   If α and ω belong to $\mathcal{L}(I)$, and we put $\varphi := \max\{|\alpha|, |\omega|\}$, then $\varphi \in \mathcal{L}(I)$ and we can replace the condition (7) by the condition

$$\tag{7'} |f_k(x)| \le \varphi(x) \qquad \text{for almost every} \quad x \in I.$$

Some Examples _____

10.4.6 Examples   **(a)**   If $k \in \mathbb{N}$, let $f_k(x) := 1/k$ for $x \in [0, k]$ and $f_k(x) := 0$ elsewhere in $[0, \infty)$.

Then the sequence converges uniformly on $[0, \infty)$ to the 0-function. However $\int_0^\infty f_k = 1$ for all $k \in \mathbb{N}$, while the integral of the 0-function equals 0. It is an exercise to show that the function $\sup\{f_k(x) : k \in \mathbb{N}\}$ does not belong to $\mathcal{R}^*[0, \infty)$, so the domination condition (7) is not satisfied.

(b)   We have $\displaystyle\lim_{k \to \infty} \int_0^1 \dfrac{x^k + 1}{x^k + 3} \, dx = \tfrac{1}{3}$.

For, if $g_k(x) := (x^k + 1)/(x^k + 3)$, then $0 \le g_k(x) \le 1$ and $g_k(x) \to 1/3$ for $x \in [0, 1)$. Thus the Dominated Convergence Theorem 10.4.5 applies.

(c)   We have $\displaystyle\lim_{k \to \infty} \int_0^k \left(1 + \dfrac{x}{k}\right)^k e^{-ax} \, dx = \dfrac{1}{a - 1}$ if $a > 1$.

Let $h_k(x) := (1 + x/k)^k e^{-ax}$ for $x \in [0, k]$ and $h_k(x) := 0$ elsewhere on $[0, \infty)$. The argument in Example 3.3.6 shows that (h_k) is an increasing sequence and converges to $e^x e^{-ax} = e^{(1-a)x}$ on $[0, \infty)$. If $a > 1$ this limit function belongs to $\mathcal{L}[0, \infty)$. Moreover, if $F(x) := e^{(1-a)x}/(1 - a)$, then $F'(x) = e^{(1-a)x}$ so that the Monotone Convergence Theorem 10.4.4 and the Fundamental Theorem 10.3.5 imply that

$$\lim_{k \to \infty} \int_0^\infty h_k = \int_0^\infty e^{(1-a)x} \, dx = F(x) \Big|_0^\infty = \frac{1}{a - 1}.$$

(d) If f is bounded and continuous on $[0, \infty)$ and if $a > 0$, then the function defined by $L(t) := \int_0^\infty e^{-tx} f(x)\, dx$ is continuous for $t \in J_a := (a, \infty)$.

Since $|e^{-tx} f(x)| \leq M e^{-ax}$ for $t \in J_a$, if (t_k) is any sequence in J_a converging to $t_0 \in J_a$, the Dominated Convergence Theorem implies that $L(t_k) \to L(t_0)$. But since the sequence $(t_k) \to t_0$ is arbitrary, then L is continuous at t_0.

(e) The integral in (d) is differentiable for $t > a$ and

$$(10) \qquad L'(t) = \int_0^\infty (-x) e^{-tx} f(x)\, dx,$$

which is the result obtained by "differentiating under the integral sign" with respect to t.

Fix a number $t_0 \in J_a$. If $t \in J_a$, then by the Mean Value Theorem applied to the function $t \mapsto e^{-tx}$, there exists a point t_x between t_0 and t such that we have $e^{-tx} - e^{-t_0 x} = -x e^{-t_x x}(t - t_0)$, whence

$$\left| \frac{e^{-tx} - e^{-t_0 x}}{t - t_0} \right| \leq x e^{-t_x x} \leq x e^{-ax}.$$

Since $\omega(x) := x e^{-ax} f(x)$ belongs to $\mathcal{L}[0, \infty)$, then for any sequence (t_k) in J_a with $t_0 \neq t_k \to t_0$, the Dominated Convergence Theorem implies that

$$\lim_{k \to \infty} \left[\frac{L(t_k) - L(t_0)}{t_k - t_0} \right] = \int_0^\infty \lim_{k \to \infty} \left[\frac{e^{-t_k x} - e^{-t_0 x}}{t_k - t_0} \right] f(x)\, dx$$

$$= \int_0^\infty (-x) e^{-t_0 x} f(x)\, dx.$$

Since (t_k) is an arbitrary sequence, then $L'(t_0)$ exists and (10) is proved.

(f) Let $D_k(t) := \int_0^k e^{-tx} \left(\frac{\sin x}{x} \right) dx$ for $k \in \mathbb{N}, t \geq 0$.

Since $|(e^{-tx} \sin x)/x| \leq e^{-tx} \leq 1$ for $t \geq 0, x \geq 0$, the integral defining D_k exists. In particular, we have

$$D_k(0) = \int_0^k \frac{\sin x}{x}\, dx.$$

We want to show that $D_k(0) \to \frac{1}{2}\pi$ as $k \to \infty$. By Example 10.3.4(d), this will show that $\int_0^\infty (\sin x)/x\, dx = \frac{1}{2}\pi$. The argument is rather complex, and uses the Dominated Convergence Theorem several times.

Since the partial derivative satisfies $\left| \dfrac{\partial}{\partial t} \left(\dfrac{e^{-tx} \sin x}{x} \right) \right| = |-e^{-tx} \sin x| \leq 1$ for $t \geq 0$, $x \geq 0$, an argument as in (e) and the Dominated Convergence Theorem imply that

$$D'_k(t) = -\int_0^k e^{-tx} \sin x\, dx \qquad \text{for} \quad k \in \mathbb{N}, t \geq 0.$$

Since a routine calculation shows that $\dfrac{\partial}{\partial x} \left(\dfrac{e^{-tx}(t \sin x + \cos x)}{t^2 + 1} \right) = -e^{-tx} \sin x$, then an application of the Fundamental Theorem gives

$$D'_k(t) = \frac{e^{-tk}(t \sin k + \cos k)}{t^2 + 1} - \frac{1}{t^2 + 1}.$$

If we put $g_k(t) := \dfrac{e^{-tk}(t \sin k + \cos k)}{t^2 + 1}$ for $0 \leq t \leq k$ and $g_k(t) := 0$ for $t > k$, then another application of the Fundamental Theorem gives

(11) $$D_k(k) - D_k(0) = \int_0^k D_k'(t)\, dt = \int_0^k g_k(t)\, dt - \int_0^k \frac{dt}{t^2 + 1}$$

$$= \int_0^\infty g_k(t)\, dt - \text{Arctan}\, k.$$

If we note that $g_k(t) \to 0$ for $t > 0$ as $k \to \infty$ and that (since $k \geq 1$)

$$|g_k(t)| \leq \frac{e^{-tk}(t + 1)}{t^2 + 1} \leq 2e^{-t} \qquad \text{for} \quad t \geq 0,$$

then the Dominated Convergence Theorem gives $\int_0^\infty g_k(t)\, dt \to 0$.

In addition, since $|(\sin x)/x| \leq 1$, we have

$$\left| D_k(k) \right| = \left| \int_0^k e^{-kx}\, \frac{\sin x}{x}\, dx \right| \leq \int_0^k e^{-kx}\, dx = \frac{e^{-kx}}{-k} \Big|_{x=0}^{x=k}$$

$$= \frac{1 - e^{-k^2}}{k} \leq \frac{1}{k} \to 0.$$

Therefore, as $k \to \infty$, formula (11) becomes

$$0 - \lim_{k \to \infty} D_k(0) = 0 - \lim_{k \to \infty} \text{Arctan}\, k = -\tfrac{1}{2}\pi.$$

As we have noted before, this gives an evaluation of **Dirichlet's Integral**:

(12) $$\int_0^\infty \frac{\sin x}{x}\, dx = \tfrac{1}{2}\pi. \qquad \square$$

Measurable Functions _____

We wish to characterize the collection of functions in $\mathcal{R}^*(I)$. In order to bypass a few minor details, we will limit our discussion to the case $I := [a, b]$. We need to introduce the notion of a "measurable function"; this class of functions contains all the functions the reader is ever likely to encounter. Measurable functions are often defined in terms of the notion of a "measurable set". However, the approach we will use is somewhat simpler and does not require a theory of measurable sets to have been developed first. (In fact, the theory of measure can be derived from properties of the integral; see Exercises 15 and 16.)

We recall from Definition 5.4.9 that a function $s : [a, b] \to \mathbb{R}$ is a **step function** if it has only a finite number of values, each value being assumed on a finite number of subintervals of $[a, b]$.

10.4.7 Definition A function $f : [a, b] \to \mathbb{R}$ is said to be **(Lebesgue) measurable** if there exists a sequence (s_k) of step functions on $[a, b]$ such that

(13) $$f(x) = \lim_{k \to \infty} s_k(x) \qquad \text{for almost every} \quad x \in [a, b].$$

We denote the collection of all measurable functions on $[a, b]$ by $\mathcal{M}[a, b]$.

We can reformulate the definition as: A function f is in $\mathcal{M}[a, b]$ if there exists a null set $Z \subset [a, b]$ and a sequence (s_k) of step functions such that

(14) $$f(x) = \lim_{k \to \infty} s_k(x) \qquad \text{for all} \quad x \in [a, b] \setminus Z.$$

It is trivial that every step function on $[a, b]$ is a measurable function. By Theorem 5.4.10, a continuous function on $[a, b]$ is a uniform limit of a sequence of step functions; therefore, every continuous function on an interval $[a, b]$ is measurable. Similarly, every monotone function on $[a, b]$ is a uniform limit of step functions (see the proof of Theorem 7.2.7); therefore, every monotone function on an interval is measurable.

At first glance, it might seem that the collection of measurable functions might not be so very large. However, the requirement that the limit (13) is required to hold only *almost everywhere* (and not everywhere), enables one to obtain much more general functions. We now give a few examples.

10.4.8 Examples **(a)** The Dirichlet function, $f(x) := 1$ for $x \in [0, 1]$ rational and $f(x) := 0$ for $x \in [0, 1]$ irrational, is a measurable function.

Since $\mathbb{Q} \cap [0, 1]$ is a null set, we can take each s_k to be the 0-function. We then obtain $s_k(x) \to f(x)$ for $x \in [0, 1] \setminus \mathbb{Q}$.

(b) Thomae's function h (see Examples 5.1.5(h) and 7.1.6) is a measurable function.

Again, take s_k to be the 0-function. Then $s_k(x) \to h(x)$ for $x \in [0, 1] \setminus \mathbb{Q}$.

(c) The function $g(x) := 1/x$ for $x \in (0, 1]$ and $g(0) := 0$ is a measurable function.

This can be seen by taking a step function $s_k(x) := 0$ for $x \in [0, 1/k)$ and (using 5.4.10) such that $|s_k(x) - 1/x| < 1/k$ for $x \in [1/k, 1]$. Then $s_k(x) \to g(x)$ for all $x \in [0, 1]$.

(d) If $f \in \mathcal{M}[a, b]$ and if $\psi : [a, b] \to \mathbb{R}$ is such that $\psi(x) = f(x)$ a.e., then $\psi \in \mathcal{M}[a, b]$.

For, if $f(x) = \lim s_k(x)$ for $x \in [a, b] \setminus Z_1$ and if $\psi(x) = f(x)$ for all $x \in [a, b] \setminus Z_2$, then $\psi(x) = \lim s_k(x)$ for all $x \in [a, b] \setminus (Z_1 \cup Z_2)$. Since $Z_1 \cup Z_2$ is a null set when Z_1 and Z_2 are, the conclusion follows.   □

The next result shows that elementary combinations of measurable functions lead to measurable functions.

10.4.9 Theorem *Let f and g belong to $\mathcal{M}[a, b]$ and let $c \in \mathbb{R}$.*

(a) *Then the functions cf, $|f|$, $f + g$, $f - g$ and $f \cdot g$ also belong to $\mathcal{M}[a, b]$.*

(b) *If $\varphi : \mathbb{R} \to \mathbb{R}$ is continuous, then the composition $\varphi \circ f \in \mathcal{M}[a, b]$.*

(c) *If (f_n) is a sequence in $\mathcal{M}[a, b]$ and $f(x) = \lim f_n(x)$ almost everywhere on I, then $f \in \mathcal{M}[a, b]$.*

Proof. (a) We will prove that $|f|$ is measurable. Let $Z \subset [a, b]$ be a null set such that (14) holds. Since $|s_k|$ is a step function, the Triangle Inequality implies that

$$0 \leq \big||f(x)| - |s_k(x)|\big| \leq \big|f(x) - s_k(x)\big| \to 0$$

for all $x \in [a, b] \setminus Z$. Therefore $|f| \in \mathcal{M}[a, b]$.

The other assertions in (a) follow from the basic properties of limits.

(b) If s_k is a step function on $[a, b]$, it is easily seen that $\varphi \circ s_k$ is also a step function on $[a, b]$. Since φ is continuous on $\mathbb{R}$ and $f(x) = \lim s_k(x)$ for all $x \in [a, b] \setminus Z$, it follows that $(\varphi \circ f)(x) = \varphi(f(x)) = \lim \varphi(s_k(x)) = \lim(\varphi \circ s_k)(x)$ for all $x \in [a, b] \setminus Z$. Therefore $\varphi \circ f$ is measurable.

(c) This conclusion is not obvious; a proof is outlined in Exercise 14.   Q.E.D.

The next result is that we can replace the step functions in Definition 10.4.7 by continuous functions. Since we will use only one part of this result, we content ourselves with a sketch of the proof of the other part.

10.4.10 Theorem   *A function $f : [a, b] \to \mathbb{R}$ is in $\mathcal{M}[a, b]$ if and only if there exists a sequence (g_k) of continuous functions such that*

(15) $f(x) = \lim_{k \to \infty} g_k(x)$ *for almost every $x \in [a, b]$.*

Proof.   ($\Leftarrow$) Let $Z \subset [a, b]$ be a null set and (g_k) be a sequence of continuous functions such that $f(x) = \lim g_k(x)$ for $x \in [a, b] \setminus Z$. Since g_k is continuous, by 5.4.10 there exists a step function s_k such that

$$|g_k(x) - s_k(x)| \leq 1/k \qquad \text{for all} \quad x \in [a, b].$$

Therefore we have

$$0 \leq |f(x) - s_k(x)| \leq |f(x) - g_k(x)| + |g_k(x) - s_k(x)|$$
$$\leq |f(x) - g_k(x)| + 1/k,$$

whence it follows that $f(x) = \lim g_k(x)$ for all $x \in [a, b] \setminus Z$.

Sketch of ($\Rightarrow$) Let Z be a null set and (s_k) be a sequence of step functions such that $f(x) = \lim s_k(x)$ for all $x \in [a, b] \setminus Z$. Without loss of generality, we may assume that each s_k is continuous at the endpoints a, b. Since s_k is discontinuous at only a finite number of points in (a, b), which can be enclosed in a finite union J_k of intervals with total length $\leq 1/k$, we can construct a piecewise linear and continuous function g_k which coincides with s_k on $[a, b] \setminus J_k$. It can be shown that $g_k(x) \to f(x)$ a.e. on I. (See [MTI] for the details.)   Q.E.D.

Functions in $\mathcal{R}^*[a, b]$ are Measurable

We now show that a generalized Riemann integrable function is measurable.

10.4.11 Measurability Theorem   *If $f \in \mathcal{R}^*[a, b]$, then $f \in \mathcal{M}[a, b]$.*

Proof.   Let $F : [a, b + 1] \to \mathbb{R}$ be the indefinite integral

$$F(x) := \int_a^x f \qquad \text{if} \quad x \in [a, b],$$

and let $F(x) := F(b)$ for $x \in (b, b + 1]$. It follows from the Fundamental Theorem (Second Form) 10.1.11(a) that F is continuous on $[a, b]$. From 10.1.11(c), there exists a null set Z such that the derivative $F'(x) = f(x)$ exists for $x \in [a, b] \setminus Z$. Therefore, if we introduce the difference quotient functions

$$g_k(x) := \frac{F(x + 1/k) - F(x)}{1/k} \qquad \text{for} \quad x \in [a, b), k \in \mathbb{N},$$

then $g_k(x) \to f(x)$ for all $x \in [a, b] \setminus Z$. Since the g_k are continuous, it follows from the part of Theorem 10.4.10 we have proved that $f \in \mathcal{M}[a, b]$.   Q.E.D.

Are Measurable Functions Integrable?

Not *every* measurable function is generalized Riemann integrable. For example, the function $g(x) := 1/x$ for $x \in (0, 1]$ and $g(0) := 0$ was seen in Example 10.4.8(c) to be measurable;

however it is not in $\mathcal{R}^*[a, b]$ because it is "too large" (as $x \to 0+$). However, if the graph of a measurable function on $[a, b]$ lies between two functions in $\mathcal{R}^*[a, b]$, then it also belongs to $\mathcal{R}^*[a, b]$.

10.4.12 Integrability Theorem *Let $f \in \mathcal{M}[a, b]$. Then $f \in \mathcal{R}^*[a, b]$ if and only if there exist functions $\alpha, \omega \in \mathcal{R}^*[a, b]$ such that*

(16) $$\alpha(x) \le f(x) \le \omega(x) \qquad for \ almost \ every \quad x \in [a, b].$$

Moreover, if either α or ω belongs to $\mathcal{L}[a, b]$, then $f \in \mathcal{L}[a, b]$.

Proof. ($\Rightarrow$) This implication is trivial, since one can take $\alpha = \omega = f$.

($\Leftarrow$) Since $f \in \mathcal{M}[a, b]$, there exists a sequence (s_k) of step functions on $[a, b]$ such that (13) holds. We define $\bar{s}_k := \mathrm{mid}\{\alpha, s_k, \omega\}$ for $k \in \mathbb{N}$, so that $\bar{s}_k(x)$ is the middle of the numbers $\alpha(x)$, $s_k(x)$ and $\omega(x)$ for each $x \in [a, b]$. It follows from Theorem 10.2.8 and the facts

$$\mathrm{mid}\{a, b, c\} = \min\{\max\{a, b\}, \max\{b, c\}, \max\{c, a\}\},$$
$$\min\{a', b', c'\} = \min\{\min\{a', b'\}, c'\},$$

that $\bar{s}_k \in \mathcal{R}^*[a, b]$ and that $\alpha \le \bar{s}_k \le \omega$. Since $f = \lim s_k = \lim \bar{s}_k$ a.e., the Dominated Convergence Theorem now implies that $f \in \mathcal{R}^*[a, b]$.

If either α or ω belongs to $\mathcal{L}[a, b]$, then we can apply Theorem 10.2.6 to conclude that f belongs to $\mathcal{L}[a, b]$. Q.E.D.

A Final Word _____

In this chapter we have made frequent reference to Lebesgue integrable functions on an interval I, which we have introduced as functions in $\mathcal{R}^*(I)$ whose absolute value also belongs to $\mathcal{R}^*(I)$. While there is no single "standard approach" to the Lebesgue integral, our approach is very different from any that are customary. A critic might say that our approach is not useful because our definition of a function in $\mathcal{L}(I)$ is not standard, but that would be wrong.

After all, one seldom uses the *definition* to confirm that a specific function is Lebesgue integrable. Instead, one uses the fact that certain simpler functions (such as step functions, polynomials, continuous functions, bounded measurable functions) belong to $\mathcal{L}(I)$, and that more complicated functions belong to $\mathcal{L}(I)$ by taking algebraic combinations or various limiting operations (e.g., Hake's Theorem or the Dominated Convergence Theorem). A famous analyst once said, "No one ever calculates a Lebesgue integral; instead, one calculates Riemann integrals and takes limits".

It is the same as with real numbers: we listed certain properties as axioms for $\mathbb{R}$ and then derived consequences of these properties which enable us to work quite effectively with the real numbers, often by taking limits.

Exercises for Section 10.4 _____

1. Consider the following sequences of functions with the indicated domains. Does the sequence converge? If so, to what? Is the convergence uniform? Is it bounded? If not bounded, is it dominated? Is it monotone? Evaluate the limit of the sequence of integrals.

(a) $\dfrac{kx}{1+kx}$ $[0, 1]$,   (b) $\dfrac{x^k}{1+x^k}$ $[0, 2]$,

(c) $\dfrac{1}{1+x^k}$ $[0, 1]$,   (d) $\dfrac{1}{1+x^k}$ $[0, 2]$.

2. Answer the questions posed in Exercise 1 for the following sequences (when properly defined).

   (a) $\dfrac{kx}{1+k\sqrt{x}}$ $[0, 1]$,   (b) $\dfrac{1}{\sqrt{x}(1+x^k)}$ $[0, 1]$,

   (c) $\dfrac{1}{\sqrt{x}(1+x^k)}$ $[1, 2]$,   (d) $\dfrac{1}{\sqrt{x}(2-x^k)}$ $[0, 1]$.

3. Discuss the following sequences of functions and their integrals on $[0, 1]$. Evaluate the limit of the integrals, when possible.

   (a) e^{-kx},   (b) e^{-kx}/x,

   (c) kxe^{-kx},   (d) k^2xe^{-kx},

   (e) $kxe^{-k^2x^2}$,   (f) kxe^{-kx^2}.

4. (a) Show that $\displaystyle\lim_{k\to\infty}\int_0^1 \dfrac{x^k\,dx}{(1+x)^2}=0$.   (b) Show that $\displaystyle\lim_{k\to\infty}\int_0^1 \dfrac{kx^k\,dx}{1+x}=\tfrac{1}{2}$.

5. If $f_k(x):=k$ for $x\in[1/k,2/k]$ and $f_k(x):=0$ elsewhere on $[0, 2]$, show that $f_k(x)\to 0$ but that $\int_0^2 f_k=1$.

6. Let (f_k) be a sequence on $[a, b]$ such that each f_k is differentiable on $[a, b]$ and $f_k'(x)\to g(x)$ with $|f_k'(x)|\le K$ for all $x\in[a, b]$. Show that the sequence $(f_k(x))$ either converges for all $x\in[a, b]$ or it diverges for all $x\in[a, b]$.

7. If f_k are the functions in Example 10.4.6(a), show that $\sup\{f_k\}$ does not belong to $\mathcal{R}^*[0, \infty)$.

8. Show directly that $\int_0^\infty e^{-tx}\,dx=1/t$ and $\int_0^\infty xe^{-tx}\,dx=1/t^2$ for $t>0$, thus confirming the results in Examples 10.4.6(d,e) when $f(x):=1$.

9. Use the differentiation formula in 10.4.6(f) to obtain $\int_0^\infty e^{-tx}\sin x\,dx=1/(t^2+1)$.

10. If $t>0$, define $E(t):=\int_0^\infty[(e^{-tx}\sin x)/x]\,dx$.
   (a) Show that E exists and is continuous for $t>a>0$. Moreover, $E(t)\to 0$ as $t\to\infty$.
   (b) Since $\left|\dfrac{\partial}{\partial t}\left(\dfrac{e^{-tx}\sin x}{x}\right)\right|\le e^{-ax}$ for $t\ge a>0$, show that $E'(t)=\dfrac{-1}{t^2+1}$ for $t>0$.
   (c) Deduce that $E(t)=\tfrac{1}{2}\pi-\operatorname{Arctan}t$ for $t>0$.
   (d) Explain why we cannot use the formula in (c) to obtain equation (12).

11. In this exercise we will establish the important formula:

$$(17)\qquad \int_0^\infty e^{-x^2}\,dx=\tfrac{1}{2}\sqrt{\pi}.$$

   (a) Let $G(t):=\int_0^1[e^{-t^2(x^2+1)}/(x^2+1)]\,dx$ for $t\ge 0$. Since the integrand is dominated by $1/(x^2+1)$ for $t\ge 0$, then G is continuous on $[0, \infty)$. Moreover, $G(0)=\operatorname{Arctan}1=\tfrac{1}{4}\pi$ and it follows from the Dominated Convergence Theorem that $G(t)\to 0$ as $t\to\infty$.
   (b) The partial derivative of the integrand with respect to t is bounded for $t\ge 0$, $x\in[0, 1]$, so $G'(t)=-2te^{-t^2}\int_0^1 e^{-t^2x^2}\,dx=-2e^{-t^2}\int_0^t e^{-u^2}\,du$.
   (c) If we set $F(t):=\left[\int_0^t e^{-x^2}\,dx\right]^2$, then the Fundamental Theorem 10.1.11 yields $F'(t)=2e^{-t^2}\int_0^t e^{-x^2}\,dx$ for $t\ge 0$, whence $F'(t)+G'(t)=0$ for all $t\ge 0$. Therefore, $F(t)+G(t)=C$ for all $t\ge 0$.
   (d) Using $F(0)=0$, $G(0)=\tfrac{1}{4}\pi$ and $\lim_{t\to\infty}G(t)=0$, we conclude that $\lim_{t\to\infty}F(t)=\tfrac{1}{4}\pi$, so that formula (17) holds.

12. Suppose $I\subseteq\mathbb{R}$ is a closed interval and that $f:[a, b]\times I\to\mathbb{R}$ is such that $\partial f/\partial t$ exists on $[a, b]\times I$, and for each $t\in[a, b]$ the function $x\mapsto f(t, x)$ is in $\mathcal{R}^*(I)$ and there exist $\alpha, \omega\in\mathcal{R}^*(I)$ such that the partial derivative satisfies $\alpha(x)\le\partial f(t, x)/\partial t\le\omega(x)$ for

a.e. $x \in I$. If $F(t) := \int_I f(t, x)\, dx$, show that F is differentiable on $[a, b]$ and that $F'(t) = \int_I \partial f(t, x)/\partial t\, dx$.

13. (a) If $f, g \in \mathcal{M}[a, b]$, show that $\max\{f, g\}$ and $\min\{f, g\}$ belong to $\mathcal{M}[a, b]$.
   (b) If $f, g, h \in \mathcal{M}[a, b]$, show that $\mathrm{mid}\{f, g, h\} \in \mathcal{M}[a, b]$.

14. (a) If (f_k) is a bounded sequence in $\mathcal{M}[a, b]$ and $f_k \to f$ a.e., show that $f \in \mathcal{M}[a, b]$. [*Hint:* Use the Dominated Convergence Theorem.]
   (b) If (g_k) is any sequence in $\mathcal{M}[a, b]$ and if $f_k := \mathrm{Arctan} \circ g_k$, show that (f_k) is a bounded sequence in $\mathcal{M}[a, b]$.
   (c) If (g_k) is a sequence in $\mathcal{M}[a, b]$ and if $g_k \to g$ a.e., show that $g \in \mathcal{M}[a, b]$.

15. A set E in $[a, b]$ is said to be **(Lebesgue) measurable** if its characteristic function $\mathbf{1}_E$ (defined by $\mathbf{1}_E(x) := 1$ if $x \in E$ and $\mathbf{1}_E(x) := 0$ if $x \in [a, b] \setminus E$) belongs to $\mathcal{M}[a, b]$. We will denote the collection of measurable sets in $[a, b]$ by $\mathbb{M}[a, b]$. In this exercise, we develop a number of properties of $\mathbb{M}[a, b]$.
   (a) Show that $E \in \mathbb{M}[a, b]$ if and only if $\mathbf{1}_E$ belongs to $\mathcal{R}^*[a, b]$.
   (b) Show that $\emptyset \in \mathbb{M}[a, b]$ and that if $[c, d] \subseteq [a, b]$, then the intervals $[c, d], [c, d), (c, d]$ and (c, d) are in $\mathbb{M}[a, b]$.
   (c) Show that $E \in \mathbb{M}[a, b]$ if and only if $E' := [a, b] \setminus E$ is in $\mathbb{M}[a, b]$.
   (d) If E and F are in $\mathbb{M}[a, b]$, then $E \cup F, E \cap F$ and $E \setminus F$ are also in $\mathbb{M}[a, b]$. [*Hint:* Show that $\mathbf{1}_{E \cup F} = \max\{\mathbf{1}_E, \mathbf{1}_F\}$, etc.]
   (e) If (E_k) is an increasing sequence in $\mathbb{M}[a, b]$, show that $E := \bigcup_{k=1}^\infty E_k$ is in $\mathbb{M}[a, b]$. Also, if (F_k) is a decreasing sequence in $\mathbb{M}[a, b]$ show that $F := \bigcap_{k=1}^\infty F_k$ is in $\mathbb{M}[a, b]$. [*Hint:* Apply Theorem 10.4.9(c).]
   (f) If (E_k) is any sequence in $\mathbb{M}[a, b]$, show that $\bigcup_{k=1}^\infty E_k$ and $\bigcap_{k=1}^\infty E_k$ are in $\mathbb{M}[a, b]$.

16. If $E \in \mathbb{M}[a, b]$, we define the **(Lebesgue) measure** of E to be the number $m(E) := \int_a^b \mathbf{1}_E$. In this exercise, we develop a number of properties of the **measure function** $m : \mathbb{M}[a, b] \to \mathbb{R}$.
   (a) Show that $m(\emptyset) = 0$ and $0 \le m(E) \le b - a$.
   (b) Show that $m([c, d]) = m([c, d)) = m((c, d]) = m((c, d)) = d - c$.
   (c) Show that $m(E') = (b - a) - m(E)$.
   (d) Show that $m(E \cup F) + m(E \cap F) = m(E) + m(F)$.
   (e) If $E \cap F = \emptyset$, show that $m(E \cup F) = m(E) + m(F)$. (This is the **additivity property** of the measure function.)
   (f) If (E_k) is an increasing sequence in $\mathbb{M}[a, b]$, show that $m(\bigcup_{k=1}^\infty E_k) = \lim_k m(E_k)$. [*Hint:* Use the Monotone Convergence Theorem.]
   (g) If (C_k) is a sequence in $\mathbb{M}[a, b]$ that is pairwise disjoint (in the sense that $C_j \cap C_k = \emptyset$ whenever $j \ne k$), show that

$$(18) \qquad m\left(\bigcup_{k=1}^\infty C_k\right) = \sum_{k=1}^\infty m(C_k).$$

(This is the **countable additivity property** of the measure function.)

CHAPTER 11

A GLIMPSE INTO TOPOLOGY

For the most part, we have considered only functions that were defined on intervals. Indeed, for certain important results on continuous functions, the intervals were also assumed to be closed and bounded. We shall now examine functions defined on more general types of sets, with the goal of establishing certain important properties of continuous functions in a more general setting. For example, we proved in Section 5.3 that a function that is continuous on a closed and bounded interval attains a maximum value. However, we will see that the hypothesis that the set is an interval is not essential, and in the proper context it can be dropped.

In Section 11.1 we define the notions of an open set, and a closed set. The study of open sets and the concepts that can be defined in terms of open sets is the study of point-set topology, so we are in fact discussing certain aspects of the topology of $\mathbb{R}$. (The mathematical area called "topology" is very abstract and goes far beyond the study of the real line, but the key ideas are to be found in real analysis. In fact, it is the study of continuous functions on $\mathbb{R}$ that motivated many of the concepts developed in topology.)

The notion of compact set is defined in Section 11.2 in terms of open coverings. In advanced analysis, compactness is a powerful and widely used concept. The compact subsets of $\mathbb{R}$ are fully characterized by the Heine-Borel Theorem, so the full strength of the idea is not as apparent as it would be in more general settings. Nevertheless, as we establish the basic properties of continuous functions on compact sets in Section 11.3, the reader should begin to appreciate how compactness arguments are used.

In Section 11.4 we take the essential features of distance on the real line and introduce a generalization of distance called a "metric". The much-used triangle inequality is the key property in this general concept of distance. We present examples and show how theorems on the real line can be extended to the context of a metric space.

The ideas in this chapter are somewhat more abstract than those in earlier chapters; however, abstraction can often lead to a deeper and more refined understanding. In this case, it leads to a more general setting for the study of analysis.

Section 11.1   Open and Closed Sets in $\mathbb{R}$

There are special types of sets that play a distinguished role in analysis—these are the open and the closed sets in $\mathbb{R}$. To expedite the discussion, it is convenient to have an extended notion of a neighborhood of a point.

11.1.1 Definition   A **neighborhood** of a point $x \in \mathbb{R}$ is any set V that contains an ε-neighborhood $V_\varepsilon(x) := (x - \varepsilon, x + \varepsilon)$ of x for some $\varepsilon > 0$.

While an ε-neighborhood of a point is required to be "symmetric about the point", the idea of a (general) neighborhood relaxes this particular feature, but often serves the same purpose.

11.1.2 Definition   (i)   A subset G of $\mathbb{R}$ is **open** in $\mathbb{R}$ if for each $x \in G$ there exists a neighborhood V of x such that $V \subseteq G$.

(ii)   A subset F of $\mathbb{R}$ is **closed in** $\mathbb{R}$ if the complement $\mathcal{C}(F) := \mathbb{R} \backslash F$ is open in $\mathbb{R}$.

To show that a set $G \subseteq \mathbb{R}$ is open, it suffices to show that each point in G has an ε-neighborhood contained in G. In fact, G is open if and only if for each $x \in G$, there exists $\varepsilon_x > 0$ such that $(x - \varepsilon_x, x + \varepsilon_x)$ is contained in G.

To show that a set $F \subseteq \mathbb{R}$ is closed, it suffices to show that each point $y \notin F$ has an ε-neighborhood disjoint from F. In fact, F is closed if and only if for each $y \notin F$ there exists $\varepsilon_y > 0$ such that $F \cap (y - \varepsilon_y, y + \varepsilon_y) = \emptyset$.

11.1.3 Examples   (a)   The entire set $\mathbb{R} = (-\infty, \infty)$ is open.

For any $x \in \mathbb{R}$, we may take $\varepsilon := 1$.

(b)   The set $G := \{x \in \mathbb{R} : 0 < x < 1\}$ is open.

For any $x \in G$ we may take ε_x to be the smaller of the numbers x, $1 - x$. We leave it to the reader to show that if $|u - x| < \varepsilon_x$ then $u \in G$.

(c)   Any open interval $I := (a, b)$ is an open set.

In fact, if $x \in I$, we can take ε_x to be the smaller of the numbers $x - a, b - x$. The reader can then show that $(x - \varepsilon_x, x + \varepsilon_x) \subseteq I$. Similarly, the intervals $(-\infty, b)$ and (a, ∞) are open sets.

(d)   The set $I := [0, 1]$ is not open.

This follows since every neighborhood of $0 \in I$ contains points not in I.

(e)   The set $I := [0, 1]$ is closed.

To see this let $y \notin I$; then either $y < 0$ or $y > 1$. If $y < 0$, we take $\varepsilon_y := |y|$, and if $y > 1$ we take $\varepsilon_y := y - 1$. We leave it to the reader to show that in either case we have $I \cap (y - \varepsilon_y, y + \varepsilon_y) = \emptyset$.

(f)   The set $H := \{x : 0 \leq x < 1\}$ is neither open nor closed. (Why?)

(g)   The empty set $\emptyset$ is open in $\mathbb{R}$.

In fact, the empty set contains no points at all, so the requirement in Definition 11.1.2(i) is vacuously satisfied. The empty set is also closed since its complement $\mathbb{R}$ is open, as was seen in part (a).   □

In ordinary parlance, when applied to doors, windows, and minds, the words "open" and "closed" are antonyms. However, when applied to subsets of $\mathbb{R}$, these words are not antonyms. For example, we noted above that the sets $\emptyset$, $\mathbb{R}$ are *both* open and closed in $\mathbb{R}$. (The reader will probably be relieved to learn that there are no other subsets of $\mathbb{R}$ that have both properties.) In addition, there are many subsets of $\mathbb{R}$ that are *neither* open nor closed; in fact, most subsets of $\mathbb{R}$ have this neutral character.

The following basic result describes the manner in which open sets relate to the operations of the union and intersection of sets in $\mathbb{R}$.

11.1.4 Open Set Properties   (a)   *The union of an arbitrary collection of open subsets in $\mathbb{R}$ is open.*

(b)   *The intersection of any finite collection of open sets in $\mathbb{R}$ is open.*

Proof. (a) Let $\{G_\lambda : \lambda \in \Lambda\}$ be a family of sets in $\mathbb{R}$ that are open, and let G be their union. Consider an element $x \in G$; by the definition of union, x must belong to G_{λ_0} for some $\lambda_0 \in \Lambda$. Since G_{λ_0} is open, there exists a neighborhood V of x such that $V \subseteq G_{\lambda_0}$. But $G_{\lambda_0} \subseteq G$, so that $V \subseteq G$. Since x is an arbitrary element of G, we conclude that G is open in $\mathbb{R}$.

(b) Suppose G_1 and G_2 are open and let $G := G_1 \cap G_2$. To show that G is open, we consider any $x \in G$; then $x \in G_1$ and $x \in G_2$. Since G_1 is open, there exists $\varepsilon_1 > 0$ such that $(x - \varepsilon_1, x + \varepsilon_1)$ is contained in G_1. Similarly, since G_2 is open, there exists $\varepsilon_2 > 0$ such that $(x - \varepsilon_2, x + \varepsilon_2)$ is contained in G_2. If we now take ε to be the smaller of ε_1 and ε_2, then the ε-neighborhood $U := (x - \varepsilon, x + \varepsilon)$ satisfies both $U \subseteq G_1$ and $U \subseteq G_2$. Thus, $x \in U \subseteq G$. Since x is an arbitrary element of G, we conclude that G is open in $\mathbb{R}$.

It now follows by an Induction argument (which we leave to the reader to write out) that the intersection of any finite collection of open sets is open.   Q.E.D.

The corresponding properties for closed sets will be established by using the general De Morgan identities for sets and their components. (See Theorem 1.1.4.)

11.1.5 Closed Set Properties (a) *The intersection of an arbitrary collection of closed sets in $\mathbb{R}$ is closed.*

(b) *The union of any finite collection of closed sets in $\mathbb{R}$ is closed.*

Proof. (a) If $\{F_\lambda : \lambda \in \Lambda\}$ is a family of closed sets in $\mathbb{R}$ and $F := \bigcap_{\lambda \in \Lambda} F_\lambda$, then $\mathcal{C}(F) = \bigcup_{\lambda \in \Lambda} \mathcal{C}(F_\lambda)$ is the union of open sets. Hence, $\mathcal{C}(F)$ is open by Theorem 11.1.4(a), and consequently, F is closed.

(b) Suppose $F_1, F_2, \cdots, F_n$ are closed in $\mathbb{R}$ and let $F := F_1 \cup F_2 \cup \cdots \cup F_n$. By the De Morgan identity the complement of F is given by

$$\mathcal{C}(F) = \mathcal{C}(F_1) \cap \cdots \cap \mathcal{C}(F_n)$$

Since each set $\mathcal{C}(F_i)$ is open, it follows from Theorem 11.1.4(b) that $\mathcal{C}(F)$ is open. Hence F is closed.   Q.E.D.

The finiteness restrictions in 11.1.4(b) and 11.1.5(b) cannot be removed. Consider the following examples:

11.1.6 Examples (a) Let $G_n := (0, 1 + 1/n)$ for $n \in \mathbb{N}$. Then G_n is open for each $n \in \mathbb{N}$, by Example 11.1.3(c). However, the intersection $G := \bigcap_{n=1}^{\infty} G_n$ is the interval $(0, 1]$ which is not open. Thus, *the intersection of infinitely many open sets in $\mathbb{R}$ need not be open.*

(b) Let $F_n := [1/n, 1]$ for $n \in \mathbb{N}$. Each F_n is closed, but the union $F := \bigcup_{n=1}^{\infty} F_n$ is the set $(0, 1]$ which is not closed. Thus, *the union of infinitely many closed sets in $\mathbb{R}$ need not be closed.* $\square$

The Characterization of Closed Sets

We shall now give a characterization of closed subsets of $\mathbb{R}$ in terms of sequences. As we shall see, closed sets are precisely those sets F that contain the limits of all convergent sequences whose elements are taken from F.

11.1.7 Characterization of Closed Sets   *Let $F \subseteq \mathbb{R}$; then the following assertions are equivalent.*

(i)   *F is a closed subset of $\mathbb{R}$.*

(ii) *If $X = (x_n)$ is any convergent sequence of elements in F, then $\lim X$ belongs to F.*

Proof.   (i) $\Rightarrow$ (ii) Let $X = (x_n)$ be a sequence of elements in F and let $x := \lim X$; we wish to show that $x \in F$. Suppose, on the contrary, that $x \notin F$; that is, that $x \in \mathcal{C}(F)$ the complement of F. Since $\mathcal{C}(F)$ is open and $x \in \mathcal{C}(F)$, it follows that there exists an ε-neighborhood V_ε of x such that V_ε is contained in $\mathcal{C}(F)$. Since $x = \lim(x_n)$, it follows that there exists a natural number $K = K(\varepsilon)$ such that $x_K \in V_\varepsilon$. Therefore we must have $x_K \in \mathcal{C}(F)$; but this contradicts the assumption that $x_n \in F$ for all $n \in \mathbb{N}$. Therefore, we conclude that $x \in F$.

(ii) $\Rightarrow$ (i) Suppose, on the contrary, that F is not closed, so that $G := \mathcal{C}(F)$ is not open. Then there exists a point $y_0 \in G$ such that for each $n \in \mathbb{N}$, there is a number $y_n \in \mathcal{C}(G) = F$ such that $|y_n - y_0| < 1/n$. It follows that $y_0 := \lim(y_n)$, and since $y_n \in F$ for all $n \in \mathbb{N}$, the hypothesis (ii) implies that $y_0 \in F$, contrary to the assumption $y_0 \in G = \mathcal{C}(F)$. Thus the hypothesis that F is not closed implies that (ii) is not true. Consequently (ii) implies (i), as asserted.   Q.E.D.

The next result is closely related to the preceding theorem. It states that a set F is closed if and only if it contains all of its cluster points. Recall from Section 4.1 that a point x is a **cluster point** of a set F if every ε-neighborhood of x contains a point of F different from x. Since by Theorem 4.1.2 each cluster point of a set F is the limit of a sequence of points in F, the result follows immediately from Theorem 11.1.7 above. We provide a second proof that uses only the relevant definitions.

11.1.8 Theorem   *A subset of $\mathbb{R}$ is closed if and only if it contains all of its cluster points.*

Proof.   Let F be a closed set in $\mathbb{R}$ and let x be a cluster point of F; we will show that $x \in F$. If not, then x belongs to the open set $\mathcal{C}(F)$. Therefore there exists an ε-neighborhood V_ε of x such that $V_\varepsilon \subseteq \mathcal{C}(F)$. Consequently $V_\varepsilon \cap F = \emptyset$, which contradicts the assumption that x is a cluster point of F.

Conversely, let F be a subset of $\mathbb{R}$ that contains all of its cluster points; we will show that $\mathcal{C}(F)$ is open. For if $y \in \mathcal{C}(F)$, then y is not a cluster point of F. It follows that there exists an ε-neighborhood V_ε of y that does not contain a point of F (except possibly y). But since $y \in \mathcal{C}(F)$, it follows that $V_\varepsilon \subseteq \mathcal{C}(F)$. Since y is an arbitrary element of $\mathcal{C}(F)$, we deduce that for every point in $\mathcal{C}(F)$ there is an ε-neighborhood that is entirely contained in $\mathcal{C}(F)$. But this means that $\mathcal{C}(F)$ is open in $\mathbb{R}$. Therefore F is closed in $\mathbb{R}$.   Q.E.D.

The Characterization of Open Sets

The idea of an open set in $\mathbb{R}$ is a generalization of the notion of an open interval. That this generalization does not lead to extremely exotic sets that are open is revealed by the next result.

11.1.9 Theorem   *A subset of $\mathbb{R}$ is open if and only if it is the union of countably many disjoint open intervals in $\mathbb{R}$.*

Proof.   Suppose that $G \neq \emptyset$ is an open set in $\mathbb{R}$. For each $x \in G$, let $A_x := \{a \in \mathbb{R} : (a, x] \subseteq G\}$ and let $B_x := \{b \in \mathbb{R} : [x, b) \subseteq G\}$. Since G is open, it follows that A_x and B_x

are not empty. (Why?) If the set A_x is bounded below, we set $a_x := \inf A_x$; if A_x is not bounded below, we set $a_x := -\infty$. Note that in either case $a_x \notin G$. If the set B_x is bounded above, we set $b_x := \sup B_x$; if B_x is not bounded above, we set $b_x := \infty$. Note that in either case $b_x \notin G$.

We now define $I_x := (a_x, b_x)$; clearly I_x is an open interval containing x. We claim that $I_x \subseteq G$. To see this, let $y \in I_x$ and suppose that $y < x$. It follows from the definition of a_x that there exists $a' \in A_x$ with $a' < y$, whence $y \in (a', x] \subseteq G$. Similarly, if $y \in I_x$ and $x < y$, there exists $b' \in B_x$ with $y < b'$, whence it follows that $y \in [x, b') \subseteq G$. Since $y \in I_x$ is arbitrary, we have that $I_x \subseteq G$.

Since $x \in G$ is arbitrary, we conclude that $\bigcup_{x \in G} I_x \subseteq G$. On the other hand, since for each $x \in G$ there is an open interval I_x with $x \in I_x \subseteq G$, we also have $G \subseteq \bigcup_{x \in G} I_x$. Therefore we conclude that $G = \bigcup_{x \in G} I_x$.

We claim that if $x, y \in G$ and $x \neq y$, then either $I_x = I_y$ or $I_x \cap I_y = \emptyset$. To prove this suppose that $z \in I_x \cap I_y$, whence it follows that $a_x < z < b_x$ and $a_y < z < b_y$. (Why?) We will show that $a_x = a_y$. If not, it follows from the Trichotomy Property that either (i) $a_x < a_y$, or (ii) $a_y < a_x$. In case (i), then $a_y \in I_x = (a_x, b_x) \subseteq G$, which contradicts the fact that $a_y \notin G$. Similarly, in case (ii), then $a_x \in I_y = (a_y, b_y) \subseteq G$, which contradicts the fact that $a_x \notin G$. Therefore we must have $a_x = a_y$ and a similar argument implies that $b_x = b_y$. Therefore, we conclude that if $I_x \cap I_y \neq \emptyset$, then $I_x = I_y$.

It remains to show that the collection of distinct intervals $\{I_x : x \in G\}$ is countable. To do this, we enumerate the set $\mathbb{Q}$ of rational numbers $\mathbb{Q} = \{r_1, r_2, \cdots, r_n, \cdots\}$ (see Theorem 1.3.11). It follows from the Density Theorem 2.4.8 that each interval I_x contains rational numbers; we select the rational number in I_x that has the smallest index n in this enumeration of $\mathbb{Q}$. That is, we choose $r_{n(x)} \in \mathbb{Q}$ such that $I_{r_{n(x)}} = I_x$ and $n(x)$ is the smallest index n such that $I_{r_n} = I_x$. Thus the set of distinct intervals $I_x, x \in G$, is put into correspondence with a subset of $\mathbb{N}$. Hence this set of distinct intervals is countable.   Q.E.D.

It is left as an exercise to show that the representation of G as a disjoint union of open intervals is uniquely determined.

It does *not* follow from the preceding theorem that a subset of $\mathbb{R}$ is closed if and only if it is the intersection of a countable collection of closed *intervals* (why not?). In fact, there are closed sets in $\mathbb{R}$ that cannot be expressed as the intersection of a countable collection of closed intervals in $\mathbb{R}$. A set consisting of two points is one example. (Why?) We will now describe the construction of a much more interesting example called the Cantor set.

The Cantor Set _____

The Cantor set, which we will denote by $\mathbb{F}$, is a very interesting example of a (somewhat complicated) set that is unlike any set we have seen up to this point. It reveals how inadequate our intuition can sometimes be in trying to picture subsets of $\mathbb{R}$.

The Cantor set $\mathbb{F}$ can be described by removing a sequence of open intervals from the closed unit interval $I := [0, 1]$. We first remove the open middle third $\left(\frac{1}{3}, \frac{2}{3}\right)$ of $[0, 1]$ to obtain the set

$$F_1 := \left[0, \tfrac{1}{3}\right] \cup \left[\tfrac{2}{3}, 1\right].$$

We next remove the open middle third of each of the two closed intervals in F_1 to obtain the set

$$F_2 := \left[0, \tfrac{1}{9}\right] \cup \left[\tfrac{2}{9}, \tfrac{1}{3}\right] \cup \left[\tfrac{2}{3}, \tfrac{7}{9}\right] \cup \left[\tfrac{8}{9}, 1\right].$$

We see that F_2 is the union of $2^2 = 4$ closed intervals, each of which is of the form $\left[k/3^2, (k+1)/3^2\right]$. We next remove the open middle thirds of each of these sets to get F_3, which is union of $2^3 = 8$ closed intervals. We continue in this way. In general, if F_n has been constructed and consists of the union of 2^n intervals of the form $\left[k/3^n, (k+1)/3^n\right]$, then we obtain the set F_{n+1} by removing the open middle third of each of these intervals. The Cantor set $\mathbb{F}$ is what remains after this process has been carried out for every $n \in \mathbb{N}$. (See Figure 11.1.1.)

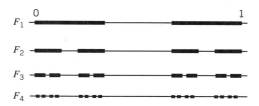

Figure 11.1.1   Construction of the Cantor set.

11.1.10 Definition   The **Cantor set** $\mathbb{F}$ is the intersection of the sets F_n, $n \in \mathbb{N}$, obtained by successive removal of open middle thirds, starting with $[0, 1]$.

Since it is the intersection of closed sets, $\mathbb{F}$ is itself a closed set by 11.1.5(a). We now list some of the properties of $\mathbb{F}$ that make it such an interesting set.

(1) The total length of the removed intervals is 1.

We note that the first middle third has length $1/3$, the next two middle thirds have lengths that add up to $2/3^2$, the next four middle thirds have lengths that add up to $2^2/3^3$, and so on. The total length L of the removed intervals is given by

$$L = \frac{1}{3} + \frac{2}{3^2} + \cdots + \frac{2^n}{3^{n+1}} + \cdots = \frac{1}{3}\sum_{n=0}^{\infty}\left(\frac{2}{3}\right)^n.$$

Using the formula for the sum of a geometric series, we obtain

$$L = \frac{1}{3} \cdot \frac{1}{1 - (2/3)} = 1.$$

Thus $\mathbb{F}$ is a subset of the unit interval $[0, 1]$ whose complement in $[0, 1]$ has total length 1.

Note also that the total length of the intervals that make up F_n is $(2/3)^n$, which has limit 0 as $n \to \infty$. Since $\mathbb{F} \subseteq F_n$ for all $n \in \mathbb{N}$, we see that *if* $\mathbb{F}$ can be said to have "length", it must have length 0.

(2) The set $\mathbb{F}$ contains no nonempty open interval as a subset.

Indeed, if $\mathbb{F}$ contains a nonempty open interval $J := (a, b)$, then since $J \subseteq F_n$ for all $n \in \mathbb{N}$, we must have $0 < b - a \le (2/3)^n$ for all $n \in \mathbb{N}$. Therefore $b - a = 0$, whence J is empty, a contradiction.

(3) The Cantor set $\mathbb{F}$ has infinitely (even uncountably) many points.

The Cantor set contains all of the endpoints of the removed open intervals, and these are all points of the form $2^k/3^n$ where $k = 0, 1, \cdots, n$ for each $n \in \mathbb{N}$. There are infinitely many points of this form.

The Cantor set actually contains many more points than those of the form $2^k/3^n$; in fact, $\mathbb{F}$ is an uncountable set. We give an outline of the argument. We note that each $x \in [0, 1]$ can be written in a ternary (base 3) expansion

$$x = \sum_{n=1}^{\infty} \frac{a_n}{3^n} = (.a_1 a_2 \cdots a_n \cdots)_3$$

where each a_n is either 0 or 1 or 2. (See the discussion at the end of Section 2.5.) Indeed, each x that lies in one of the removed open intervals has $a_n = 1$ for some n; for example, each point in $(\frac{1}{3}, \frac{2}{3})$ has $a_1 = 1$. The endpoints of the removed intervals have two possible ternary expansions, one having no 1s; for example, $3 = (.100\cdots)_3 = (.022\cdots)_3$. If we choose the expansion without 1s for these points, then $\mathbb{F}$ consists of all $x \in [0, 1]$ that have ternary expansions with no 1s; that is, a_n is 0 or 2 for all $n \in \mathbb{N}$. We now define a mapping φ of $\mathbb{F}$ onto $[0, 1]$ as follows:

$$\varphi\left(\sum_{n=1}^{\infty} \frac{a_n}{3^n}\right) := \sum_{n=1}^{\infty} \frac{(a_n/2)}{2^n} \qquad \text{for} \quad x \in \mathbb{F}.$$

That is, $\varphi\big((.a_1 a_2 \cdots)_3\big) = (.b_1 b_2 \cdots)_2$ where $b_n = a_n/2$ for all $n \in \mathbb{N}$ and $(.b_1 b_2 \cdots)_2$ denotes the *binary* representation of a number. Thus φ is a surjection of $\mathbb{F}$ onto $[0, 1]$. Assuming that $\mathbb{F}$ is countable, Theorem 1.3.10 implies that there exists a surjection ψ of $\mathbb{N}$ onto $\mathbb{F}$, so that $\varphi \circ \psi$ is a surjection of $\mathbb{N}$ onto $[0, 1]$. Another application of Theorem 1.3.10 implies that $[0, 1]$ is a countable set, which contradicts Theorem 2.5.5. Therefore $\mathbb{F}$ is an uncountable set.

Exercises for Section 11.1

1. If $x \in (0, 1)$, let ε_x be as in Example 11.1.3(b). Show that if $|u - x| < \varepsilon_x$, then $u \in (0, 1)$.

2. Show that the intervals (a, ∞) and $(-\infty, a)$ are open sets, and that the intervals $[b, \infty)$ and $(-\infty, b]$ are closed sets.

3. Write out the Induction argument in the proof of part (b) of the Open Set Properties 11.1.4.

4. Prove that $(0, 1] = \bigcap_{n=1}^{\infty} (0, 1 + 1/n)$, as asserted in Example 11.1.6(a).

5. Show that the set $\mathbb{N}$ of natural numbers is a closed set in $\mathbb{R}$.

6. Show that $A = \{1/n : n \in \mathbb{N}\}$ is not a closed set, but that $A \cup \{0\}$ is a closed set.

7. Show that the set $\mathbb{Q}$ of rational numbers is neither open nor closed.

8. Show that if G is an open set and F is a closed set, then $G \backslash F$ is an open set and $F \backslash G$ is a closed set.

9. A point $x \in \mathbb{R}$ is said to be an **interior point** of $A \subseteq \mathbb{R}$ in case there is a neighborhood V of x such that $V \subseteq A$. Show that a set $A \subseteq \mathbb{R}$ is open if and only if every point of A is an interior point of A.

10. A point $x \in \mathbb{R}$ is said to be a **boundary point** of $A \subseteq \mathbb{R}$ in case every neighborhood V of x contains points in A and points in $\mathcal{C}(A)$. Show that a set A and its complement $\mathcal{C}(A)$ have exactly the same boundary points.

11. Show that a set $G \subseteq \mathbb{R}$ is open if and only if it does not contain any of its boundary points.

12. Show that a set $F \subseteq \mathbb{R}$ is closed if and only if it contains all of its boundary points.

13. If $A \subseteq \mathbb{R}$, let A° be the union of all open sets that are contained in A; the set A° is called the **interior** of A. Show that A° is an open set, that it is the largest open set contained in A, and that a point z belongs to A° if and only if z is an interior point of A.

14. Using the notation of the preceding exercise, let A, B be sets in $\mathbb{R}$. Show that $A^\circ \subseteq A$, $(A^\circ)^\circ = A^\circ$, and that $(A \cap B)^\circ = A^\circ \cap B^\circ$. Show also that $A^\circ \cup B^\circ \subseteq (A \cup B)^\circ$, and give an example to show that the inclusion may be proper.

15. If $A \subseteq \mathbb{R}$, let A^- be the intersection of all closed sets containing A; the set A^- is called the **closure** of A. Show that A^- is a closed set, that it is the smallest closed set containing A, and that a point w belongs to A^- if and only if w is either an interior point or a boundary point of A.

16. Using the notation of the preceding exercise, let A, B be sets in $\mathbb{R}$. Show that we have $A \subseteq A^-$, $(A^-)^- = A^-$, and that $(A \cup B)^- = A^- \cup B^-$. Show that $(A \cap B)^- \subseteq A^- \cap B^-$, and give an example to show that the inclusion may be proper.

17. Give an example of a set $A \subseteq \mathbb{R}$ such that $A^\circ = \emptyset$ and $A^- = \mathbb{R}$.

18. Show that if $F \subseteq \mathbb{R}$ is a closed nonempty set that is bounded above, then $\sup F$ belongs to F.

19. If G is open and $x \in G$, show that the sets A_x and B_x in the proof of Theorem 11.1.9 are not empty.

20. If the set A_x in the proof of Theorem 11.1.9 is bounded below, show that $a_x := \inf A_x$ does not belong to G.

21. If in the notation used in the proof of Theorem 11.1.9, we have $a_x < y < x$, show that $y \in G$.

22. If in the notation used in the proof of Theorem 11.1.9, we have $I_x \cap I_y \neq \emptyset$, show that $b_x = b_y$.

23. Show that each point of the Cantor set $\mathbb{F}$ is a cluster point of $\mathbb{F}$.

24. Show that each point of the Cantor set $\mathbb{F}$ is a cluster point of $\mathcal{C}(\mathbb{F})$.

Section 11.2 Compact Sets

In advanced analysis and topology, the notion of a "compact" set is of enormous importance. This is less true in $\mathbb{R}$ because the Heine-Borel Theorem gives a very simple characterization of compact sets in $\mathbb{R}$. Nevertheless, the definition and the techniques used in connection with compactness are very important, and the real line provides an appropriate place to see the idea of compactness for the first time.

The definition of compactness uses the notion of an open cover, which we now define.

11.2.1 Definition Let A be a subset of $\mathbb{R}$. An **open cover** of A is a collection $\mathcal{G} = \{G_\alpha\}$ of open sets in $\mathbb{R}$ whose union contains A; that is,

$$A \subseteq \bigcup_\alpha G_\alpha.$$

If $\mathcal{G}'$ is a subcollection of sets from $\mathcal{G}$ such that the union of the sets in $\mathcal{G}'$ also contains A, then $\mathcal{G}'$ is called a **subcover** of $\mathcal{G}$. If $\mathcal{G}'$ consists of finitely many sets, then we call $\mathcal{G}'$ a **finite subcover** of $\mathcal{G}$.

There can be many different open covers for a given set. For example, if $A := [1, \infty)$, then the reader can verify that the following collections of sets are all open covers of A:

$$\mathcal{G}_0 := \{(0, \infty)\},$$
$$\mathcal{G}_1 := \{(r-1, r+1) : r \in \mathbb{Q}, r > 0\},$$
$$\mathcal{G}_2 := \{(n-1, n+1) : n \in \mathbb{N}\},$$
$$\mathcal{G}_3 := \{(0, n) : n \in \mathbb{N}\},$$
$$\mathcal{G}_4 := \{(0, n) : n \in \mathbb{N}, n \geq 23\}.$$

We note that $\mathcal{G}_2$ is a subcover of $\mathcal{G}_1$, and that $\mathcal{G}_4$ is a subcover of $\mathcal{G}_3$. Of course, many other open covers of A can be described.

11.2.2 Definition   A subset K of $\mathbb{R}$ is said to be **compact** if *every* open cover of K has a finite subcover.

In other words, a set K is compact if, whenever it is contained in the union of a collection $\mathcal{G} = \{G_\alpha\}$ of open sets in $\mathbb{R}$, then it is contained in the union of some *finite* number of sets in $\mathcal{G}$.

It is very important to note that, in order to apply the definition to prove that a set K is compact, we must examine an *arbitrary* collection of open sets whose union contains K, and show that K is contained in the union of some finite number of sets in the given collection. That is, it must be shown that *any* open cover of K has a finite subcover. On the other hand, to prove that a set H is *not* compact, it is sufficient to exhibit one *specific* collection $\mathcal{G}$ of open sets whose union contains H, but such that the union of any finite number of sets in $\mathcal{G}$ fails to contain H. That is, H is not compact if there exists some open cover of H that has no finite subcover.

11.2.3 Examples   **(a)**   Let $K := \{x_1, x_2, \cdots, x_n\}$ be a finite subset of $\mathbb{R}$. If $\mathcal{G} = \{G_\alpha\}$ is an open cover of K, then each x_i is contained in some set G_{α_i} in $\mathcal{G}$. Then the union of the sets in the collection $\{G_{\alpha_1}, G_{\alpha_2}, \cdots, G_{\alpha_n}\}$ contains K, so that it is a finite subcover of $\mathcal{G}$. Since $\mathcal{G}$ was arbitrary, it follows that the finite set K is compact.

(b)   Let $H := [0, \infty)$. To prove that H is not compact, we will exhibit an open cover that has no finite subcover. If we let $G_n := (-1, n)$ for each $n \in \mathbb{N}$, then $H \subseteq \bigcup_{n=1}^{\infty} G_n$, so that $\mathcal{G} := \{G_n : n \in \mathbb{N}\}$ is an open cover of H. However, if $\{G_{\alpha_1}, G_{\alpha_2}, \cdots, G_{\alpha_n}\}$ is any finite subcollection of $\mathcal{G}$, and if we let $m := \sup\{n_1, n_2, \cdots, n_k\}$, then

$$G_{n_1} \cup G_{n_2} \cup \cdots \cup G_{n_k} = G_m = (-1, m).$$

Evidently, this union fails to contain $H = [0, \infty)$. Thus no finite subcollection of $\mathcal{G}$ will have its union contain H, and therefore H is not compact.

(c)   Let $J := (0, 1)$. If we let $G_n := (1/n, 1)$ for each $n \in \mathbb{N}$, then it is readily seen that $J = \bigcup_{n=1}^{\infty} G_n$. Thus $\mathcal{G} := \{G_n : n \in \mathbb{N}\}$ is an open cover of J. If $\{G_{n_1}, G_{n_2}, \cdots, G_{n_r}\}$ is any finite subcollection of $\mathcal{G}$, and if we set $s := \sup\{n_1, n_2, \cdots, n_r\}$ then

$$G_{n_1} \cup G_{n_2} \cup \cdots \cup G_{n_r} = G_s = (1/s, 1).$$

Since $1/s$ is in J but not in G_s, we see that the union does not contain J. Therefore, J is not compact. $\square$

We now wish to describe all compact subsets of $\mathbb{R}$. First we will establish by rather straightforward arguments that any compact set in $\mathbb{R}$ must be both closed and bounded. Then we will show that these properties in fact characterize the compact sets in $\mathbb{R}$. This is the content of the Heine-Borel Theorem.

11.2.4 Theorem   *If K is a compact subset of $\mathbb{R}$, then K is closed and bounded.*

Proof. We will first show that K is bounded. For each $m \in \mathbb{N}$, let $H_m := (-m, m)$. Since each H_m is open and since $K \subseteq \bigcup_{m=1}^{\infty} H_m = \mathbb{R}$, we see that the collection $\{H_m : m \in \mathbb{N}\}$ is an open cover of K. Since K is compact, this collection has a finite subcover, so there exists $M \in \mathbb{N}$ such that

$$K \subseteq \bigcup_{m=1}^{M} H_m = H_M = (-M, M).$$

Therefore K is bounded, since it is contained in the bounded interval $(-M, M)$.

We now show that K is closed, by showing that its complement $\mathcal{C}(K)$ is open. To do so, let $u \in \mathcal{C}(K)$ be arbitrary and for each $n \in \mathbb{N}$, we let $G_n := \{y \in \mathbb{R} : |y - u| > 1/n\}$. It is an exercise to show that each set G_n is open and that $\mathbb{R} \backslash \{u\} = \bigcup_{n=1}^{\infty} G_n$. Since $u \notin K$, we have $K \subseteq \bigcup_{n=1}^{\infty} G_n$. Since K is compact, there exists $m \in \mathbb{N}$ such that

$$K \subseteq \bigcup_{n=1}^{m} G_n = G_m.$$

Now it follows from this that $K \cap (u - 1/m, u + 1/m) = \emptyset$, so that the interval $(u - 1/m, u + 1/m) \subseteq \mathcal{C}(K)$. But since u was an arbitrary point in $\mathcal{C}(K)$, we infer that $\mathcal{C}(K)$ is open. Q.E.D.

We now prove that the conditions of Theorem 11.2.4 are both necessary and sufficient for a subset of $\mathbb{R}$ to be compact.

11.2.5 Heine-Borel Theorem *A subset K of $\mathbb{R}$ is compact if and only if it is closed and bounded.*

Proof. We have shown in Theorem 11.2.4 that a compact set in $\mathbb{R}$ must be closed and bounded. To establish the converse, suppose that K is closed and bounded, and let $\mathcal{G} = \{G_\alpha\}$ be an open cover of K. We wish to show that K must be contained in the union of some finite subcollection from $\mathcal{G}$. The proof will be by contradiction. We assume that:

(1) K is not contained in the union of any finite number of sets in $\mathcal{G}$.

By hypothesis, K is bounded, so there exists $r > 0$ such that $K \subseteq [-r, r]$. We let $I_1 := [-r, r]$ and bisect I_1 into two closed subintervals $I_1' := [-r, 0]$ and $I_1'' := [0, r]$. At least one of the two subsets $K \cap I_1'$, and $K \cap I_1''$ must be nonvoid and have the property that it is not contained in the union of any finite number of sets in $\mathcal{G}$. [For if both of the sets $K \cap I_1'$ and $K \cap I_1''$ are contained in the union of some finite number of sets in $\mathcal{G}$, then $K = (K \cap I_1') \cup (K \cap I_1'')$ is contained in the union of some finite number of sets in $\mathcal{G}$, contrary to the assumption (1).] If $K \cap I_1'$ is not contained in the union of some finite number of sets in $\mathcal{G}$, we let $I_2 := I_1'$; otherwise $K \cap I_1''$ has this property and we let $I_2 := I_1''$.

We now bisect I_2 into two closed subintervals I_2' and I_2''. If $K \cap I_2'$ is nonvoid and is not contained in the union of some finite number of sets in $\mathcal{G}$, we let $I_3 := I_2'$; otherwise $K \cap I_2''$ has this property and we let $I_3 := I_2''$.

Continuing this process, we obtain a nested sequence of intervals (I_n). By the Nested Intervals Property 2.5.2, there is a point z that belongs to all of the I_n, $n \in \mathbb{N}$. Since each interval I_n contains infinitely many points in K (why?), the point z is a cluster point of K. Moreover, since K is assumed to be closed, it follows from Theorem 11.1.8 that $z \in K$.

Therefore there exists a set G_λ in $\mathcal{G}$ with $z \in G_\lambda$. Since G_λ is open, there exists $\varepsilon > 0$ such that

$$(z - \varepsilon, z + \varepsilon) \subseteq G_\lambda.$$

On the other hand, since the intervals I_n are obtained by repeated bisections of $I_1 = [-r, r]$, the length of I_n is $r/2^{n-2}$. It follows that if n is so large that $r/2^{n-2} < \varepsilon$, then $I_n \subseteq (z - \varepsilon, z + \varepsilon) \subseteq G_\lambda$. But this means that if n is such that $r/2^{n-2} < \varepsilon$, then $K \cap I_n$ is contained in the *single* set G_λ in $\mathcal{G}$, contrary to our construction of I_n. This contradiction shows that the assumption (1) that the closed bounded set K requires an infinite number of sets in $\mathcal{G}$ to cover it is untenable. We conclude that K is compact.   Q.E.D.

Remark   It was seen in Example 11.2.3(b) that the closed set $H := [0, \infty)$ is *not* compact; note that H is not bounded. It was also seen in Example 11.2.3(c) that the bounded set $J := (0, 1)$ is *not* compact; note that J is not closed. Thus, we cannot drop either hypothesis of the Heine-Borel Theorem.

We can combine the Heine-Borel Theorem with the Bolzano-Weierstrass Theorem 3.4.8 to obtain a sequential characterization of the compact subsets of $\mathbb{R}$.

11.2.6 Theorem   *A subset K of $\mathbb{R}$ is compact if and only if every sequence in K has a subsequence that converges to a point in K.*

Proof.   Suppose that K is compact and let (x_n) be a sequence with $x_n \in K$ for all $n \in \mathbb{N}$. By the Heine-Borel Theorem, the set K is bounded so that the sequence (x_n) is bounded; by the Bolzano-Weierstrass Theorem 3.4.8, there exists a subsequence (x_{n_k}) that converges. Since K is closed (by Theorem 11.2.4), the limit $x := \lim(x_{n_k})$ is in K. Thus every sequence in K has a subsequence that converges to a point of K.

To establish the converse, we will show that if K is either not closed or not bounded, then there must exist a sequence in K that has no subsequence converging to a point of K. First, if K is not closed, then there is a cluster point c of K that does not belong to K. Since c is a cluster point of K, there is a sequence (x_n) with $x_n \in K$ and $x_n \neq c$ for all $n \in \mathbb{N}$ such that $\lim(x_n) = c$. Then every subsequence of (x_n) also converges to c, and since $c \notin K$, there is no subsequence that converges to a point of K.

Second, if K is not bounded, then there exists a sequence (x_n) in K such that $|x_n| > n$ for all $n \in \mathbb{N}$. (Why?) Then every subsequence of (x_n) is unbounded, so that no subsequence of it can converge to a point of K.   Q.E.D.

Remark   The reader has probably noticed that there is a similarity between the compactness of the interval $[a, b]$ and the existence of δ-fine partitions for $[a, b]$. In fact, these properties are equivalent, each being deducible from the other. However, compactness applies to sets that are more general than intervals.

Exercises for Section 11.2

1. Exhibit an open cover of the interval $(1, 2]$ that has no finite subcover.

2. Exhibit an open cover of $\mathbb{N}$ that has no finite subcover.

3. Exhibit an open cover of the set $\{1/n: n \in \mathbb{N}\}$ that has no finite subcover.

4. Prove, using Definition 11.2.2, that if F is a closed subset of a compact set K in $\mathbb{R}$, then F is compact.

5. Prove, using Definition 11.2.2, that if K_1 and K_2 are compact sets in $\mathbb{R}$, then their union $K_1 \cup K_2$ is compact.

6. Use the Heine-Borel Theorem to prove the following version of the Bolzano-Weierstrass Theorem: Every bounded infinite subset of $\mathbb{R}$ has a cluster point in $\mathbb{R}$. (Note that if a set has no cluster points, then it is closed by Theorem 11.1.8.)

7. Find an infinite collection $\{K_n : n \in \mathbb{N}\}$ of compact sets in $\mathbb{R}$ such that the union $\bigcup_{n=1}^{\infty} K_n$ is not compact.

8. Prove that the intersection of an arbitrary collection of compact sets in $\mathbb{R}$ is compact.

9. Let $(K_n : n \in \mathbb{N})$ be a sequence of nonempty compact sets in $\mathbb{R}$ such that $K_1 \supseteq K_2 \supseteq \cdots \supseteq K_n \supseteq \cdots$. Prove that there exists at least one point $x \in \mathbb{R}$ such that $x \in K_n$ for all $n \in \mathbb{N}$; that is, the intersection $\bigcap_{n=1}^{\infty} K_n$ is not empty.

10. Let $K \neq \emptyset$ be a compact set in $\mathbb{R}$. Show that $\inf K$ and $\sup K$ exist and belong to K.

11. Let $K \neq \emptyset$ be compact in $\mathbb{R}$ and let $c \in \mathbb{R}$. Prove that there exists a point a in K such that $|c - a| = \inf\{|c - x| : x \in K\}$.

12. Let $K \neq \emptyset$ be compact in $\mathbb{R}$ and let $c \in \mathbb{R}$. Prove that there exists a point b in K such that $|c - b| = \sup\{|c - x| : x \in K\}$.

13. Use the notion of compactness to give an alternative proof of Exercise 5.3.18.

14. If K_1 and K_2 are disjoint nonempty compact sets, show that there exist $k_i \in K_i$ such that $0 < |k_1 - k_2| = \inf\{|x_1 - x_2| : x_i \in K_i\}$

15. Give an example of disjoint closed sets F_1, F_2 such that $0 = \inf\{|x_1 - x_2| : x_i \in F_i\}$.

Section 11.3   Continuous Functions

In this section we will examine the way in which the concept of continuity of functions can be related to the topological ideas of open sets and compact sets. Some of the fundamental properties of continuous functions on intervals presented in Section 5.3 will be established in this context. Among other things, these new arguments will show that the concept of continuity and many of its important properties can be carried to a greater level of abstraction. This will be discussed briefly in the next section on metric spaces.

Continuity ———————————————————————————

In Section 5.1 we were concerned with continuity at a point, that is, with the "local" continuity of functions. We will now be mainly concerned with "global" continuity in the sense that we will assume that the functions are continuous on their entire domains.

The continuity of a function $f : A \to \mathbb{R}$ at a point $c \in A$ was defined in Section 5.1. Theorem 5.1.2 stated that f is continuous at c if and only if for every ε-neighborhood $V_\varepsilon(f(c))$ of $f(c)$ there exists a δ-neighborhood $V_\delta(c)$ of c such that if $x \in V_\delta(c) \cap A$, then $f(x) \in V_\varepsilon(f(c))$. We wish to restate this condition for continuity at a point in terms of general neighborhoods. (Recall from 11.1.1 that a neighborhood of a point c is any set U that contains an ε-neighborhood of c for some $\varepsilon > 0$.)

11.3.1 Lemma   *A function $f : A \to \mathbb{R}$ is continuous at the point c in A if and only if for every neighborhood U of $f(c)$, there exists a neighborhood V of c such that if $x \in V \cap A$, then $f(x) \in U$.*

Proof.   Suppose f satisfies the stated condition. Then given $\varepsilon > 0$, we let $U = V_\varepsilon(f(c))$ and then obtain a neighborhood V for which $x \in V \cap A$ implies $f(x) \in U$. If we choose $\delta > 0$ such that $V_\delta(c) \subseteq V$, then $x \in V_\delta(c) \cap A$ implies $f(x) \in U$; therefore f is continuous at c according to Theorem 5.1.2.

Conversely, if f is continuous at c in the sense of Theorem 5.1.2, then since any neighborhood U of $f(c)$ contains an ε-neighborhood $V_\varepsilon(f(c))$, it follows that taking the δ-neighborhood $V = V_\delta(c)$ of c of Theorem 5.1.2 satisfies the condition of the lemma.
                                                                                                                Q.E.D.

We note that the statement that $x \in V \cap A$ implies $f(x) \in U$ is equivalent to the statement that $f(V \cap A) \subseteq U$; that is, that the direct image of $V \cap A$ is contained in U. Also from the definition of inverse image, this is the same as $V \cap A \subseteq f^{-1}(U)$. (See Definition 1.1.7 for the definitions of direct and inverse images.) Using this observation, we now obtain a condition for a function to be continuous on its domain in terms of open sets. In more advanced courses in topology, part (b) of the next result is often taken as the definition of (global) continuity.

11.3.2 Global Continuity Theorem   *Let $A \subseteq \mathbb{R}$ and let $f : A \to \mathbb{R}$ be a function with domain A. Then the following are equivalent:*

(a)   *f is continuous at every point of A.*

(b)   *For every open set G in $\mathbb{R}$, there exists an open set H in $\mathbb{R}$ such that $H \cap A = f^{-1}(G)$.*

Proof.   (a) $\Rightarrow$ (b). Assume that f is continuous at every point of A, and let G be a given open set in $\mathbb{R}$. If c belongs to $f^{-1}(G)$, then $f(c) \in G$, and since G is open, G is a neighborhood of $f(c)$. Therefore, by the preceding lemma, it follows from the continuity of f that there is an open set $V(c)$ such that $x \in V(c)$ implies that $f(x) \in G$; that is, $V(c)$ is contained in the inverse image $f^{-1}(G)$. Select $V(c)$ for each c in $f^{-1}(G)$, and let H be the union of all these sets $V(c)$. By the Open Set Properties 11.1.4, the set H is open, and we have $H \cap A = f^{-1}(G)$. Hence (a) implies (b).

(b) $\Rightarrow$ (a). Let c be any point A, and let G be an open neighborhood of $f(c)$. Then condition (b) implies that there exists an open set H in $\mathbb{R}$ such that $H \cap A = f^{-1}(G)$. Since $f(c) \in G$, it follows that $c \in H$, so H is a neighborhood of c. If $x \in H \cap A$, then $f(c) \in G$, and therefore f is continuous at c. Thus (b) implies (a).   Q.E.D.

In the case that $A = \mathbb{R}$, the preceding result simplifies to some extent.

11.3.3 Corollary   *A function $f : \mathbb{R} \to \mathbb{R}$ is continuous if and only if $f^{-1}(G)$ is open in $\mathbb{R}$ whenever G is open.*

It must be emphasized that the Global Continuity Theorem 11.3.2 does *not* say that if f is a continuous function, then the direct image $f(G)$ of an open set is necessarily open. In general, a continuous function will not send open sets to open sets. For example, consider the continuous function $f : \mathbb{R} \to \mathbb{R}$ defined by

$$f(x) := x^2 + 1 \qquad \text{for} \quad x \in \mathbb{R}.$$

If G is the open set $G := (-1, 1)$, then the direct image under f is $f(G) = [1, 2)$, which is not open in $\mathbb{R}$. See the exercises for additional examples.

Preservation of Compactness

In Section 5.3 we proved that a continuous function takes a closed, bounded interval $[a, b]$ onto a closed, bounded interval $[m, M]$, where m and M are the minimum and maximum values of f on $[a, b]$, respectively. By the Heine-Borel Theorem, these are compact subsets of $\mathbb{R}$, so that Theorem 5.3.8 is a special case of the following theorem.

11.3.4 Preservation of Compactness If K is a compact subset of $\mathbb{R}$ and if $f : K \to \mathbb{R}$ is continuous on K, then $f(K)$ is compact.

Proof. Let $\mathcal{G} = \{G_\lambda\}$ be an open cover of the set $f(K)$. We must show that $\mathcal{G}$ has a finite subcover. Since $f(K) \subseteq \bigcup G_\lambda$, it follows that $K \subseteq \bigcup f^{-1}(G_\lambda)$. By Theorem 11.3.2, for each G_λ there is an open set H_λ such that $H_\lambda \cap K = f^{-1}(G_\lambda)$. Then the collection $\{H_\lambda\}$ is an open cover of the set K. Since K is compact, this open cover of K contains a finite subcover $\{H_{\lambda_1}, H_{\lambda_2}, \cdots, H_{\lambda_n}\}$. Then we have

$$\bigcup_{i=1}^{n} f^{-1}\left(G_{\lambda_i}\right) = \bigcup_{i=1}^{n} H_{\lambda_i} \cap K \supseteq K.$$

From this it follows that $\bigcup_{i=1}^{n} G_{\lambda_i} \supseteq f(K)$. Hence we have found a finite subcover of $\mathcal{G}$. Since $\mathcal{G}$ was an arbitrary open cover of $f(K)$, we conclude that $f(K)$ is compact. Q.E.D.

11.3.5 Some Applications We will now show how to apply the notion of compactness (and the Heine-Borel Theorem) to obtain alternative proofs of some important results that we have proved earlier by using the Bolzano-Weierstrass Theorem. In fact, these theorems remain true if the intervals are replaced by arbitrary nonempty compact sets in $\mathbb{R}$.

(1) The Boundedness Theorem 5.3.2 is an immediate consequence of Theorem 11.3.4 and the Heine-Borel Theorem 11.2.5. Indeed, if $K \subseteq \mathbb{R}$ is compact and if $f: K \to \mathbb{R}$ is continuous on K, then $f(K)$ is compact and hence bounded.

(2) The Maximum-Minimum Theorem 5.3.4 also is an easy consequence of Theorem 11.3.4 and the Heine-Borel Theorem. As before, we find that $f(K)$ is compact and hence bounded in $\mathbb{R}$, so that $s^* := \sup f(K)$ exists. If $f(K)$ is a finite set, then $s^* \in f(K)$. If $f(K)$ is an infinite set, then s^* is a cluster point of $f(K)$ [see Exercise 11.2.6]. Since $f(K)$ is a closed set, by the Heine-Borel Theorem, it follows from Theorem 11.1.8 that $s^* \in f(K)$. We conclude that $s^* = f(x^*)$ for some $x^* \in K$.

(3) We can also give a proof of the Uniform Continuity Theorem 5.4.3 based on the notion of compactness. To do so, let $K \subseteq \mathbb{R}$ be compact and let $f : K \to \mathbb{R}$ be continuous on K. Then given $\varepsilon > 0$ and $u \in K$, there is a number $\delta_u := \delta\left(\frac{1}{2}\varepsilon, u\right) > 0$ such that if $x \in K$ and $|x - u| < \delta_u$ then $|f(x) - f(u)| < \frac{1}{2}\varepsilon$. For each $u \in K$, let $G_u := \left(u - \frac{1}{2}\delta_u, u + \frac{1}{2}\delta_u\right)$ so that G_u is open; we consider the collection $\mathcal{G} = \{G_u : u \in K\}$. Since $u \in G_u$ for $u \in K$, it is trivial that $K \subseteq \bigcup_{u \in K} G_u$. Since K is compact, there are a finite number of sets, say $G_{u_1}, \cdots, G_{u_M}$ whose union contains K. We now define

$$\delta(\varepsilon) := \tfrac{1}{2} \inf\left\{\delta_{u_1}, \cdots, \delta_{u_M}\right\},$$

so that $\delta(\varepsilon) > 0$. Now if $x, u \in K$ and $|x - u| < \delta(\varepsilon)$, then there exists some u_k with $k = 1, \cdots, M$ such that $x \in G_{u_k}$; therefore $|x - u_k| < \frac{1}{2}\delta_{u_k}$. Since we have $\delta(\varepsilon) \leq \frac{1}{2}\delta_{u_k}$ it follows that

$$|u - u_k| \leq |u - x| + |x - u_k| < \delta_{u_k}.$$

But since $\delta_{u_k} = \delta\left(\frac{1}{2}\varepsilon, u_k\right)$ it follows that both

$$|f(x) - f(u_k)| < \frac{1}{2}\varepsilon \qquad \text{and} \qquad |f(u) - f(u_k)| < \frac{1}{2}\varepsilon.$$

Therefore we have $|f(x) - f(u)| < \varepsilon$.

We have shown that if $\varepsilon > 0$, then there exists $\delta(\varepsilon) > 0$ such that if x, u are any points in K with $|x - u| < \delta(\varepsilon)$, then $|f(x) - f(u)| < \varepsilon$. Since $\varepsilon > 0$ is arbitrary, this shows that f is uniformly continuous on K, as asserted.   □

We conclude this section by extending the Continuous Inverse Theorem 5.6.5 to functions whose domains are compact subsets of $\mathbb{R}$, rather than intervals in $\mathbb{R}$.

11.3.6 Theorem   *If K is a compact subset of $\mathbb{R}$ and $f : K \to \mathbb{R}$ is injective and continuous, then f^{-1} is continuous on $f(K)$.*

Proof.   Since K is compact, then Theorem 11.3.4 implies that the image $f(K)$ is compact. Since f is injective by hypothesis, the inverse function f^{-1} is defined on $f(K)$ to K. Let (y_n) be any convergent sequence in $f(K)$, and let $y_0 = \lim(y_n)$. To establish the continuity of f^{-1}, we will show that the sequence $(f^{-1}(y_n))$ converges to $f^{-1}(y_0)$.

Let $x_n := f^{-1}(y_n)$ and, by way of contradiction, assume that (x_n) does not converge to $x_0 := f^{-1}(y_0)$. Then there exists an $\varepsilon > 0$ and a subsequence (x_k') such that $|x_k' - x_0| \geq \varepsilon$ for all k. Since K is compact, we conclude from Theorem 11.2.6 that there is a subsequence (x_r'') of the sequence (x_k') that converges to a point x^* of K. Since $|x^* - x_0| \geq \varepsilon$, we have $x^* \neq x_0$. Now since f is continuous, we have $\lim(f(x_r'')) = f(x^*)$. Also, since the subsequence (y_n'') of (y_n) that corresponds to the subsequence (x_n'') of (x_n) must converge to the same limit as (y_n) does, we have

$$\lim(f(x_r'')) = \lim(y_n'') = y_0 = f(x_0).$$

Therefore we conclude that $f(x^*) = f(x_0)$. However, since f is injective, this implies that $x^* = x_0$, which is a contradiction. Thus we conclude that f^{-1} takes convergent sequences in $f(K)$ to convergent sequences in K, and hence f^{-1} is continuous.   Q.E.D.

Exercises for Section 11.3

1. Let $f : \mathbb{R} \to \mathbb{R}$ be defined by $f(x) = x^2$ for $x \in \mathbb{R}$.
   (a) Show that the inverse image $f^{-1}(I)$ of an open interval $I := (a, b)$ is either an open interval, the union of two open intervals, or empty, depending on a and b.
   (b) Show that if I is an open interval containing 0, then the direct image $f(I)$ is not open.

2. Let $f : \mathbb{R} \to \mathbb{R}$ be defined by $f(x) := 1/(1 + x^2)$ for $x \in \mathbb{R}$.
   (a) Find an open interval (a, b) whose direct image under f is not open.
   (b) Show that the direct image of the closed interval $[0, \infty)$ is not closed.

3. Let $I := [1, \infty)$ and let $f(x) := \sqrt{x - 1}$ for $x \in I$. For each ε-neighborhood $G = (-\varepsilon, +\varepsilon)$ of 0, exhibit an open set H such that $H \cap I = f^{-1}(G)$.

4. Let $h : \mathbb{R} \to \mathbb{R}$ be defined by $h(x) := 1$ if $0 \le x \le 1$, $h(x) := 0$ otherwise. Find an open set G such that $h^{-1}(G)$ is not open, and a closed set F such that $h^{-1}(F)$ is not closed.

5. Show that if $f : \mathbb{R} \to \mathbb{R}$ is continuous, then the set $\{x \in \mathbb{R} : f(x) < \alpha\}$ is open in $\mathbb{R}$ for each $\alpha \in \mathbb{R}$.

6. Show that if $f : \mathbb{R} \to \mathbb{R}$ is continuous, then the set $\{x \in \mathbb{R} : f(x) \le \alpha\}$ is closed in $\mathbb{R}$ for each $\alpha \in \mathbb{R}$.

7. Show that if $f : \mathbb{R} \to \mathbb{R}$ is continuous, then the set $\{x \in \mathbb{R} : f(x) = k\}$ is closed in $\mathbb{R}$ for each $k \in \mathbb{R}$.

8. Give an example of a function $f : \mathbb{R} \to \mathbb{R}$ such that the set $\{x \in \mathbb{R} : f(x) = 1\}$ is neither open nor closed in $\mathbb{R}$.

9. Prove that $f : \mathbb{R} \to \mathbb{R}$ is continuous if and only if for each closed set F in $\mathbb{R}$, the inverse image $f^{-1}(F)$ is closed.

10. Let $I := [a, b]$ and let $f : I \to \mathbb{R}$ and $g : I \to \mathbb{R}$ be continuous functions on I. Show that the set $\{x \in I : f(x) = g(x)\}$ is closed in $\mathbb{R}$.

Section 11.4 Metric Spaces

This book has been devoted to a careful study of the real number system and a number of different limiting processes that can be defined for functions of a real variable. A central topic was the study of continuous functions. At this point, with a strong understanding of analysis on the real line, the study of more general spaces and the related limit concepts can begin. It is possible to generalize the fundamental concepts of real analysis in several different ways, but one of the most fruitful is in the context of metric spaces, where a metric is an abstraction of a distance function.

In this section, we will introduce the idea of metric space and then indicate how certain areas of the theory developed in this book can be extended to this new setting. We will discuss the concepts of neighborhood of a point, open and closed sets, convergence of sequences, and continuity of functions defined on metric spaces. Our purpose in this brief discussion is not to develop the theory of metric spaces to any great extent, but to reveal how the key ideas and techniques of real analysis can be put into a more abstract and general framework. The reader should note how the basic results of analysis on the real line serve to motivate and guide the study of analysis in more general contexts.

Generalization can serve two important purposes. One purpose is that theorems derived in general settings can often be applied in many particular cases without the need of a separate proof for each special case. A second purpose is that by removing the nonessential (and sometimes distracting) features of special situations, it is often possible to understand the real significance of a concept or theorem.

Metrics

On the real line, basic limit concepts were defined in terms of the distance $|x - y|$ between two points x, y in $\mathbb{R}$, and many theorems were proved using the absolute value function. Actually, a careful study reveals that only a few key properties of the absolute value were required to prove many fundamental results, and it happens that these properties can be extracted and used to define more general distance functions called "metrics".

11.4.1 Definition A **metric** on a set S is a function $d : S \times S \to \mathbb{R}$ that satisfies the following properties:

(a) $d(x, y) \geq 0$ for all $x, y \in S$ (*positivity*);

(b) $d(x, y) = 0$ if and only if $x = y$ (*definiteness*);

(c) $d(x, y) = d(y, x)$ for all $x, y \in S$ (*symmetry*);

(d) $d(x, y) \leq d(x, z) + d(z, y)$ for all $x, y, z \in S$ (*triangle inequality*).

A **metric space** (S, d) is a set S together with a metric d on S.

We consider several examples of metric spaces.

11.4.2 Examples **(a)** The familiar metric on $\mathbb{R}$ is defined by

$$d(x, y) := |x - y| \qquad \text{for} \quad x, y \in \mathbb{R}.$$

Property 11.4.1(d) for d follows from the Triangle Inequality for absolute value because we have

$$
\begin{aligned}
d(x, y) = |x - y| &= |(x - z) + (z - y)| \\
&\leq |x - z| + |z - y| = d(x, z) + d(z, y),
\end{aligned}
$$

for all $x, y, z \in \mathbb{R}$.

(b) The distance function in the plane obtained from the Pythagorean Theorem provides one example of a metric in $\mathbb{R}^2$. That is, we define the metric d on $\mathbb{R}^2$ as follows: if $P_1 := (x_1, y_1)$ and $P_2 := (x_2, y_2)$ are points in $\mathbb{R}^2$, then

$$d(P_1, P_2) := \sqrt{(x_1 - x_2)^2 + (y_1 - y_2)^2}$$

(c) It is possible to define several different metrics on the same set. On $\mathbb{R}^2$, we can also define the metric d_1 as follows:

$$d_1(P_1, P_2) := |x_1 - x_2| + |y_1 - y_2|$$

Still another metric on $\mathbb{R}^2$ is d_∞ defined by

$$d_\infty(P_1, P_2) := \sup \{|x_1 - x_2|, |y_1 - y_2|\}.$$

The verifications that d_1 and d_∞ satisfy the properties of a metric are left as exercises.

(d) Let $C[0, 1]$ denote the set of all continuous functions on the interval $[0, 1]$ to $\mathbb{R}$. For f, g in $C[0, 1]$, we define

$$d_\infty(f, g) := \sup\{|f(x) - g(x)| : x \in [0, 1]\}.$$

Then it can be verified that d_∞ is a metric on $C[0, 1]$. This metric is the uniform norm of $f - g$ on $[0, 1]$ as defined in Section 8.1; that is, $d_\infty(f, g) = \|f - g\|$, where $\|f\|$ denotes the uniform norm of f on the set $[0, 1]$.

(e) We again consider $C[0, 1]$, but we now define a different metric d_1 by

$$d_1(f, g) := \int_0^1 |f - g| \qquad \text{for} \quad f, g \in C[0, 1].$$

The properties of the integral can be used to show that this is indeed a metric on $C[0, 1]$. The details are left as an exercise.

(f) Let S be any nonempty set. For $s, t \in S$, we define

$$d(s, t) := \begin{cases} 0 & \text{if} \quad s = t, \\ 1 & \text{if} \quad s \neq t. \end{cases}$$

It is an exercise to show that d is a metric on S. This metric is called the **discrete metric** on the set S. $\square$

We note that if (S, d) is a metric space, and if $T \subseteq S$, then d' defined by $d'(x, y) := d(x, y)$ for all $x, y \in T$ gives a metric on T, which we generally denote by d. With this understanding, we say that (T, d) is also a metric space. For example, the metric d on $\mathbb{R}$ defined by the absolute value is a metric on the set $\mathbb{Q}$ of rational numbers, and thus $(\mathbb{Q}, d)$ is also a metric space.

Neighborhoods and Convergence

The basic notion needed for the introduction of limit concepts is that of neighborhood, and this is defined in metric spaces as follows.

11.4.3 Definition Let (S, d) be a metric space. Then for $\varepsilon > 0$, the ε-**neighborhood** of a point x_0 in S is the set

$$V_\varepsilon(x_0) := \{x \in S : d(x_0, x) < \varepsilon\}.$$

A **neighborhood** of x_0 is any set U that contains an ε-neighborhood of x_0 for some $\varepsilon > 0$.

Any notion defined in terms of neighborhoods can now be defined and discussed in the context of metric spaces by modifying the language appropriately. We first consider the convergence of sequences.

A sequence in a metric space (S, d) is a function $X : \mathbb{N} \to S$ with domain $\mathbb{N}$ and range in S, and the usual notations for sequence are used; we write $X = (x_n)$, but now $x_n \in S$ for all $n \in \mathbb{N}$. When we replace the absolute value by a metric in the definition of sequential convergence, we get the notion of convergence in a metric space.

11.4.4 Definition Let (x_n) be a sequence in the metric space (S, d). The sequence (x_n) is said to **converge** to x in S if for any $\varepsilon > 0$ there exists $K \in \mathbb{N}$ such that $x_n \in V_\varepsilon(x)$ for all $n \geq K$.

Note that since $x_n \in V_\varepsilon(x)$ if and only if $d(x_n, x) < \varepsilon$, a sequence (x_n) converges to x if and only if for any $\varepsilon > 0$ there exists K such that $d(x_n, x) < \varepsilon$ for all $n \geq K$. In other words, a sequence (x_n) in (S, d) converges to x if and only if the sequence of real numbers $\big(d(x_n, x)\big)$ converges to 0.

11.4.5 Examples **(a)** Consider $\mathbb{R}^2$ with the metric d defined in Example 11.4.2(b). If $P_n = (x_n, y_n) \in \mathbb{R}^2$ for each $n \in \mathbb{N}$, then we claim that the sequence (P_n) converges to $P = (x, y)$ with respect to this metric if and only if the sequences of real numbers (x_n) and (y_n) converge to x and y, respectively.

First, we note that the inequality $|x_n - x| \leq d(P_n, P)$ implies that if (P_n) converges to P with respect to the metric d, then the sequence (x_n) converges to x; the convergence of (y_n) follows in a similar way. The converse follows from the inequality $d(P_n, P) \leq |x_n - x| + |y_n - y|$, which is readily verified. The details are left to the reader.

(b) Let d_∞ be the metric on $C[0, 1]$ defined in Example 11.4.2(d). Then a sequence (f_n) in $C[0, 1]$ converges to f with respect to this metric if and only if (f_n) converges to f uniformly on the set $[0, 1]$. This is established in Lemma 8.1.8 in the discussion of the uniform norm.   □

Cauchy Sequences

The notion of Cauchy sequence is a significant concept in metric spaces. The definition is formulated as expected, with the metric replacing the absolute value.

11.4.6 Definition   Let (S, d) be a metric space. A sequence (x_n) in S is said to be a **Cauchy sequence** if for each $\varepsilon > 0$, there exists $H \in \mathbb{N}$ such that $d(x_n, x_m) < \varepsilon$ for all $n, m \geq H$.

The Cauchy Convergence Theorem 3.5.5 for sequences in $\mathbb{R}$ states that a sequence in $\mathbb{R}$ is a Cauchy sequence if and only if it converges to a point of $\mathbb{R}$. This theorem is not true for metric spaces in general, as the examples that follow will reveal. Those metric spaces for which Cauchy sequences are convergent have special importance.

11.4.7 Definition   A metric space (S, d) is said to be **complete** if each Cauchy sequence in S converges to a point of S.

In Section 2.3 the Completeness Property of $\mathbb{R}$ is stated in terms of the order properties by requiring that every nonempty subset of $\mathbb{R}$ that is bounded above has a supremum in $\mathbb{R}$. The convergence of Cauchy sequences is deduced as a theorem. In fact, it is possible to reverse the roles of these fundamental properties of $\mathbb{R}$: the Completeness Property of $\mathbb{R}$ can be stated in terms of Cauchy sequences as in 11.4.7, and the Supremum Property can then be deduced as a theorem. Since many metric spaces do not have an appropriate order structure, a concept of completeness must be described in terms of the metric, and Cauchy sequences provide the natural vehicle for this.

11.4.8 Examples   **(a)**   The metric space $(\mathbb{Q}, d)$ of rational numbers with the metric defined by the absolute value function is *not* complete.
For example, if (x_n) is a sequence of rational numbers that converges to $\sqrt{2}$, then it is Cauchy in $\mathbb{Q}$, but it does not converge to a point of $\mathbb{Q}$. Therefore $(\mathbb{Q}, d)$ is not a complete metric space.

(b)   The space $C[0, 1]$ with the metric d_∞ defined in 11.4.2(d) is complete.
To prove this, suppose that (f_n) is a Cauchy sequence in $C[0, 1]$ with respect to the metric d_∞. Then, given $\varepsilon > 0$, there exists H such that

$$\text{(1)} \qquad \qquad \left| f_n(x) - f_m(x) \right| < \varepsilon$$

for all $x \in [0, 1]$ and all $n, m \geq H$. Thus for each x, the sequence $(f_n(x))$ is Cauchy in $\mathbb{R}$, and therefore converges in $\mathbb{R}$. We define f to be the pointwise limit of the sequence; that is, $f(x) := \lim(f_n(x))$ for each $x \in [0, 1]$. It follows from (1) that for each $x \in [0, 1]$ and each $n \geq H$, we have $\left| f_n(x) - f(x) \right| \leq \varepsilon$. Consequently the sequence (f_n) converges uniformly to f on $[0, 1]$. Since the uniform limit of continuous functions is also continuous (by 8.2.2), the function f is in $C[0, 1]$. Therefore the metric space $(C[0, 1], d_\infty)$ is complete.

(c)   If d_1 is the metric on $C[0, 1]$ defined in 11.4.2(e), then the metric space $(C[0, 1], d_1)$ is *not* complete.

To prove this statement, it suffices to exhibit a Cauchy sequence that does not have a limit in the space. We define the sequence (f_n) for $n \geq 3$ as follows (see Figure 11.4.1):

$$f_n(x) := \begin{cases} 1 & \text{for} \quad 0 \leq x \leq 1/2, \\ 1 + n/2 - nx & \text{for} \quad 1/2 < x \leq 1/2 + 1/n, \\ 0 & \text{for} \quad 1/2 + 1/n < x \leq 1. \end{cases}$$

Note that the sequence (f_n) converges pointwise to the discontinuous function $f(x) := 1$ for $0 \leq x \leq 1/2$ and $f(x) := 0$ for $1/2 < x \leq 1$. Hence $f \notin C[0, 1]$; in fact, there is no function $g \in C[0, 1]$ such that $d_1(f_n, g) \to 0$.   □

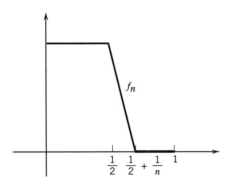

Figure 11.4.1   The sequence (f_n)

Open Sets and Continuity

With the notion of neighborhood defined, the definitions of open set and closed set read the same as for sets in $\mathbb{R}$.

11.4.9 Definition   Let (S, d) be a metric space. A subset G of S is said to be an **open** set in S if for every point $x \in S$ there is a neighborhood U of x such that $U \subseteq G$. A subset F of S is said to be a **closed** set in S if the complement $S \backslash F$ is an open set in S.

Theorems 11.1.4 and 11.1.5 concerning the unions and intersections of open sets and closed sets can be extended to metric spaces without difficulty. In fact, the proofs of those theorems carry over to metric spaces with very little change: simply replace the ε-neighborhoods $(x - \varepsilon, x + \varepsilon)$ in $\mathbb{R}$ by ε-neighborhoods $V_\varepsilon(x)$ in S.

We now can examine the concept of continuity for functions that map one metric space (S_1, d_1) into another metric space (S_2, d_2). Note that we modify the property in 5.1.2 of continuity for functions on $\mathbb{R}$ by replacing neighborhoods in $\mathbb{R}$ by neighborhoods in the metric spaces.

11.4.10 Definition   Let (S_1, d_1) and (S_2, d_2) be metric spaces, and let $f : S_1 \to S_2$ be a function from S_1 to S_2. The function f is said to be **continuous** at the point c in S_1 if for every ε-neighborhood $V_\varepsilon(f(c))$ of $f(c)$ there exists a δ-neighborhood $V_\delta(c)$ of c such that if $x \in V_\delta(c)$, then $f(x) \in V_\varepsilon(f(c))$.

The ε-δ formulation of continuity can be stated as follows: $f : S_1 \to S_2$ is continuous at c if and only if for each $\varepsilon > 0$ there exists $\delta > 0$ such that $d_1(x, c) < \delta$ implies that $d_2(f(x), f(c)) < \varepsilon$.

The Global Continuity Theorem can be established for metric spaces by appropriately modifying the argument for functions on $\mathbb{R}$.

11.4.11 Global Continuity Theorem   *If (S_1, d_1) and (S_2, d_2) are metric spaces, then a function $f : S_1 \to S_2$ is continuous on S_1 if and only if $f^{-1}(G)$ is open in S whenever G is open in S_2.*

The notion of compactness extends immediately to metric spaces. A metric space (S, d) is said to be **compact** if each open cover of S has a finite subcover. Then by modifying the proof of 11.3.4, we obtain the following result.

11.4.12 Preservation of Compactness   *If (S, d) is a compact metric space and if the function $f : S \to \mathbb{R}$ is continuous, then $f(S)$ is compact in $\mathbb{R}$.*

The important properties of continuous functions given in 11.3.5 then follow immediately. The Boundedness Theorem, the Maximum-Minimum Theorem, and the Uniform Continuity Theorem for real-valued continuous functions on a compact metric space are all established by appropriately modifying the language of the proofs given in 11.3.5.

Semimetrics _____

11.4.13 Definition   A **semimetric** on a set S is a function $d : S \times S \to \mathbb{R}$ that satisfies all of the conditions in Definition 11.4.1, except that condition (b) is replaced by the weaker condition

(b′) $d(x, y) = 0$ if $x = y$.

A **semimetric space** (S, d) is a set S together with a semimetric d on S.

Thus every metric is a semimetric, and every metric space is a semimetric space. However, the converse is not true. For example, if $P_1 := (x_1, y_1)$ and $P_2 := (x_2, y_2)$ are points in the space $\mathbb{R}^2$, the function d_1 defined by

$$d_1(P_1, P_2) := |x_1 - x_2|,$$

is easily seen to be a semimetric, but it is not a metric since any two points with the same first coordinate have "d_1-distance" equal to 0.

Somewhat more interestingly, if f, g are any functions in $\mathcal{L}[a, b]$, we have defined (in Definition 10.2.9) the distance function:

$$\text{dist}(f, g) := \int_a^b |f - g|.$$

Here it is clear that any two functions that are equal except at a countable set of points will have distance equal to 0 from each other (in fact, this is also true when the functions are equal almost everywhere).

The reader can retrace the discussion in the present section and see that most of what we have done remains true for semimetrics and semimetric spaces. The main difference is that a sequence in a semimetric space does not necessarily converge to a *unique* limit. While this seems to be rather unusual, it is actually not a very serious problem and one can learn to adjust to this situation. The other alternative is to "identify" points that have distance 0 from each other. This identification procedure is often invoked, but it means one is dealing with "equivalence classes" rather than individual points. Often this cure is worse than the malady.

Exercises for Section 11.4

1. Show that the functions d_1 and d_∞ defined in 11.4.2(c) are metrics on $\mathbb{R}^2$.

2. Show that the functions d_∞ and d_1 defined in 11.4.2(d, e) are metrics on $C[0, 1]$.

3. Verify that the discrete metric on a set S as defined in 11.4.2(f) is a metric.

4. If $P_n := (x_n, y_n) \in \mathbb{R}^2$ and d_∞ is the metric in 11.4.2(c), show that (P_n) converges to $P := (x, y)$ with respect to this metric if and only if (x_n) and (y_n) converge to x and y, respectively.

5. Verify the conclusion of Exercise 4 if d_∞ is replaced by d_1.

6. Let S be a nonempty set and let d be the discrete metric defined in 11.4.2(f). Show that in the metric space (S, d), a sequence (x_n) in S converges to x if and only if there is a $K \in \mathbb{N}$ such that $x_n = x$ for all $n \geq K$.

7. Show that if d is the discrete metric on a set S, then every subset of S is both open and closed in (S, d).

8. Let $P := (x, y)$ and $O := (0, 0)$ in $\mathbb{R}^2$. Draw the following sets in the plane:
   (a) $\{P \in \mathbb{R}^2 : d_1(O, P) \leq 1\}$,
   (b) $\{P \in \mathbb{R}^2 : d_\infty(O, P) \leq 1\}$.

9. Prove that in any metric space, an ε-neighborhood of a point is an open set.

10. Prove Theorem 11.4.11.

11. Prove Theorem 11.4.12.

12. If (S, d) is a metric space, a subset $A \subseteq S$ is said to be **bounded** if there exists $x_0 \in S$ and a number $B > 0$ such that $A \subseteq \{x \in S: d(x, x_0) \leq B\}$. Show that if A is a compact subset of S, then A is closed and bounded.

APPENDIX A

LOGIC AND PROOFS

Natural science is concerned with collecting facts and organizing these facts into a coherent body of knowledge so that one can understand nature. Originally much of science was concerned with observation, the collection of information, and its classification. This classification gradually led to the formation of various "theories" that helped the investigators to remember the individual facts and to be able to explain and sometimes predict natural phenomena. The ultimate aim of most scientists is to be able to organize their science into a coherent collection of general principles and theories so that these principles will enable them both to understand nature and to make predictions of the outcome of future experiments. Thus they want to be able to develop a system of general principles (or axioms) for their science that will enable them to *deduce* the individual facts and consequences from these general laws.

Mathematics is different from the other sciences: by its very nature, it is a deductive science. That is not to say that mathematicians do not collect facts and make observations concerning their investigations. In fact, many mathematicians spend a large amount of time performing calculations of special instances of the phenomena they are studying in the hopes that they will discover "unifying principles". (The great Gauss did a vast amount of calculation and studied much numerical data before he was able to formulate a conjecture concerning the distribution of prime numbers.) However, even after these principles and conjectures are formulated, the work is far from over, for mathematicians are not satisfied until conjectures have been derived (i.e., proved) from the axioms of mathematics, from the definitions of the terms, and from results (or theorems) that have previously been proved. Thus, a mathematical statement is not a theorem until it has been carefully derived from axioms, definitions, and previously proved theorems.

A few words about the axioms (i.e., postulates, assumptions, etc.) of mathematics are in order. There are a few axioms that apply to all of mathematics—the "axioms of set theory"—and there are specific axioms within different areas of mathematics. Sometimes these axioms are stated formally, and sometimes they are built into definitions. For example, we list properties in Chapter 2 that we assume the real number system possesses; they are really a set of axioms. As another example, the definition of a "group" in abstract algebra is basically a set of axioms that we assume a set of elements to possess, and the study of group theory is an investigation of the consequences of these axioms.

Students studying real analysis for the first time usually do not have much experience in understanding (not to mention constructing) proofs. In fact, one of the main purposes of this course (and this book) is to help the reader gain experience in the type of critical thought that is used in this deductive process. The purpose of this appendix is to help the reader gain insight about the techniques of proof.

Statements and Their Combinations

All mathematical proofs and arguments are based on **statements**, which are declarative sentences or meaningful strings of symbols that can be classified as being true or false. It is not necessary that we know whether a given statement is actually true or false, but it must be one or the other, and it cannot be both. (This is the *Principle of the Excluded Middle*.) For example, the sentence "Chickens are pretty" is a matter of opinion and not a statement in the sense of logic. Consider the following sentences:

- It rained in Kuala Lumpur on June 2, 1988.
- Thomas Jefferson was shorter than John Adams.
- There are infinitely many twin primes.
- This sentence is false.

The first three are statements: the first is true, the second is false, and the third is either true or false, but we are not sure which at this time. The fourth sentence is not a statement; it can be neither true nor false since it leads to contradictory conclusions.

Some statements (such as "$1 + 1 = 2$") are always true; they are called **tautologies**. Some statements (such as "$2 = 3$") are always false; they are called **contradictions** or **falsities**. Some statements (such as "$x^2 = 1$") are sometimes true and sometimes false (e.g., true when $x = 1$ and false when $x = 3$). Or course, for the statement to be completely clear, it is necessary that the proper context has been established and the meaning of the symbols has been properly defined (e.g., we need to know that we are referring to integer arithmetic in the preceding examples).

Two statements P and Q are said to be **logically equivalent** if P is true exactly when Q is true (and hence P is false exactly when Q is false). In this case we often write $P \equiv Q$. For example, we write

$$(x \text{ is Abraham Lincoln}) \equiv (x \text{ is the 16th president of the United States}).$$

There are several different ways of forming new statements from given ones by using logical connectives.

If P is a statement, then its **negation** is the statement denoted by

$$\text{not } P$$

which is true when P is false, and is false when P is true. (A common notation for the negation of P is $\neg P$.) A little thought shows that

$$P \equiv \text{not(not } P).$$

This is the *Principle of Double Negation*.

If P and Q are statements, then their **conjunction** is the statement denoted by

$$P \text{ and } Q$$

which is true when both P and Q are true, and is false otherwise. (A standard notation for the conjunction of P and Q is $P \wedge Q$.) It is evident that

$$(P \text{ and } Q) \equiv (Q \text{ and } P).$$

Similarly, the **disjunction** of P and Q is the statement denoted by

$$P \text{ or } Q$$

which is true when at least one of P and Q is true, and false only when they are both false. In legal documents "or" is often denoted by "and/or" to make it clear that this disjunction is also true when *both* P and Q are true. (A standard notation for the disjunction of P and Q is $P \vee Q$.) It is also evident that

$$(P \text{ or } Q) \equiv (Q \text{ or } P).$$

To contrast disjunctive and conjunctive statements, note that the statement "$2 < \sqrt{2}$ and $\sqrt{2} < 3$" is false, but the statement "$2 < \sqrt{2}$ or $\sqrt{2} < 3$" is true (since $\sqrt{2}$ is approximately equal to $1.4142\cdots$).

Some thought shows that negation, conjunction, and disjunction are related by *DeMorgan's Laws*:

$$\text{not } (P \text{ and } Q) \equiv (\text{not } P) \text{ or } (\text{not } Q),$$
$$\text{not } (P \text{ or } Q) \equiv (\text{not } P) \text{ and } (\text{not } Q).$$

The first of these equivalencies can be illustrated by considering the statements

$$P : x = 2, \qquad Q : y \in A.$$

The statement $(P \text{ and } Q)$ is true when both $(x = 2)$ and $(y \in A)$ are true, and it is false when at least one of $(x = 2)$ and $(y \in A)$ is false; that is, the statement not$(P \text{ and } Q)$ is true when at least one of the statements $(x \neq 2)$ and $(y \notin A)$ holds.

Implications

A very important way of forming a new statement from given ones is the **implication** (or **conditional**) statement, denoted by

$$(P \Rightarrow Q), \qquad (\text{if } P \text{ then } Q), \qquad \text{or} \qquad (P \text{ implies } Q).$$

Here P is called the **hypothesis**, and Q is called the **conclusion** of the implication. To help understand the truth values of the implication, consider the statement

$$\text{If I win the lottery today, then I'll buy Sam a car.}$$

Clearly this statement is false if I win the lottery and don't buy Sam a car. What if I don't win the lottery today? Under this circumstance, I haven't made any promise about buying anyone a car, and since the condition of winning the lottery did not materialize, my failing to buy Sam a car should not be considered as breaking a promise. Thus the implication is regarded as true when the hypothesis is not satisfied.

In mathematical arguments, we are very much interested in implications when the hypothesis is true, but not much interested in them when the hypothesis is false. The accepted procedure is to take the statement $P \Rightarrow Q$ to be false only when P is true and Q is false; in all other cases the statement $P \Rightarrow Q$ is true. (Consequently, if P is false, then we agree to take the statement $P \Rightarrow Q$ to be true whether or not Q is true or false. That may seem strange to the reader, but it turns out to be convenient in practice and consistent with the other rules of logic.)

We observe that the definition of $P \Rightarrow Q$ is logically equivalent to

$$\text{not } \big(P \text{ and } (\text{not } Q)\big),$$

because this statement is false only when P is true and Q is false, and it is true in all other cases. It also follows from the first DeMorgan Law and the Principle of Double Negation that $P \Rightarrow Q$ is logically equivalent to the statement

$$(\text{not } P) \text{ or } Q,$$

since this statement is true unless both (not P) and Q are false; that is, unless P is true and Q is false.

Contrapositive and Converse

As an exercise, the reader should show that the implication $P \Rightarrow Q$ is logically equivalent to the implication

$$(\text{not } Q) \Rightarrow (\text{not } P),$$

which is called the **contrapositive** of the implication $P \Rightarrow Q$. For example, if $P \Rightarrow Q$ is the implication

If I am in Chicago, then I am in Illinois,

then the contrapositive $(\text{not } Q) \Rightarrow (\text{not } P)$ is the implication

If I am not in Illinois, then I am not in Chicago.

The equivalence of these two statements is apparent after a bit of thought. In attempting to establish an implication, it is sometimes easier to establish the contrapositive, which is logically equivalent to it. (This will be discussed in more detail later.)

If an implication $P \Rightarrow Q$ is given, then one can also form the statement

$$Q \Rightarrow P,$$

which is called the **converse** of $P \Rightarrow Q$. The reader must guard against confusing the converse of an implication with its contrapositive, since they are quite different statements. While the contrapositive is logically equivalent to the given implication, the converse is not. For example, the converse of the statement

If I am in Chicago, then I am in Illinois,

is the statement

If I am in Illinois, then I am in Chicago.

Since it is possible to be in Illinois but not in Chicago, these two statements are evidently *not* logically equivalent.

There is one final way of forming statements that we will mention. It is the **double implication** (or the **biconditional**) statement, which is denoted by

$$P \Longleftrightarrow Q \qquad \text{or} \qquad P \text{ if and only if } Q,$$

and which is defined by

$$(P \Rightarrow Q) \text{ and } (Q \Rightarrow P).$$

It is a straightforward exercise to show that $P \Longleftrightarrow Q$ is true precisely when P and Q are both true, or both false.

Context and Quantifiers

In any form of communication, it is important that the individuals have an appropriate context in mind. Statements such as "I saw Mary today" may not be particularly informative if the hearer knows several persons named Mary. Similarly, if one goes into the middle of a mathematical lecture and sees the equation $x^2 = 1$ on the blackboard, it is useful for the viewer to know what the writer means by the letter x and the symbol 1. Is x an integer? A function? A matrix? A subgroup of a given group? Does 1 denote a natural number? The identity function? The identity matrix? The trivial subgroup of a group?

Often the context is well understood by the conversants, but it is always a good idea to establish it at the start of a discussion. For example, many mathematical statements involve one or more variables whose values usually affect the truth or the falsity of the statement, so we should always make clear what the possible values of the variables are.

Very often mathematical statements involve expressions such as "for all", "for every", "for some", "there exists", "there are", and so on. For example, we may have the statements

$$\text{For any integer } x, \ x^2 = 1$$

and

$$\text{There exists an integer } x \text{ such that } x^2 = 1.$$

Clearly the first statement is false, as is seen by taking $x = 3$; however, the second statement is true since we can take either $x = 1$ or $x = -1$.

If the context has been established that we are talking about integers, then the above statements can safely be abbreviated as

$$\text{For any } x, \ x^2 = 1$$

and

$$\text{There exists an } x \text{ such that } x^2 = 1.$$

The first statement involves the **universal quantifier** "for every", and is making a statement (here false) about *all* integers. The second statement involves the **existential quantifier** "there exists", and is making a statement (here true) about *at least one* integer.

These two quantifiers occur so often that mathematicians often use the symbol ∀ to stand for the universal quantifier, and the symbol ∃ to stand for the existential quantifier. That is,

∀   denotes   "for every",

∃   denotes   "there exists".

While we do not use these symbols in this book, it is important for the reader to know how to read formulas in which they appear. For example, the statement

(i) $(\forall x)(\exists y)(x + y = 0)$

(understood for integers) can be read

$$\text{For every integer } x, \ \text{there exists}$$
$$\text{an integer } y \text{ such that } x + y = 0.$$

Similarly the statement

(ii) $(\exists y)(\forall x)(x + y = 0)$

can be read

$$\text{There exists an integer } y, \ \text{such that}$$
$$\text{for every integer } x, \ \text{then } x + y = 0.$$

These two statements are very different; for example, the first one is true and the second one is false. The moral is that the *order* of the appearance of the two different types of quantifiers is *very important*. It must also be stressed that if several variables appear in a mathematical expression with quantifiers, the values of the later variables should be assumed to depend on all of the values of the variables that are mentioned earlier. Thus in the (true) statement (i) above, the value of y depends on that of x; here if $x = 2$, then $y = -2$, while if $x = 3$, then $y = -3$.

It is important that the reader understand how to negate a statement that involves quantifiers. In principle, the method is simple.

(a) To show that it is false that every element x in some set possesses a certain property $\mathcal{P}$, it is enough to produce a single **counter-example** (that is, a particular element in the set that does not possess this property); and

(b) To show that it is false that there exists an element y in some set that satisfies a certain property $\mathcal{P}$, we need to show that every element y in the set fails to have that property.

Therefore, in the process of forming a negation,

$$\text{not } (\forall x)\mathcal{P} \quad \text{becomes} \quad (\exists x) \text{ not } \mathcal{P}$$

and similarly

$$\text{not } (\exists y)\mathcal{P} \quad \text{becomes} \quad (\forall y) \text{ not } \mathcal{P}.$$

When several quantifiers are involved, these changes are repeatedly used. Thus the negation of the (true) statement (i) given previously becomes in succession

$$\text{not } (\forall x)\,(\exists y)\,(x + y = 0),$$
$$(\exists x) \text{ not } (\exists y)\,(x + y = 0),$$
$$(\exists x)\,(\forall y) \text{ not } (x + y = 0),$$
$$(\exists x)\,(\forall y)\,(x + y \neq 0).$$

The last statement can be rendered in words as:

> There exists an integer x, such that
> for every integer y, then $x + y \neq 0$.

(This statement is, of course, false.)

Similarly, the negation of the (false) statement (ii) given previously becomes in succession

$$\text{not } (\exists y)\,(\forall x)\,(x + y = 0),$$
$$(\forall y) \text{ not } (\forall x)\,(x + y = 0),$$
$$(\forall y)\,(\exists x) \text{ not } (x + y = 0),$$
$$(\forall y)\,(\exists x)\,(x + y \neq 0).$$

The last statement is rendered in words as

> For every integer y, there exists
> an integer x such that $x + y \neq 0$.

Note that this statement is true, and that the value (or values) of x that make $x + y \neq 0$ depends on y, in general.

Similarly, the statement

> For every $\delta > 0$, the interval $(-\delta, \delta)$
> contains a point belonging to the set A,

can be seen to have the negation

> There exists $\delta > 0$ such that the interval
> $(-\delta, \delta)$ does not contain any point in A.

The first statement can be symbolized

$$(\forall \delta > 0)\,(\exists\, y \in A)\,(y \in (-\delta, \delta)),$$

and its negation can be symbolized by

$$(\exists\, \delta > 0)\,(\forall y \in A)\,(y \notin (-\delta, \delta))$$

or by

$$(\exists\, \delta > 0)\,(A \cap (-\delta, \delta) = \emptyset).$$

It is the strong opinion of the authors that, while the use of this type of symbolism is often convenient, it is *not* a substitute for thought. Indeed, the readers should ordinarily reason for themselves what the negation of a statement is and not rely slavishly on symbolism. While good notation and symbolism can often be a useful aid to thought, it can never be an adequate replacement for thought and understanding.

Direct Proofs

Let P and Q be statements. The assertion that the hypothesis P of the implication $P \Rightarrow Q$ implies the conclusion Q (or that $P \Rightarrow Q$ is a theorem) is the assertion that whenever the hypothesis P is true, then Q is true.

The construction of a **direct proof** of $P \Rightarrow Q$ involves the construction of a string of statements $R_1, R_2, \cdots, R_n$ such that

$$P \Rightarrow R_1, \quad R_1 \Rightarrow R_2, \quad \cdots, \quad R_n \Rightarrow Q.$$

(The *Law of the Syllogism* states that if $R_1 \Rightarrow R_2$ and $R_2 \Rightarrow R_3$ are true, then $R_1 \Rightarrow R_3$ is true.) This construction is usually not an easy task; it may take insight, intuition, and considerable effort. Often it also requires experience and luck.

In constructing a direct proof, one often works forward from P and backward from Q. We are interested in logical consequences of P; that is, statements $Q_1, \cdots, Q_k$ such that $P \Rightarrow Q_i$. And we might also examine statements $P_1, \cdots, P_r$ such that $P_j \Rightarrow Q$. If we can work forward from P and backward from Q so the string "connects" somewhere in the middle, then we have a proof. Often in the process of trying to establish $P \Rightarrow Q$ one finds that one must strengthen the hypothesis (i.e., add assumptions to P) or weaken the conclusion (that is, replace Q by a nonequivalent consequence of Q).

Most students are familiar with "direct" proofs of the type described above, but we will give one elementary example here. Let us prove the following theorem.

Theorem 1 *The square of an odd integer is also an odd integer.*

If we let n stand for an integer, then the hypothesis is:

$$P : n \text{ is an odd integer.}$$

The conclusion of the theorem is:

$$Q : n^2 \text{ is an odd integer.}$$

We need the definition of odd integer, so we introduce the statement

$$R_1 : n = 2k - 1 \text{ for some integer } k.$$

Then we have $P \Rightarrow R_1$. We want to deduce the statement $n^2 = 2m - 1$ for some integer m, since this would imply Q. We can obtain this statement by using algebra:

$$R_2 : n^2 = (2k - 1)^2 = 4k^2 - 4k + 1,$$
$$R_3 : n^2 = (4k^2 - 4k + 2) - 1,$$
$$R_4 : n^2 = 2(2k^2 - 2k + 1) - 1.$$

If we let $m = 2k^2 - 2k + 1$, then m is an integer (why?), and we have deduced the statement

$$R_5 : n^2 = 2m - 1.$$

Thus we have $P \Rightarrow R_1 \Rightarrow R_2 \Rightarrow R_3 \Rightarrow R_4 \Rightarrow R_5 \Rightarrow Q$, and the theorem is proved.

Of course, this is a clumsy way to present a proof. Normally, the formal logic is suppressed and the argument is given in a more conversational style with complete English sentences. We can rewrite the preceding proof as follows.

Proof of Theorem 1.   If n is an odd integer, then $n = 2k - 1$ for some integer k. Then the square of n is given by $n^2 = 4k^2 - 4k + 1 = 2(2k^2 - 2k + 1) - 1$. If we let $m = 2k^2 - 2k + 1$, then m is an integer (why?) and $n^2 = 2m - 1$. Therefore, n^2 is an odd integer.   Q.E.D.

At this stage, we see that we may want to make a preliminary argument to prove that $2k^2 - 2k + 1$ is an integer whenever k is an integer. In this case, we could state and prove this fact as a **Lemma**, which is ordinarily a preliminary result that is needed to prove a theorem, but has little interest by itself.

Incidentally, the letters Q.E.D. stand for *quod erat demonstrandum*, which is Latin for "which was to be demonstrated".

Indirect Proofs

There are basically two types of indirect proofs: (i) contrapositive proofs, and (ii) proofs by contradiction. Both types start with the assumption that the conclusion Q is false, in other words, that the statement "not Q" is true.

(i)   Contrapositive proofs. Instead of proving $P \Rightarrow Q$, we may prove its logically equivalent contrapositive: not $Q \Rightarrow$ not P.

Consider the following theorem.

Theorem 2   *If n is an integer and n^2 is even, then n is even.*

The negation of "$Q : n$ is even" is the statement "not $Q : n$ is odd". The hypothesis "$P : n^2$ is even" has a similar negation, so that the contrapositive is the implication: If n is odd, then n^2 is odd. But this is exactly Theorem 1, which was proved above. Therefore this provides a proof of Theorem 2.

The contrapositive proof is often convenient when the universal quantifier is involved, for the contrapositive form will then involve the existential quantifier. The following theorem is an example of this situation.

Theorem 3   *Let $a \geq 0$ be a real number. If, for every $\varepsilon > 0$, we have $0 \leq a < \varepsilon$, then $a = 0$.*

Proof.   If $a = 0$ is false, then since $a \geq 0$, we must have $a > 0$. In this case, if we choose $\varepsilon_0 = \frac{1}{2} a$, then we have $\varepsilon_0 > 0$ and $\varepsilon_0 < a$, so that the hypothesis $0 \leq a < \varepsilon$ for all $\varepsilon > 0$ is false.   Q.E.D.

Here is one more example of a contrapositive proof.

Theorem 4   *If m, n are natural numbers such that $m + n \geq 20$, then either $m \geq 10$ or $n \geq 10$.*

Proof.   If the conclusion is false, then we have both $m < 10$ and $n < 10$. (Recall De-Morgan's Law.) Then addition gives us $m + n < 10 + 10 = 20$, so that the hypothesis is false.   Q.E.D.

(ii)   Proof by contradiction. This method of proof employs the fact that if C is a contradiction (i.e., a statement that is always false, such as "$1 = 0$"), then the two statements

$$\big(P \text{ and } (\text{not } Q)\big) \Rightarrow C, \qquad P \Rightarrow Q$$

are logically equivalent. Thus we establish $P \Rightarrow Q$ by showing that the statement $\big(P \text{ and } (\text{not } Q)\big)$ implies a contradiction.

Theorem 5   *Let $a > 0$ be a real number. If $a > 0$, then $1/a > 0$.*

Proof.   We suppose that the statement $a > 0$ is true and that the statement $1/a > 0$ is false. Therefore, $1/a \leq 0$. But since $a > 0$ is true, it follows from the order properties of $\mathbb{R}$ that $a(1/a) \leq 0$. Since $1 = a(1/a)$, we deduce that $1 \leq 0$. However, this conclusion contradicts the known result that $1 > 0$.   Q.E.D.

There are several classic proofs by contradiction (also known as *reductio ad absurdum*) in the mathematical literature. One is the proof that there is no rational number r that satisfies $r^2 = 2$. (This is Theorem 2.1.4 in the text.) Another is the proof of the infinitude of primes, found in Euclid's *Elements*. Recall that a natural number p is prime if its only integer divisors are 1 and p itself. We will assume the basic results that each prime number is greater than 1 and each natural number greater than 1 is either prime or divisible by a prime.

Theorem 6   (Euclid's *Elements*, Book IX, Proposition 20.)   *There are infinitely many prime numbers.*

Proof.   If we suppose by way of contradiction that there are finitely many prime numbers, then we may assume that $S = \{p_1, \cdots, p_n\}$ is the set of *all* prime numbers. We let $m = p_1 \cdots p_n$, the product of all the primes, and we let $q = m + 1$. Since $q > p_i$ for all i, we see that q is not in S, and therefore q is not prime. Then there exists a prime p that is a divisor of q. Since p is prime, then $p = p_j$ for some j, so that p is a divisor of m. But if p divides both m and $q = m + 1$, then p divides the difference $q - m = 1$. However, this is impossible, so we have obtained a contradiction.   Q.E.D.

APPENDIX B

FINITE AND COUNTABLE SETS

We will establish the results that were stated in Section 1.3 without proof. The reader should refer to that section for the definitions.

The first result is sometimes called the "Pigeonhole Principle". It may be interpreted as saying that if m pigeons are put into n pigeonholes and if $m > n$, then at least two pigeons must share one of the pigeonholes. This is a frequently-used result in combinatorial analysis. It yields many useful consequences.

B.1 Theorem   *Let $m, n \in \mathbb{N}$ with $m > n$. Then there does not exist an injection from $\mathbb{N}_m$ into $\mathbb{N}_n$.*

Proof.   We will prove this by induction on n.

If $n = 1$ and if g is any map of $\mathbb{N}_m$ ($m > 1$) into $\mathbb{N}_1$, then it is clear that $g(1) = \cdots = g(m) = 1$, so that g is not injective.

Assume that $k > 1$ is such that if $m > k$, there is no injection from $\mathbb{N}_m$ into $\mathbb{N}_k$. We will show that if $m > k + 1$, there is no function $h : \mathbb{N}_m \to \mathbb{N}_{k+1}$ that is an injection.

Case 1:   If the range $h(\mathbb{N}_m) \subseteq \mathbb{N}_k \subset \mathbb{N}_{k+1}$, then the induction hypothesis implies that h is not an injection of $\mathbb{N}_m$ into $\mathbb{N}_k$, and therefore into $\mathbb{N}_{k+1}$.

Case 2:   Suppose that $h(\mathbb{N}_m)$ is not contained in $\mathbb{N}_k$. If more than one element in $\mathbb{N}_m$ is mapped into $k + 1$, then h is not an injection. Therefore, we may assume that a single $p \in \mathbb{N}_m$ is mapped into $k + 1$ by h. We now define $h_1 : \mathbb{N}_{m-1} \to \mathbb{N}_k$ by

$$h_1(q) := \begin{cases} h(q) & \text{if } q = 1, \cdots, p - 1, \\ h(q + 1) & \text{if } q = p, \cdots, m - 1. \end{cases}$$

Since the induction hypothesis implies that h_1 is not an injection into $\mathbb{N}_k$, it is easily seen that h is not an injection into $\mathbb{N}_{k+1}$.   Q.E.D.

We now show that a finite set determines a unique number in $\mathbb{N}$.

1.3.2 Uniqueness Theorem   *If S is a finite set, then the number of elements in S is a unique number in $\mathbb{N}$.*

Proof.   If the set S has m elements, there exists a bijection f_1 of $\mathbb{N}_m$ onto S. If S also has n elements, there exists a bijection f_2 of $\mathbb{N}_m$ onto S. If $m > n$, then (by Exercise 19 of Section 1.1) $f_2^{-1} \circ f_1$ is a bijection of $\mathbb{N}_m$ onto $\mathbb{N}_n$, which contradicts Theorem B.1. If $n > m$, then $f_1^{-1} \circ f_2$ is a bijection of $\mathbb{N}_n$ onto $\mathbb{N}_m$, which contradicts Theorem B.1. Therefore we have $m = n$.   Q.E.D.

B.2 Theorem   *If $n \in \mathbb{N}$, there does not exist an injection from $\mathbb{N}$ into $\mathbb{N}_n$.*

Proof.   Assume that $f : \mathbb{N} \to \mathbb{N}_n$ is an injection, and let $m := n + 1$. Then the restriction of f to $\mathbb{N}_m \subset \mathbb{N}$ is also in injection into $\mathbb{N}_n$. But this contradicts Theorem B.1.   Q.E.D.

1.3.3 Theorem   *The set $\mathbb{N}$ of natural numbers is an infinite set.*

Proof.   If $\mathbb{N}$ is a finite set, there exists some $n \in \mathbb{N}$ and a bijection f of $\mathbb{N}_n$ onto $\mathbb{N}$. In this case the inverse function f^{-1} is a bijection (and hence an injection) of $\mathbb{N}$ onto $\mathbb{N}_n$. But this contradicts Theorem B.2.   Q.E.D.

We will next establish Theorem 1.3.8 by defining a bijection of $\mathbb{N} \times \mathbb{N}$ onto $\mathbb{N}$. We will obtain an explicit formula for the counting procedure of $\mathbb{N} \times \mathbb{N}$ that is displayed in Figure 1.3.1; the reader should refer to that figure during the ensuing discussion. The set $\mathbb{N} \times \mathbb{N}$ is viewed as a collection of diagonals; the first diagonal has 1 point, the second has 2 points, $\cdots$, and the kth diagonal has k points. In view of Example 1.2.4(a), the total number of points in diagonals 1 through k is therefore given by

$$\psi(k) := 1 + 2 + \cdots + k = \tfrac{1}{2} k(k+1).$$

The fact that ψ is strictly increasing follows from Mathematical Induction and

(1) $\psi(k+1) = \psi(k) + (k+1) \qquad \text{for} \quad k \in \mathbb{N}.$

The point (m, n) in $\mathbb{N} \times \mathbb{N}$ lies in the kth diagonal when $k = m + n - 1$, and it is the mth point in that diagonal as we move downward from left to right. (For example, the point $(3, 2)$ lies in the 4th diagonal (since $3 + 2 - 1 = 4$) and is the 3rd point in that diagonal.) Therefore, in the counting scheme shown in Figure 1.3.1, we count the point (m, n) by first counting the points in the first $k - 1 = m + n - 2$ diagonals and then adding m. According to this analysis, our counting function $h : \mathbb{N} \times \mathbb{N} \to \mathbb{N}$ is given by

(2) $h(m, n) := \psi(m + n - 2) + m \qquad \text{for} \quad (m, n) \in \mathbb{N} \times \mathbb{N}.$

(For example, the point $(3, 2)$ is counted as number $h(3, 2) = \psi(5 - 2) + 3 = \psi(3) + 3 = 6 + 3 = 9$, as in Figure 1.3.1. Also, the point $(17, 25)$ is counted as number $h(17, 25) = \psi(40) + 17 = 837$.) While this geometric argument has been suggestive and has led to the counting formula (2), we must now prove that h is in fact a bijection of $\mathbb{N} \times \mathbb{N}$ onto $\mathbb{N}$.

1.3.8 Theorem   *The set $\mathbb{N} \times \mathbb{N}$ is denumerable.*

Proof.   We will show that the function h defined in (2) is a bijection.

(a) We first show that h is injective. If $(m, n) \neq (m', n')$, then either (i) $m + n \neq m' + n'$, or (ii) $m + n = m' + n'$ and $m \neq m'$.

In case (i), we may suppose $m + n < m' + n'$. Then, using formula (1), the fact that ψ is increasing, and $m' > 0$, we have

$$
\begin{aligned}
h(m, n) &= \psi(m + n - 2) + m \leq \psi(m + n - 2) + (m + n - 1) \\
&= \psi(m + n - 1) \leq \psi(m' + n' - 2) \\
&< \psi(m' + n' - 2) + m' = h(m', n').
\end{aligned}
$$

In case (ii), if $m + n = m' + n'$ and $m \neq m'$, then

$$h(m, n) - m = \psi(m + n - 2) = \psi(m' + n' - 2) = h(m', n') - m',$$

whence $h(m, n) \neq h(m', n')$.

(b) Next we show that h is surjective.

Clearly $h(1, 1) = 1$. If $p \in \mathbb{N}$ with $p \geq 2$, we will find a pair $(m_p, n_p) \in \mathbb{N} \times \mathbb{N}$ with $h(m_p, n_p) = p$. Since $p < \psi(p)$, then the set $E_p := \{k \in \mathbb{N} : p \leq \psi(k)\}$ is nonempty.

Using the Well-Ordering Property 1.2.1, we let $k_p > 1$ be the least element in E_p. (This means that p lies in the k_pth diagonal.) Since $p \geq 2$, it follows from equation (1) that

$$\psi(k_p - 1) < p \leq \psi(k_p) = \psi(k_p - 1) + k_p.$$

Let $m_p := p - \psi(k_p - 1)$ so that $1 \leq m_p \leq k_p$, and let $n_p := k_p - m_p + 1$ so that $1 \leq n_p \leq k_p$ and $m_p + n_p - 1 = k_n$. Therefore,

$$h(m_p, n_p) = \psi(m_p + n_p - 2) + m_p = \psi(k_p - 1) + m_p = p.$$

Thus h is a bijection and $\mathbb{N} \times \mathbb{N}$ is denumerable.   Q.E.D.

The next result is crucial in proving Theorems 1.3.9 and 1.3.10.

B.3 Theorem   *If $A \subseteq \mathbb{N}$ and A is infinite, there exists a function $\varphi : \mathbb{N} \to A$ such that $\varphi(n + 1) > \varphi(n) \geq n$ for all $n \in \mathbb{N}$. Moreover, φ is a bijection of $\mathbb{N}$ onto A.*

Proof.   Since A is infinite, it is not empty. We will use the Well-Ordering Property 1.2.1 of $\mathbb{N}$ to give a recursive definition of φ.

Since $A \neq \emptyset$, there is a least element of A, which we define to be $\varphi(1)$; therefore, $\varphi(1) \geq 1$.

Since A is infinite, the set $A_1 := A\backslash\{\varphi(1)\}$ is not empty, and we define $\varphi(2)$ to be least element of A_1. Therefore $\varphi(2) > \varphi(1) \geq 1$, so that $\varphi(2) \geq 2$.

Suppose that φ has been defined to satisfy $\varphi(n + 1) > \varphi(n) \geq n$ for $n = 1, \cdots, k - 1$, whence $\varphi(k) > \varphi(k - 1) \geq k - 1$ so that $\varphi(k) \geq k$. Since the set A is infinite, the set

$$A_k := A\backslash\{\varphi(1), \cdots, \varphi(k)\}$$

is not empty and we define $\varphi(k + 1)$ to be the least element in A_k. Therefore $\varphi(k + 1) > \varphi(k)$, and since $\varphi(k) \geq k$, we also have $\varphi(k + 1) \geq k + 1$. Therefore, φ is defined on all of $\mathbb{N}$.

We claim that φ is an injection. If $m > n$, then $m = n + r$ for some $r \in \mathbb{N}$. If $r = 1$, then $\varphi(m) = \varphi(n + 1) > \varphi(n)$. Suppose that $\varphi(n + k) > \varphi(n)$; we will show that $\varphi(n + (k + 1)) > \varphi(n)$. Indeed, this follows from the fact that $\varphi(n + (k + 1)) = \varphi((n + k) + 1) > \varphi(n + k) > \varphi(n)$. Since $\varphi(m) > \varphi(n)$ whenever $m > n$, it follows that φ is an injection.

We claim that φ is a surjection of $\mathbb{N}$ onto A. If not, the set $\tilde{A} := A\backslash\varphi(\mathbb{N})$ is not empty, and we let p be the least element in $\tilde{A}$. We claim that p belongs to the set $\{\varphi(1), \cdots, \varphi(p)\}$. Indeed, if this is not true, then

$$p \in A \backslash \{\varphi(1), \cdots, \varphi(p)\} = A_p,$$

so that $\varphi(p + 1)$, being the least element in A_p, must satisfy $\varphi(p + 1) \leq p$. But this contradicts the fact that $\varphi(p + 1) > \varphi(p) \geq p$. Therefore $\tilde{A}$ is empty and φ is a surjection onto A.   Q.E.D.

B.4 Theorem   *If $A \subseteq \mathbb{N}$, then A is countable.*

Proof.   If A is finite, then it is countable, so it suffices to consider the case that A is infinite. In this case, Theorem B.3 implies that there exists a bijection φ of $\mathbb{N}$ onto A, so that A is denumerable and, therefore, countable.   Q.E.D.

1.3.9 Theorem   *Suppose that S and T are sets and that $T \subseteq S$.*

(a)   *If S is a countable set, then T is a countable set.*

(b)   *If T is an uncountable set, then S is an uncountable set.*

Proof.   (a) If S is a finite set, it follows from Theorem 1.3.5(a) that T is finite, and therefore countable. If S is denumerable, then there exists a bijection ψ of S onto $\mathbb{N}$. Since $\psi(S) \subseteq \mathbb{N}$, Theorem B.4 implies that $\psi(S)$ is countable. Since the restriction of ψ to T is a bijection onto $\psi(T)$ and $\psi(T) \subseteq \mathbb{N}$ is countable, it follows that T is also countable.

(b) This assertion is the contrapositive of the assertion in (a).   Q.E.D.

APPENDIX C

THE RIEMANN AND
LEBESGUE CRITERIA

We will give here proofs of the Riemann and Lebesgue Criteria for a function to be Riemann integrable. First we will give the Riemann Criterion, which is interesting in itself, and also leads to the more incisive Lebesgue Criterion.

C.1 Riemann Integrability Criterion   *Let $f : [a, b] \to \mathbb{R}$ be bounded. Then the following assertions are equivalent:*

(a) $f \in \mathcal{R}[a, b]$.

(b)   *For every $\varepsilon > 0$ there exists a partition $\mathcal{P}_\varepsilon$ such that if $\dot{\mathcal{P}}_1, \dot{\mathcal{P}}_2$ are any tagged partitions having the same subintervals as $\mathcal{P}_\varepsilon$, then*

(1)
$$|S(f; \dot{\mathcal{P}}_1) - S(f; \dot{\mathcal{P}}_2)| < \varepsilon.$$

(c)   *For every $\varepsilon > 0$ there exists a partition $\mathcal{P}_\varepsilon = \{I_i\}_{i=1}^n = \{[x_{i-1}, x_i]\}_{i=1}^n$ such that if $m_i := \inf\{f(x) : x \in I_i\}$ and $M_i := \sup\{f(x) : x \in I_i\}$ then*

(2)
$$\sum_{i=1}^n (M_i - m_i)(x_i - x_{i-1}) < 2\varepsilon.$$

Proof.   (a) $\Rightarrow$ (b)   Given $\varepsilon > 0$, let $\eta_\varepsilon > 0$ be as in the Cauchy Criterion 7.2.1, and let $\mathcal{P}_\varepsilon$ be any partition with $\|\mathcal{P}_\varepsilon\| < \eta_\varepsilon$. Then if $\dot{\mathcal{P}}_1, \dot{\mathcal{P}}_2$ are any tagged partitions with the same subintervals as $\mathcal{P}_\varepsilon$, then $\|\dot{\mathcal{P}}_1\| < \eta_\varepsilon$ and $\|\dot{\mathcal{P}}_2\| < \eta_\varepsilon$ and so (1) holds.

(b) $\Rightarrow$ (c)   Given $\varepsilon > 0$, let $\mathcal{P}_\varepsilon = \{I_i\}_{i=1}^n$ be a partition as in (b) and let m_i and M_i be as in the statement of (c). Since m_i is an infimum and M_i is a supremum, there exist points u_i and v_i in I_i with

$$f(u_i) < m_i + \frac{\varepsilon}{2(b - a)} \qquad \text{and} \qquad M_i - \frac{\varepsilon}{2(b - a)} < f(v_i),$$

so that we have

$$M_i - m_i < f(v_i) - f(u_i) + \frac{\varepsilon}{(b - a)} \qquad \text{for} \quad i = 1, \cdots, n.$$

If we multiply these inequalities by $(x_i - x_{i-1})$ and sum, we obtain

$$\sum_{i=1}^n (M_i - m_i)(x_i - x_{i-1}) < \sum_{i=1}^n (f(v_i) - f(u_i))(x_i - x_{i-1}) + \varepsilon.$$

We let $\dot{\mathcal{Q}}_1 := \{(I_i, u_i)\}_{i=1}^n$ and $\dot{\mathcal{Q}}_2 := \{(I_i, v_i)\}_{i=1}^n$, so that these tagged partitions have the same subintervals as $\mathcal{P}_\varepsilon$ does. Also, the sum on the right side equals $S(f; \dot{\mathcal{Q}}_2) - S(f; \dot{\mathcal{Q}}_1)$. Hence it follows from (1) that inequality (2) holds.

(c) $\Rightarrow$ (a)   Define the step functions α_ε and ω_ε on $[a, b]$ by

$$\alpha_\varepsilon(x) := m_i \qquad \text{and} \qquad \omega_\varepsilon(x) := M_i \qquad \text{for} \quad x \in (x_{i-1}, x_i),$$

and $\alpha_\varepsilon(x_i) := f(x_i) =: \omega_\varepsilon(x_i)$ for $i = 0, 1, \cdots, n$; then $\alpha_\varepsilon(x) \le f(x) \le \omega_\varepsilon(x)$ for $x \in [a, b]$. Since α_ε and ω_ε are step functions, they are Riemann integrable and

$$\int_a^b \alpha_\varepsilon = \sum_{i=1}^n m_i(x_i - x_{i-1}) \qquad \text{and} \qquad \int_a^b \omega_\varepsilon = \sum_{i=1}^n M_i(x_i - x_{i-1}).$$

Therefore it follows that

$$\int_a^b (\omega_\varepsilon - \alpha_\varepsilon) = \sum_{i=1}^n (M_i - m_i)(x_i - x_{i-1}).$$

If we apply (2), we have that

$$\int_a^b (\omega_\varepsilon - \alpha_\varepsilon) < 2\varepsilon.$$

Since $\varepsilon > 0$ is arbitrary, the Squeeze Theorem implies that $f \in \mathcal{R}[a, b]$.   Q.E.D.

We have already seen that every continuous function on $[a, b]$ is Riemann integrable. We also saw in Example 7.1.6 that Thomae's function is Riemann integrable. Since Thomae's function has a countable set of points of discontinuity, it is evident that continuity is not a necessary condition for Riemann integrability. Indeed, it is reasonable to ask "how discontinuous" a function may be, yet still be Riemann integrable. The Riemann Criterion throws some light on that question in showing that sums of the form (2) must be arbitrarily small. Since the terms $(M_i - m_i)(x_i - x_{i-1})$ in this sum are all ≥ 0, it follows that each of these terms must be small. Such a term will be small if (i) the difference $M_i - m_i$ is small (which will be the case if the function is continuous on the interval $[x_{i-1}, x_i]$), or if (ii) an interval where the difference $M_i - m_i$ is not small has small length.

The Lebesgue Criterion, which we will discuss next, makes these ideas more precise. But first it is convenient to have the notion of the oscillation of a function.

C.2 Definition   Let $f : A \to \mathbb{R}$ be a bounded function. If $S \subseteq A \subseteq \mathbb{R}$, we define the **oscillation of f on S** to be

(3) $$W(f; S) := \sup\{|f(x) - f(y)| : x, y \in S\}.$$

It is easily seen that we can also write

$$W(f; S) = \sup\{f(x) - f(y) : x, y \in S\}$$
$$= \sup\{f(x) : x \in S\} - \inf\{f(x) : x \in S\}.$$

It is also trivial that if $S \subseteq T \subseteq A$, then

$$0 \le W(f; S) \le W(f; T) \le 2 \cdot \sup\{|f(x)| : x \in A\}.$$

If $r > 0$, we recall that the r-neighborhood of $c \in A$ is the set

$$V_r(c) := \{x \in A : |x - c| < r\}.$$

C.3 Definition   If $c \in A$, we define the **oscillation of f at c** by

(4) $$w(f; c) := \inf\{W(f; V_r(c)) : r > 0\} = \lim_{r \to 0+} W(f; V_r(c)).$$

Since $r \mapsto W(f; V_r(c))$ is an increasing function for $r > 0$, this right-hand limit exists and equals the indicated infimum.

C.4 Lemma *If $f : A \to \mathbb{R}$ is bounded and $c \in A$, then f is continuous at c if and only if the oscillation $w(f; c) = 0$.*

Proof. ($\Rightarrow$) If f is continuous at c, given $\varepsilon > 0$ there exists $\delta > 0$ such that if $x \in V_r(c)$, then $|f(x) - f(c)| < \varepsilon/2$. Therefore, if $x, y \in V_r(c)$, we have $|f(x) - f(y)| < \varepsilon$, whence $0 \le w(f; c) \le W(f; V_r(c)) \le \varepsilon$. Since $\varepsilon > 0$ is arbitrary, this implies that $w(f; c) = 0$.
    ($\Leftarrow$) If $w(f; c) = 0$ and $\varepsilon > 0$, there exists $s > 0$ with $W(f; V_s(c)) < \varepsilon$. Thus, if $|x - c| < s$ then $|f(x) - f(c)| < \varepsilon$, and f is continuous at c.   Q.E.D.

We will now give the details of the proof of the Lebesgue Integrability Criterion. First we recall the statement of the theorem.

Lebesgue's Integrability Criterion *A bounded function $f : [a, b] \to \mathbb{R}$ is Riemann integrable if and only if it is continuous almost everywhere on $[a, b]$.*

Proof. ($\Rightarrow$) Let $\varepsilon > 0$ be given and, for each $k \in \mathbb{N}$, let $H_k := \{x \in [a, b] : w(f; x) > 1/2^k\}$. We will show that H_k is contained in the union of a finite number of intervals having total length $< \varepsilon/2^k$.
    By the Riemann Criterion, there is a partition $\mathcal{P}_k = \{[x_{i-1}^k, x_i^k]\}_{i=1}^{n(k)}$ such that if m_i^k (respectively, M_i^k) is the infimum (resp., supremum) of f on the interval $[x_{i-1}^k, x_i^k]$, then

$$\sum_{i=1}^{n(k)} (M_i^k - m_i^k)(x_i^k - x_{i-1}^k) < \varepsilon/4^k.$$

If $x \in H_k \cap (x_{i-1}^k, x_i^k)$, there exists $r > 0$ such that $V_r(x) \subseteq (x_{i-1}^k, x_i^k)$, whence

$$1/2^k \le w(f; x) \le W(f; V_r(x)) \le M_i^k - m_i^k.$$

If we denote a summation over those i with $H_k \cap (x_{i-1}^k, x_i^k) \ne \emptyset$ by $\sum'$, then

$$(1/2^k) \sum{}' (x_i^k - x_{i-1}^k) \le \sum_{i=1}^{n(k)} (M_i^k - m_i^k)(x_i^k - x_{i-1}^k) \le \varepsilon/4^k,$$

whence it follows that

$$\sum{}' (x_i^k - x_{i-1}^k) \le \varepsilon/2^k.$$

Since H_k differs from the union of sets $H_k \cap (x_i^k - x_{i-1}^k)$ by at most a finite number of the partition points, we conclude that H_k is contained in the union of a finite number of intervals with total length $< \varepsilon/2^k$.
    Finally, since $D := \{x \in [a, b] : w(f; x) > 0\} = \bigcup_{k=1}^{\infty} H_k$, it follows that the set D of points of discontinuity of $f \in \mathcal{R}[a, b]$ is a null set.
    ($\Leftarrow$) Let $|f(x)| \le M$ for $x \in [a, b]$ and suppose that the set D of points of discontinuity of f is a null set. Then, given $\varepsilon > 0$ there exists a countable set $\{J_k\}_{k=1}^{\infty}$ of open intervals with $D \subseteq \bigcup_{k=1}^{\infty} J_k$ and $\sum_{k=1}^{\infty} l(J_k) < \varepsilon/2M$. Following R. A. Gordon, we will define a gauge on $[a, b]$ that will be useful.

(i) If $t \notin D$, then f is continuous at t and there exists $\delta(t) > 0$ such that if $x \in V_{\delta(t)}(t)$ then $|f(x) - f(t)| < \varepsilon/2$, whence

$$0 \le M_t - m_t := \sup\{f(x) : x \in V_{\delta(t)}(t)\} - \inf\{f(x) : x \in V_{\delta(t)}(t)\} \le \varepsilon.$$

(ii) If $t \in D$, we choose $\delta(t) > 0$ such that $V_{\delta(t)}(t) \subseteq J_k$ for some k. For these values of t, we have $0 \le M_t - m_t \le 2M$.

Thus we have defined a gauge δ on $[a, b]$. If $\dot{\mathcal{P}} = \{([x_{i-1}, x_i], t_i)\}_{i=1}^n$ is a δ-fine partition of $[a, b]$, we divide the indices i into two disjoint sets

$$S_c := \{i : t_i \notin D\} \qquad \text{and} \qquad S_d := \{i : t_i \in D\}.$$

If $\dot{\mathcal{P}}$ is δ-fine, we have $[x_{i-1}, x_i] \subseteq V_{\delta(t_i)}(t_i)$, whence it follows that $M_i - m_i \le M_{t_i} - m_{t_i}$. Consequently, if $i \in S_c$ then $M_i - m_i \le \varepsilon$, while if $i \in S_d$ we have $M_i - m_i \le 2M$. However, the collection of intervals $[x_{i-1}, x_i]$ with $i \in S_d$ are contained in the union of the intervals $\{J_k\}$ whose total length is $< \varepsilon/2M$. Therefore

$$\sum_{i=1}^n (M_i - m_i)(x_i - x_{i-1})$$

$$= \sum_{i \in S_c} (M_i - m_i)(x_i - x_{i-1}) + \sum_{i \in S_d} (M_i - m_i)(x_i - x_{i-1})$$

$$\le \sum_{i \in S_c} \varepsilon(x_i - x_{i-1}) + \sum_{i \in S_d} 2M(x_i - x_{i-1})$$

$$\le \varepsilon(b - a) + 2M \cdot (\varepsilon/2M) \le \varepsilon(b - a + 1).$$

Since $\varepsilon > 0$ is arbitrary, we conclude that $f \in \mathcal{R}[a, b]$.   Q.E.D.

APPENDIX D

APPROXIMATE INTEGRATION

We will supply here the proofs of Theorems 7.4.3, 7.4.6 and 7.4.8. We will not repeat the statement of these results, and we will use the notations introduced in Section 7.4 and refer to numbered equations there. It will be seen that some important results from Chapters 5 and 6 are used in these proofs.

Proof of Theorem 7.4.3. If $k = 1, 2, \cdots, n$, let $a_k := a + (k-1)h$ and let $\varphi_k : [0, h] \to \mathbb{R}$ be defined by

$$\varphi_k(t) := \tfrac{1}{2} t \left[f(a_k) + f(a_k + t) \right] - \int_{a_k}^{a_k + t} f(x)\, dx$$

for $t \in [0, h]$. Note that $\varphi_k(0) = 0$ and that (by Theorem 7.3.6)

$$\varphi_k'(t) = \tfrac{1}{2} \left[f(a_k) + f(a_k + t) \right] + \tfrac{1}{2} t f'(a_k + t) - f(a_k + t)$$
$$= \tfrac{1}{2} \left[f(a_k) - f(a_k + t) \right] + \tfrac{1}{2} t f'(a_k + t).$$

Consequently $\varphi_k'(0) = 0$ and

$$\varphi_k''(t) = -\tfrac{1}{2} f'(a_k + t) + \tfrac{1}{2} f'(a_k + t) + \tfrac{1}{2} t f''(a_k + t)$$
$$= \tfrac{1}{2} t f''(a_k + t).$$

Now let A, B be defined by

$$A := \inf\{ f''(x) : x \in [a, b] \}, \qquad B := \sup\{ f''(x) : x \in [a, b] \}$$

so that we have $\tfrac{1}{2} At \le \varphi_k''(t) \le \tfrac{1}{2} Bt$ for $t \in [0, h]$, $k = 1, 2, \cdots, n$. Integrating and applying Theorem 7.3.1, we obtain (since $\varphi_k'(0) = 0$) that $\tfrac{1}{4} At^2 \le \varphi_k'(t) \le \tfrac{1}{4} Bt^2$ for $t \in [0, h]$, $k = 1, 2, \cdots, n$. Integrating again and taking $t = h$, we obtain (since $\varphi_k(0) = 0$) that

$$\tfrac{1}{12} Ah^3 \le \varphi_k(h) \le \tfrac{1}{12} Bh^3$$

for $k = 1, 2, \cdots, n$. If we add these inequalities and note that

$$\sum_{k=1}^{n} \varphi_k(h) = T_k(f) - \int_a^b f(x)\, dx,$$

we conclude that $\tfrac{1}{12} Ah^3 n \le T_n(f) - \int_a^b f(x)\, dx \le \tfrac{1}{12} Bh^3 n$. Since $h = (b-a)/n$, we have

$$\tfrac{1}{12} A(b-a)h^2 \le T_n(f) - \int_a^b f(x)\, dx \le \tfrac{1}{12} B(b-a)h^2.$$

Since f'' is continuous on $[a, b]$, it follows from the definitions of A and B and Bolzano's Intermediate Value Theorem 5.3.7 that there exists a point c in $[a, b]$ such that equation (4) in Section 7.4 holds.   Q.E.D.

Proof of Theorem 7.4.6. If $k = 1, 2, \cdots, n$, let $c_k := a + (k - \frac{1}{2})h$, and $\psi_k : [0, \frac{1}{2}h] \to \mathbb{R}$ be defined by

$$\psi_k(t) := \int_{c_k-t}^{c_k+t} f(x)\,dx - f(c_k)2t$$

for $t \in [0, \frac{1}{2}h]$. Note that $\psi_k(0) = 0$ and that since

$$\psi_k(t) := \int_{c_k}^{c_k+t} f(x)\,dx - \int_{c_k}^{c_k-t} f(x)\,dx - f(c_k)2t,$$

we have

$$\psi_k'(t) = f(c_k + t) - f(c_k - t)(-1) - 2f(c_k)$$
$$= \big[f(c_k + t) + f(c_k - t)\big] - 2f(c_k).$$

Consequently $\psi_k'(0) = 0$ and

$$\psi_k''(t) = f'(c_k + t) + f'(c_k - t)(-1)$$
$$= f'(c_k + t) - f'(c_k - t).$$

By the Mean Value Theorem 6.2.4, there exists a point $c_{k,t}$ with $|c_k - c_{k,t}| < t$ such that $\psi_k''(t) = 2tf''(c_{k,t})$. If we let A and B be as in the proof of Theorem 7.4.3, we have $2tA \le \psi_k''(t) \le 2tB$ for $t \in [0, h/2]$, $k = 1, 2, \cdots, n$. It follows as before that

$$\tfrac{1}{3}At^3 \le \psi_k(t) \le \tfrac{1}{3}Bt^3$$

for all $t \in [0, \frac{1}{2}h]$, $k = 1, 2, \cdots, n$. If we put $t = \frac{1}{2}h$, we get

$$\frac{1}{24}Ah^3 \le \psi_k\left(\tfrac{1}{2}h\right) \le \frac{1}{24}Bh^3.$$

If we add these inequalities and note that

$$\sum_{k=1}^{n} \psi_k\left(\tfrac{1}{2}h\right) = \int_a^b f(x)\,dx - M_n(f),$$

we conclude that

$$\frac{1}{24}Ah^3n \le \int_a^b f(x)\,dx - M_n(f) \le \frac{1}{24}Bh^3n.$$

If we use the fact that $h = (b - a)/n$ and apply Bolzano's Intermediate Value Theorem 5.3.7 to f'' on $[a, b]$ we conclude that there exists a point $\gamma \in [a, b]$ such that (7) in Section 7.4 holds.   Q.E.D.

Proof of Theorem 7.4.8. If $k = 0, 1, 2, \cdots, \frac{1}{2}n - 1$, let $c_k := a + (2k + 1)h$, and let $\varphi_k : [0, h] \to \mathbb{R}$ be defined by

$$\varphi_k(t) := \tfrac{1}{3}t\big[f(c_k - t) + 4f(c_k) + f(c_k + t)\big] - \int_{c_k-t}^{c_k+t} f(x)\,dx.$$

Evidently $\varphi_k(0) = 0$ and

$$\varphi_k'(t) = \tfrac{1}{3}t\big[-f'(c_k - t) + f'(c_k + t)\big] - \tfrac{2}{3}\big[f(c_k - t) - 2f(c_k) + f(c_k + t)\big],$$

so that $\varphi_k'(0) = 0$ and

$$\varphi_k''(t) = \tfrac{1}{3}t\big[f''(c_k - t) + f''(c_k + t)\big] - \tfrac{1}{3}\big[-f'(c_k - t) + f'(c_k + t)\big],$$

so that $\varphi_k''(0) = 0$ and

$$\varphi_k'''(t) = \tfrac{1}{3} t \left[f'''(c_k + t) - f'''(c_k - t) \right].$$

Hence it follows from the Mean Value Theorem 6.2.4 that there is a $\gamma_{k,t}$ with $|c_k - \gamma_{k,t}| \le t$ such that $\varphi_k'''(t) = \tfrac{2}{3} t^2 f^{(4)}(\gamma_{k,t})$. If we let A and B be defined by

$$A := \inf\{ f^{(4)}(x) : x \in [a, b]\} \qquad \text{and} \qquad B := \sup\{ f^{(4)}(x) : x \in [a, b]\},$$

then we have

$$\tfrac{2}{3} A t^2 \le \varphi_k'''(t) \le \tfrac{2}{3} B t^2$$

for $t \in [0, h]$, $k = 0, 1, \cdots, \tfrac{1}{2} n - 1$. After three integrations, this inequality becomes

$$\frac{1}{90} A t^5 \le \varphi_k(t) \le \frac{1}{90} B t^5$$

for all $t \in [0, h]$, $k = 0, 1, \cdots, \tfrac{1}{2} n - 1$. If we put $t = h$, we get

$$\frac{1}{90} A h^5 \le \varphi_k(h) \le \frac{1}{90} B h^5$$

for $k = 0, 1, \cdots, \tfrac{1}{2} n - 1$. If we add these $\tfrac{1}{2} n$ inequalities and note that

$$\sum_{k=0}^{\frac{1}{2}n-1} \varphi_k(h) = S_n(f) - \int_a^b f(x)\, dx,$$

we conclude that

$$\frac{1}{90} A h^5 \frac{n}{2} \le S_n(f) - \int_a^b f(x)\, dx \le \frac{1}{90} B h^5 \frac{n}{2}.$$

Since $h = (b - a)/n$, it follows from Bolzano's Intermediate Value Theorem 5.3.7 (applied to $f^{(4)}$) that there exists a point $c \in [a, b]$ such that the relation (10) in Section 7.4 holds.

Q.E.D.

TWO EXAMPLES

In this appendix we will give an example of a continuous function that has a derivative at no point and of a continuous curve in $\mathbb{R}^2$ whose range contains the entire unit square of $\mathbb{R}^2$. Both proofs use the Weierstrass M-Test 9.4.6.

A Continuous Nowhere Differentiable Function

The example we will give is a modification of one due to B. L. van der Waerden in 1930. Let $f_0 : \mathbb{R} \to \mathbb{R}$ be defined by $f_0(x) := \text{dist}(x, \mathbb{Z}) = \inf\{|x - k| : k \in \mathbb{Z}\}$, so that f_0 is a continuous "sawtooth" function whose graph consists of lines with slope ± 1 on the intervals $[k/2, (k + 1)/2]$, $k \in \mathbb{Z}$. For each $m \in \mathbb{N}$, let $f_m(x) := (1/4^m) f_0(4^m x)$, so that f_m is also a continuous sawtooth function whose graph consists of lines with slope ± 1 and with $0 \le f_m(x) \le 1/(2 \cdot 4^m)$. (See Figure E.1.)

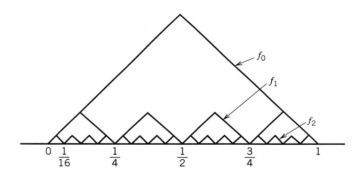

Figure E.1   Graphs of f_0, f_1, and f_2.

We now define $g : \mathbb{R} \to \mathbb{R}$ by $g(x) := \sum_{m=0}^{\infty} f_m(x)$. The Weierstrass M-Test implies that the series is uniformly convergent on $\mathbb{R}$; hence g is continuous on $\mathbb{R}$. We will now show that g is *not* differentiable at any point of $\mathbb{R}$.

Fix $x \in \mathbb{R}$. For each $n \in \mathbb{N}$, let $h_n := \pm 1/4^{n+1}$, with the sign chosen so that both $4^n x$ and $4^n(x + h_n)$ lie in the same interval $[k/2, (k + 1)/2]$. Since f_0 has slope ± 1 on this interval, then

$$\varepsilon_n := \frac{f_n(x + h_n) - f_n(x)}{h_n} = \frac{f_0(4^n x + 4^n h_n) - f_0(4^n x)}{4^n h_n} = \pm 1.$$

In fact if $m < n$, then the graph of f_m also has slope ± 1 on the interval between x and $x + h_n$ and so

$$\varepsilon_m := \frac{f_m(x + h_n) - f_m(x)}{h_n} = \pm 1 \qquad \text{for} \quad m < n.$$

On the other hand, if $m > n$, then $4^m(x + h_n) - 4^m x = \pm 4^{m-n-1}$ is an integer, and since f_0 has period equal to 1, it follows that

$$f_m(x + h_n) - f_m(x) = 0.$$

Consequently, we have

$$\frac{g(x + h_n) - g(x)}{h_n} = \sum_{m=0}^{n} \frac{f_m(x + h_n) - f_m(x)}{h_n} = \sum_{m=0}^{n} \varepsilon_m,$$

whence the difference quotient $(g(x + h_n) - g(x))/h_n$ is an odd integer if n is even, and an even integer if n is odd. Therefore, the limit

$$\lim_{h \to 0} \frac{g(x + h) - g(x)}{h}$$

does not exist, so g is not differentiable at the arbitrary point $x \in \mathbb{R}$.

A Space-Filling Curve

We will now give an example of a space-filling curve that was constructed by I. J. Schoenberg in 1936. Let $\varphi : \mathbb{R} \to \mathbb{R}$ be the continuous, even function with period 2 given by

$$\varphi(t) := \begin{cases} 0 & \text{for} \quad 0 \le t \le 1/3, \\ 3t - 1 & \text{for} \quad 1/3 < t < 2/3, \\ 1 & \text{for} \quad 2/3 \le t \le 1. \end{cases}$$

(See Figure E.2.) For $t \in [0, 1]$, we define the functions

$$f(t) := \sum_{k=0}^{\infty} \frac{\varphi(3^{2k}t)}{2^{k+1}} \qquad \text{and} \qquad g(t) := \sum_{k=0}^{\infty} \frac{\varphi(3^{2k+1}t)}{2^{k+1}}.$$

Since $0 \le \varphi(x) \le 1$ and is continuous, the Weierstrass M-Test implies that f and g are continuous on $[0, 1]$; moreover, $0 \le f(t) \le 1$ and $0 \le g(t) \le 1$. We will now show that an arbitrary point (x_0, y_0) in $[0, 1] \times [0, 1]$ is the image under (f, g) of some point $t_0 \in [0, 1]$. Indeed, let x_0 and y_0 have the binary (= base 2) expansions:

$$x_0 = \frac{a_0}{2} + \frac{a_2}{2^2} + \frac{a_4}{2^3} + \cdots \qquad \text{and} \qquad y_0 = \frac{a_1}{2} + \frac{a_3}{2^2} + \frac{a_5}{2^3} + \cdots,$$

where each a_k equals 0 or 1. It will be shown that $x_0 = f(t_0)$ and $y_0 = g(t_0)$, where t_0 has the ternary (= base 3) expansion

$$t_0 = \sum_{k=0}^{\infty} \frac{2a_k}{3^{k+1}} = \frac{2a_0}{3} + \frac{2a_1}{3^2} + \frac{2a_2}{3^3} + \frac{2a_3}{3^4} + \cdots.$$

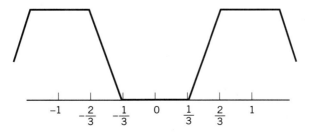

Figure E.2   Graph of φ.

First, we note that the above formula does yield a number in $[0, 1]$. We also note that if $a_0 = 0$, then $0 \le t_0 \le 1/3$ so that $\varphi(t_0) = 0$, and if $a_0 = 1$, then $2/3 \le t_0 \le 1$ so that $\varphi(t_0) = 1$; therefore, in both cases $\varphi(a_0) = a_0$. Similarly, it is seen that for each $n \in \mathbb{N}$ there exists $m_n \in \mathbb{N}$ such that

$$3^n t_0 = 2m_n + \frac{2a_n}{3} + \frac{2a_{n+1}}{3^2} + \cdots,$$

whence it follows from the fact that φ has period 2 that $\varphi(3^n t_0) = a_n$. Finally, we conclude that

$$f(t_0) = \sum_{k=0}^{\infty} \frac{\varphi(3^{2k} t_0)}{2^{k+1}} = \sum_{k=0}^{\infty} \frac{a_{2k}}{2^{k+1}} = x_0,$$

and

$$g(t_0) = \sum_{k=0}^{\infty} \frac{\varphi(3^{2k+1} t_0)}{2^{k+1}} = \sum_{k=0}^{\infty} \frac{a_{2k+1}}{2^{k+1}} = y_0.$$

Therefore $x_0 = f(t_0)$ and $y_0 = g(t_0)$ as claimed.

REFERENCES

Apostol, T. M., *Mathematical Analysis*, Second Edition, Addison-Wesley, Reading, MA, 1974.

Bartle, R. G., *The Elements of Real Analysis*, Second Edition, referred to as [ERA], John Wiley & Sons, New York, 1976.

___, *The Elements of Integration and Lebesgue Measure*, Wiley Classics Edition, John Wiley & Sons, New York, 1995.

___, *Return to the Riemann Integral*, Amer. Math. Monthly, **103** (1996), 625–632.

___, *A Modern Theory of Integration*, referred to as [MTI], Grad. Studies in Math., Amer. Math. Society, Providence, RI, 2000.

Bartle, R. G. and D. R. Sherbert, *Introduction to Real Analysis*, Second Edition, John Wiley & Sons, New York, 1992.

Barwise, J. and J. Etchemendy, *The Language of First Order Logic*, Univ. of Chicago Press, Chicago, 1990.

Birkhoff, G. and S. MacLane, *A Survey of Modern Algebra*, Fourth Edition, Macmillan Publishing Co., New York, 1977.

Boas, R. P., Jr., *A Primer of Real Functions*, Fourth Edition, Carus Monograph No. 13, Math. Assn. Amer., Washington, D.C. 1996.

DePree, J. D. and C. W. Swartz, *Introduction to Real Analysis*, John Wiley & Sons, New York, 1988.

Gelbaum, B. R. and J. M. H. Olmsted, *Counterexamples in Analysis*, Holden-Day, San Francisco, 1964.

Gordon, R. A., *The Integrals of Lebesgue, Denjoy, Perron, and Henstock*, Grad. Studies in Math., vol. 4, Amer. Math. Soc., Providence, 1994.

___, *The Use of Tagged Partitions in Elementary Real Analysis*, Amer. Math. Monthly, **105** (1998), 105–117 and 886.

Hawkins, T., *Lebesgue's Theory of Integration. Its Origins and Developments*, Univ. of Wisconsin Press, Madison, WI, 1970. Reprint, Amer. Math. Soc., Chelsea Series, 1998.

Kline, M., *Mathematical Thought from Ancient to Modern Times*, Oxford Univ. Press, New York, 1972.

McLeod, R. M., *The Generalized Riemann Integral*, Carus Monograph, No. 20, Math. Assn. Amer., Washington, D.C., 1980.

Wilder, R. L., *The Foundations of Mathematics*, Second Edition, John Wiley & Sons, New York, 1965.

PHOTO CREDITS

Chapter 3
Page 74: Courtesy The New York Public Library. Page 93: Corbis-Bettmann.

Chapter 4
Page 97: Courtesy David Eugene Smith Collection, Columbia University.

Chapter 5
Page 119: Courtesy The New York Public Library.

Chapter 6
Page 157: Courtesy National Portrait Gallery, London.

Chapter 7
Page 193: Baldwin Ward/Corbis-Bettmann.

Chapter 10
Page 275: Courtesy Patrick Muldowney, University of Ulster.

HINTS FOR SELECTED EXERCISES

Reader: Do not look at these hints unless you are stymied. However, after putting a considerable amount of thought into a problem, sometimes just a little hint is all that is needed. Many of the exercises call for proofs, and there is usually no single approach that is correct, so even if you have a totally different argument, yours may be correct. Very few of the following hints give much detail, and some may seem downright cryptic at first. Somewhat more detail is presented for the earlier material.

Section 1.1

1. Show that if $A \subseteq B$, then $A = A \cap B$. Next show that if $A = A \cap B$, then $A \subseteq B$.

2. Show that if $x \in A \setminus (B \cap C)$, then $x \in (A \setminus B) \cup (A \setminus C)$. Next show that if $y \in (A \setminus B) \cup (A \setminus C)$, then $y \in A \setminus (B \cap C)$. Since the sets $A \setminus (B \cap C)$ and $A \setminus (B \cap C)$ contain the same elements, they are equal.

5. (a) $A_1 \cap A_2 = \{6, 12, 18, 24, \cdots\} = \{6k : k \in \mathbb{N}\} = A_5$.
   (b) $\bigcup A_n = \mathbb{N} \setminus \{1\}$ and $\bigcap A_n = \emptyset$.

7. No. For example, both $(0, 1)$ and $(0, -1)$ belong to C.

9. (a) $f(E) = [2, 3]$, so $h(E) = g(f(E)) = g([2, 3]) = [4, 9]$.
   (b) $g^{-1}(G) = [-2, 2]$, so $h^{-1}(G) = [-4, 0]$.

13. If $x \in f^{-1}(G) \cap f^{-1}(H)$, then $x \in f^{-1}(G)$ and $x \in f^{-1}(H)$, so that $f(x) \in G$ and $f(x) \in H$. Then $f(x) \in G \cap H$, and hence $x \in f^{-1}(G \cap H)$. This shows that $f^{-1}(G) \cap f^{-1}(H) \subseteq f^{-1}(G \cap H)$.

15. One possibility is $f(x) := (x - a)/(b - a)$.

19. If $g(f(x_1)) = g(f(x_2))$, then $f(x_1) = f(x_2)$, so that $x_1 = x_2$, which implies that $g \circ f$ is injective. If $w \in C$, there exists $y \in B$ such that $g(y) = w$, and there exists $x \in A$ such that $f(x) = y$. Then $g(f(x)) = w$, so that $g \circ f$ is surjective. Thus $g \circ f$ is a bijection.

20. (a) If $f(x_1) = f(x_2)$, then $g(f(x_1)) = g(f(x_2))$, which implies $x_1 = x_2$, since $g \circ f$ is injective. Thus f is injective.

Section 1.2

1. Note that $1/(1 \cdot 2) = 1/(1 + 1)$. Also $k/(k + 1) + 1/[(k + 1)(k + 2)] = (k + 1)/(k + 2)$.

2. $[\frac{1}{2}k(k + 1)]^2 + (k + 1)^3 = [\frac{1}{2}(k + 1)(k + 2)]^2$.

4. $\frac{1}{3}(4k^3 - k) + (2k + 1)^2 = \frac{1}{3}[4(k + 1)^3 - (k + 1)]$.

6. $(k + 1)^3 + 5(k + 1) = (k^3 + 5k) + 3k(k + 1) + 6$ and $k(k + 1)$ is always even.

8. $5^{k+1} - 4(k + 1) - 1 = 5 \cdot 5^k - 4k - 5 = (5^k - 4k - 1) + 4(5^k - 1)$.

13. If $k < 2^k$, then $k + 1 < 2^k + 1 < 2^k + 2^k = 2(2^k) = 2^{k+1}$.

16. It is true for $n = 1$ and $n \geq 5$, but false for $n = 2, 3, 4$.

18. $\sqrt{k} + 1/\sqrt{k + 1} = (\sqrt{k}\sqrt{k + 1} + 1)/\sqrt{k + 1} > (k + 1)/\sqrt{k + 1} = \sqrt{k + 1}$.

Section 1.3

1. Use Exercise 1.1.19 (= Exercise 19 of Section 1.1).

2. Part (b) Let f be a bijection of $\mathbb{N}_m$ onto A and let $C = \{f(k)\}$ for some $k \in \mathbb{N}_m$. Define g on $\mathbb{N}_{m-1}$ by $g(i) := f(i)$ for $i = 1, \cdots, k-1$, and $g(i) := f(i+1)$ for $i = k, \cdots, m-1$. Then g is a bijection of $\mathbb{N}_{m-1}$ onto $A \setminus C$.

3. (a) There are $6 = 3 \cdot 2 \cdot 1$ different injections of S into T.
   (b) There are 3 surjections that map a into 1, and there are 3 other surjections that map a into 2.

7. If T_1 is denumerable, take $T_2 = \mathbb{N}$. If f is a bijection of T_1 onto T_2, and if g is a bijection of T_2 onto $\mathbb{N}$, then (by Exercise 1.1.19) $g \circ f$ is a bijection of T_1 onto $\mathbb{N}$, so that T_1 is denumerable.

9. If $S \cap T = \emptyset$ and $f : \mathbb{N} \to S$, $g : \mathbb{N} \to T$ are bijections onto S and T, respectively, let $h(n) := f((n+1)/2)$ if n is odd and $h(n) := g(n/2)$ if n is even.

10. (a) $\mathcal{P}(\{1, 2\}) = \{\emptyset, \{1\}, \{2\}, \{1, 2\}\}$ has $2^2 = 4$ elements.
    (c) $\mathcal{P}(\{1, 2, 3, 4\})$ has $2^4 = 16$ elements.

11. Let $S_{n+1} := \{x_1, \cdots, x_n, x_{n+1}\} = S_n \cup \{x_{n+1}\}$ have $n+1$ elements. Then a subset of S_{n+1} either (i) contains x_{n+1}, or (ii) does not contain x_{n+1}. There is a total of $2^n + 2^n = 2 \cdot 2^n = 2^{n+1}$ subsets of S_{n+1}.

12. For each $m \in \mathbb{N}$, the collection of all subsets of $\mathbb{N}_m$ is finite. Note that $\mathcal{F}(\mathbb{N}) = \bigcup_{m=1}^{\infty} \mathcal{P}(\mathbb{N}_m)$.

Section 2.1

1. (a) Justify the steps in: $b = 0 + b = (-a + a) + b = -a + (a + b) = -a + 0 = -a$.
   (c) Apply (a) to the equation $a + (-1)a = a(1 + (-1)) = a \cdot 0 = 0$.

2. (a) $-(a + b) = (-1)(a + b) = (-1)a + (-1)b = (-a) + (-b)$.
   (c) Note that $(-a)(-(1/a)) = a(1/a) = 1$.

3. (a) $3/2$ (b) $0, 2$
   (c) $2, -2$ (d) $1, -2$

6. Note that if $q \in \mathbb{Z}$ and if $3q^2$ is even, then q^2 is even, so that q is even.

7. If $p \in \mathbb{N}$, then there are three possibilities: for some $m \in \mathbb{N} \cup \{0\}$, (i) $p = 3m$, (ii) $p = 3m + 1$, or (iii) $p = 3m + 2$.

10. (a) If $c = d$, then 2.1.7(b) implies $a + c < b + d$. If $c < d$, then $a + c < b + c < b + d$.

13. If $a \neq 0$, then 2.1.8(a) implies that $a^2 > 0$; since $b^2 \geq 0$, it follows that $a^2 + b^2 > 0$.

15. (a) If $0 < a < b$, then 2.1.7(c) implies that $0 < a^2 < ab < b^2$. Then by Example 2.1.13(a), we infer that $a = \sqrt{a^2} < \sqrt{ab} < \sqrt{b^2} = b$.

16. (a) $\{x : x > 4 \text{ or } x < -1\}$.   (b) $\{x : 1 < x < 2 \text{ or } -2 < x < -1\}$.
    (c) $\{x : -1 < x < 0 \text{ or } x > 1\}$.   (d) $\{x : x < 0 \text{ or } x > 1\}$.

19. The inequality is equivalent to $0 \leq a^2 - 2ab + b^2 = (a - b)^2$.

20. (a) Use 2.1.7(c).

21. (a) Let $S := \{n \in \mathbb{N} : 0 < n < 1\}$. If S is not empty, the Well-Ordering Property of $\mathbb{N}$ implies there is a least element m in S. However, $0 < m < 1$ implies that $0 < m^2 < m$, and since m^2 is also in S, this is a contradiction of the fact that m is the least element of S.

22. (a) Let $x := c - 1 > 0$ and apply Bernoulli's Inequality 2.1.13(c).

24. (a) If $m > n$, then $k := m - n \in \mathbb{N}$, and $c^k \geq c > 1$ which implies that $c^m > c^n$. Conversely, the hypotheses that $c^m > c^n$ and $m \leq n$ lead to a contradiction.

25. Let $b := c^{1/mn}$ and show that $b > 1$. Exercise 24(a) implies that $c^{1/n} = b^m > b^n = c^{1/m}$ if and only if $m > n$.

26. Fix $m \in \mathbb{N}$ and use Mathematical Induction to prove that $a^{m+n} = a^m a^n$ and $(a^m)^n = a^{mn}$ for all $n \in \mathbb{N}$. Then, for a given $n \in \mathbb{N}$, prove that the equalities are valid for all $m \in \mathbb{N}$.

Section 2.2

1. (a) If $a \geq 0$, then $|a| = a = \sqrt{a^2}$; if $a < 0$, then $|a| = -a = \sqrt{a^2}$.
    (b) It suffices to show that $|1/b| = 1/|b|$ for $b \neq 0$ (why?). Consider the cases $b > 0$ and $b < 0$.

3. If $x \leq y \leq z$, then $|x - y| + |y - z| = (y - x) + (z - y) = z - x = |z - x|$. To establish the converse, show that $y < x$ and $y > z$ are impossible. For example, if $y < x \leq z$, it follows from what we have shown and the given relationship that $|x - y| = 0$, so that $y = x$, a contradiction.

6. (a) $-2 \leq x \leq 9/2$ (b) $-2 \leq x \leq 2$.

7. $x = 4$ or $x = -3$.

8. (a) $x < 0$ (b) $-3/2 < x < 1/2$.

10. $\{x : -3 < x < -5/2 \text{ or } 3/2 < x < 2\}$.

11. $\{x : 1 < x < 4\}$.

12. (a) $\{(x, y) : y = \pm x\}$.   (c) The hyperbolas $y = 2/x$ and $y = -2/x$.

13. (a) If $y \geq 0$, then $-y \leq x \leq y$ and we get the region in the upper half-plane on or between the lines $y = x$ and $y = -x$.

16. (a) Suppose that $a \leq b$.

17. If $a \leq b \leq c$, then $\text{mid}\{a, b, c\} = b = \min\{b, c, c\} = \min\{\max\{a, b\}, \max\{b, c\}, \max\{c, a\}\}$. The other cases are similar.

Section 2.3

1. Since $0 \leq x$ for all $x \in S_1$, then $u = 0$ is a lower bound of S_1. If $v > 0$, then v is not a lower bound of S_1 because $v/2 \in S_1$ and $v/2 < v$. Therefore $\inf S_1 = 0$.

3. Since $1/n \leq 1$ for all $n \in \mathbb{N}$, then 1 is an upper bound for S_3.

4. $\sup S_4 = 2$ and $\inf S_4 = 1/2$.

6. Let $u \in S$ be an upper bound of S. If v is another upper bound of S, then $u \leq v$. Hence $u = \sup S$.

9. Let $u := \sup A$, $v := \sup B$ and $w := \sup\{u, v\}$. Then w is an upper bound of $A \cup B$, because if $x \in A$, then $x \leq u \leq w$, and if $x \in B$, then $x \leq v \leq w$. If z is any upper bound of $A \cup B$, then z is an upper bound of A and of B, so that $u \leq z$ and $v \leq z$. Hence $w \leq z$. Therefore, $w = \sup(A \cup B)$.

11. Consider two cases: $u \geq s^*$ and $u < s^*$.

Section 2.4

1. Since $1 - 1/n < 1$ for all $n \in \mathbb{N}$, 1 is an upper bound. To show that 1 is the supremum, it must be shown that for each $\varepsilon > 0$ there exists $n \in \mathbb{N}$ such that $1 - 1/n > 1 - \varepsilon$, which is equivalent to $1/n < \varepsilon$. Apply the Archimedean Property 2.4.3 or 2.4.5.

2. $\inf S = -1$ and $\sup S = 1$.

4. (a) Let $u := \sup S$ and $a > 0$. Then $x \leq u$ for all $x \in S$, whence $ax \leq au$ for all $x \in S$, whence it follows that au is an upper bound of aS. If v is another upper bound of aS, then $ax \leq v$

for all $x \in S$, whence $x \le v/a$ for all $x \in S$, showing that v/a is an upper bound for S so that $u \le v/a$, from which we conclude that $au \le v$. Therefore $au = \sup(aS)$.

5. Let $u := \sup f(X)$. Then $f(x) \le u$ for all $x \in X$, so that $a + f(x) \le a + u$ for all $x \in X$, whence $\sup\{a + f(x) : x \in X\} \le a + u$. If $w < a + u$, then $w - a < u$, so that there exists $x_w \in X$ with $w - a < f(x_w)$, whence $w < a + f(x_w)$, and thus w is not an upper bound for $\{a + f(x) : x \in X\}$.

7. If $u := \sup f(X)$ and $v := \sup g(X)$, then $f(x) \le u$ and $g(x) \le v$ for all $x \in X$, whence $f(x) + g(x) \le u + v$ for all $x \in X$.

9. (a) $f(x) = 1$ for $x \in X$. \qquad (b) $g(y) = 0$ for $y \in Y$.

11. Let $S := \{h(x, y) : x \in X, y \in Y\}$. We have $h(x, y) \le F(x)$ for all $x \in X, y \in Y$ so that $\sup S \le \sup\{F(x) : x \in X\}$. If $w < \sup\{F(x) : x \in X\}$, then there exists $x_0 \in X$ with $w < F(x_0) = \sup\{h(x_0, y) : y \in Y\}$, whence there exists $y_0 \in Y$ with $w < h(x_0, y_0)$. Thus w is not an upper bound of S, and so $w < \sup S$. Since this is true for any w such that $w < \sup\{F(x) : x \in X\}$, we conclude that $\sup\{F(x) : x \in X\} \le \sup S$.

13. Note that $n < 2^n$ (whence $1/2^n < 1/n$) for any $n \in \mathbb{N}$.

14. Let $S_3 := \{s \in \mathbb{R} : 0 \le s, s^2 < 3\}$. Show that S_3 is nonempty and bounded by 3 and let $y := \sup S_3$. If $y^2 < 3$ and $1/n < (3 - y^2)/(2y + 1)$ show that $y + 1/n \in S_3$. If $y^2 > 3$ and $1/m < (y^2 - 3)/2y$ show that $y - 1/m \in S_3$. Therefore $y^2 = 3$.

17. If $x < 0 < y$, then we can take $r = 0$. If $x < y < 0$, we apply 2.4.8 to obtain a rational number between $-y$ and $-x$.

Section 2.5

2. S has an upper bound b and a lower bound a if and only if S is contained in the interval $[a, b]$.

4. Because z is neither a lower bound nor an upper bound of S.

5. If $z \in \mathbb{R}$, then z is not a lower bound of S so there exists $x_z \in S$ such that $x_z \le z$. Similarly, there exists $y_z \in S$ such that $z \le y_z$.

8. If $x > 0$, then there exists $n \in \mathbb{N}$ with $1/n < x$, so that $x \notin J_n$. If $y \le 0$, then $y \notin J_1$.

10. Let $\eta := \inf\{b_n : n \in \mathbb{N}\}$; we claim that $a_n \le \eta$ for all n. Fix $n \in \mathbb{N}$; we will show that a_n is a lower bound for the set $\{b_k : k \in \mathbb{N}\}$. We consider two cases. (j) If $n \le k$, then since $I_n \supseteq I_k$, we have $a_n \le a_k \le b_k$. (jj) If $k < n$, then since $I_k \supseteq I_n$, we have $a_n \le b_n \le b_k$. Therefore $a_n \le b_k$ for all $k \in \mathbb{N}$, so that a_n is a lower bound for $\{b_k : k \in \mathbb{N}\}$ and so $a_n \le \eta$. In particular, this shows that $\eta \in [a_n, b_n]$ for all n, so that $\eta \in \bigcap I_n$.

12. $\frac{3}{8} = (.011000 \cdots)_2 = (.010111 \cdots)_2$. $\frac{7}{16} = (.0111000 \cdots)_2 = (.0110111 \cdots)_2$.

13. (a) $\frac{1}{3} \approx (.0101)_2$. \qquad (b) $\frac{1}{3} = (.010101 \cdots)_2$, the block 01 repeats.

16. $1/7 = .142\,857 \cdots$, the block repeats. $2/19 = .105\,263\,157\,894\,736\,842 \cdots$, the block repeats.

17. $1.25\,137 \cdots 137 \cdots = 31253/24975$, \quad $35.14653 \cdots 653 \cdots = 3511139/99900$.

Section 3.1

1. (a) $0, 2, 0, 2, 0$ \qquad\qquad (c) $1/2, 1/6, 1/12, 1/20, 1/30$

3. (a) $1, 4, 13, 40, 121$ \qquad\qquad (c) $1, 2, 3, 5, 4$.

5. (a) We have $0 < n/(n^2 + 1) < n/n^2 = 1/n$. Given $\varepsilon > 0$, let $K(\varepsilon) \ge 1/\varepsilon$.
   (c) We have $|(3n + 1)/(2n + 5) - 3/2| = 13/(4n + 10) < 13/4n$. Given $\varepsilon > 0$, let $K(\varepsilon) \ge 13/4\varepsilon$.

6. (a) $1/\sqrt{n + 7} < 1/\sqrt{n}$ \qquad (b) $|2n/(n + 2) - 2| = 4/(n + 2) < 4/n$
   (c) $\sqrt{n}/(n + 1) < 1/\sqrt{n}$ \qquad (d) $|(-1)^n n/(n^2 + 1)| \le 1/n$.

9. $0 < \sqrt{x_n} < \varepsilon \iff 0 < x_n < \varepsilon^2$.

11. $|1/n - 1/(n + 1)| = 1/n(n + 1) < 1/n^2 \le 1/n$.

13. Let $b := 1/(1 + a)$ where $a > 0$. Since $(1 + a)^n > \frac{1}{2}n(n - 1)a^2$, we have that $0 < nb^n \le n/[\frac{1}{2}n(n - 1)a^2] \le 2/[(n - 1)a^2]$. Thus $\lim(nb^n) = 0$.

15. If $n > 3$, then $0 < n^2/n! < n/(n - 2)(n - 1) < 1/(n - 3)$.

Section 3.2

1. (a) $\lim(x_n) = 1$ (c) $x_n \ge n/2$, so the sequence diverges.

3. $Y = (X + Y) - X$.

6. (a) 4   (b) 0   (c) 1   (d) 0.

8. In (3) the exponent k is fixed, but in $(1 + 1/n)^n$ the exponent varies.

9. $\lim(y_n) = 0$ and $\lim(\sqrt{n}y_n) = \frac{1}{2}$.

11. b.

13. (a) 1   (b) 1.

15. (a) $L = a$ (b) $L = b/2$ (c) $L = 1/b$ (d) $L = 8/9$.

18. (a) Converges to 0   (c) Converges to 0.

20. (a) (1)   (b) (n).

21. Yes. (Why?)

22. From Exercise 2.2.16, $u_n = \frac{1}{2}(x_n + y_n + |x_n - y_n|)$.

23. Use Exercises 2.2.16(b), 2.2.17, and the preceding exercise.

Section 3.3

1. (x_n) is a bounded decreasing sequence. The limit is 4.

2. The limit is 1.   3. The limit is 2.   4. The limit is 2.

5. (y_n) is increasing. The limit is $y = \frac{1}{2}(1 + \sqrt{1 + 4p})$.

7. (x_n) is increasing.

10. (s_n) is decreasing and (t_n) is increasing. Also $t_n \le x_n \le s_n$ for $n \in \mathbb{N}$.

11. Note $y_n = 1/(n + 1) + 1/(n + 2) + \cdots + 1/2n < 1/(n + 1) + 1/(n + 1) + \cdots + 1/(n + 1)$ $= n/(n + 1) < 1$.

13. (a) e (b) e^2 (c) e (d) $1/e$.

14. Note that if $n \ge 2$, then $0 \le s_n - \sqrt{2} \le s_n^2 - 2$.

15. Note that $0 \le s_n - \sqrt{5} \le (s_n^2 - 5)/\sqrt{5} \le (s_n^2 - 5)/2$.

16. $e_2 = 2.25$, $e_4 = 2.441\,406$, $e_8 = 2.565\,785$, $e_{16} = 2.637\,928$.

17. $e_{50} = 2.691\,588$, $e_{100} = 2.704\,814$, $e_{1000} = 2.716\,924$.

Section 3.4

1. For example $x_{2n-1} := 2n - 1$ and $x_{2n} := 1/2n$.

3. $L = \frac{1}{2}(1 + \sqrt{5})$.

7. (a) e (b) $e^{1/2}$ (c) e^2 (d) e^2.

8. (a) 1   (b) $e^{3/2}$.

12. Choose $n_1 \geq 1$ so that $|x_{n_1}| > 1$, then choose $n_2 > n_1$ so that $|x_{n_2}| > 2$, and, in general, choose $n_k > n_{k-1}$ so that $|x_{n_k}| > k$.

13. $(x_{2n-1}) = (-1, -1/3, -1/5, \cdots)$.

14. Choose $n_1 \geq 1$ so that $x_{n_1} \geq s - 1$, then choose $n_2 > n_1$ so that $x_{n_2} > s - 1/2$, and, in general, choose $n_k > n_{k-1}$ so that $x_{n_k} > s - 1/k$.

Section 3.5

1. For example, $((-1)^n)$.

3. (a) Note that $|(-1)^n - (-1)^{n+1}| = 2$ for all $n \in \mathbb{N}$.
    (c) Take $m = 2n$, so $x_m - x_n = x_{2n} - x_n = \ln 2n - \ln n = \ln 2$ for all n.

5. $\lim(\sqrt{n+1} - \sqrt{n}) = 0$. But, if $m = 4n$, then $\sqrt{4n} - \sqrt{n} = \sqrt{n}$ for all n.

8. Let $u := \sup\{x_n : n \in \mathbb{N}\}$. If $\varepsilon > 0$, let H be such that $u - \varepsilon < x_H \leq u$. If $m \geq n \geq H$, then $u - \varepsilon < x_n \leq x_m \leq u$ so that $|x_m - x_n| < \varepsilon$.

10. $\lim(x_n) = (1/3)x_1 + (2/3)x_2$.   12. The limit is $\sqrt{2} - 1$.

13. The limit is $1 + \sqrt{2}$.

14. Four iterations give $r = 0.201\,64$ to 5 places.

Section 3.6

1. If $\{x_n : n \in \mathbb{N}\}$ is not bounded above, choose $n_{k+1} > n_k$ such that $x_{n_k} \geq k$ for $k \in \mathbb{N}$.

3. Note that $|x_n - 0| < \varepsilon$ if and only if $1/x_n > 1/\varepsilon$.

4. (a) $[\sqrt{n} > a] \iff [n > a^2]$ (c) $\sqrt{n-1} \geq \sqrt{n/2}$ when $n \geq 2$.

8. (a) $n < (n^2 + 2)^{1/2}$.
    (c) Since $n < (n^2 + 1)^{1/2}$, then $n^{1/2} < (n^2 + 1)^{1/2}/n^{1/2}$.

9. (a) Since $x_n/y_n \to \infty$, there exists K_1 such that if $n \geq K_1$, then $x_n \geq y_n$. Now apply Theorem 3.6.4(a).

Section 3.7

1. The partial sums of $\sum b_n$ are a subsequence of the partial sums of $\sum a_n$.

3. (a) Since $1/(n+1)(n+2) = 1/(n+1) - 1/(n+2)$, the series is telescoping.

6. (a) The sequence $(\cos n)$ does not converge to 0.
    (b) Since $|(\cos n)/n^2| \leq 1/n^2$, the convergence of $\sum(\cos n)/n^2$ follows from Example 3.7.6(c) and Theorem 3.7.7.

7. The "even" sequence (s_{2n}) is decreasing, the "odd" sequence (s_{2n+1}) is increasing, and $-1 \leq s_n \leq 0$. Also $0 \leq s_{2n} - s_{2n+1} = 1/\sqrt{2n+1}$.

9. $\sum 1/n^2$ is convergent, but $\sum 1/n$ is not.

11. Show that $b_k \geq a_1/k$ for $k \in \mathbb{N}$, whence $b_1 + \cdots + b_n \geq a_1(1 + \cdots + 1/n)$.

12. Evidently $2a(4) \leq a(3) + a(4)$ and $2^2 a(8) \leq a(5) + \cdots + a(8)$, etc. Also $a(2) + a(3) \leq 2a(2)$ and $a(4) + \cdots + a(7) \leq 2^2 a(2^2)$, etc. The stated inequality follows by addition. Now apply the Comparison Test 3.7.7.

14. (a) The terms are decreasing and $2^n/2^n \ln(2^n) = 1/(n \ln 2)$. Since $\sum 1/n$ diverges, so does $\sum 1/(n \ln n)$.

15. (a) The terms are decreasing and $2^n/2^n (\ln 2^n)^c = (1/n^c) \cdot (1/\ln 2)^c$. Now use the fact that $\sum(1/n^c)$ converges when $c > 1$.

Section 4.1

1. (a-c) If $|x - 1| \leq 1$, then $|x + 1| \leq 3$ so that $|x^2 - 1| \leq 3|x - 1|$. Thus, $|x - 1| < 1/6$ assures that $|x^2 - 1| < 1/2$, etc.
   (d) If $|x - 1| < 1$, then $|x^3 - 1| \leq 7|x - 1|$.

2. (a) Since $|\sqrt{x} - 2| = |x - 4|/(\sqrt{x} + 2) \leq \frac{1}{2}|x - 4|$, then $|x - 4| < 1$ implies that we have $|\sqrt{x} - 2| < \frac{1}{2}$.
   (b) If $|x - 4| < 2 \times 10^{-2} = .02$, then $|\sqrt{x} - 2| < .01$.

5. If $0 < x < a$, then $0 < x + c < a + c < 2a$, so that $|x^2 - c^2| = |x + c||x - c| \leq 2a|x - c|$. Given $\varepsilon > 0$, take $\delta := \varepsilon/2a$.

8. If $c \neq 0$, show that $|\sqrt{x} - \sqrt{c}| \leq (1/\sqrt{c})|x - c|$, so we can take $\delta := \varepsilon\sqrt{c}$. If $c = 0$, we can take $\delta := \varepsilon^2$.

9. (a) If $|x - 2| < 1/2$ show that $|1/(1 - x) + 1| = |(x - 2)/(x - 1)| \leq 2|x - 2|$. Thus we can take $\delta := \inf\{1/2, \varepsilon/2\}$.
   (c) If $x \neq 0$, then $|x^2/|x| - 0| = |x|$. Take $\delta := \varepsilon$.

10. (a) If $|x - 2| < 1$, then $|x^2 + 4x - 12| = |x + 6||x - 2| < 9|x - 2|$. We may take $\delta := \inf\{1, \varepsilon/9\}$.
    (b) If $|x + 1| < 1/4$, then $|(x + 5)/(3x + 2) - 4| = 7|x + 1|/|2x + 3| < 14|x + 1|$, and we may take $\delta := \inf\{1/4, \varepsilon/14\}$.

11. (a) Let $x_n := 1/n$.   (c) Let $x_n := 1/n$ and $y_n := -1/n$.

13. (b) If $f(x) := \text{sgn}(x)$, then $\lim_{x \to 0}(f(x))^2 = 1$, but $\lim_{x \to 0} f(x)$ does not exist.

14. (a) Since $|f(x) - 0| \leq |x|$, we have $\lim_{x \to 0} f(x) = 0$.
    (b) If $c \neq 0$ is rational, let (x_n) be a sequence of irrational numbers that converges to c; then $f(c) = c \neq 0 = \lim(f(x_n))$. What if c is irrational?

16. The restriction of sgn to $[0, 1]$ has a limit at 0.

Section 4.2

1. (a) 10   (b) -3 (c) 1/12   (d) 1/2.

2. (a) 1   (b) 4   (c) 2   (d) 1/2.

3. Multiply the numerator and denominator by $\sqrt{1 + 2x} + \sqrt{1 + 3x}$.

4. Consider $x_n := 1/2\pi n$ and $\cos(1/x_n) = 1$. Use the Squeeze Theorem 4.2.7.

8. If $|x| \leq 1, k \in \mathbb{N}$, then $|x^k| = |x|^k \leq 1$, whence $-x^2 \leq x^{k+2} \leq x^2$.

11. (a) No limit   (b) 0   (c) No limit   (d) 0.

Section 4.3

2. Let $f(x) := \sin(1/x)$ for $x < 0$ and $f(x) := 0$ for $x > 0$.

3. Given $\alpha > 0$, if $0 < x < 1/\alpha^2$, then $\sqrt{x} < 1/\alpha$, and so $f(x) > \alpha$.

5. (a) If $\alpha > 1$ and $1 < x < \alpha/(\alpha - 1)$, then $\alpha < x/(x - 1)$, hence we have $\lim_{x \to 1+} x/(x - 1) = \infty$.
   (c) Since $(x + 2)/\sqrt{x} > 2/\sqrt{x}$, the limit is ∞.
   (e) If $x > 0$, then $1/\sqrt{x} < (\sqrt{x + 1})/x$, so the right-hand limit is ∞.
   (g) 1   (h) -1.

8. Note that $|f(x) - L| < \varepsilon$ for $x > K$ if and only if $|f(1/z) - L| < \varepsilon$ for $0 < z < 1/K$.

9. There exists $\alpha > 0$ such that $|xf(x) - L| < 1$ whenever $x > \alpha$. Hence $|f(x)| < (|L| + 1)/x$ for $x > \alpha$.

12. No. If $h(x) := f(x) - g(x)$, then $\lim_{x \to \infty} h(x) = 0$ and we have
$$f(x)/g(x) = 1 + h(x)/g(x) \to 1.$$

13. Suppose that $|f(x) - L| < \varepsilon$ for $x > K$, and that $g(y) > K$ for $y > H$. Then $|f \circ g(y) - L| < \varepsilon$ for $y > H$.

Section 5.1

4. (a) Continuous if $x \neq 0, \pm 1, \pm 2, \cdots$ (b) Continuous if $x \neq \pm 1, \pm 2, \cdots$
   (c) Continuous if $\sin x \neq 0, 1$ (d) Continuous if $x \neq 0, \pm 1, \pm 1/2, \cdots$.

7. Let $\varepsilon := f(c)/2$, and let $\delta > 0$ be such that if $|x - c| < \delta$, then $|f(x) - f(c)| < \varepsilon$, which implies that $f(x) > f(c) - \varepsilon = f(c)/2 > 0$.

8. Since f is continuous at x, we have $f(x) = \lim(f(x_n)) = 0$. Thus $x \in S$.

10. Note that $\left| |x| - |c| \right| \leq |x - c|$.

13. Since $|g(x) - 6| \leq \sup\{|2x - 6|, |x - 3|\} = 2|x - 3|$, g is continuous at $x = 3$. If $c \neq 3$, let (x_n) be a sequence of rational numbers converging to c and let (y_n) be a sequence of irrational numbers converging to c. Then $\lim(g(x_n)) \neq \lim(g(y_n))$

Section 5.2

1. (a) Continuous on $\mathbb{R}$ (c) Continuous for $x \neq 0$.

2. Use 5.2.1(a) and Induction; or, use 5.2.8 with $g(x) := x^n$.

4. Continuous at every noninteger.

7. Let $f(x) := 1$ if x is rational, and $f(x) := -1$ if x is irrational.

12. First show that $f(0) = 0$ and $f(-x) = -f(x)$ for all $x \in \mathbb{R}$; then note that $f(x - x_0) = f(x) - f(x_0)$. Consequently f is continuous at the point x_0 if and only if it is continuous at 0. Thus, if f is continuous at x_0, then it is continuous at 0, and hence everywhere.

13. First show that $f(0) = 0$ and (by Induction) that $f(x) = cx$ for $x \in \mathbb{N}$, and hence also for $x \in \mathbb{Z}$. Next show that $f(x) = cx$ for $x \in \mathbb{Q}$. Finally, if $x \notin \mathbb{Q}$, let $x = \lim(r_n)$ for some sequence in $\mathbb{Q}$.

15. If $f(x) \geq g(x)$, then both expressions give $h(x) = f(x)$; and if $f(x) \leq g(x)$, then $h(x) = g(x)$ in both cases.

Section 5.3

1. Apply either the Boundedness Theorem 5.3.2 to $1/f$, or the Maximum-Minimum Theorem 5.3.4 to conclude that $\inf f(I) > 0$.

3. Choose a sequence (x_n) such that $|f(x_{n+1})| \leq \frac{1}{2}|f(x_n)| \leq (\frac{1}{2})^n |f(x_1)|$. Apply the Bolzano-Weierstrass Theorem to obtain a convergent subsequence.

4. Suppose that p has odd degree n and that the coefficient a_n of x^n is positive. By 4.3.16, $\lim_{x \to \infty} p(x) = \infty$ and $\lim_{x \to -\infty} p(x) = -\infty$.

5. In the intervals $[1.035, 1.040]$ and $[-7.026, -7.025]$.

7. In the interval $[0.7390, 0.7391]$.

8. In the interval $[1.4687, 1.4765]$.

9. (a) 1   (b) 6.

10. $1/2^n < 10^{-5}$ implies that $n > (5 \ln 10)/\ln 2 \approx 16.61$. Take $n = 17$.

11. If $f(w) < 0$, then it follows from Theorem 4.2.9 that there exists a δ-neighborhood $V_\delta(w)$ such that $f(x) < 0$ for all $x \in V_\delta(w)$.

14. Apply Theorem 4.2.9 to $\beta - f(x)$.

15. If $0 < a < b \le \infty$, then $f((a, b)) = (a^2, b^2)$; if $-\infty \le a < b < 0$, then $f((a, b)) = (b^2, a^2)$. If $a < 0 < b$, then $f((a, b))$ is not an open interval, but equals $[0, c)$ where $c := \sup\{a^2, b^2\}$. Images of closed intervals are treated similarly.

16. For example, if $a < 0 < b$ and $c := \inf\{1/(a^2 + 1), 1/(b^2 + 1)\}$, then $g((a, b)) = (c, 1]$. If $0 < a < b$, then $g((a, b)) = (1/(b^2 + 1), 1/(a^2 + 1))$. Also $g([-1, 1]) = [1/2, 1]$. If $a < b$, then $h((a, b)) = (a^3, b^3)$ and $h((a, b]) = (a^3, b^3]$.

17. Yes. Use the Density Theorem 2.4.8.

19. Consider $g(x) := 1/x$ for $x \in J := (0, 1)$.

Section 5.4

1. Since $1/x - 1/u = (u - x)/xu$, it follows that $|1/x - 1/u| \le (1/a^2)|x - u|$ for $x, u \in [a, \infty)$.

3. (a) Let $x_n := n + 1/n$, $u_n := n$.
   (b) Let $x_n := 1/2n\pi$, $u_n := 1/(2n\pi + \pi/2)$.

6. If M is a bound for both f and g on A, show that $|f(x)g(x) - f(u)g(u)| \le M|f(x) - f(u)| + M|g(x) - g(u)|$ for all $x, u \in A$.

8. Given $\varepsilon > 0$ there exists $\delta_f > 0$ such that $|y - v| < \delta_f$ implies $|f(y) - f(v)| < \varepsilon$. Now choose $\delta_g > 0$ so that $|x - u| < \delta_g$ implies $|g(x) - g(u)| < \delta_f$.

11. If $|g(x) - g(0)| \le K|x - 0|$ for all $x \in [0, 1]$, then $\sqrt{x} \le Kx$ for $x \in [0, 1]$. But if $x_n := 1/n^2$, then K must satisfy $n \le K$ for all $n \in \mathbb{N}$, which is impossible.

14. Since f is bounded on $[0, p]$, it follows that it is bounded on $\mathbb{R}$. Since f is continuous on $J := [-1, p + 1]$, it is uniformly continuous on J. Now show that this implies that f is uniformly continuous on $\mathbb{R}$.

Section 5.5

1. (a) The δ-intervals are $[-\frac{1}{4}, \frac{1}{4}]$, $[\frac{1}{4}, \frac{3}{4}]$, and $[\frac{3}{8}, \frac{9}{8}]$.
   (b) The third δ-interval does not contain $[\frac{1}{2}, 1]$.

2. (a) Yes.   (b) Yes.

3. No. The first δ_2-interval is $[-\frac{1}{10}, \frac{1}{10}]$ and does not contain $[0, \frac{1}{4}]$.

4. (b) If $t \in (\frac{1}{2}, 1)$ then $[t - \delta(t), t + \delta(t)] = [-\frac{1}{2} + \frac{3}{2}t, \frac{1}{2} + \frac{1}{2}t] \subset (\frac{1}{4}, 1)$.

6. We could have two subintervals having c as a tag with one of them not contained in the δ-interval around c.

7. If $\dot{\mathcal{P}} := \{([a, x_1], t_1), \cdots ([x_{k-1}, c], t_k), ([c, x_{k+1}], t_{k+1}), \cdots, ([x_n, b], t_n)\}$ is δ^*-fine, then $\dot{\mathcal{P}}' := \{([a, x_1], t_1), \cdots, ([x_{k-1}, c], t_k)\}$ is a δ'-fine partition of $[a, c]$ and $\dot{\mathcal{P}}'' := \{([c, x_{k+1}], t_{k+1}), \cdots, ([x_n, b], t_n)\}$ is a δ''-fine partition of $[c, b]$.

9. The hypothesis that f is locally bounded presents us with a gauge δ. If $\{([x_{i-1}, x_i], t_i)\}_{i=1}^n$ is a δ-fine partition of $[a, b]$ and M_i is a bound for $|f|$ on $[x_{i-1}, x_i]$, let $M := \sup\{M_i : i = 1, \cdots, n\}$.

Section 5.6

1. If $x \in [a, b]$, then $f(a) \le f(x)$.

4. If $0 \le f(x_1) \le f(x_2)$ and $0 \le g(x_1) \le g(x_2)$, then $f(x_1)g(x_1) \le f(x_2)g(x_1) \le f(x_2)g(x_2)$.

6. If f is continuous at c, then $\lim(f(x_n)) = f(c)$, since $c = \lim(x_n)$. Conversely, since $0 \le j_f(c) \le f(x_{2n}) - f(x_{2n+1})$, it follows that $j_f(c) = 0$, so f is continuous at c.

7. Apply Exercises 2.4.4, 2.4.5 and the Principle of the Iterated Infima (analogous to the result in Exercise 2.4.11).

8. Let $x_1 \in I$ be such that $y = f(x_1)$ and $x_2 \in I$ be such that $y = g(x_2)$. If $x_2 \le x_1$, then $y = g(y_2) < f(x_2) \le f(x_1) = y$, a contradiction.

11. Note that f^{-1} is continuous at every point of its domain $[0, 1] \cup (2, 3]$.

14. Let $y := x^{1/n}$ and $z := x^{1/q}$ so that $y^n = x = z^q$, whence (by Exercise 2.1.26) $y^{np} = x^p = z^{qp}$. Since $np = mq$, show that $(x^{1/n})^m = (x^{1/q})^p$, or $x^{m/n} = x^{p/q}$. Now consider the case where $m, p \in \mathbb{Z}$.

15. Use the preceding exercise and Exercise 2.1.26.

Section 6.1

1. (a) $f'(x) = \lim_{h \to 0}[(x + h)^3 - x^3]/h = \lim_{h \to 0}(3x^2 + 3xh + h^2) = 3x^2$,

    (c) $h'(x) = \lim_{h \to 0}\dfrac{\sqrt{x + h} - \sqrt{x}}{h} = \lim_{h \to 0}\dfrac{1}{\sqrt{x + h} + \sqrt{x}} = \dfrac{1}{2\sqrt{x}}$.

4. Note that $|f(x)/x| \le |x|$ for $x \in \mathbb{R}$.

5. (a) $f'(x) = (1 - x^2)/(1 + x^2)^2$ (b) $g'(x) = (x - 1)/\sqrt{5 - 2x + x^2}$
    (c) $h'(x) = mkx^{k-1}(\cos x^k)(\sin x^k)^{m-1}$ (d) $k'(x) = 2x \sec^2(x^2)$.

6. The function f' is continuous for $n \ge 2$ and is differentiable for $n \ge 3$.

8. (a) $f'(x) = 2$ for $x > 0$, $f'(x) = 0$ for $-1 < x < 0$, and $f'(x) = -2$ for $x < -1$,
    (c) $h'(x) = 2|x|$ for all $x \in \mathbb{R}$,

10. If $x \ne 0$, then $g'(x) = 2x \sin(1/x^2) - (2/x)\cos(1/x^2)$. Moreover, $g'(0) = \lim_{h \to 0} h \sin(1/h^2) = 0$. Consider $x_n := 1/\sqrt{2n\pi}$.

11. (a) $f'(x) = 2/(2x + 3)$ (b) $g'(x) = 6(L(x^2))^2/x$
    (c) $h'(x) = 1/x$ (d) $k'(x) = 1/(xL(x))$.

14. $1/h'(0) = 1/2$, $1/h'(1) = 1/5$, and $1/h'(-1) = 1/5$.

16. $D[\text{Arctan } y] = 1/D[\tan x] = 1/\sec^2 x = 1/(1 + y^2)$.

Section 6.2

1. (a) Increasing on $[3/2, \infty)$, decreasing on $(-\infty, 3/2]$,
    (c) Increasing on $(-\infty, -1]$ and $[1, \infty)$

2. (a) Relative minimum at $x = 1$; relative maximum at $x = -1$,
    (c) Relative maximum at $x = 2/3$.

3. (a) Relative minima at $x = \pm 1$; relative maxima at $x = 0, \pm 4$,
    (c) Relative minima at $x = -2, 3$; relative maximum at $x = 2$.

6. If $x < y$ there exists c in (x, y) such that $|\sin x - \sin y| = |\cos c||y - x|$.

9. $f(x) = x^4(2 + \sin(1/x)) > 0$ for $x \ne 0$, so f has an absolute minimum at $x = 0$. Show that $f'(1/2n\pi) < 0$ for $n \ge 2$ and $f'(2/(4n + 1)\pi) > 0$ for $n \ge 1$.

10. $g'(0) = \lim_{x \to 0}(1 + 2x \sin(1/x)) = 1 + 0 = 1$, and if $x \neq 0$, then $g'(x) = 1 + 4x \sin(1/x) - 2\cos(1/x)$. Now show that $g'(1/2n\pi) < 0$ and that we have $g'(2/(4n+1)\pi) > 0$ for $n \in \mathbb{N}$.

14. Apply Darboux's Theorem 6.2.12.

17. Apply the Mean Value Theorem to the function $g - f$ on $[0, x]$.

20. (a, b) Apply the Mean Value Theorem.
      (c) Apply Darboux's Theorem to the results of (a) and (b).

Section 6.3

1. $A = B(\lim_{x \to c} f(x)/g(x)) = 0$.

4. Note that $f'(0) = 0$, but that $f'(x)$ does not exist if $x \neq 0$.

6. (a) 1 | (b) 1 | (c) 0 | (d) 1/3.

7. (a) 1 | (b) ∞ | (c) 0 | (d) 0.

8. (a) 0 | (b) 0 | (c) 0 | (d) 0.

9. (a) 1 | (b) 1 | (c) e^3 | (d) 0.

10. (a) 1 | (b) 1 | (c) 1 | (d) 0.

Section 6.4

1. $f^{(2n-1)}(x) = (-1)^n a^{2n-1} \sin ax$ and $f^{(2n)}(x) = (-1)^n a^{2n} \cos ax$ for $n \in \mathbb{N}$.

4. Apply Taylor's Theorem to $f(x) := \sqrt{1+x}$ at $x_0 := 0$ and note that $R_1(x) < 0$ and $R_2(x) > 0$ for $x > 0$.

5. $1.095 < \sqrt{1.2} < 1.1$ and $1.375 < \sqrt{2} < 1.5$.

6. $R_2(0.2) < 0.0005$ and $R_2(1) < 0.0625$.

11. With $n = 4$, $\ln 1.5 = 0.40$; with $n = 7$, $\ln 1.5 = 0.405$.

17. Apply Taylor's Theorem to f at $x_0 = c$ to show that $f(x) \geq f(c) + f'(c)(x - c)$.

19. Since $f(2) < 0$ and $f(2.2) > 0$, there is a zero of f in $[2.0, 2.2]$. The value of x_4 is approximately 2.094 551 5.

20. $r_1 \approx 1.452\,626\,88$ and $r_2 \approx -1.164\,035\,14$.   21. $r \approx 1.324\,717\,96$.

22. $r_1 \approx 0.158\,594\,34$ and $r_2 \approx 3.146\,193\,22$.   23. $r_1 \approx 0.5$ and $r_2 \approx 0.809\,016\,99$.

24. $r \approx 0.739\,085\,13$.

Section 7.1

1. (a) $\|P_1\| = 2$ | (b) $\|P_2\| = 2$ | (c) $\|P_3\| = 1.4$ | (d) $\|P_4\| = 2$.

2. (a) $0^2 \cdot 1 + 1^2 \cdot 1 + 2^2 \cdot 2 = 0 + 1 + 8 = 9$
      (b) 37   (c) 13   (d) 33.

5. (a) If $u \in [x_{i-1}, x_i]$, then $x_{i-1} \leq u$ so that $c_1 \leq t_i \leq x_i \leq x_{i-1} + \|\dot{P}\|$ whence $c_1 - \|\dot{P}\| \leq x_{i-1} \leq u$. Also $u \leq x_i$ so that $x_i - \|\dot{P}\| \leq x_{i-1} \leq t_i \leq c_2$, whence $u \leq x_i \leq c_2 + \|\dot{P}\|$.

10. g is not bounded. Take rational tags.

12. Let P_n be the partition of $[0, 1]$ into n equal parts. If $\dot{P}_n$ is this partition with rational tags, then $S(f; \dot{P}_n) = 1$, while if $\dot{Q}_n$ is this partition with irrational tags, then $S(f; \dot{Q}_n) = 0$.

13. Argue as in Example 7.1.3(d).

15. If $\|\dot{\mathcal{P}}\| < \delta_\varepsilon := \varepsilon/4\alpha$, then the union of the subintervals in $\dot{\mathcal{P}}$ with tags in $[c, d]$ contains the interval $[c + \delta_\varepsilon, d - \delta_\varepsilon]$ and is contained in $[c - \delta_\varepsilon, d + \delta_\varepsilon]$. Therefore $\alpha(d - c - 2\delta_\varepsilon) \leq S(\varphi; \dot{\mathcal{P}}) \leq \alpha(d - c + 2\delta_\varepsilon)$, whence $|S(\varphi; \dot{\mathcal{P}}) - \alpha(d - c)| \leq 2\alpha\delta_\varepsilon < \varepsilon$.

16. (b) In fact, $(x_i^2 + x_i x_{i-1} + x_{i-1}^2) \cdot (x_i - x_{i-1}) = x_i^3 - x_{i-1}^3$.
    (c) The terms in $S(Q; \dot{Q})$ telescope.

18. Let $\dot{\mathcal{P}} = \{([x_{i-1}, x_i], t_i)\}_{i=1}^n$ be a tagged partition of $[a, b]$ and let $\dot{Q} := \{([x_{i-1} + c, x_i + c], t_i + c)\}_{i=1}^n$ so that $\dot{Q}$ is a tagged partition of $[a + c, b + c]$ and $\|\dot{Q}\| = \|\dot{\mathcal{P}}\|$. Moreover, $S(g; \dot{Q}) = S(f; \dot{\mathcal{P}})$ so that $|S(g; \dot{Q}) - \int_a^b f| = |S(f; \dot{\mathcal{P}}) - \int_a^b f| < \varepsilon$ when $\|\dot{Q}\| < \delta_\varepsilon$.

Section 7.2

2. If the tags are all rational, then $S(h; \dot{\mathcal{P}}) \geq 1$, while if the tags are all irrational, then $S(h; \dot{\mathcal{P}}) = 0$.

3. Let $\dot{\mathcal{P}}_n$ be the partition of $[0, 1]$ into n equal subintervals with $t_1 = 1/n$ and $\dot{Q}_n$ be the same subintervals tagged by irrational points.

5. If $c_1, \cdots, c_n$ are the distinct values taken by φ, then $\varphi^{-1}(c_j)$ is the union of a finite collection $\{J_{j1}, \cdots, J_{jr_j}\}$ of disjoint subintervals of $[a, b]$. We can write $\varphi = \sum_{j=1}^n \sum_{k=1}^{r_j} c_j \varphi_{J_{jk}}$.

6. Not necessarily.

8. If $f(c) > 0$ for some $c \in (a, b)$, there exists $\delta > 0$ such that $f(x) > \frac{1}{2}f(c)$ for $|x - c| \leq \delta$. Then $\int_a^b f \geq \int_{c-\delta}^{c+\delta} f \geq (2\delta)\frac{1}{2}f(c) > 0$. If c is an endpoint, a similar argument applies.

10. Use Bolzano's Theorem 5.3.7.

12. Indeed, $|g(x)| \leq 1$ and is continuous on every interval $[c, 1]$ where $0 < c < 1$. The preceding exercise applies.

13. Let $f(x) := 1/x$ for $x \in (0, 1]$ and $f(0) := 0$.

16. Let $m := \inf f(x)$ and $M := \sup f$. By Theorem 7.1.4(c), we have $m(b - a) \leq \int_a^b f \leq M(b - a)$. By Bolzano's Theorem 5.3.7, there exists $c \in [a, b]$ such that $f(c) = (\int_a^b f)/(b - a)$.

19. (a) Let $\dot{\mathcal{P}}_n$ be a sequence of tagged partitions of $[0, a]$ with $\|\dot{\mathcal{P}}_n\| \to 0$ and let $\dot{\mathcal{P}}_n^*$ be the corresponding "symmetric" partition of $[-a, a]$. Show that $S(f; \dot{\mathcal{P}}_n^*) = 2S(f; \dot{\mathcal{P}}_n) \to 2\int_0^a f$.

21. Note that $x \mapsto f(x^2)$ is an even continuous function.

22. Let $x_i := i(\pi/2)$ for $i = 0, 1, \cdots, n$. Then we have that $(\pi/2n)\sum_{i=0}^{n-1} f(\cos x_i) = (\pi/2n)\sum_{k=1}^n f(\sin x_k)$.

Section 7.3

1. Suppose that $E := \{a = c_0 < c_1 < \cdots < c_m = b\}$ contains the points in $[a, b]$ where the derivative $F'(x)$ either does not exist, or does not equal $f(x)$. Then $f \in \mathcal{R}[c_{i-1}, c_i]$ and $\int_{c_{i-1}}^{c_i} f = F(c_i) - F(c_{i-1})$. Exercise 7.2.14 and Corollary 7.2.10 imply that $f \in \mathcal{R}[a, b]$ and that $\int_a^b f = \sum_{i=1}^m (F(c_i) - F(c_{i-1})) = F(b) - F(a)$.

2. $E = \emptyset$.   3. Let $E := \{-1, 1\}$. If $x \notin E$, $G'(x) = g(x)$.

4. Indeed, $B'(x) = |x|$ for all x.   6. $F_c = F_a - \int_a^c f$.

7. Let h be Thomae's function. There is no function $H : [0, 1] \to \mathbb{R}$ such that $H'(x) = h(x)$ for x in some nondegenerate open interval; otherwise Darboux's Theorem 6.2.12 would be contradicted on this interval.

9. (a) $G(x) = F(x) - F(c)$,   (b) $H(x) = F(b) - F(x)$,   (c) $S(x) = F(\sin x) - F(x)$.

10. Use Theorem 7.3.6 and the Chain Rule 6.1.6.

11. (a) $F'(x) = 2x(1 + x^6)^{-1}$ (b) $F'(x) = (1 + x^2)^{1/2} - 2x(1 + x^4)^{1/2}$.

13. $g'(x) = f(x + c) - f(x - c)$.

16. (a) Take $\varphi(t) = 1 + t^2$ to get $\frac{1}{3}(2^{3/2} - 1)$.
     (b) Take $\varphi(t) = 1 + t^3$ to get $\frac{4}{3}$.
     (c) Take $\varphi(t) = 1 + \sqrt{t}$ to get $\frac{4}{3}(3^{3/2} - 2^{3/2})$.
     (d) Take $\varphi(t) = t^{1/2}$ to get $2(\sin 2 - \sin 1)$.

18. (a) Take $x = \varphi(t) = t^{1/2}$, so $t = \psi(x) = x^2$ to get $4(1 - \ln(5/3))$.
     (b) Take $x = \varphi(t) = (t + 1)^{1/2}$, so $t = \psi(x) = x^2 - 1$ to get $\ln(3 + 2\sqrt{2}) - \ln 3$.
     (c) Take $x = \varphi(t) = t^{1/2}$ to get $2(3/2 + \ln 3/2)$.
     (d) Take $x = \varphi(t) = t^{1/2}$ to get $\operatorname{Arctan} 1 - \operatorname{Arctan}(1/2)$.

19. In (a) – (c) $\varphi'(0)$ does not exist. For (a), integrate over $[c, 4]$ and let $c \to 0+$. For (c), the integrand is even so the integral equals $2 \int_0^1 (1 + t)^{1/2}\, dt$.

20. (b) $\bigcup_n Z_n$ is contained in $\bigcup_{n,k} J_k^n$ and the sum of the lengths of these intervals is $\leq \sum_n \varepsilon/2^n = \varepsilon$.

21. (a) The Product Theorem 7.3.16 applies.
     (b) We have $\mp 2t \int_a^b fg \leq t^2 \int_a^b f^2 + \int_a^b g^2$.
     (c) Let $t \to \infty$ in (b).
     (d) If $\int_a^b f^2 \neq 0$, let $t = \left(\int_a^b g^2 / \int_a^b f^2\right)^{1/2}$ in (b).

22. Note that sgn $\circ\, h$ is Dirichlet's function, which is not Riemann integrable.

Section 7.4

1. Use (4) with $n = 4, a = 1, b = 2, h = 1/4$. Here $1/4 \leq f''(c) \leq 2$, so $T_4 \approx 0.697\,02$.

3. $T_4 \approx 0.782\,79$.

4. The index n must satisfy $2/12n^2 < 10^{-6}$; hence $n > 1000/\sqrt{6} \approx 408.25$.

5. $S_4 \approx 0.785\,39$.

6. The index n must satisfy $96/180n^4 < 10^{-6}$; hence $n \geq 28$.

12. The integral is equal to the area of one quarter of the unit circle. The derivatives of h are unbounded on $[0, 1]$. Since $h''(x) \leq 0$, the inequality is $T_n(h) < \pi/4 < M_n(h)$. See Exercise 8.

13. Interpret K as an area. Show that $h''(x) = -(1 - x^2)^{3/2}$ and that $h^{(4)}(x) = -3(1 + 4x^2)(1 - x^2)^{-7/2}$. To eight decimal places, $\pi = 3.141\,592\,65$.

14. Approximately 3.653 484 49.   15. Approximately 4.821 159 32.

16. Approximately 0.835 648 85.   17. Approximately 1.851 937 05.

18. 1.   19. Approximately 1.198 140 23.

20. Approximately 0.904 524 24.

Section 8.1

1. Note that $0 \leq f_n(x) \leq x/n \to 0$ as $n \to \infty$.

3. If $x > 0$, then $|f_n(x) - 1| < 1/(nx)$.

5. If $x > 0$, then $|f_n(x)| \leq 1/(nx) \to 0$.

7. If $x > 0$, then $0 < e^{-x} < 1$.

9. If $x > 0$, then $0 \leq x^2 e^{-nx} = x^2 (e^{-x})^n \to 0$, since $0 < e^{-x} < 1$.

10. If $x \in \mathbb{Z}$, the limit equals 1. If $x \notin \mathbb{Z}$, the limit equals 0.

11. If $x \in [0, a]$, then $|f_n(x)| \leq a/n$. However, $f_n(n) = 1/2$.

14. If $x \in [0, b]$, then $|f_n(x)| \leq b^n$. However, $f_n(2^{-1/n}) = 1/3$.

15. If $x \in [a, \infty)$, then $|f_n(x)| \leq 1/(na)$. However, $f_n(1/n) = \frac{1}{2} \sin 1 > 0$.

18. The maximum of f_n on $[0, \infty)$ is at $x = 1/n$, so $\|f_n\|_{[0,\infty)} = 1/(ne)$.

20. If n is sufficiently large, $\|f_n\|_{[a,\infty)} = n^2 a^2 / e^{na}$. However, $\|f_n\|_{[0,\infty)} = 4/e^2$.

23. Let M be a bound for $(f_n(x))$ and $(g_n(x))$ on A, whence also $|f(x)| \leq M$. The Triangle Inequality gives $|f_n(x) g_n(x) - f(x) g(x)| \leq M(|f_n(x) - f(x)| + |g_n(x) - g(x)|)$ for $x \in A$.

Section 8.2

1. The limit function is $f(x) := 0$ for $0 \leq x < 1$, $f(1) := 1/2$, and $f(x) := 1$ for $1 < x \leq 2$.

4. If $\varepsilon > 0$ is given, let K be such that if $n \geq K$, then $\|f_n - f\|_I < \varepsilon/2$. Then $|f_n(x_n) - f(x_0)| \leq |f_n(x_n) - f(x_n)| + |f(x_n) - f(x_0)| \leq \varepsilon/2 + |f(x_n) - f(x_0)|$. Since f is continuous (by Theorem 8.2.2) and $x_n \to x_0$, then $|f(x_n) - f(x_0)| < \varepsilon/2$ for $n \geq K'$, so that $|f_n(x_n) - f(x_0)| < \varepsilon$ for $n \geq \max\{K, K'\}$.

6. Here $f(0) = 1$ and $f(x) = 0$ for $x \in (0, 1]$. The convergence is not uniform on $[0, 1]$.

7. Given $\varepsilon := 1$, there exists $K > 0$ such that if $n \geq K$ and $x \in A$, then $|f_n(x) - f(x)| < 1$, so that $|f_n(x)| \leq |f_K(x)| + 1$ for all $x \in A$. Let $M := \max\{\|f_1\|_A, \cdots, \|f_{K-1}\|_A, \|f_K\|_A + 1\}$.

8. $f_n(1/\sqrt{n}) = \sqrt{n}/2$.

10. Here (g_n) converges uniformly to the zero function. The sequence (g_n') does not converge uniformly.

11. Use the Fundamental Theorem 7.3.1 and Theorem 8.2.4.

13. If $a > 0$, then $\|f_n\|_{[a,\pi]} \leq 1/(na)$ and Theorem 8.2.4 applies.

15. Here $\|g_n\|_{[0,1]} \leq 1$ for all n. Now apply Theorem 8.2.5.

20. Let $f_n(x) := x^n$ on $[0, 1)$.

Section 8.3

1. Let $A := x > 0$ and let $m \to \infty$ in (5). For the upper estimate on e, take $x = 1$ and $n = 3$ to obtain $|e - 2\frac{2}{3}| < 1/12$, so $e < 2\frac{3}{4}$.

2. Note that if $n \geq 9$, then $2/(n+1)! < 6 \times 10^{-7} < 5 \times 10^{-6}$. Hence $e \approx 2.71828$.

3. Evidently $E_n(x) \leq e^x$ for $x \geq 0$. To obtain the other inequality, apply Taylor's Theorem 6.4.1 to $[0, a]$.

5. Note that $0 \leq t^n/(1+t) \leq t^n$ for $t \in [0, x]$.

6. $\ln 1.1 \approx 0.0953$ and $\ln 1.4 \approx 0.3365$. Take $n > 19,999$.

7. $\ln 2 \approx 0.6931$.

10. $L'(1) = \lim[L(1 + 1/n) - L(1)]/(1/n) = \lim L((1 + 1/n)^n) = L(\lim(1 + 1/n)^n) = L(e) = 1$.

11. (c) $(xy)^\alpha = E(\alpha L(xy)) = E(\alpha L(x) + \alpha L(y)) = E(\alpha L(x)) \cdot E(\alpha L(y)) = x^\alpha \cdot y^\alpha$.

12. (b) $(x^\alpha)^\beta = E(\beta L(x^\alpha)) = E(\beta \alpha L(x)) = x^{\alpha\beta}$, and similarly for $(x^\beta)^\alpha$.

15. Use 8.3.14 and 8.3.9(vii).

17. Indeed, we have $\log_a x = (\ln x)/(\ln a) = [(\ln x)/(\ln b)] \cdot [(\ln b)/(\ln a)]$ if $a \neq 1, b \neq 1$. Now take $a = 10, b = e$.

Section 8.4

1. If $n > 2|x|$, then $|\cos x - C_n(x)| \leq (16/15)|x|^{2n}/(2n)!$, so $\cos(0.2) \approx 0.980\,067$, $\cos 1 \approx 0.549\,302$. Similarly, $\sin(0.2) \approx 0.198\,669$ and $\sin 1 \approx 0.841\,471$.

4. We integrate 8.4.8(x) twice on $[0, x]$. Note that the polynomial on the left has a zero in the interval $[1.56, 1.57]$, so $1.56 \leq \pi/2$.

5. Exercise 8.4.4 shows that $C_4(x) \leq \cos x \leq C_3(x)$ for all $x \in \mathbb{R}$. Integrating several times, we get $S_4(x) \leq \sin x \leq S_5(x)$ for all $x > 0$. Show that $S_4(3.05) > 0$ and $S_5(3.15) < 0$. (This procedure can be sharpened.)

6. If $|x| \leq A$ and $m > n > 2A$, then $|c_m(x) - c_n(x)| < (16/15)A^{2n}/(2n)!$, whence the convergence of (c_n) to c is uniform on each interval $[-A, A]$.

7. $D[(c(x))^2 - (s(x))^2] = 0$ for all $x \in \mathbb{R}$. For uniqueness, argue as in 8.4.4.

8. Let $g(x) := f(0)c(x) + f'(0)s(x)$ for $x \in \mathbb{R}$, so that $g''(x) = g(x)$, $g(0) = f(0)$ and $g'(0) = f'(0)$. Therefore $h(x) := f(x) - g(x)$ has the property that $h''(x) = h(x)$ for all $x \in \mathbb{R}$ and $h(0) = 0, h'(0) = 0$. Thus $g(x) = f(x)$ for all $x \in \mathbb{R}$, so that $f(x) = f(0)c(x) + f'(0)s(x)$.

9. If $\varphi(x) := c(-x)$, show that $\varphi''(x) = \varphi(x)$ and $\varphi(0) = 1$, $\varphi'(0) = 0$, so that $\varphi(x) = c(x)$ for all $x \in \mathbb{R}$. Therefore c is even.

Section 9.1

1. Let s_n be the nth partial sum of $\sum_1^\infty a_n$, let t_n be the nth partial sum of $\sum_1^\infty |a_n|$, and suppose that $a_n \geq 0$ for $n > P$. If $m > n > P$, show that $t_m - t_n = s_m - s_n$. Now apply the Cauchy Criterion.

3. Take positive terms until the partial sum exceeds 1, then take negative terms until the partial sum is less than 1, then take positive terms until the partial sum exceeds 2, etc.

5. Yes.

6. If $n \geq 2$, then $s_n = -\ln 2 - \ln n + \ln(n + 1)$. Yes.

9. We have $s_{2n} - s_n \geq na_{2n} = \frac{1}{2}(2na_{2n})$, and $s_{2n+1} - s_n \geq \frac{1}{2}(2n + 1)a_{2n+1}$. Consequently $\lim(na_n) = 0$.

11. Indeed, if $|n^2 a_n| \leq M$ for all n, then $|a_n| \leq M/n^2$.

13. (a) Rationalize to obtain $\sum x_n$ where $x_n := [\sqrt{n}(\sqrt{n+1} + \sqrt{n})]^{-1}$ and note that $x_n \approx y_n := 1/(2n)$. Now apply the Limit Comparison Test 3.7.8.
     (b) Rationalize and compare with $\sum 1/n^{3/2}$.

14. If $\sum a_n$ is absolutely convergent, the partial sums of $\sum |a_n|$ are bounded, say by M. Evidently the absolute value of the partial sums of any subseries of a_n are also bounded by M.
     Conversely, if every subseries of $\sum a_n$ is convergent, then the subseries consisting of the strictly positive (and strictly negative) terms are absolutely convergent, whence it follows that $\sum a_n$ is absolutely convergent.

Section 9.2

1. (a) Convergent; compare with $\sum 1/n^2$.   (c) Divergent; note that $2^{1/n} \to 1$.

2. (a) Divergent; apply 9.2.1 with $b_n := 1/n$.
     (c) Convergent; use 9.2.4 and note that $(n/(n + 1))^n \to 1/e < 1$.

3. (a) $(\ln n)^p < n$ for large n, by L'Hospital's Rule.

(c)   Convergent; note that $(\ln n)^{\ln n} > n^2$ for large n.

(e)   Divergent; apply 9.2.6 or Exercise 3.7.12.

4. (a)   Convergent   (b)   Divergent   (c)   Divergent

(d)   Convergent; note that $(\ln n)\exp(-n^{1/2}) < n\exp(-n^{1/2}) < 1/n^2$ for large n, by L'Hospital's Rule.

(e)   Divergent   (f)   Divergent.

6. Apply the Integral Test 9.2.6.

7. (a, b)   Convergent   (c)   Divergent   (d)   Convergent.

9. If $m > n \geq K$, then $|s_m - s_n| \leq |x_{n+1}| + \cdots + |x_m| < r^{n+1}/(1 - r)$. Now let $m \to \infty$.

12. (a)   A crude estimate of the remainder is given by $s - s_4 < \int_5^\infty x^{-2}\,dx = 1/5$. Similarly $s - s_{10} < 1/11$ and $s - s_n < 1/(n + 1)$, so that 999 terms suffice to get $s - s_{999} < 1/1000$.

(d)   If $n \geq 4$, then $x_{n+1}/x_n \leq 5/8$ so (by Exercise 10) $|s - s_4| \leq 5/12$. If $n \geq 10$, then $x_{n+1}/x_n \leq 11/20$ so that $|s - s_{10}| \leq (10/2^{10})(11/9) < 0.012$. If $n = 14$, then $|s - s_{14}| < 0.000\,99$.

13. (b)   Here $\sum_{n+1}^\infty < \int_n^\infty x^{-3/2}\,dx = 2/\sqrt{n}$, so $|s - s_{10}| < 0.633$ and $|s - s_n| < 0.001$ when $n > 4 \times 10^6$.

(c)   If $n \geq 4$, then $|s - s_n| \leq (0.694)x_n$ so that $|s - s_4| < 0.065$. If $n \geq 10$, then $|s - s_n| \leq (0.628)x_n$ so that $|s - s_{10}| < 0.000\,023$.

14. Note that (s_{3n}) is not bounded.

16. Note that, for an integer with n digits, there are 9 ways of picking the first digit and 10 ways of picking each of the other $n - 1$ digits. There is one value of m_k from 1 to 9, there is one value from 10 to 19, one from 20 to 29, etc.

18. Here $\lim(n(1 - x_{n+1}/x_n)) = (c - a - b) + 1$, so the series is convergent if $c > a + b$ and is divergent if $c < a + b$.

Section 9.3

1. (a)   Absolutely convergent   (b)   Conditionally convergent

(c)   Divergent   (d)   Conditionally convergent.

2. Show by induction that $s_2 < s_4 < s_6 < \cdots < s_5 < s_3 < s_1$. Hence the limit lies between s_n and s_{n+1} so that $|s - s_n| < |s_{n+1} - s_n| = z_{n+1}$.

5. Use Dirichlet's Test with $(y_n) := (+1, -1, -1, +1, +1, -1, -1, \cdots)$. Or, group the terms in pairs (after the first) and use the Alternating Series Test.

7. If $f(x) := (\ln x)^p/x^q$, then $f'(x) < 0$ for x sufficiently large. L'Hospital's Rule shows that the terms in the alternating series approach 0.

8. (a)   Convergent   (b)   Divergent   (c)   Divergent   (d)   Divergent.

11. Dirichlet's Test does not apply (directly, at least), since the partial sums of the series generated by $(1, -1, -1, 1, 1, 1, \cdots)$ are not bounded.

15. (a)   Use Abel's Test with $x_n := 1/n$.

(b)   Use the Cauchy Inequality with $x_n := \sqrt{a_n}$, $y_n := 1/n$, to get $\sum \sqrt{a_n}/n \leq (\sum a_n)^{1/2}(\sum 1/n^2)^{1/2}$, establishing convergence.

(d)   Let $a_n := [n(\ln n)^2]^{-1}$, which converges by the Integral Test. However, $b_n := [\sqrt{n}\ln n]^{-1}$, which diverges.

Section 9.4

1. (a)   Take $M_n := 1/n^2$ in the Weierstrass M-Test.

(c) Since $|\sin y| \leq |y|$, the series converges for all x. But it is not uniformly convergent on $\mathbb{R}$. If $a > 0$, the series is uniformly convergent for $|x| \leq a$.

(d) If $0 \leq x \leq 1$, the series is divergent. If $1 < x < \infty$, the series is convergent. It is uniformly convergent on $[a, \infty)$ for $a > 1$. However, it is not uniformly convergent on $(1, \infty)$.

4. If $\rho = \infty$, then the sequence $(|a_n|^{1/n})$ is not bounded. Hence if $|x_0| > 0$, then there are infinitely many $k \in \mathbb{N}$ with $|a_k| > 1/|x_0|$ so that $|a_k x_0^k| > 1$. Thus the series is not convergent when $x_0 \neq 0$.

5. Suppose that $L := \lim(|a_n|/|a_{n+1}|)$ exists and that $0 < L < \infty$. It follows from the Ratio Test that $\sum a_n x^n$ converges for $|x| < L$ and diverges for $|x| > L$. The Cauchy-Hadamard Theorem implies that $L = R$.

6. (a) $R = \infty$ (b) $R = \infty$ (c) $R = 1/e$
    (d) 1 (e) $R = 4$ (f) $R = 1$.

8. Use $\lim(n^{1/n}) = 1$.

10. By the Uniqueness Theorem 9.4.13, $a_n = (-1)^n a_n$ for all n.

12. If $n \in \mathbb{N}$, there exists a polynomial P_n such that $f^{(n)}(x) = e^{-1/x^2} P_n(1/x)$ for $x \neq 0$.

13. Let $g(x) := 0$ for $x \geq 0$ and $g(x) := e^{-1/x^2}$ for $x < 0$. Show that $g^{(n)}(0) = 0$ for all n.

16. Substitute $-y$ for x in Exercise 15 and integrate from $y = 0$ to $y = x$ for $|x| < 1$, which is justified by Theorem 9.4.11.

19. $\int_0^x e^{-t^2} dt = \sum_{n=0}^{\infty} (-1)^n x^{2n+1}/n!(2n + 1)$ for $x \in \mathbb{R}$.

20. Apply Exercise 14 and $\int_0^{\pi/2} (\sin x)^{2n} dx = \dfrac{\pi}{2} \cdot \dfrac{1 \cdot 3 \cdot 5 \cdots (2n - 1)}{2 \cdot 4 \cdot 6 \cdots 2n}$.

Section 10.1

1. (a) Since $t_i - \delta(t_i) \leq x_{i-1}$ and $x_i \leq t_i + \delta(t_i)$, then $0 \leq x_i - x_{i-1} \leq 2\delta(t_i)$.
    (b) Apply (a) to each subinterval.

2. (b) Consider the tagged partition $\{([0, 1], 1), ([1, 2], 1), ([2, 3], 3), ([3, 4], 3)\}$.

3. (a) If $\dot{\mathcal{P}} = \{([x_{i-1}, x_i], t_i)\}_{i=1}^n$ and if t_k is a tag for both subintervals $[x_{k-1}, x_k]$ and $[x_k, x_{k+1}]$, we must have $t_k = x_k$. We replace these two subintervals by the subinterval $[x_{k-1}, x_{k+1}]$ with the tag t_k, keeping the δ-fineness property.
    (b) No.
    (c) If $t_k \in (x_{k-1}, x_k)$, then we replace $[x_{k-1}, x_k]$ by the two intervals $[x_{k-1}, t_k]$ and $[t_k, x_k]$ both tagged by t_k, keeping the δ-fineness property.

4. If $x_{k-1} \leq 1 \leq x_k$ and if t_k is the tag for $[x_{k-1}, x_k]$, then we cannot have $t_k > 1$, since then $t_k - \delta(t_k) = \frac{1}{2}(t_k + 1) > 1$. Similarly, we cannot have $t_k < 1$, since then $t_k + \delta(t_k) = \frac{1}{2}(t_k + 1) < 1$. Therefore $t_k = 1$.

5. (a) Let $\delta(t) := \frac{1}{2} \min\{|t - 1|, |t - 2|, |t - 3|\}$ if $t \neq 1, 2, 3$ and $\delta(t) := 1$ for $t = 1, 2, 3$.
    (b) Let $\delta_2(t) := \min\{\delta(t), \delta_1(t)\}$, where δ is as in part (a).

7. (a) $F_1(x) := (2/3)x^{3/2} + 2x^{1/2}$,
    (b) $F_2(x) := (2/3)(1 - x)^{3/2} - 2(1 - x)^{1/2}$,
    (c) $F_3(x) := (2/3)x^{3/2}(\ln x - 2/3)$ for $x \in (0, 1]$ and $F_3(0) := 0$,
    (d) $F_4(x) := 2x^{1/2}(\ln x - 2)$ for $x \in (0, 1]$ and $F_4(0) := 0$,
    (e) $F_5(x) := -\sqrt{1 - x^2} + \text{Arcsin } x$.
    (f) $F_6(x) := \text{Arcsin}(x - 1)$.

8. The tagged partition $\dot{\mathcal{P}}_z$ need not be δ_ε-fine, since the value $\delta_\varepsilon(z)$ may be much smaller than $\delta_\varepsilon(x_j)$.

9. If f were integrable, then $\int_0^1 f \geq \int_0^1 s_n = 1/2 + 1/3 + \cdots + 1/(n+1)$.

10. We enumerate the nonzero rational numbers as $r_k = m_k/n_k$ and define $\delta_\varepsilon(m_k/n_k) := \varepsilon/(n_k 2^{k+1})$ and $\delta_\varepsilon(x) := 1$ otherwise.

12. The function M is not continuous on $[-2, 2]$.

13. L_1 is continuous and $L_1'(x) = l_1(x)$ for $x \neq 0$, so Theorem 10.1.9 applies.

15. We have $C_1'(x) = (3/2)x^{1/2}\cos(1/x) + x^{-1/2}\sin(1/x)$ for $x > 0$. Since the first term in C_1' has a continuous extension to $[0, 1]$, it is integrable.

16. We have $C_2'(x) = \cos(1/x) + (1/x)\sin(1/x)$ for $x > 0$. By the analogue of Exercise 7.2.12, the first term belongs to $\mathcal{R}[0, 1]$.

17. (a) Take $\varphi(t) := t^2 + t - 2$ so $E_\varphi = \emptyset$ to get 6.
    (b) Take $\varphi(t) := \sqrt{t}$ so $E_\varphi = \{0\}$ to get $2(2 + \ln 3)$.
    (c) Take $\varphi(t) := \sqrt{t-1}$ so $E_\varphi = \{1\}$ to get 2 Arctan 2.
    (d) Take $\varphi(t) := \text{Arcsin}\, t$ so $E_\varphi = \{1\}$ to get $\frac{1}{4}\pi$.

19. (a) In fact $f(x) := F'(x) = \cos(\pi/x) + (\pi/x)\sin(\pi/x)$ for $x > 0$. We set $f(0) := 0$, $F'(0) := 0$. Note that f is continuous on $(0, 1]$.
    (b) $F(a_k) = 0$ and $F(b_k) = (-1)^k/k$. Apply Theorem 10.1.9.
    (c) If $|f| \in \mathcal{R}^*[0, 1]$, then $\sum_{k=1}^n 1/k \leq \sum_{k=1}^n \int_{a_k}^{b_k} |f| \leq \int_0^1 |f|$ for all $n \in \mathbb{N}$.

20. Indeed, $\text{sgn}(f(x)) = (-1)^k = m(x)$ on $[a_k, b_k]$ so $m(x) \cdot f(x) = |m(x)f(x)|$ for $x \in [0, 1]$. Since the restrictions of m and $|m|$ to every interval $[c, 1]$ for $0 < c < 1$ are step functions, they belong to $\mathcal{R}[c, 1]$. By Exercise 7.2.11, m and $|m|$ belong to $\mathcal{R}[0, 1]$ and $\int_0^1 m = \sum_{k=1}^\infty (-1)^k/k(2k+1)$ and $\int_0^1 |m| = \sum_{k=1}^\infty 1/k(2k+1)$.

21. Indeed, $\varphi(x) = \Phi'(x) = |\cos(\pi/x)| + (\pi/x)\sin(\pi/x) \cdot \text{sgn}(\cos(\pi/x))$ for $x \notin E$ by Example 6.1.7(c). Evidently φ is not bounded near 0. If $x \in [a_k, b_k]$, then $\varphi(x) = |\cos(\pi/x)| + (\pi/x)|\sin(\pi/x)|$ so that $\int_{a_k}^{b_k} |\varphi| = \Phi(b_k) - \Phi(a_k) = 1/k$, whence $|\varphi| \notin \mathcal{R}^*[0, 1]$.

22. Here $\psi(x) = \Psi'(x) = 2x|\cos(\pi/x)| + \pi \sin(\pi/x) \cdot \text{sgn}(\cos(\pi/x))$ for $x \notin \{0\} \cup E_1$ by Example 6.1.7(b). Since ψ is bounded, Exercise 7.2.11 applies. We cannot apply Theorem 7.3.1 to evaluate $\int_0^b \psi$ since E is not finite, but Theorem 10.1.9 applies and $\psi \in \mathcal{R}[0, 1]$. Corollary 7.3.15 implies that $|\psi| \in \mathcal{R}[0, 1]$.

23. If $p \geq 0$, then $mp \leq fp \leq Mp$, where m and M denote the infimum and the supremum of f on $[a, b]$, so that $m \int_a^b p \leq \int_a^b fp \leq M \int_a^b p$. If $\int_a^b p = 0$, the result is trivial; otherwise, the conclusion follows from Bolzano's Intermediate Value Theorem 5.3.7.

24. By the Multiplication Theorem 10.1.14, $fg \in \mathcal{R}^*[a, b]$. If g is increasing, then $g(a)f \leq fg \leq g(b)f$ so that $g(a)\int_a^b f \leq \int_a^b fg \leq g(b)\int_a^b f$. Let $K(x) := g(a)\int_a^x f + g(b)\int_x^b f$, so that K is continuous and takes all values between $K(b)$ and $K(a)$.

Section 10.2

2. (a) If $G(x) := 3x^{1/3}$ for $x \in [0, 1]$ then $\int_c^1 g = G(1) - G(c) \to G(1) = 3$.
   (b) We have $\int_c^1 (1/x)\, dx = \ln c$, which does not have a limit in $\mathbb{R}$ as $c \to 0$.

3. Here $\int_0^c (1-x)^{-1/2}\, dx = 2 - 2(1-c)^{1/2} \to 2$ as $c \to 1-$.

5. Because of continuity, $g_1 \in \mathcal{R}^*[c, 1]$ for all $c \in (0, 1)$. If $\omega(x) := x^{-1/2}$, then $|g_1(x)| \leq \omega(x)$ for all $x \in [0, 1]$. The "left version" of the preceding exercise implies that $g_1 \in \mathcal{R}^*[0, 1]$ and the above inequality and the Comparison Test 10.2.4 imply that $g_1 \in \mathcal{L}[0, 1]$.

6. (a) The function is bounded on $[0, 1]$ (use l'Hospital) and continuous in $(0, 1)$.
   (c) If $x \in (0, \frac{1}{2}]$ the integrand is dominated by $|(\ln \frac{1}{2})\ln x|$. If $x \in [\frac{1}{2}, 1)$ the integrand is dominated by $|(\ln \frac{1}{2})\ln(1-x)|$.

7. (a) Convergent   (b, c) Divergent   (d, e) Convergent   (f) Divergent.

10. By the Multiplication Theorem 10.1.4, $fg \in \mathcal{R}^*[a, b]$. Since $|f(x)g(x)| \leq B|f(x)|$, then $fg \in \mathcal{L}[a, b]$ and $\|fg\| \leq B\|f\|$.

11. (a) Let $f(x) := (-1)^k 2^k / k$ for $x \in [c_{k-1}, c_k)$ and $f(1) := 0$, where the c_k are as in Example 10.2.2(a). Then $f^+ := \max\{f, 0\} \notin \mathcal{R}^*[0, 1]$.
     (b) Use the first formula in the proof of Theorem 10.2.7.

13. (jj) If $f(x) = g(x)$ for all $x \in [a, b]$, then dist$(f, g) = \int_a^b |f - g| = 0$.
     (jjj) dist$(f, g) = \int_a^b |f - g| = \int_a^b |g - f| = $ dist(g, f).
     (jk) dist$(f, h) = \int_a^b |f - h| \leq \int_a^b |f - g| + \int_a^b |g - h| = $ dist$(f, g) + $ dist(g, h).

16. If (f_n) converges to f in $\mathcal{L}[a, b]$, given $\varepsilon > 0$ there exists $K(\varepsilon/2)$ such that if $m, n \geq K(\varepsilon/2)$ then $\|f_m - f\| < \varepsilon/2$ and $\|f_n - f\| < \varepsilon/2$. Therefore $\|f_m - f_n\| \leq \|f_m - f\| + \|f - f_n\| < \varepsilon/2 + \varepsilon/2 = \varepsilon$. Thus we may take $H(\varepsilon) := K(\varepsilon/2)$.

18. If $m > n$, then $\|g_m - g_n\| \leq 1/n + 1/m \to 0$. One can take $g := $ sgn.

19. No.

20. We can take k to be the 0-function.

Section 10.3

1. Let $b \geq \max\{a, 1/\delta(\infty)\}$. If $\dot{\mathcal{P}}$ is a δ-fine partition of $[a, b]$, show that $\dot{\mathcal{P}}$ is a δ-fine subpartition of $[a, \infty)$.

3. If $f \in \mathcal{L}[a, \infty)$, apply the preceding exercise to $|f|$. Conversely, if $\int_p^q |f| < \varepsilon$ for $q > p \geq K(\varepsilon)$, then $|\int_a^q f - \int_a^p f| \leq \int_p^q |f| < \varepsilon$ so both $\lim_\gamma \int_a^\gamma f$ and $\lim_\gamma \int_a^\gamma |f|$ exist; therefore $f, |f| \in \mathcal{R}^*[a, \infty)$ and so $f \in \mathcal{L}[a, \infty)$.

5. If $f, g \in \mathcal{L}[a, \infty)$, then $f, |f|, g$ and $|g|$ belong to $\mathcal{R}^*[a, \infty)$, so Example 10.3.3(a) implies that $f + g$ and $|f| + |g|$ belong to $\mathcal{R}^*[a, \infty)$ and that $\int_a^\infty (|f| + |g|) = \int_a^\infty |f| + \int_a^\infty |g|$. Since $|f + g| \leq |f| + |g|$, it follows that $\int_a^\gamma |f + g| \leq \int_a^\gamma |f| + \int_a^\gamma |g| \leq \int_a^\infty |f| + \int_a^\infty |g|$, whence $\|f + g\| \leq \|f\| + \|g\|$.

6. Indeed, $\int_1^\gamma (1/x)\, dx = \ln \gamma$, which does not have a limit as $\gamma \to \infty$. Or, use Exercise 2 and the fact that $\int_p^{2p} (1/x)\, dx = \ln 2 > 0$ for all $p \geq 1$.

8. If $\gamma > 0$, then $\int_0^\gamma \cos x\, dx = \sin \gamma$, which does not have a limit as $\gamma \to \infty$.

9. (a) We have $\int_0^\gamma e^{-sx}\, dx = (1/s)(1 - e^{-s\gamma}) \to 1/s$.
     (b) Let $G(x) := -(1/s)e^{-sx}$ for $x \in [0, \infty)$, so G is continuous on $[0, \infty)$ and $G(x) \to 0$ as $x \to \infty$. By the Fundamental Theorem 10.3.5, we have $\int_0^\infty g = -G(0) = 1/s$.

12. (a) If $x \geq e$, then $(\ln x)/x \geq 1/x$.
     (b) Integrate by parts on $[1, \gamma]$ and then let $\gamma \to \infty$.

13. (a) $|\sin x| \geq 1/\sqrt{2} > 1/2$ and $1/x > 1/(n+1)\pi$ for $x \in (n\pi + \pi/4, n\pi + 3\pi/4)$.
     (b) If $\gamma > (n+1)\pi$, then $\int_0^\gamma |D| \geq (1/4)(1/1 + 1/2 + \cdots + 1/(n+1))$.

15. Let $u = \varphi(x) = x^2$. Now apply Exercise 14.

16. (a) Convergent   (b, c) Divergent   (d) Convergent   (e) Divergent
     (f) Convergent.

17. (a) If $f_1(x) := \sin x$, then $f_1 \notin \mathcal{R}^*[0, \infty)$. In Exercise 14, take $f_2(x) := x^{-1/2} \sin x$ and $\varphi_2(x) := 1/\sqrt{x}$.
     (c) Take $f(x) := x^{-1/2} \sin x$ and $\varphi(x) := (x+1)/x$.

18. (a) $f(x) := \sin x$ is in $\mathcal{R}^*[0, \gamma]$, and $F(x) := \int_0^x \sin t\, dt = 1 - \cos x$ is bounded on $[0, \infty)$, and $\varphi(x) := 1/x$ decreases monotonely to 0.
     (c) $F(x) := \int_0^x \cos t\, dt = \sin x$ is bounded on $[0, \infty)$ and $\varphi(x) := x^{-1/2}$ decreases monotonely to 0.

19. Let $u = \varphi(x) := x^2$.

20. (a) If $\gamma > 0$, then $\int_0^\gamma e^{-x} \, dx = 1 - e^{-\gamma} \to 1$ so $e^{-x} \in \mathcal{R}^*[0, \infty)$. Similarly $e^{-|x|} = e^x \in \mathcal{R}^*(-\infty, 0]$.

   (c) $0 \le e^{-x^2} \le e^{-x}$ for $|x| \ge 1$, so $e^{-x^2} \in \mathcal{R}^*[0, \infty)$. Similarly on $(-\infty, 0]$.

Section 10.4

1. (a) Converges to 0 at $x = 0$, to 1 on $(0, 1]$. Not uniform. Bounded by 1. Increasing. Limit $= 1$.
    (c) Converges to 1 on $[0, 1)$, to $\frac{1}{2}$ at $x = 1$. Not uniform. Bounded by 1. Increasing. Limit $= 1$.

2. (a) Converges to $\sqrt{x}$ on $[0, 1]$. Uniform. Bounded by 1. Increasing. Limit $= 2/3$.
    (c) Converges to $\frac{1}{2}$ at $x = 1$, to 0 on $(1, 2]$. Not uniform. Bounded by 1. Decreasing. Limit $= 0$.

3. (a) Converges to 1 at $x = 0$, to 0 on $(0, 1]$. Not uniform. Bounded by 1. Decreasing. Limit $= 0$.
    (c) Converges to 0. Not uniform. Bounded by $1/e$. Not monotone. Limit $= 0$.
    (e) Converges to 0. Not uniform. Bounded by $1/\sqrt{2e}$. Not monotone. Limit $= 0$.

4. (a) The Dominated Convergence Theorem applies.
    (b) $f_k(x) \to 0$ for $x \in [0, 1)$, but $(f_k(1))$ is not bounded. No obvious dominating function. Integrate by parts and use (a). The result shows that the Dominated Convergence Theorem does not apply.

6. Suppose that $(f_k(c))$ converges for some $c \in [a, b]$. By the Fundamental Theorem, $f_k(x) - f_k(c) = \int_c^x f_k'$. By the Dominated Convergence Theorem, $\int_c^x f_k' \to \int_c^x g$, whence $(f_k(x))$ converges for all $x \in [a, b]$. Note that if $f_k(x) := (-1)^k$, then $(f_k(x))$ does not converge for any $x \in [a, b]$.

7. Indeed, $g(x) := \sup\{f_k(x) : k \in \mathbb{N}\}$ equals $1/k$ on $(k - 1, k]$, so that $\int_0^n g = 1 + \frac{1}{2} + \cdots + \frac{1}{n}$. Hence $g \notin \mathcal{R}^*[0, \infty)$.

10. (a) If $a > 0$, then $|(e^{-tx} \sin x)/x| \le e^{-ax}$ for $t \in J_a := (a, \infty)$. If $t_k \in J_a$ and $t_k \to t_0 \in J_a$, then the argument in 10.4.6(d) shows that E is continuous at t_0. Also, if $t_k \ge 1$, then $|(e^{-t_k x} \sin x)/x| \le e^{-x}$ and the Dominated Convergence Theorem implies that $E(t_k) \to 0$. Thus $E(t) \to 0$ as $t \to \infty$.
    (b) It follows as in 10.4.6(e) that $E'(t_0) = -\int_0^\infty e^{-t_0 x} \sin x \, dx = -1/(t_0^2 + 1)$.
    (c) By 10.1.9, $E(s) - E(t) = \int_t^s E'(t) \, dt = -\int_t^s (t^2 + 1)^{-1} dt = \operatorname{Arctan} t - \operatorname{Arctan} s$ for $s, t > 0$. But $E(s) \to 0$ and $\operatorname{Arctan} s \to \pi/2$ as $s \to \infty$.
    (d) We do not know that E is continuous as $t \to 0+$.

12. Fix $x \in I$. As in 10.4.6(e), if $t, t_0 \in [a, b]$, there exists t_x between t, t_0 such that $f(t, x) - f(t_0, x) = (t - t_0)\frac{\partial f}{\partial t}(t_x, x)$. Therefore $\alpha(x) \le [f(t, x) - f(t_0, x)]/(t - t_0) \le \omega(x)$ when $t \ne t_0$. Now argue as before and use the Dominated Convergence Theorem 10.4.5.

13. (a) If (s_k) is a sequence of step functions converging to f a.e., and (t_k) is a sequence of step functions converging to g a.e., Theorem 10.4.9(a) and Exercise 2.2.16 imply that $(\max\{s_k, t_k\})$ is a sequence of step functions that converges to $\max\{f, g\}$ a.e. Similarly, for $\min\{f, g\}$.

14. (a) Since $f_k \in \mathcal{M}[a, b]$ is bounded, it belongs to $\mathcal{R}^*[a, b]$. The Dominated Convergence Theorem implies that $f \in \mathcal{R}^*[a, b]$. The Measurability Theorem 10.4.11 now implies that $f \in \mathcal{M}[a, b]$.
    (b) Since $t \mapsto \operatorname{Arctan} t$ is continuous, Theorem 10.4.9(b) implies that $f_k := \operatorname{Arctan} \circ g_k \in \mathcal{M}[a, b]$. Further, $|f_k(x)| \le \frac{1}{2}\pi$ for $x \in [a, b]$.
    (c) If $g_k \to g$ a.e., it follows from the continuity of Arctan that $f_k \to f$ a.e. Parts (a,b) imply that $f \in \mathcal{M}[a, b]$ and Theorem 10.4.9(b) applied to $\varphi = \tan$ implies that $g = \tan \circ f \in \mathcal{M}[a, b]$.

15. (a) Since $\mathbf{1}_E$ is bounded, it is in $\mathcal{R}^*[a, b]$ if and only if it is in $\mathcal{M}[a, b]$.

(c) $\mathbf{1}_{E'} = 1 - \mathbf{1}_E$.

(d) $\mathbf{1}_{E \cup F}(x) = \max\{\mathbf{1}_E(x), \mathbf{1}_F(x)\}$ and $\mathbf{1}_{E \cap F}(x) = \min\{\mathbf{1}_E(x), \mathbf{1}_F(x)\}$. Further, $E \setminus F = E \cap F'$.

(e) If (E_k) is an increasing sequence in $\mathbb{M}[a, b]$, then $(\mathbf{1}_{E_k})$ is an increasing sequence in $\mathcal{M}[a, b]$ with $\mathbf{1}_E(x) = \lim \mathbf{1}_{E_k}(x)$, and we can apply Theorem 10.4.9(c). Similarly, $(\mathbf{1}_{F_k})$ is a decreasing sequence in $\mathcal{M}[a, b]$ and $\mathbf{1}_F(x) = \lim \mathbf{1}_{F_k}(x)$.

(f) Let $A_n := \bigcup_{k=1}^n E_k$, so that (A_n) is an increasing sequence in $\mathbb{M}[a, b]$ with $\bigcup_{n=1}^\infty A_n = E$, so (e) applies. Similarly, if $B_n := \bigcap_{k=1}^n F_k$, then (B_n) is a decreasing sequence in $\mathbb{M}[a, b]$ with $\bigcap_{n=1}^\infty B_n = F$.

16. (a) $m(\emptyset) = \int_a^b 0 = 0$ and $0 \le \mathbf{1}_E \le 1$ implies $0 \le m(E) = \int_a^b \mathbf{1}_E \le b - a$.

(b) Since $\mathbf{1}_{[c,d]}$ is a step function, then $m([c, d]) = d - c$.

(c) Since $\mathbf{1}_{E'} = 1 - \mathbf{1}_E$, we have $m(E') = \int_a^b (1 - \mathbf{1}_E) = (b - a) - m(E)$.

(d) Note that $\mathbf{1}_{E \cup F} + \mathbf{1}_{E \cap F} = \mathbf{1}_E + \mathbf{1}_F$.

(f) If (E_k) is increasing in $\mathbb{M}[a, b]$ to E, then $(\mathbf{1}_{E_k})$ is increasing in $\mathcal{M}[a, b]$ to $\mathbf{1}_E$. The Monotone Convergence Theorem 10.4.4 applies.

(g) If (C_k) is pairwise disjoint and $E_n := \bigcup_{k=1}^n C_k$ for $n \in \mathbb{N}$, then $m(E_n) = m(C_1) + \cdots + m(C_n)$. Since $\bigcup_{k=1}^\infty C_k = \bigcup_{n=1}^\infty E_n$ and (E_n) is increasing, (f) implies that $m(\bigcup_{k=1}^\infty C_k) = \lim_n m(E_n) = \lim_n \sum_{k=1}^n m(C_k) = \sum_{n=1}^\infty m(C_k)$.

Section 11.1

1. If $|x - u| < \inf\{x, 1 - x\}$, then $u < x + (1 - x) = 1$ and $u > x - x = 0$, so that $0 < u < 1$.

3. Since the union of two open sets is open, then $G_1 \cup \cdots \cup G_k \cup G_{k+1} = (G_1 \cup \cdots \cup G_k) \cup G_{k+1}$ is open.

5. The complement of $\mathbb{N}$ is the union $(-\infty, 1) \cup (1, 2) \cup \cdots$ of open intervals.

7. Corollary 2.4.9 implies that every neighborhood of x in $\mathbb{Q}$ contains a point not in $\mathbb{Q}$.

10. x is a boundary point of A $\iff$ every neighborhood V of x contains points in A and points in $\mathcal{C}(a)$ $\iff$ x is a boundary point of $\mathcal{C}(a)$.

12. The sets F and $\mathcal{C}(F)$ have the same boundary points. Therefore F contains all of its boundary points $\iff$ $\mathcal{C}(F)$ does not contain any of its boundary points $\iff$ $\mathcal{C}(F)$ is open.

13. $x \in A^\circ$ $\iff$ x belongs to an open set $V \subseteq A$ $\iff$ x is an interior point of A.

15. Since A^- is the intersection of all closed sets containing A, then by 11.1.5(a) it is a closed set containing A. Since $\mathcal{C}(A^-)$ is open, then $z \in \mathcal{C}(A^-)$ $\iff$ z has a neighborhood $V_\varepsilon(z)$ in $\mathcal{C}(A^-)$ $\iff$ z is neither an interior point nor a boundary point of A.

19. If $G \ne \emptyset$ is open and $x \in G$, then there exists $\varepsilon > 0$ such that $V_\varepsilon(x) \subseteq G$, whence it follows that $a := x - \varepsilon$ is in A_x.

21. If $a_x < y < x$ then since $a_x := \inf A_x$ there exists $a' \in A_x$ such that $a_x < a' \le y$. Therefore $(y, x] \subseteq (a', x] \subseteq G$ and $y \in G$.

23. If $x \in \mathbb{F}$ and $n \in \mathbb{N}$, the interval I_n in F_n containing x has length $1/3^n$. Let y_n be an endpoint of I_n with $y_n \ne x$. Then $y_n \in \mathbb{F}$ (why?) and $y_n \to x$.

24. As in the preceding exercise, take z_n to be the midpoint of I_n. Then $z_n \notin \mathbb{F}$ (why?) and $z_n \to x$.

Section 11.2

1. Let $G_n := (1 + 1/n, 3)$ for $n \in \mathbb{N}$.

3. Let $G_n := (1/2n, 2)$ for $n \in \mathbb{N}$.

5. If $\mathcal{G}_1$ is an open cover of K_1 and $\mathcal{G}_2$ is an open cover of K_2, then $\mathcal{G}_1 \cup \mathcal{G}_2$ is an open cover of $K_1 \cup K_2$.

7. Let $K_n := [0, n]$ for $n \in \mathbb{N}$.

10. Since $K \neq \emptyset$ is bounded, it follows that inf K exists in $\mathbb{R}$. If $K_n := \{k \in K : k \leq (\inf K) + 1/n\}$, then K_n is closed and bounded, hence compact. By the preceding exercise $\bigcap K_n \neq \emptyset$, but if $x_0 \in \bigcap K_n$, then $x_0 \in K$ and it is readily seen that $x_0 = \inf K$. [Alternatively, use Theorem 11.2.6.]

12. Let $\emptyset \neq K \subseteq \mathbb{R}$ be compact and let $c \in \mathbb{R}$. If $n \in \mathbb{N}$, there exists $x_n \in K$ such that $\sup\{|c - x| : x \in K\} - 1/n < |c - x_n|$. Now apply the Bolzano-Weierstrass Theorem.

15. Let $F_1 := \{n : n \in \mathbb{N}\}$ and $F_2 := \{n + 1/n : n \in \mathbb{N}, n \geq 2\}$.

Section 11.3

1. (a) If $a < b \leq 0$, then $f^{-1}(I) = \emptyset$. If $a < 0 < b$, then $f^{-1}(I) = (-\sqrt{b}, \sqrt{b})$. If $0 \leq a < b$, then $f^{-1}(I) = (-\sqrt{b}, -\sqrt{a}) \cup (\sqrt{a}, \sqrt{b})$.

3. $f^{-1}(G) = f^{-1}([0, \varepsilon)) = [1, 1 + \varepsilon^2) = (0, 1 + \varepsilon^2) \cap I$.

4. Let $G := (1/2, 3/2)$. Let $F := [-1/2, 1/2]$.

8. Let f be the Dirichlet Discontinuous Function.

9. First note that if $A \subseteq \mathbb{R}$ and $x \in \mathbb{R}$, then we have $x \in f^{-1}(\mathbb{R} \setminus A) \iff f(x) \in \mathbb{R} \setminus A \iff f(x) \notin A \iff x \notin f^{-1}(A) \iff x \in \mathbb{R} \setminus f^{-1}(A)$; therefore, $f^{-1}(\mathbb{R} \setminus A) = \mathbb{R} \setminus f^{-1}(A)$. Now use the fact that a set $F \subseteq \mathbb{R}$ is closed if and only if $\mathbb{R} \setminus F$ is open, together with Corollary 11.3.3.

Section 11.4

1. If $P_i := (x_i, y_i)$ for $i = 1, 2, 3$, then $d_1(P_1, P_2) \leq (|x_1 - x_3| + |x_3 - x_2|) + (|y_1 - y_3| + |y_3 - y_2|) = d_1(P_1, P_3) + d_1(P_3, P_2)$. Thus d_1 satisfies the Triangle Inequality.

2. Since $|f(x) - g(x)| \leq |f(x) - h(x)| + |h(x) - g(x)| \leq d_\infty(f, h) + d_\infty(h, g)$ for all $x \in [0, 1]$, it follows that $d_\infty(f, g) \leq d_\infty(f, h) + d_\infty(h, g)$ and d_∞ satisfies the Triangle Inequality.

3. We have $s \neq t$ if and only if $d(s, t) = 1$. If $s \neq t$, the value of $d(s, u) + d(u, t)$ is either 1 or 2 depending on whether u equals s or t, or neither.

4. Since $d_\infty(P_n, P) = \sup\{|x_n - x|, |y_n - y|\}$, if $d_\infty(P_n, P) \to 0$ then it follows that both $|x_n - x| \to 0$ and $|y_n - y| \to 0$, whence $x_n \to x$ and $y_n \to y$. Conversely, if $x_n \to x$ and $y_n \to y$, then $|x_n - x| \to 0$ and $|y_n - y| \to 0$, whence $d_\infty(P_n, P) \to 0$.

6. If a sequence (x_n) in S converges to x relative to the discrete metric d, then $d(x_n, x) \to 0$ which implies that $x_n = x$ for all sufficiently large n. The converse is trivial.

7. Show that a set consisting of a single point is open. Then it follows that every set is an open set, so that every set is also a closed set. (Why?)

10. Let $G \subseteq S_2$ be open in (S_2, d_2) and let $x \in f^{-1}(G)$ so that $f(x) \in G$. Then there exists an ε-neighborhood $V_\varepsilon(f(x)) \subseteq G$. Since f is continuous at x, there exists a δ-neighborhood $V_\delta(x)$ such that $f(V_\delta(x)) \subseteq V_\varepsilon(f(x))$. Since $x \in f^{-1}(G)$ is arbitrary, we conclude that $f^{-1}(G)$ is open in (S_1, d_1). The proof of the converse is similar.

11. Let $\mathcal{G} = \{G_\alpha\}$ be a cover of $f(S) \subseteq \mathbb{R}$ by open sets in $\mathbb{R}$. It follows from 11.4.11 that each set $f^{-1}(G_\alpha)$ is open in (S, d). Therefore, the collection $\{f^{-1}(G_\alpha)\}$ is an open cover of S. Since (S, d) is compact, a finite subcollection $\{f^{-1}(G_{\alpha_1}), \cdots, f^{-1}(G_{\alpha_N})\}$ covers S, whence it follows that the sets $\{G_{\alpha_1}, \cdots, G_{\alpha_N}\}$ must form a finite subcover of $\mathcal{G}$ for $f(S)$. Since $\mathcal{G}$ was an arbitrary open cover of $f(S)$, we conclude that $f(S)$ is compact.

INDEX